Towards Sustainable Global Food Systems

Towards Sustainable Global Food Systems: Conceptual and Policy Analysis of Agriculture, Food and Environment Linkages

Special Issue Editors

Ruerd Ruben
Jan Verhagen

MDPI • Basel • Beijing • Wuhan • Barcelona • Belgrade

Special Issue Editors
Ruerd Ruben
Wageningen University & Research
The Netherlands

Jan Verhagen
Wageningen University & Research
The Netherlands

Editorial Office
MDPI
St. Alban-Anlage 66
4052 Basel, Switzerland

This is a reprint of articles from the Special Issue published online in the open access journal *Sustainability* (ISSN 2071-1050) from 2018 to 2019 (available at: https://www.mdpi.com/journal/sustainability/special_issues/Agriculture_Food_Environment)

For citation purposes, cite each article independently as indicated on the article page online and as indicated below:

LastName, A.A.; LastName, B.B.; LastName, C.C. Article Title. *Journal Name* **Year**, *Article Number*, Page Range.

ISBN 978-3-03897-814-5 (Pbk)
ISBN 978-3-03897-815-2 (PDF)

Contents

About the Special Issue Editors

Ruerd Ruben holds the special chair of Impact Analysis for Food Systems at Wageningen University and is research coordinator for the knowledge base program on "Global Food and Nutrition Security" at Wageningen Economic Research. He is a member of the planning and management committee for the Consortium of International Agricultural Research Centers (CGIAR) in research program "Agriculture for Nutrition and Health" (A4NH) and a member of the steering committee for the the International Food Policy Research Institute- International Institute for Environment and Development (IFPRI-IIED) program "CERES 2030". He is a board member of the Dutch FoodFirst coalition and of the Agriprofocus network. He is actively involved in policy research and advice on issues related to the dynamics of food systems, smallholder involvement in tropical agri-food value chains, the effectiveness of incentives for improving healthier diets, and the impact of standards and certification for enhancing food chain upgrading. He has published more than 100 articles in peer-reviewed journals and co-edited 10 books with academic publishers. Ruerd lived and worked for 14 years in Central America, where he was engaged in programs of land reform, cooperative development, and smallholder agriculture. Thereafter, he coordinated multidisciplinary research and training programs on food security and sustainable land use in several East and West African countries and in China. He was the director of the independent Policy and Operations Evaluation (IOB) department at the Netherlands Ministry of Foreign Affairs and Professor in Development Effectiveness and Director of the Centre for International Development Issues (CIDIN) at Radboud University Nijmegen.

Jan Verhagen is a generalist in soil science, agronomy, and ecology and experienced in providing practical solutions to complex issues requiring the integration of knowledge on various fields. He works on methodology development and the design of sustainable agricultural production systems to address development and climate change. Jan has more than 20 years of experience working in Europe, South-East Asia, and Sub-Saharan Africa. Jan has represented the Netherlands at various national and international climate-change-related events. In 2001 and 2007, he was a member of the Dutch delegation to the plenary to approve the Intergovernmental Panel on Climate Change (IPCC) working group II report. From 2007 to 2015, he represented the Western European and Others Group (WEOG) in the United Nations Framework Convention on Climate Change (UNFCCC) Least-Developed Countries Expert Group (LEG), providing technical guidance and advice to the least-developed countries on adaptation issues. In this capacity, he has been involved in the execution of the National Adaptation Plans of Action and co-designed the National Adaptation Plan process. From 2011 to 2015, he was the Dutch co-chair, with Canada, of the Inventories and Measurements group of the Global Research Alliance on Agricultural Greenhouse Gases, addressing the role of agriculture in reducing greenhouse gas emissions. For Wageningen University & Research, he has co-led several research programs related to food security and climate change. Jan currently leads the Wageningen policy support program on food security for the Dutch Ministry of Agriculture.

sustainability

MDPI

Article

The Challenge of Food Systems Research: What Difference Does It Make?

Ruerd Ruben [1,*], Jan Verhagen [2] and Christine Plaisier [1]

1 Wageningen Economic Research, Wageningen University & Research, P.O. Box 29703, 2502 LS The Hague, The Netherlands; christine.plaisier@wur.nl
2 Wageningen Plant Research, Wageningen University & Research, P.O. Box 35, 6700 AA Wageningen, The Netherlands; jan.verhagen@wur.nl
* Correspondence: ruerd.ruben@wur.nl; Tel.: +31 (0)70-361-5624

Received: 30 November 2018; Accepted: 23 December 2018; Published: 31 December 2018

Abstract: Recent discussions on the results of food security programs devote key attention to complex interactions between policy interventions and business innovation for improving nutrition outcomes. This shift from linear approaches of food and nutrition security towards a more interlinked and nested analysis of food systems dynamics has profound implications for the design and organization of research and innovation processes. In this article we outline our experience with interdisciplinary and interactive processes of food systems analysis at different scale levels, paying systematic attention to three critical system interfaces: intersections with other systems, interactions within the food system, and incentives for food system innovations (the so-called: 3I approach). We discuss the importance of these interfaces for leveraging food system adaptation and managing food system transformation. We also provide illustrative examples of the relevance of food systems analysis for the identification of appropriate and effective programs for reinforcing the resilience, responsiveness and inclusiveness of novel food and nutrition programs.

Keywords: food systems; interdisciplinary research; feedbacks & interlinkages; food policy; 3I Approach

1. Introduction

Research and policy on food security has long been dominated by questions regarding availability, access and utilization. Major attention has been given to the identification of key factors that influence the availability of food at different levels (i.e., individual, household, region, country), and the likely implication of the growth of world population for food security [1]. Other studies focus on understanding of processes that reinforce access to food and/or improve food utilization. Less attention is devoted to the multidimensional nature of food security as influenced by interactions between technical, economic, social and cultural factors. Moreover, the rather linear nature of many food security analyses (focusing mainly on intensification of food production) is increasingly challenged by more complex causal mechanisms that focus on competing goals, emerging system properties and dynamic feedback mechanisms [2].

Food systems include all elements and activities related to the production, processing, distribution, preparation and consumption of food, the market and institutional networks for their governance, and the socio-economic and environmental outcomes of these activities [3–5]. Food systems analysis is based on systematic appraisal of different underlying processes that influence food availability, access and utilization, as well as a detailed analysis of the roles of different stakeholders involved, notably the role of the consumer in nutrition-oriented food systems. It requires a thorough understanding of the structure of a food system and the dynamics of food system changes over time and space in relation to predefined societal, environmental or distributional goals. Important pillars for adequately linking food system analysis to nutrition policy are [6–8]:

- Household targeting: focus on nutritional outcomes for different categories of consumers (differentiated by wealth, gender and age) that have particular types of dietary preferences;
- Multiple delivery pathways of food: food access is satisfied through a combination of home production, open markets purchase, supply by retail and supermarkets, and out of home consumption from restaurants and food services;
- Interactive governance of material flows and information exchange networks between different stakeholders and steering of decision-making processes by the food systems environment;
- Diet implication: effects on dietary intake and possible nutritional imbalances resulting from the combination of diverse baskets of food products.

In this article we aim to assess the conceptual challenges and practical opportunities for analysing the structure and performance of food systems, and we identify how food systems analysis could deliver new and innovative insights for nutrition policy in developing countries. The main objectives of the article are (a) to identify the strategic interfaces between different levels of the food system, and (b) to assess food system responses to business innovations or policy interventions. Therefore, the article responds both to the analytical challenge of understanding food systems performance as well as the empirical challenge of identifying appropriate public and private actions for supporting food systems change in line with societal goals. Moreover, we outline a framework to explore pathways of food system adaptation and to assess the dynamics of food systems transition that goes beyond the mere description of alternative options. This permits us to explicitly acknowledge trade-offs between production and nutrition goals, and may support active engagement of public and private stakeholders [9].

In a conceptual sense, food system analysis is usually conducted in a context where different goals and ambitions are simultaneously pursued and trade-offs between system objectives are likely to occur. Clear understanding of the interdependencies between different stakeholders (i.e., producers, traders, processors, consumers, policy makers) is required for adequately tracing how activities (material and human inputs) translate into desired outcomes (food security results and impact on nutrition). System boundaries should be acknowledged to identify the feasible solutions space. Changes at one system level might lead to undesirable results elsewhere in the food system, and improved knowledge on these interactions could possibly give rise to other types of interventions.

In a practical sense, food systems analysis asks for support from a wide variety of disciplines and also requires the involvement of multiple stakeholders. The willingness to cooperate is usually based on beliefs and expectations that such a research process provides innovative outcomes and more relevant insights. Moreover, the engagement of other (non-science) parties in the research process enables better feedback and linkages to policy and practice [10]. This type of broad multi-stakeholder cooperation and knowledge exchange is considered particularly important for understanding adaptive processes that depend on interaction between technical and behavioral drivers of food system change.

The remainder of this article is organized as follows. In Section 2 we outline the analytical framework for adequately understanding food system structure and dynamics, showing the importance of interlinkages and feedback mechanisms. We briefly assess different methodological strands within food systems analysis and indicate their prospects for food policy appraisal. Section 3 discusses the requirements for collaborative research around critical system interfaces that enable the analysis of food systems performance. In Section 4 we translate this framework for food systems analysis towards food and nutrition policy outcomes and identify some key areas where major differences with traditional food security programs become visible. Finally, in Section 5 we present a theory of change for analysing food systems dynamics to pursue different development trends, and we summarize in Section 6 the advantages of our food systems approach for better dovetailing public and private actions and for capturing potential trade-offs between different societal goals in a timely way.

2. Understanding Food Systems Performance

Most analyses of agricultural development start by addressing opportunities for increasing the production of food either by increasing the cultivated areas (extensive growth) and/or through higher yields (intensive growth). Much attention is usually given to agronomic research around the design and extension of farming systems that have the capacity to generate higher returns from land. The latter approach has been complemented by economic research focusing on the identification of appropriate incentives and policies to support input use by local smallholders (seeds, fertilizer, credit) towards higher returns from family labour and greater overall factor productivity (i.e., output generated by all production factors together).

At the end of the 1990s, intensive international debates around the Millennium Development Goals (MDGs)—later on followed by 'Zero Hunger' challenge as part of the Sustainable Development Goals (SDGs)—asked for wider understanding of the different pathways towards food security. This was not only because a single goal (food availability) was complemented with additional dimensions (food access; food utilization; food safety), but also because aspects of time (stability) and space (environmental resilience) increasingly required attention.

Food systems research took off when strategies for supporting food supply (production) and food demand (consumption) began to be simultaneously analysed within an integrated analytical framework. Trade-offs between these goals and tensions between instruments have been frequently registered, for instance with respect to prices (i.e., high prices to support producers affect consumer demand for food) and for supporting investments (i.e., low interest rates support input use but may reduce employment). This becomes even more complicated when specific outcomes for well-defined target groups (i.e., vulnerable households; people in remote areas; urban populations; women) are pursued, or when agriculture is supposed to contribute to wider societal goals (nutrition, health, employment, environment and climate).

Food systems analysis has undergone important changes over the last few decades. Three different strands and related narratives can be distinguished: (a) descriptive analysis of the structure of food systems with emphasis on the identification of key components [11–13]; (b) explorative analysis of different policy options and opportunities for improving food systems performance [14–17], and (c) interactive analysis of food system transitions and adaptive innovation strategies for creating synergies and coherence between key agents [18–20]. While there is still limited communication between these narratives—also related to the different academic disciplines underlying each of them (i.e., agro-ecology, economics, nutrition and sociology)—it is urgently needed to reinforce our understanding of feasible and effective strategies for supporting transitions towards healthier and sustainable diets [21,22]. This calls for approaches that allow to bridge the gaps between hard and soft systems analysis and that are capable of blending multi-level and multi-stakeholder dynamics [2].

Whereas many recent studies focus on the characteristics and features of food systems [23–25], we consider it more useful to analyse which are the main dilemmas for making the food systems framework useful for overcoming dilemmas in public policy and business practice [26]. Therefore, we introduce a distinction between food system adaptation and food system transformation to highlight different types of responses and interventions that underpin the design of food policies and innovations [27,28]. Adaptation of interactions between food system components may enable timely adjustment to key bottlenecks, whereas external shocks call for major structural transformations. Such a dynamic approach to food systems change can be particularly helpful to engage multiple stakeholders into a common and coherent strategy that satisfies their long-term objectives [29].

Food systems analysis has been applied in two different arenas: for scaling public nutrition policies and to support food business learning platforms and innovations networks. Therefore, different leverage points need to be identified that generate enduring improvements in food systems performance. The relevance and effectiveness of this framework can be demonstrated by the emergence of new forms of multi-stakeholder coalitions that support food system adaptation (see Section 4) and the multi-level food systems transformations and responses to different types of external shocks (see Section 5). The initial representation of food systems was based on a fairly linear understanding of the linkages between food supply and demand activities (See Figure 1). It reflected an increasing awareness that different stakeholders in the supply chain perform specific functions for providing access to key inputs, processing primary outputs, and the marketing and distribution of food towards final consumers. The original material and energy flow approach has been complemented in more economic terms by analyses of the value chain [30] that looks at the transactions between stakeholders and analyses the price and non-price properties of exchange conditions (trust, reliability, frequency, etc.).

Figure 1. Linear food supply system.

Whereas the supply/value chain framework improved the understanding about (horizontal) interdependencies, it still provides little insight into system interactions and feedbacks. Also, externalities for the society and the environment need more attention. The inclusion of waste recovery and nutrient recycling into the model means that the system is better described with a circular representation (see Figure 2). This also enables us to capture better the linkages between (intermediary) inputs and outputs that are relevant for improving food system efficiency.

Figure 2. Circular food system.

It was rapidly acknowledged that these linear and circular frameworks cannot do full justice to the multiple levels and dynamic interlinkages between food production and nutrition, especially because they disregard important spatial and temporal interactions [31]. Therefore, different efforts were made to develop a more 'nested' approach to food systems as part of a set of wider sub-systems. This includes both downstream linkages towards soil dynamics and their environmental and climate effects, as well as upstream linkages within village and regional governance regimes and the linkages of food to labour and capital markets that influence to a large extent the potential inclusiveness of food systems.

The recognition of the importance of systems dynamics and feedback mechanisms incited a range of complex multi-level graphic representations and related modelling exercises [32,33]. Food systems are increasingly visualised as networks that provide environmental services, enhance human welfare and promote community-based socio-economic development, and thus contribute to sustainability, resilience and equity [34]. Food choices and dietary outcomes are embedded in household/family dynamics and village/regional conditions, whereas the availability and the supply and demand of food is strongly governed by the food environment (see Figure 3).

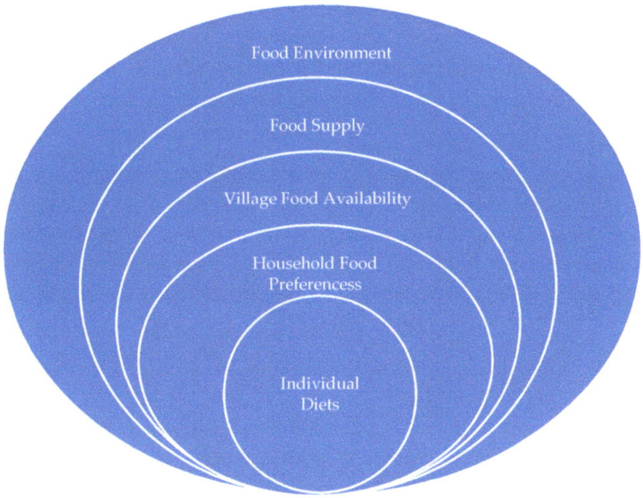

Figure 3. Nested food system.

This integrated framework of food systems has been embraced by major international fora, like by the Committee on World Food Security (CFS), the High Level Panel of Experts on Food Security and Nutrition (HLPE) and the Global Panel on Agriculture and Food Systems for Nutrition (GLOPAN). It is considered a useful framework for understanding major interactions at five different levels of the food system:

- Physical food supply chains responding to behavioural preferences;
- Material food flows generating information on availability and prices;
- Price and non-price incentives that influence household demand;
- Resources (inputs) that are required for enabling reliable food supply;
- Policy and institutional environment that shape individual food choices.

Using a food systems approach for policy analysis has profound implications. This is not only because multiple—sometimes conflicting—goals need to be considered, but also because several external factors (demography, urbanization, infrastructure, economic growth, climate change) influence to a great extent the internal food systems dynamics [35]. This calls for an in-depth analysis of potential trade-offs and/or possible synergies between healthier diets, sustainable resource management, resilient food systems and inclusive development [36]. Therefore, a combination of modelling tools (scenario development) and lab-in-the field (behavioural experiments) approaches need to be used to enhance our insights into the interactions and feedback mechanisms of food systems.

3. Interfaces for Food System Analysis

Based on our understanding of the structural features of food systems, we can now identify the analytical framework to conduct empirical research on food systems dynamics. Given the complexity of nested interactions between system levels and stakeholders, it is considered most useful to organize research as part of a concrete intervention framework. This implies that we advocate for 'Research in Development', not as a sequential process but as an interactive approach to understand system interfaces through direct stakeholder involvement.

Taking stock of the experiences from the Wageningen Research knowledge base (KB) program on global food and nutrition systems, we discuss three strategic interfaces—the so-called 3 I framework—that are critical for identifying and understanding leverage points towards food systems change:

a) Intersections with other non-food systems that influence the supply and demand of food;
b) Interactions between stakeholders that are engaged at different levels of the food system;
c) Incentives that influence the adaptive behaviour and response of food system stakeholders.

(a) Intersections with Other Systems

In the first place, any adequate analysis of the food systems dynamics requires a thorough understanding of intersections with other systems. This refers both to horizontal linkages at space or landscape level, as well as to vertical linkages with non-food systems (such as local labour markets or international trade perspectives). It implies that the performance of the food system will be influenced by several types of 'external' events that shape to a certain extent the internal dynamics of the food system. Some critical events in other systems that are likely to occasion major food system shocks are:

- Population dynamics such as migration, leading to spatial shifts in food demand between rural and urban populations;
- Economic (income) growth, that will influence food preferences towards demand for more processed food and more food purchased at supermarkets;
- Technological development (i.e., ICT) that may open new, sometimes unforeseen, opportunities for food production, distribution and sales;
- Climate change that may lead to spatial adjustment of appropriate production locations and/or requirements for reorganization of the supply chain.

These external trends shape the opportunities for modifying food production and food distribution, and also create new spaces for diversifying food demand. These external trends may either reinforce or balance the changes in internal dynamics of the food systems. Urbanization and economic growth tend to reinforce demand for more animal-based diets and processed food, whereas technological change has the potential to improve the efficiency in food production and distribution. Climate change, on the other hand, could modify the production potential and even lead to changes in feasible production locations, and thus has direct effects on food supply and food prices.

(b) Interactions within the Food System

In the second place, interactions of activities exist between different stakeholders within the food system. Food systems outcomes are shaped by activities at different system levels, and desired outcomes sometimes require 'remote' interventions that generate results elsewhere in the system. A food systems approach can be helpful to identify and promote activities that are beneficial to desired outcomes, and should also consider alternative activities based on possible trade-offs with other outcomes. Typical examples of such multi-level interactions that generate different outcomes at different system levels are:

- Food waste reduction efforts made by primary producers tend to pay-off elsewhere in the supply chain through improved margins for retailers (due to longer shelf life);
- Healthier dietary choices may be supported through direct consumer targeting (vouchers; subsidies), but this could be done in a more cost-effective manner by reinforcing the food supply environment (retail design, peer norms);
- Food safety assessments by consumers are largely based on trust that adequate risk management practices are taken in upstream activities (processing, trade, storage).

Adequate understanding of these interactions within the food system enables a more strategic selection of focal points for influencing food system outcomes. Given these interdependencies between stakeholders, healthier or more sustainable diets can be promoted by modifying exchange conditions for food transactions. This refers both to price and non-price characteristics for supply and demand of food that may lead to changes in production practices and/or consumers' preferences. Stakeholder cooperation could reinforce alliances between food system stakeholders (convergence) but also conflicting outcomes are possible.

(c) Incentives towards Food Systems Stakeholders

In the third place, food system analysis asks for a clear understanding of the effectiveness of incentives for stakeholders to change behaviour and to make choices or implement activities that are needed to achieve the defined goals. Since food systems change can be supported by different types of incentives (prices, taxes/subsidies, information, laws) that influence stakeholders into different directions, it is quite possible that conflicting signals are given and, therefore, that not all system outcomes can be reached simultaneously. Some of these common food system trade-offs refer to:

- Cooling during transport and storage of fast-moving perishable consumer goods such as fresh vegetables and dairy (healthier food) is more demanding in terms of energy use (less sustainable);
- Healthier food choices are hard to influence with market incentives and depend to a large extent on social norms;
- Investments in better waste management tend to increase overall market availability that leads to lower producer prices, thus taking away the initial incentive for engagement.

For the identification of suitable incentives towards healthier diets and sustainable food systems, it is important to understand direct response reactions from producers and consumers, but also spillover effects to other food system stakeholders and the general equilibrium effects on market prices. These secondary implications may either reinforce or counterbalance the original incentive structure and thus determine to a great extent the overall system outcome.

This so-called 3I framework can be very useful to support information exchange between academic disciplines and to foster active engagement of multiple stakeholders [37]. It is also relevant for steering food system policy analysis where insights regarding technical feasibility and behavioural responses need to be simultaneously considered. In addition, different food system goals (such as sustainability, inclusiveness, fairness and resilience) may be compared with respect to their potential trade-offs. Finally, adequate understanding of interactions between different system levels (food production, value chains, food consumption) is helpful to identify the role of different stakeholders (public policy, civic advocacy, business networks) for supporting food system innovations [38].

The food systems analysis thus considers some of the key variables and their major potential interactions as underlying factors influencing dietary choices (by consumers) and allocative choices (by producers and processors). It outlines clearly that effective policies for food system adaptation and transformation require coordinated actions by different public and private stakeholders.

The food system approach might also be a useful framework to identify game changers for improving systems performance. It enables better understanding of both positive and negative feedbacks and considers behavioural interactions that shape the stakeholder's responses to external incentives or shocks. Systems change can thus be based on understanding thresholds and managing the tipping points [39]. The latter are sometimes real regime shifts: large, often abrupt and non-linear changes in food systems behaviour, usually triggered by conflicts between social and environmental factors (poverty traps; ecological shocks) that directly affect people's livelihoods [40]. This understanding is important to identify possible instruments and innovations that are proposed to support food system improvements.

4. New Insights on Food Systems Performance

Food systems analysis is expected to deliver new insights for a more accurate process of public policy formulation and private sector investment appraisal. It is increasingly seen as a suitable way to improve food systems' outcomes (in terms of inclusiveness and sustainability) in order to deal with competing priorities (healthy and sustainable diets) and to harmonize complex interfaces that exist between the stakeholders within food systems [41–43].

Therefore, it is useful to ask the question: *what differences does it make* to rely on food systems analysis for the selection of public policy and private investment priorities, and which instruments and incentives can be used to support these objectives? We focus on five food systems levels that are of critical importance for internal interaction, and distinguish between different types of interventions (or leverage points [44]) that focus on (a) supply-led production logic and (b) demand-led food consumption logic (see Table 1). For each approach we also indicate the most likely stakeholder coalition to harmonize multi-level policy interventions.

Table 1. Interventions to improve food system performance.

Leverage Point	Conventional Logic	Food System Logic	Multi-Stakeholder Coalition
Production systems	(a) Focus on improving inputs (seed, fertilizers) and training in good agricultural practices (GAP) for yield improvements, shifting towards more commercial agriculture	(b) Focus on (local and external) market demand that generates incentives to pull-in farm-level investments and reinforce farmers' bargaining position	(a) Producer-based with (private and public) support services (b) Marketing cooperatives and contract farming
Value chains	(a) Income generation and value added generation by upgrading and sales with private processors and traders networks	(b) Reinforce local processing and strengthen (female) bargaining to generate higher value added and improve intra-household welfare	(a) Midstream input providers and traders (b) Farmer organization & trade contracts
Distribution networks	(a) Guarantee stable market outlets through guaranteed quality and delivery loyalty	(b) Develop long-term upstream/downstream trust relationships and stable revenue streams	(a) Trader-oriented contracts (b) Producer-trader interface
Household livelihoods	(a) Support to higher land productivity and income generation trough specialisation with higher farm employment	(b) Support to higher labour productivity with education and risk reduction through (on/off farm) activity diversification	(a) Public land titling and private input providers (b) Village networks and civic agencies
Food choices	(a) Attention to food security through supply and stable access to sufficient, affordable (and safe) food	(b) Attention for informed choices for healthy and varied diets through home production and market purchase	(a) Public infrastructure and private retail (b) Food environment and behavioural change communication (NGOs)

Source: author's elaboration.

Programs that focus on improving farm **production systems** usually devote major attention to the input side (better seeds and fertilizers to support the adoption of improved agricultural practices) enabling producers to earn higher incomes through engagement in commercial farming. Better results might be achieved if due attention is also given to incentives from the market environment (e.g., de-risking) that incite farmers to make the necessary investments and reinforce their bargaining position in major markets. The latter interventions are likely to deliver better results in terms of household consumption and dietary intake that are strongly driven by certainty on revenue streams and risk diversification from engagement in different market outlets.

Interventions at the level of **value chains** are critically important to guarantee that farmers can reap the product of their investment efforts. In practice, however, a large proportion of the generated value added accrues to upstream chain actors. In addition to value chain integration and upgrading, it is important to guarantee that women receive a fair share of the generated revenues. This is a key condition to guarantee that higher revenues are translated into better nutrition and improved household welfare [45].

The organization of food **distribution networks** also influences dietary outcomes. Traditional marketing programs focus on reaching scale, reinforcing quality and strengthening loyalty between producers and processors (avoiding free riding and sides sales). Food systems approaches recognize also the importance of social and cultural interfaces and thus support trust and loyalty as initial conditions that enable smallholders to make farm-level investments for improved production systems.

In a similar vein, many programs that focus on **household livelihoods** tend to search for higher land productivity, usually through a higher degree of activity specialization and more on-farm employment. However, rural households look to optimize utility and thus search for higher labour productivity, also off-farm and outside agriculture (i.e., more revenues and higher leisure), and benefit particularly from engagement into risk-reducing activities. To enable smallholders' adoption of high-yielding activities, social and commercial networks are of vital importance.

Finally, improving individual **food choices** requires first of all better availability of (different types of) food and stable access to affordable, safe and healthy food items. While this is a necessary condition for supporting food choice, it is certainly not a sufficient condition for realizing such an outcome. Therefore, demand-side interventions are also required that reinforce the food environment and provide information and incentives to individuals and households for improving behavioural choices.

In summary, the food system approach provides an analytical framework that gives new insights in intervention pathways which enrich the 'menu of opportunities' for linking key food policy instruments and for involving different stakeholders. This will enable a better understanding of the interactions between the material and behavioural drivers of food systems change that are vital for linking food production, exchange and consumption, to identify effective food governance mechanisms and to assist stakeholders to make better informed choices on resource allocation and investment.

The involvement of different (public, private and civic) stakeholders is required to guarantee that supply and demand-side requirements of food systems are balanced. Moreover, public support programs need to be embedded into business drivers for innovation to safeguard food systems responsiveness. These new insights are generated by using systematically the analytical food systems lens that identifies more effective pathways towards food systems transformation in line with stakeholders' perceptions and societal development goals.

5. Drivers for Food System Change

The usefulness of food systems analysis can also be assessed within a more dynamic framework by looking at how food systems respond to some major external drivers of change (urbanization; economic growth; climate change; information and communication technology (ICT) connectivity). This analysis of pathways for food system transformation is based on understanding of how external change may lead to adjustments within the food system, and what type of adaptations are likely to take place in different layers (i.e., individual, household, village, region and value chain) of the food system as outlined in Figure 3 [15,18]. Food system responses are thus of critical importance to guarantee that trade-offs between conflicting aims can be reconciled and competing claims on resources are overcome [46,47].

We distinguish three areas of responses to major trends that together determine the adaptive capacity of food systems: resilience capacity, inclusiveness, and sustainability (see Table 2). As shown in Table 2, the suggested responses to external shocks are not isolated and limited to particular stages in the food supply chain, but make use of the interlinkages and feedback within the nested food system. Looking at intersections with other systems, interactions within the food system, and incentives towards food system stakeholders leads to other types of responses and invites different types of interventions (see also Section 3) that might be overlooked in more conventional food chain and network analysis.

Table 2. Examples of food systems responses to major trends.

Outcome Areas Change Drivers	Resilience	Inclusiveness	Sustainability
Urbanisation and Migration	Regular supply, loss reduction and stable prices of perishable food (drying, cooling, packaging) through consumer market information (i.e., mobile app)	Access to healthy (fresh) affordable food in (poor) neighbourhoods with high dietary gaps (i.e., open markets, home delivery)	Food supply chains based on transparent & responsive food environment (grading; food certification, etc.)
Economic Growth	Infrastructure investments to enhance food system adaptation (i.e., smart information and communication technology (ICT) solutions)	Pro-poor targeting of public and private investments and demand-led food delivery options (consumer coops)	Guiding dietary change in home and out-of-home consumption (convenient & healthy diets)
Climate Change	Farmer training and inputs finance for mitigation and adaptation	De-risking of smallholder finance (i.e., pre-finance mechanisms)	Early investments to anticipate shifts in appropriate production areas and techniques
Connectivity	Informed choice on market outlets (street food, corner shop, supermarket)	Dietary knowledge through behavioural change practices (i.e., school meals)	Providing product information (labels) and through public campaigns

Source: author's elaboration.

Important external food systems challenges are related to rapid migration towards (peri-)urban settlements that may increase the likelihood of food losses and waste in longer-distance food chains, require more storage and processing facilities, and lead to changes in diets (more processed foods) and shifts in market outlets (supermarkets, convenience shops and restaurants) that might affect inclusiveness and resilience. Economic growth and rising incomes tend to modify the food demand of the middle classes towards convenience consumption, whereas poorer population segments search for protection against price fluctuation (e.g., through affiliation to consumer cooperatives). Food system responses to climate change require either fixed investments for adaptation of production systems (provided by international banks) and/or climate-smart finance for climate mitigation purposes (provided by NGOs). Finally, improved connectedness through ICT systems enables consumers to make more informed food choices and could eventually support demand-led shifts in consumer behaviour towards more sustainable products and healthier diets.

Identifying such cross-cutting solutions to enhance inclusive, responsive and resilient food systems performance requires concerted efforts by interdisciplinary teams that are committed to engage in an informed dialogue with food system stakeholders. Finding out-of-the-box alternatives implies that most attention should be given to an understanding of the food system interlinkages and feedbacks, and less to each of the separate components.

The integration of such transdisciplinary teams is usually built on joint training and knowledge exchange, strong commitment to learning and innovation, transparent information exchange and common understanding of the shared purpose that can only be reached through committed multi-stakeholder collaboration. Important new insights on suitable policy interventions and business innovations in response to external shocks tend to be generated at the interface between explorative analysis (inventory of change options) and interactive appraisal (portfolio of change pathways) that enable to overcome critical food system bottlenecks.

Therefore, it is of critical importance that food systems analysis is undertaken as part of a wider agenda of agrarian and social transformation. While adjustments of individual components could certainly be helpful to solve specific local problems, they do not change the overall dynamics of food systems interactions [48]. Moreover, identifying tools for creating synergies between food systems goals (i.e., healthy and sustainable food) can be better addressed when nested food system levels are simultaneously considered and payoffs to each of the stakeholders can be clearly acknowledged.

6. Conclusions and Outlook

Policies for influencing food systems' performance need to be based on an adequate understanding of both the relationships between key stakeholders as well as the interactions with the external environment. Many different instruments can be used to support safe and healthier food choices and to improve dietary outcomes, but their effectiveness cannot be generally acknowledged [49]. This is primarily due to the wide diversity in strategic responses amongst food system stakeholders and the dynamic feedbacks between different food system levels.

We outlined in this article some major analytical challenges for describing the structural components of food systems and for analysing opportunities for food systems change empirically. Given the complex interactions between different system levels and the strategic responses of each of the stakeholders, it is difficult to offer adequate foresight on possible pathways towards food systems transition [50]. Better understanding of (internal) leverage points and (external) drivers of system change, as well as timely identification of potential trade-offs between food system goals permits us to prioritize key policy interventions as enablers for business innovation practices.

Important insights from food systems analysis indicate that solutions to major food and nutrition challenges can be found in other parts of the system, sometimes far from the area where the problem became manifest. This may lead to another type of interventions that strategically rely on intersections with other systems, the interactions within the food system, or the incentives towards stakeholders, in order to identify actions that can improve food systems performance and ultimately support food systems transformation. Improving insights in dynamic adjustments pathways and strategic stakeholder responses can be very helpful for creating public–private coalitions that enable food systems change.

Based on experiences in food systems research and the parallel adjustment in leading paradigms for operational food policy analysis, we can identify three critical conditions that should be considered for an interactive analysis of food system transitions:

1. Multi-level interdependencies between food system activities permit focused actions towards leverage points that may result in coherent outcomes at aggregate system level;
2. Multiple goals optimization that are based on adaptive innovation practices and learning loops towards scaling of food systems' change strategies;
3. Multiple stakeholder activities that together are able to create synergies and multipliers that permit the bridging of trade-offs.

Therefore, it is of paramount importance to develop analytical tools that enable the assessment of the likely outcomes of nutrition-oriented public policies and investment priorities to evaluate empirically the effectiveness of different (sets of) instruments for satisfying key stakeholders' goals and for reaching strategic development objectives. Given the diverse and multi-level responses to policy incentives and simultaneously occurring changes in external conditions, broad coalitions between different (public and private/civic) stakeholders are necessary to overcome possible trade-offs.

Author Contributions: R.R.: first author, coordinator KB-22 theme "Global food and nutrition security" and CGIAR-A4NH Management Committee member. J.V., C.P.: coordinator KB-22 theme "Global food and nutrition security, Wageningen Research.

Funding: This research received was funded by the Dutch Ministry of Agriculture, Nature and Food Quality, grant number KB-22.

Acknowledgments: The authors wish to thank the CGIAR-A4NH Flagship "Food Systems for Healthier Diets" funded by the Dutch Ministry of Foreign Affairs, and the KB-22 theme "Global food and nutrition security" of Wageningen University and Research funded by the Dutch Ministry of Agriculture, Nature and Food Quality for funding this study conducted under the Food Systems umbrella.

Conflicts of Interest: The authors declare no conflict of interest.

References

1. Van Ittersum, M.K.; Van Bussel, L.G.; Wolf, J.; Grassini, P.; Van Wart, J.; Guilpart, N.; Claessens, L.; de Groot, H.; Wiebe, K.; Mason-D'Croz, D. Can sub-saharan africa feed itself? *Proc. Natl. Acad. Sci. USA* **2016**, *113*, 14964–14969. [CrossRef] [PubMed]
2. Foran, T.; Butler, J.R.; Williams, L.J.; Wanjura, W.J.; Hall, A.; Carter, L.; Carberry, P.S. Taking complexity in food systems seriously: An interdisciplinary analysis. *World Dev.* **2014**, *61*, 85–101. [CrossRef]
3. Goodman, D.; DuPuis, E.M.; Goodman, M.K. *Alternative Food Networks: Knowledge, Practice, and Politics*; Routledge: Abingdon, UK, 2012.
4. Morley, A.; Marsden, T. Current food questions and their scholarly challenges: Creating and framing a sustainable food paradigm. In *Sustainable Food Systems*; Routledge: Abingdon, UK, 2014; pp. 17–45.
5. Pitt, H.; Jones, M. Scaling up and out as a pathway for food system transitions. *Sustainability* **2016**, *8*, 1025. [CrossRef]
6. Grant, M. A food systems approach for food and nutrition security. *Sight Life* **2015**, *29*, 87–90.
7. Herforth, A.; Ahmed, S. The food environment, its effects on dietary consumption, and potential for measurement within agriculture-nutrition interventions. *Food Secur.* **2015**, *7*, 505–520. [CrossRef]
8. Tschirley, D.; Haggblade, S.; Reardon, T. Africa's emerging food system transformation. In *Global Center for Food Systems Innovation*; White Paper; Michigan State University: East Lansing, MI, USA, 2013.
9. Dunning, R.; Bloom, J.D.; Creamer, N. The local food movement, public-private partnerships, and food system resiliency. *J. Environ. Stud. Sci. Springer Assoc. Environ. Stud. Sci.* **2015**, *5*, 661–670. [CrossRef]
10. Kjellberg, A.; Sohlenius, G. Principles of multidisciplinary cooperation in research, especially behavioural science and manufacturing. *CIRP Ann.* **1993**, *42*, 541–543. [CrossRef]
11. Scott, P. Global panel on agriculture and food systems for nutrition: Food systems and diets: Facing the challenges of the 21st century. *Food Secur. Sci. Sociol. Econ. Food Prod. Access Food* **2017**, *9*, 653–654. [CrossRef]
12. Ingram, J. A food systems approach to researching food security and its interactions with global environmental change. *Food Secur.* **2011**, *3*, 417–431. [CrossRef]
13. Westhoek, H.; Ingram, J.; Van Berkum, S.; Özay, L.; Hajer, M. Food systems and natural resources. In *A Report of the Working Group on Food Systems of the International Resource Panel*; United Nations Environment Programme: Paris, France, 2016.
14. Van Berkum, S.; Dengerink, J.; Ruben, R. *The Food Systems Approach: Sustainable Solutions for a Sufficient Supply of Healthy Food*; Wageningen Economic Research: Wageningen, The Netherlands, 2018.
15. Allen, T.; Prosperi, P. Modeling sustainable food systems. *Environ. Manag.* **2016**, *57*, 956–975. [CrossRef]
16. *Shaping the Future of Global Food Systems: A Scenarios Analysis*; World Economic Forum: Cologny, Switzerland, 2017.
17. Caron, P.; y de Loma-Osorio, G.F.; Nabarro, D.; Hainzelin, E.; Guillou, M.; Andersen, I.; Arnold, T.; Astralaga, M.; Beukeboom, M.; Bickersteth, S. Food systems for sustainable development: Proposals for a profound four-part transformation. *Agron. Sustain. Dev.* **2018**, *38*, 41. [CrossRef]
18. Bui, S.; Cardona, A.; Lamine, C.; Cerf, M. Sustainability transitions: Insights on processes of niche-regime interaction and regime reconfiguration in agri-food systems. *J. Rural Stud.* **2016**, *48*, 92–103. [CrossRef]
19. De Schutter, O. The political economy of food systems reform. *Eur. Rev. Agric. Econ.* **2017**, *44*, 705–731. [CrossRef]
20. Meadows, D. Places to intervene in a system. *Whole Earth* **1997**, *91*, 78–84.
21. Béné, C.; Oosterveer, P.; Lamotte, L.; Brouwer, I.D.; de Haan, S.; Prager, S.D.; Talsma, E.F.; Khoury, C.K. When food systems meet sustainability–current narratives and implications for actions. *World Dev.* **2019**, *113*, 116–130. [CrossRef]

22. Fresco, L.O.; Ruben, R.; Herens, M. Challenges and perspectives for supporting sustainable and inclusive food systems. *GREAT Insights Magazine*, 2017; 13–15.
23. Eakin, H.; Connors, J.P.; Wharton, C.; Bertmann, F.; Xiong, A.; Stoltzfus, J. Identifying attributes of food system sustainability: Emerging themes and consensus. *Agric. Hum.Values* **2016**, *34*, 57–773. [CrossRef]
24. Sundaram, J. Tackling the nutrition challenge: A food systems approach. *J. Dev.* **2014**, *2*, 141. [CrossRef]
25. Gladek, E.; Fraser, M.; Roemers, G.; Sabag Muñoz, O.; Kennedy, E.; Hirsch, P. *The Global Food System: An Analysis*; Metabolic: Zeist, The Netherlands, 2016; p. 180.
26. Gillespie, S.; van den Bold, M. Agriculture, Food Systems, and Nutrition: Meeting the Challenge. 2017. Available online: https://onlinelibrary.wiley.com/doi/pdf/10.1002/gch2.201600002 (accessed on 05 December 2017).
27. Lang, T. Food security and food sustainability: Reformulating the debate. *Geogr. J.* **2012**, *178*, 313–332. [CrossRef]
28. Miller, D.; Welch, R.M. Food system strategies for preventing micronutrient malnutrition. *Food Policy* **2013**, *42*, 115–128. [CrossRef]
29. Hawkes, C. Policy coherence across the food system for nutrition: From challenge to opportunity. *GREAT Insights Mag.* **2017**, *6*. Available online: https://ecdpm.org/great-insights/sustainable-food-systems/policy-coherence-across-food-system-nutrition-challenge-opportunity/ (accessed on 12 August 2018).
30. Gereffi, G.; Fernandez-Stark, K. *Global Value Chain Analysis: A Primer*; Duke University: Durham, NC, USA, 2016.
31. Hunt, K.P. A nested approach to the right to food: Food security, gender violence, and human rights. *J. Agric. Food Syst. Commun. Dev.* **2016**, *7*, 141–143. [CrossRef]
32. Vandermeer, J.; Aga, A.; Allgeier, J.E.; Badgley, C.; Baucom, R.; Blesh, J.; Shapiro, L.F.; Jones, A.D.; Hoey, L.; Perfecto, I. Feeding prometheus: An interdisciplinary approach for solving the global food crisis. *Front. Sustain. Food Syst.* **2018**, *2*, 39. [CrossRef]
33. Miles, A.; DeLonge, M.S.; Carlisle, L. Triggering a positive research and policy feedback cycle to support a transition to agroecology and sustainable food systems. *Agroecol. Sustain. Food Syst.* **2017**, *41*, 855–879. [CrossRef]
34. Haddad, L.; Hawkes, C.; Waage, J.; Webb, P.; Godfray, C.; Toulmin, C. Food Systems and Diets: Facing the Challenges of the 21st Century; 2016. Available online: http://openaccess.city.ac.uk/19323/ (accessed on 24 May 2018).
35. Tendall, D.; Joerin, J.; Kopainsky, B.; Edwards, P.; Shreck, A.; Le, Q.; Krütli, P.; Grant, M.; Six, J. Food system resilience: Defining the concept. *Glob. Food Secur.* **2015**, *6*, 17–23. [CrossRef]
36. Fresco, L.O. Challenges for food system adaptation today and tomorrow. *Environ. Sci. Policy* **2009**, *12*, 378–385. [CrossRef]
37. Drewnowski, A.; Team, T.E.I. The Chicago consensus on sustainable food systems science. *Front. Nutr.* **2017**, *4*, 74. [CrossRef] [PubMed]
38. Kim, D.; Anderson, V. *System Archetype Basics, from Story to Structure*; Pegasus communication. Inc.: Waltham, MA, USA, 1998.
39. Benton, T.; Fairweather, D.; Graves, A.; Harris, J.; Jones, A.; Lenton, T.; Norman, R.; O'Riordan, T.; Pope, E.; Tiffin, R. *Environmental Tipping Points and Food System Dynamics*; Global Food Security: Swindon, UK, 2017.
40. Biggs, R.; Carpenter, S.R.; Brock, W.A. Turning back from the brink: Detecting an impending regime shift in time to avert it. *Proc. Natl. Acad. Sci. USA* **2009**, *106*, 826–831. [CrossRef]
41. Ericksen, P.; Stewart, B.; Dixon, J.; Barling, D.; Loring, P.; Anderson, M.; Ingram, J. The value of a food system approach. *Food Secur. Glob. Environ. Change* **2010**, *25*, 24–25.
42. Ingram, J.; Ericksen, P.; Liverman, D. *Food Security and Global Environmental Change*; Routledge: Abingdon, UK, 2012.
43. Garnett, T. Three perspectives on sustainable food security: Efficiency, demand restraint, food system transformation. What role for lca. *J. Clean. Prod.* **2013**, *30*, 1e9.
44. West, P.C.; Gerber, J.S.; Engstrom, P.M.; Mueller, N.D.; Brauman, K.A.; Carlson, K.M.; Cassidy, E.S.; Johnston, M.; MacDonald, G.K.; Ray, D.K. Leverage points for improving global food security and the environment. *Science* **2014**, *345*, 325–328. [CrossRef]
45. Brown, L.; Deshpande, C.; Hill, C.; Lambrou, Y.; Laudazi, M.; Ragasa, C. Module 1: Gender and food security. In *Gender and Agriculture Sourcebook: Investing in Women as Drivers of Agricultural Growth*; 2009;

Available online: http://siteresources.worldbank.org/INTGENAGRLIVSOUBOOK/Resources/Module1.pdf (accessed on 12 October 2018).

46. Garnett, T.; Godfray, C.H. *Sustainable Intensification in Agriculture. Navigating a Course through Competing Food System Priorities*; University of Oxford Programme on the Future of Food, University of Oxford: Oxford, UK, 2012.
47. Lang, T.; Barling, D. Nutrition and sustainability: An emerging food policy discourse. *Proc. Nutr. Soc.* **2013**, *72*, 1–12. [CrossRef] [PubMed]
48. Freedgood, J.; Pierce-Quiñonez, M.; Meter, K. Emerging assessment tools to inform food system planning. *J. Agric. Food Syst. Commun. Dev.* **2016**, *2*, 83–104. [CrossRef]
49. Pinstrup-Andersen, P.; Watson, D.D., II. *Food Policy for Developing Countries: The Role of Government in Global, National, and Local Food Systems*; Cornell University Press: Ithaca, NY, USA, 2011.
50. Thompson, J.; Millstone, E.; Scoones, I.; Ely, A.; Marshall, F.; Shah, E.; Stagl, S.; Wilkinson, J. *Agri-Food System Dynamics: Pathways to Sustainability in an Era of Uncertainty*. 2007. Available online: https://www.ids.ac.uk/files/agriculture.pdf (accessed on 8 April 2016).

sustainability

MDPI

Article

Does Agricultural Commercialization Affect Food Security: The Case of Crop-Producing Households in the Regions of Post-Reform Vietnam?

Vincent Linderhof *, Valerie Janssen and Thom Achterbosch

Wageningen Economic Research, Prinses Beatrixlaan 582-528, 2595 BM The Hague, The Netherlands; valerie.janssen@wur.nl (V.J.); thom.achterbosch@wur.nl (T.A.)

* Correspondence: vincent.linderhof@wur.nl; Tel.: +31-70-335-8396

Received: 31 December 2018; Accepted: 20 February 2019; Published: 27 February 2019

check for updates

Abstract: Agricultural commercialization remains a widely pursued approach in development projects to improve food security in low-income countries, although there is no clear scientific evidence for it. This study examines the impact of agricultural commercialization on the food security status of crop-producing households in the regions of Vietnam in the 1990s. We used the food system framework including output and input markets. We explore three indicators of commercialization: Cash crop production share (CCPS), crop output market participation share (COMPS), and crop input market participation share (CIMPS) based on fertilizer use. For food security, we looked at caloric intake and dietary diversity (Food Variety Score). We use a balanced panel data sample from the Vietnam Living Standards Survey (VLSS) of 1992/93 and 1997/98. We apply four specifications for all combinations of commercialization indicators and food security indicators for seven regions: OLS 1992/93, OLS 1997/98, pooled sample, and difference estimator. The results show that the effect of commercialization on food security is widely heterogeneous. It depends upon the commercialization indicator and the region in Vietnam. In general, there is no clear evidence for the direction of commercialization on either caloric intake or dietary diversity; however, it is clear that the impacts are generally more positive for southern regions than for northern regions of Vietnam.

Keywords: commercialization; Vietnam; food system; fertilizer use; caloric intake; dietary diversity Vietnam Living Standards Survey (VLSS)

1. Introduction

Commercialization has been presented as a way out of poverty and as a way to improve food security for poor farming households in low-income countries since the 1980s [1]. So far, the promotion of commercialization has been targeted at the agricultural outputs of poor farming households. There are two reasons to re-examine this relationship. Firstly, commercialization should be regarded from a food-system perspective, which means that different elements of the food system, as defined in earlier studies [2], can contribute to commercialization. From the farmer perspective, commercialization is not limited to increased market outputs or the production of cash crops. It can also refer to the purchasing of fertilizer or the hiring of labor, for instance. Secondly, not all low- and middle-income countries have available detailed surveys with which the impact of commercialization on farming households can be analyzed. For instance, Myanmar is a centrally planned economy that is opening up to a more market-oriented economy. However, Myanmar lacks reliable surveys for analyzing the impacts of commercialization on poverty and food security. Therefore, we sought a country with similar characteristics, which has experienced political and economic transition, and which has available household surveys to analyze the impact of commercialization on food security. Therefore, we chose to look at the impact of commercialization on food security in Vietnam in the 1990s, as it reflects a similar degree of political and economic change to that which Myanmar is experiencing now. Therefore, this study will focus on the impact of commercialization from a food-system perspective on food security of farming households in Vietnam in the 1990s.

In the 1990s, Vietnam experienced high levels of economic growth [3] caused by a set of economic reforms, the so-called *doi moi* [4]. These reforms started around 1986 and focused on transforming the centrally planned economy gradually towards a more market-oriented system [5]. At the same time, Vietnam achieved a significant reduction in poverty [3]. The share of the population living in poverty dropped from 58% in 1993 to 37% in 1998 [6], while the share of undernourishment declined from 45.6% to 35.4% between 1991 and 1995 [6]. The privatization of the agricultural sector was one of the key elements of the *doi moi*. Farm households were, from that moment onwards, allowed to make their own decisions on the allocation of land, the type of crops produced, and whether or not they sell their produce at markets. Additionally, the market prices of crops and inputs were liberalized [5], and collective farms were privatized—similar to that observed in China a couple of years earlier [4]. In 1996, ten years after the *doi moi* started, the agricultural sector was still employing 70% of the population [6].

There are studies that analyzed the impact of economic reforms and trade liberalization on agriculture and the income distribution in rural Vietnam [7,8]. However, it is still unclear how economic prosperity, caused by these political and economic changes, has affected food security in this period of poverty in Vietnam. In the literature on commercialization and food security, there has been limited empirical research on the topic despite much discussion on the topic in the literature in the 1980s and 1990s [9]. Most empirical studies found a positive effect of commercialization on income, but only a marginal effect on nutrition or food security [9]. Most studies were applied in the African context, with hardly any examples for the South East Asia region published.

This study will analyze the impact of commercialization on food security from a food system's perspective. In particular, we will analyze the relationship between indicators of commercialization and food security using three commercialization indicators, namely, the share of cash crops in production, the share of market participation, and share of inputs used from the market. For food security, we distinguish two indicators, namely caloric intake and the Food Variety Score (FVS). As we explore the relationship between the commercialization of agriculture and food security, we will solely consider farm households involved in crop production in our analysis. Moreover, we only include farm households for which we have two observations in time so that we can explore the change over time at the farm household level. As the food systems differ across the seven administrative regions [10], we explore the relationships for each of the seven regions separately.

The set of explanatory variables include commercialization, socio-economic, and farm characteristics. Using panel data from the Vietnam Living Standards Survey (VLSS) for 1992–1993 and 1997-1998, we explore different model specifications for explaining food security such as an Ordinary Least Squared (OLS) model for the individual cross-sectional data sets of 1992/93 and 1997/98, an OLS for the pooled sample of two data sets, and a fixed effects (FE) difference estimator.

The structure of the paper is as follows. Section 2 discusses the literature on commercialization and the impact on income and food security of smallholder farmers. Section 3 presents the methodology, and Section 3 discusses the data of the VLSS in more detail. Section 4 examines the results and, finally, Section 4 presents the conclusions of the study and provides further discussion.

2. Literature and Methodology

2.1. Literature Review

Agriculture in low-income countries is known to be a crucial provider of income, livelihoods, and environmental services [11]. Moreover, agriculture and its commercialization are seen as particularly promising ways out of poverty for poor farming households in low-income countries [1]. In theory, specialization and commercialization of agriculture are much more efficient than subsistence farming. Specialization and commercialization of agriculture can improve the productivity and competitiveness of smallholder farmers. Gains in income could occur through comparative advantages, economies of scale, and increased productivity caused by social learning effects [12]. In addition, the improved agricultural productivity reduces the amount of labor required on farms, which implies mobility of labor from agriculture towards other sectors of the economy [12]. However, the commercialization of agriculture can also lead to a decline in crop production diversity at the farm household level [13]. This would mean that households can become less self-sufficient and more dependent on local food markets. In regions where markets are not well-integrated, volatile market prices of crops and inputs, inefficient marketing institutions, and poor infrastructure pose risks to household income [14,15]. Moreover, due to the lack of access to credit, households are unable to mitigate these risks [14]. In such regions, subsistence farming serves as a kind of insurance against the risks and costs of the market [1].

In order to achieve improved agricultural productivity, attention should be given to increasing access to assets and diversifying income sources other than from agriculture [11]. Since we are focusing on the impact of commercialization on food security of farming households, other impacts of commercialization are beyond the scope of this study.

Farming households have different ways in which they can improve their food security status. We adopt the framework which distinguishes three different pathways [16], see Figure 1. The market pathway represents the most direct impact of commercialization of agriculture from an output perspective, i.e., higher quantities of agricultural commodities sold at the market. However, commercialization of agriculture might also affect the own-production pathway, as it implies changes in input use which affect agricultural productivity, and potentially results in higher own production.

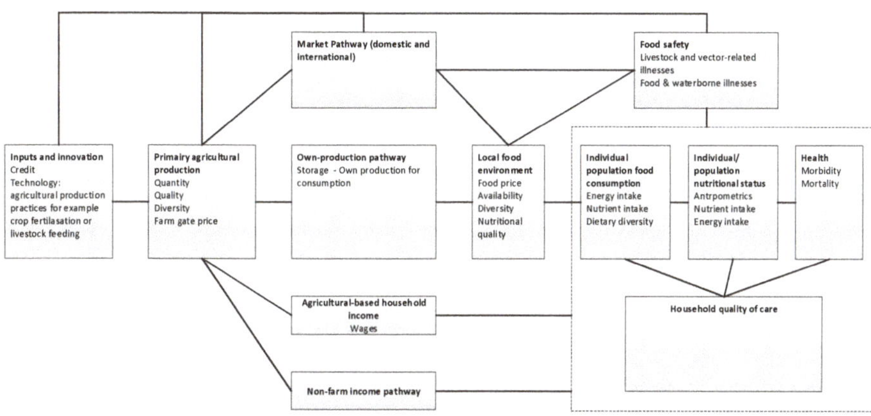

Figure 1. Conceptual framework of the links between agriculture, the food system, nutrition, and public health. Source: Adopted from [16].

In the literature, we observed two streams of thoughts, which are interesting to take into account. The first explored the impact of commercialization on farmers' income and poverty. These studies hypothesized that commercialization has a positive impact on a farmer's income. In addition, some of these studies assumed that improved income will also affect food security in a positive way. The results on farmers' income from these studies ranged from negative to positive depending on the local conditions, while the effects of increased income on food security were either positive or neutral, depending on household decisions. The decisions tended to vary based on culture and social groups [17]. Increased income could increase the demand for more diversified and nutritious diets, namely an increase in expenditures on animal products, fruits, and vegetables to replace cereals and pulses [18]. Although increased diversification tends to yield higher levels of micronutrient content in diets, this might not be the case for caloric intake [19]. When income increases, households do not spend everything on obtaining more calories. Instead, they often buy better-tasting and more expensive calories [20]. Moreover, non-food expenditures are also an important factor at play. Households might prioritize to purchase non-food items over increasing their caloric intake [20].

For example, in a study in the Philippines, cash crops production significantly increased household income, but due to the purchasing of more expensive calories and non-food items, this increase did not translate into a higher preschooler nutritional status [21]. In a study concerning Southwestern Kenya, similar results were found [22]. Cash crop production increased income and showed a small positive effect on household caloric intake. The additional income, however, was mainly spent on non-food items such as housing and school fees.

The second stream of literature focused on the impact of commercialization and food security directly. The impact of commercialization on food security could also function through changes in farmers' own production [16], rather than only through income.

On the one hand, increases in income provide farmers with the opportunity to make investments that could lead to higher productivity, which would improve food security [23,24]. On the other hand, commercialization can lead to less diversification of crops and more specialization at the smallholder level, but in general, diversification tends to increase at the sector level [25]. In the case of the Malawian domestic food crisis for instance, the effect of commercialization on food security was negative [26]. During the period of food price shocks, cash crop production was associated with negative health effects on children in the utero state [26].

A more recent study based on data from three African countries confirmed the earlier findings that there is little evidence for a relationship between commercialization and food security [9]. In contrast to many earlier studies, the study did not investigate cash crop production as an indicator of commercialization but used the share of output sold at the market of total production [9]. However, the commercialization of input factors, i.e., participation at input markets of fertilizer and pesticides, for instance, as included in this study, was not considered.

The relationship between agricultural commercialization and food security can also be considered at a more macro-economic level. Agricultural commercialization causes households in different areas with different resources to specialize in different crops as the agricultural transformation takes place. This leads to greater diversification on the level of the agricultural sector as a whole. Finally, the highest level of aggregation, the economy as a whole, eventually shows the highest level of diversification. Originally, this diversity is expected to be low, but the increased importance and accessibility of international trade will fuel the inherent desire of people for more diverse diets [25].

2.2. Commercialization

One of the main focuses of this research is comparing different ways of operationalizing commercialization. Most research on commercialization has been explored from the perspective of agricultural development [9]. These studies use a very simple definition of commercialization with an indicator that only focuses on whether a farmer grows cash crops or not. When farmers are growing cash crops, it means that they are market-oriented for selling their production. Cash-crop production is frequently accompanied by the modernization and intensification of cultivation through improved inputs or investments [27].

In this paper, we will look at the impact of commercialization at the farm level from a more holistic perspective, namely the food system perspective. This means that we do not only link commercialization to what farmers produce but also link it to what farmers require for their productions, such as the different factor inputs (e.g., land, labor, and capital, but also inputs like seed, fertilizer, etc.). Even when cash-crop growing is considered to be the definition of commercialization, commercialization of agriculture involves multiple aspects including the input and the output side of production [1,12].

In total, we will consider two separate measures of commercialization, each of which represents a crucial element of agricultural commercialization, namely output and input markets. We will consider the effect of each of these measures on food security separately. The most commonly used measure of commercialization is that of output markets. The cash crop indicator that was mentioned above is also an example of this, as this indicator would consider whether or not households are participating in cash crop production—which in the case of cash crops is equal to output markets, as they are not destined for own consumption at all.

2.2.1. Cash crop production share (*CCPS*)

In practice, smallholder farmers that are involved in cash-crop growing are likely to be involved in trading non-cash crops as well. Therefore, we will not use the dichotomous indicator for involvement in cash-crop production but we will use the cash-crop share in the total production value. The advantage of this share is that it is more comparable with other indicators that we will explore in this study.

Suppose that a farmer can grow K different types of crops. We define a subset K_c of cash crops. Then the *CCPS* indicator is defined as:

$$CCPS_i = \frac{\sum_{c=1}^{K_c} \overline{P_c} Q_{ic}}{\sum_{k=1}^{K} \overline{P_k} Q_{ik}} \tag{1}$$

where Q_{ic} is the quantity of cash crop c produced by farm household i evaluated at an average community level price $\overline{P_c}$, and Q_{ik} is the total quantity of crop k produced by farm household i, evaluated at an average community level price $\overline{P_k}$. K_c is the set of crops identified as cash crops and c is the index of cash crops with $c \in K_c$. The set of cash crops K_c is a subset of the set of all crops K.

So, if a farm household i only sells cash crops, $CCPS_i = 1$. If a farm household does not sell any cash crops, $CCPS_i = 0$. Note that this farmer could sell non-cash crops at the market, which will not be reflected by the *CCPS* indicator.

2.2.2. Crop output market participation share (*COMPS*)

In order to take into account all market sales of crops by a farmer household, we use the *COMPS* indicator, which is calculated as the proportion of the value of crops sold at the market and the total value of crop production [28].

$$COMPS_i = \frac{\sum_{k=1}^{K} \overline{P_k} S_{ik}}{\sum_{k=1}^{K} \overline{P_k} Q_{ik}} \tag{2}$$

where S_{ik} is the quantity of crops k sold at the market by farm household i evaluated at an average community level price $\overline{P_k}$. Note that $S_{ik} \leq Q_{ik}$. Therefore, when a farm household i sells the whole crop production at the market, $S_{ik} = Q_{ik}$ and $COMPS_i = 1$. When a farm household does not sell any crop production at the market, $S_{ik} = 0$ and $COMPS_i = 0$.

2.2.3. Crop input market participation share (*CIMPS*)

Both the CCPS and COMPS indicators above are based on the market sales of a household farm, which only partly comply with the food-system perspective. In order to obtain a more comprehensive picture, we propose the crop input market participation share (CIMPS) indicator. It is defined as the share of purchased inputs value to the total value of inputs used for production.

$$CIMPS_i = \frac{\sum_{r=1}^{R} \overline{W_r} X_{ir}}{\sum_{r=1}^{R} \overline{W_r} I_{ir}} \tag{3}$$

where X_{ir} is the amount of input r purchased (or hired in the case of labor) by the farm household i at the average input price $\overline{W_r}$, I_{ir} is the total amount of input r used in the production of the household, R is the set of different inputs, and r is the index of inputs with $r \in R$. So, when the farm households only uses inputs from the market, $X_{ik} = I_{ik}$, and $CIMPS_i = 1$. Conversely, when the farm household does not purchase any inputs from the market, $X_{ik} = 0$ and $CIMPS_i = 1$. In the case of the CIMPS indicator, we use the (calculated) value of inputs so that we can sum different inputs, which is infeasible when using physical amounts. Moreover, the use of physical amounts could be problematic in the case of fertilizer use because different crops require different amounts of fertilizer [12].

2.3. Food Security

A common indicator of food security in the literature is the energy intake of food consumption [21,28,29]. We specify this measure in our study as the caloric energy intake per day per adult male equivalent.

However, even when households have sufficient levels of caloric intake, they might still lack diversity in the intake of their nutrients [29]. To explore dietary diversity, we apply the Food Variety Score (FVS), which reflects the diversity of diets of households, as the VLSS lacks data on micronutrient consumption. The FVS is a count of the number of food items consumed, which is calculated for all households separately. In the FVS, all food items are equally weighted.

With the two food security indicators together, a more comprehensive outlook on food security can be encapsulated. For instance, farm households might have sufficient caloric intake, but their diet still might lack diversity in nutrients, as indicated by FVS [29]. Conversely, farm households might have insufficient caloric energy intake but a high variety of their diet. The two indicators need to be examined together to reach an accurate and balanced conclusion.

2.4. Regression Specifications

To avoid multi-collinearity, three separate regression models are specified, each with a different indicator for commercialization. The specifications are:

$$Y_{it} = \beta_0 + \beta_1 C_{it} + \beta_2 Z_{it} + \varepsilon_{it} \tag{4}$$

where Y_{it} is the food security status of household i at time t, C_{it} is the farm household's commercialization indicator, Z_{it} is a set of explanatory variables, and ε_{it} is the error term.

The explanatory variables include socio-economic and farm characteristics of the farm household. Socio-economic characteristics comprise of age, gender, and education level of the household head, as well as the household size and the dependency ratio. The dependency ratio is the ratio of the number of children and elderly in a household over the number of household members in the labor force. Farm characteristics include land holdings, the value of farm equipment, and the livestock holdings. Moreover, region-specific dummies to correct for unobserved heterogeneity across regions or use panel data to correct for unobserved heterogeneity across individuals are also used.

When regressing the effect of commercialization on food security at the household level, there may be differences in access to credit or access to markets that influence the household's transaction costs, and these are captured by household- and region-specific factors [30]. Thus, the circumstances of a farm household partly pre-determine the effect of commercialization. Farmers in remote areas with large distances to markets are less likely to participate in market activities (selling crop yields or buying crop inputs).

In similar studies, the food security model specified in Equation (4) is likely to suffer from misspecification because of a potential causal relationship between food security and commercialization, or unobserved heterogeneity. As Equation (4) reflects the impact of commercialization on food security, the status of food security might also affect the degree of commercialization in the next growing season. As we observe the food security status after a harvesting period, we expect that the commercialization indicators affect the food security status but not the other way around. In other words, it is unlikely that endogeneity of commercialization factors is present.

As indicated in Equation (4), we apply a panel data specification in our analyses. However, the panel data estimation results with the fixed effects (FE) estimator indicated that there is only minor variation in our samples and subsamples over time. As a result, we only looked at the first difference estimator, based on the specification in Equation (5).

$$\Delta Y_{it} = \beta_0 + \beta_1 \Delta C_{it} + \beta_2 \Delta Z_{it} + \varepsilon_{it} \tag{5}$$

In Equation (5), there is the same set of explanatory variables as in Equation (1). For all combinations of two food security indicators (Y_{it}) and three commercialization indicators (C_{it}), we estimated four specifications: OLS 1992/93, 1997/98, pooled OLS, and first difference estimations.

3. Data

This study uses two cross-sections of the Vietnam Living Standards Survey (VLSS), namely for the periods 1992/93 and 1997/98. The VLSS was conducted by Vietnam's General Statistics Office (GSO) in collaboration with the World Bank [31,32]. Both surveys are representative at the national and regional level. The surveys include 4800 and 6002 households for 1992/93 and 1997/98, respectively (Table 1). A total of approximately 4300 households participated in both surveys [3].

Table 1. Households types included in the Vietnam Living Standards Survey (VLSS) 1992/93 and 1997/98.

Household Type	1992/93		1997/98	
	#	Share (%)	#	Share (%)
Total number of households	4800	100.0	6002	100.0
No agricultural production	846	17.6	1647	27.4
Only livestock production	196	4.1	161	2.7
Involved in crop production	3758	78.3	4194	69.9
Involved in cropping	3758	100.0	4194	100.0
Involved in cropping in both years	3231	86.0	3231	77.0
Only one of the years	527	14.0	963	23.0
Sample	3231	100.0	3231	100.0
Included observations	2943	91.1	2943	91.1
Excluded observations	288	8.9	288	8.9

The questionnaires of the surveys included questions on households' food consumption, agriculture (production and equipment), demographics, and socio-economic aspects. In addition, community questionnaires of the VLSS were administered in 120 rural communities included in the sample. This community questionnaire consists of questions on demographics, economy and infrastructure, education, health and agriculture, and prices.

According to the VLSS in 1992/93, 82.4% of the Vietnamese households were involved in agriculture, and this share declined to 72.6% in 1998, see Table 1. For households involved in crop production, the share declined from 78.3% in 1992/93 to 69.9% in 1997/98. The declining trend in agricultural involvement in Vietnamese households was also observed by the World Bank, which reported a strong decline in the employment in agriculture from 70% of the total employment to 65.3% between 1996 and 1998 [11].

For analyzing the impact of commercialization on food security, we select a sample of households which are involved in crop production and were present in both surveys. In this way, we can see the development of household farms with respect to both commercialization activities and food security. Farm households with missing data or extreme/outlying values on relevant indicators on commercialization and food security were excluded. We trimmed caloric intake per adult male equivalent per day to the range of 500–5000 kilocalories. The final sample contained 2943 farm households.

3.1. Food Security

Both the caloric intake and the FVS were derived from the food consumption section of the VLSS. It registers food consumption for households rather than individuals. Respondents were asked to recall food consumption for two lengths of periods, annually with information on annual food expenditures and physical units and fortnightly with information on whether or not food items were consumed recently. For caloric intake, we used the annual information to calculate the total amount consumed for each food item, as we will explain later on. The FVS is not provided by the VLSS, but we derived it from the survey. First, we counted the food items bought in the previous fortnight. This ignores the food items (crops or animal products) produced or stored by the household. Then, we calculated the number of food items which were produced and stored for consumption. Both lists were then combined into a single list for each household, and the number of food items on the combined list was counted.

The use of derived or constructed food security indicators has some issues. On the one hand, food security indicators might be overestimated. Firstly, recall periods of actual food consumption are preferably one day or one week. Secondly, the consequence of long recall periods is that consumers achieve higher levels of dietary diversity by definition. Thirdly, the diversity of food consumption is likely to be overestimated. On the other hand, the food consumption data has a category "food away from home", i.e., lunch or dinner eaten outside the home. For this food category, it is unclear what kind of food items or how much food was eaten. This category was significant. By ignoring this category, food security indicators might be underestimated. As a consequence, the values of the FVSs derived from the VLSS may be rather high.

The energy intake of food items was not directly included in the VLSS, and we used the annual food consumption data in combination with energy conversion factors for food items, see Table A1 in Appendix A. This approach implicitly assumes that no food is wasted [33]. The caloric intake will fluctuate with the size and the type of household members. In order to make the caloric intake comparable between different types of household sizes, we calculated the caloric intake per day per Adult Male Equivalent (AME). The AME indicator was derived from the household composition based on the conversion factors for age and gender of the household members, see Table A2 in Appendix A. The mean AME in 1992/93 was 4.05 and in 1997/98, it was 3.95. This is a decline of 0.5%, while the average household size declined by more than 3%, see Table A3 in Appendix A.

The average households' caloric intake in the sample increased from 2514 kcal per AME per day in 1992/93 to 2531 kcal in 1997/98, which is an increase of 0.7% (Table 2). In 1992/93, Central Highlands showed the highest average energy intake (2854 kcal) and North Central Coast the lowest (2307 kcal). The other regions all have an average caloric intake that is similar to the national average. In 1997/98, the differences in energy intake across regions are negligible—the Red River Delta showed the highest energy intake (2620 kcal) and South Central Coast the lowest (2434 kcal) in 1997/98. The highest increase observed was 6.8% in the North Central Coast region, and the highest decline was—13.2% in the Central Highlands.

Table 2. Mean, standard deviation, and relative change of the caloric intake per adult male equivalents (AME) per day (in kcal) per region and per period.

	1992/93		1997/98		Change
Region	**Mean**	**St.dev**	**Mean**	**St.dev**	**%**
North mountains and midlands	2587	580	2594	523	0.28
Red River Delta	2540	531	2620	575	3.11
North Central Coast	2307	523	2463	513	6.76
South Central Coast	2506	745	2434	531	−2.90
Central Highlands	2854	785	2479	446	−13.16
Southeast	2561	783	2563	562	0.06
Mekong Delta	2542	723	2455	614	−3.42
Total	2514	634	2531	557	0.69

Table 3 shows the constructed Food Variety Score for the different regions in Vietnam. The FVS increased from 9.4 in 1992/93 to 10.8 1997/98, which is an increase of 14.6%. The highest FVS found was in the Central Highlands for both cross-sections of the sample, namely 11.6 and 12.1 in 1992/93 and 1997/98, respectively. The north mountains and midlands region showed the lowest FVS in 1992/93 (8.2) and in 1997/98 (9.7). The FVSs of all regions increased except for the Mekong Delta, where the FVS declined from 11.0 in 1992/93 to 10.1 in 1997/98. The highest increase in the period was observed in the Southeast region with an increase of 32.3%.

Table 3. Mean, standard deviation, and relative change of the Food Variety Score (FVS) per region and per period.

	1992/93		1997/98		Change
Region	**Mean**	**Standard Deviation**	**Mean**	**Standard Deviation**	**%**
North mountains and midlands	8.18	3.73	9.69	4.15	18.45
Red River Delta	8.37	4.16	10.07	4.96	20.25
North Central Coast	9.32	3.72	10.84	4.57	16.38
South Central Coast	10.31	4.43	13.15	3.98	27.56
Central Highlands	11.58	5.88	12.11	3.51	4.55
Southeast	10.95	4.20	14.49	4.63	32.29
Mekong Delta	10.96	4.79	10.05	4.14	−8.35
Total	9.39	4.35	10.77	4.65	14.64

Therefore, four regions (i.e., North mountains and midlands, Red River Delta, North Central Coast, and Southeast) showed higher caloric intake and diversity, while Mekong Delta showed a decline in both indicators. Central Highlands and South Central Coast had lower caloric intakes but larger diversity. It is important to note that we only compared the average of the food security indicators. The individual score of farm households may differ from this average trend in food security, see Figure 2.

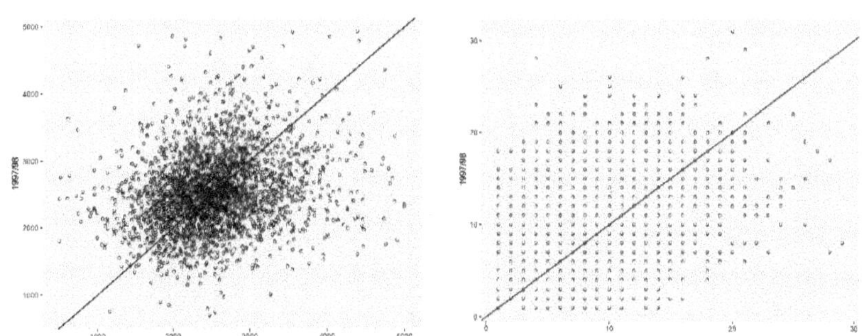

(a) (b)

Figure 2. Change of food security indicators between 1992/93 and 1997/98 at the farm household level: this is (**a**) caloric intake per adult male equivalent (AME) per day (kcal) and (**b**) Food Variety Score (FVS).

3.2. *Commercialization*

After the political and economic changes that took place in Vietnam, it is reasonable to expect farmers and farm households to have more opportunities to access markets, which, from a food-system perspective, refers to both for selling their outputs, as well as for purchasing their agricultural inputs.

Farmers grow cash crops usually for market sales or exports. In Vietnam, the crops produced for export are cashew, coffee, pepper, rubber, and tea [34]. Rice is both produced for export and domestic consumption, but the share of rice that is sold on the domestic markets is much lower than the shares of the cash crops named above [34]. Moreover, rice is not only the main crop produced in Vietnam, but there are also land restrictions that enforce the production of rice in certain areas [35]. For these reasons, rice is not included in the list of cash crops as was done in some earlier studies [34].

Table 4 shows the descriptive statistics of the CCPS indicator, which measures the share of cash-crop value in the total value of production. At the national level, the share of cash crops is 3–4% in both cross-sections. Table 4 shows that there is a large difference in the CCPS indicator values across regions. Central Highlands and the Southeast region are regions known for their cash-crop production.

Table 4. Mean and standard deviation of the cash crop participation share (CCPS) per region and per survey.

	1992/93		1997/98	
Region	**Mean**	**Standard Deviation**	**Mean**	**Standard Deviation**
North mountains and midlands	1.95	6.50	0.87	3.83
Red River Delta	0.24	2.12	0.24	2.18
North Central Coast	1.33	6.53	1.95	7.92
South Central Coast	1.68	8.36	1.00	4.74
Central Highlands	47.16	36.14	74.90	38.42
Southeast	19.28	33.05	19.86	35.85
Mekong Delta	0.21	1.79	0.11	0.99
Total	3.28	13.92	3.82	16.88

In Central Highlands and the Southeast region, the CCPS indicator is significant and has increased over time, while in the other regions the CCPS indicator is lower than 2%. In Central Highlands, the CCPS indicator increased from 47% to almost 75% in the period of analysis, which represents almost a 59% increase. The CCPS indicator for the Southeast region increased more marginally, from 19.3% to 19.9% (a 3% increase).

Table 5 shows the descriptive statistics of the COMPS indicator, which reflects the total share of the marketed output in the total production value. The national average was 28.2% in 1992/93 and 40.2% in 1997/98. This represents an increase of 42.4%. All regions showed a substantial increase of the COMPS over time. As shown in Table 5, there are large differences across regions. Central Highlands and the Southeast region have high values of COMPS in 1992/93, as well as in 1997/98. Both regions also have significant shares of cash crops which is also reflected in the COMPS. The increase of the COMPS for both regions between 1992/93 and 1997/98 was between 20% and 25%, and this is lower than the national average (42.4%), see Table 7. The Mekong Delta region showed a high value for the COMPS compared to the northern, non-cash crop regions.

Table 5. Mean and standard deviation of the crop output market participation share (COMPS) per region and per survey.

	1992/93		1997/98	
Region	**Mean**	**Standard Deviation**	**Mean**	**Standard Deviation**
North mountains and midlands	21.16	18.41	30.47	23.93
Red River Delta	20.00	16.22	28.78	21.10
North Central Coast	20.12	16.24	30.09	20.98
South Central Coast	21.71	21.53	35.34	23.47
Central Highlands	69.50	24.17	84.89	22.61
Southeast	51.09	33.86	63.60	31.80
Mekong Delta	46.73	26.40	66.85	26.10
Total	28.23	24.78	40.20	28.95

The CIMPS reflects the share of inputs purchased in the total input use for production. Although we could have taken into account multiple types of input such as labor, seeds, etc., we limited this indicator to fertilizer use alone because there were no reliable indicators for other inputs purchased at the market such as pesticides or irrigation water. There are two types of fertilizers in the VLSS, namely chemical fertilizer and organic fertilizer. Chemical fertilizer was always purchased at the market, while organic fertilizer can be produced by the farmer or purchased at the market. This was registered as part of the agricultural production of the VLSS. The CIMPS indicator is defined as the sum of chemical and organic fertilizer purchased over the total value of fertilizer used for production.

Table 6 shows the CIMPS indicator across regions and over time. On average, 47.7% of the total fertilizer used was purchased at the market in 1992/93, this share increased to 51.6% in 1997/98. This is an increase of more than 8% in 5 years. Note that this increase does not indicate whether fertilizer use has increased or not. Furthermore, Table 6 shows large differences in the values of the CIMPS, although the differences are stable over time. In the Mekong Delta, farmers purchased most of the fertilizer used at the markets. Also, in the Central Highlands and the Southeast region, the CIMPS values were equal to or over 80%. In the other four regions, the CIMPS value was in the range of 25-50%. All regions showed an increase of the CIMPS over time. In the north mountains and midlands, the increase of the CIMPS was 28% across the 2 periods examined, see Table 7.

Table 6. Mean, standard deviation, and relative change of the crop input market share (CIMPS) per region and per survey.

	1992/93		1997/98	
Region	**Mean**	**Standard Deviation**	**Mean**	**Standard Deviation**
North mountains and midlands	25.16	22.06	32.21	20.75
Red River Delta	35.68	18.55	38.21	22.62
North Central Coast	28.15	21.17	31.00	20.11
South Central Coast	44.67	29.41	50.88	30.50
Central Highlands	79.77	29.54	87.43	23.90
Southeast	81.14	26.17	82.29	25.45
Mekong Delta	97.28	10.29	99.55	2.64
Total	47.68	34.20	51.58	33.59

Table 7. Change in cash crop participation share (CCPS), crop output market participation share (COMPS), and crop input market participation share (CIMPS) over time per region.

	CCPS	COMPS	CIMPS
Region	**Change (%)**	**Change (%)**	**Change (%)**
North mountains and midlands		44.0	28.0
Red River Delta		43.9	7.1
North Central Coast		49.6	10.1
South Central Coast		62.8	13.9
Central Highlands	58.80	22.1	9.6
Southeast	3.02	24.5	1.4
Mekong Delta		43.0	2.3
Total	16.34	42.4	8.2

At the individual farm level, the picture is similar, see Figure 3.

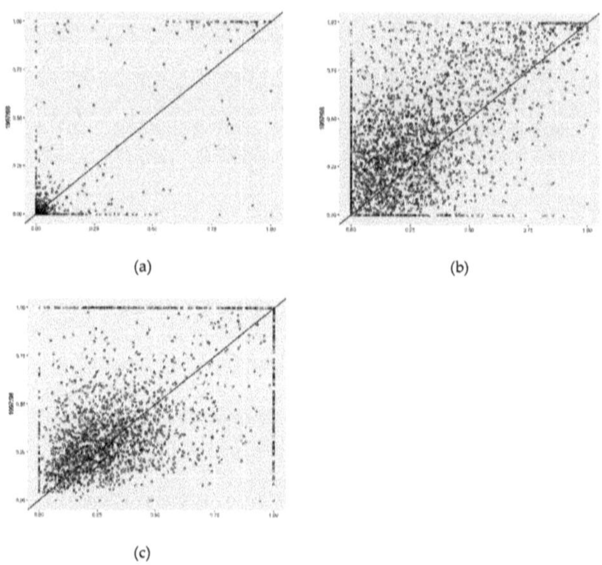

Figure 3. Change of commercialization indicators between 1992/93 and 1997/98 at farm household level: (**a**) cash crop participation share (CCPS), (**b**) crop output market participation share (COMPS) and (**c**) crop input market participation share (CIMPS).

4. Results

The results of the descriptive statistics analysis in the previous section indicate the heterogeneity of the regions with respect to food security, commercialization, and the development of both over time. In this section, we will explore whether or not there is a relationship between commercialization and food security. In particular, we are interested in whether or not commercialization after a political and economic change can contribute to an improved food security status of crop-farming households. We applied the regressions on three commercialization indicators (CCPS, COMPS, and CIMPS) to two food security indicators (caloric intake and dietary diversity measured by the FVS). The regressions were run separately for the seven regions of Vietnam.

For all combinations of commercialization indicators, food security indicators, and regions, we applied four specifications. For the first two specifications, we applied a linear model for each year separately, so that we can identify whether or not the coefficients of commercialization differ over time, per region. Then, we used a specification for a pooled regression assuming that coefficients for determinants are the same for both years. Finally, we applied a fixed effects (FE) difference estimator, because the within group variation in our sample was limited and yielded biased results.

Below, we first present the impact of commercialization on caloric intake, followed by the impact of commercialization on the dietary diversity of the households. For convenience, we only present the results of the coefficients of the commercialization indicators in the tables. Detailed estimation results are presented in supplementary tables.

4.1. Caloric Intake

The impact of the CCPS was only tested for the Central Highlands and the Southeast regions because the participation in cash-crop production was negligible in the other regions. For Central Highlands, there is no significant effect for CCPS on the caloric intake in any of the four specifications, see Table 8. In the case of the Southeast region, the impact of CCPS on FVS was positive and significant for all specifications.

Table 8. Regression results of cash crop participation share (CCPS), crop output market participation share (COMPS), and crop input market participation share (CIMPS) on caloric intake per adult male equivalent (AME) per day.

Region (Number of observations)	OLS 1992/93	OLS 1997/98	Pooled	Difference-in-Difference
CCPS				
Central Highlands (74)	−254.74	184.13	14.55	113.70
Southeast (181)	392.13 **	231.42 *	310.15 ***	314.14 **
COMPS				
North mountains and midlands (606)	58.33	−49.67	−7.66	76.01
Red River Delta (738)	−33.55	−193.98**	−119.02	−104.12
North Central Coast (530)	−100.93	66.07	33.90	161.78
South Central Coast (303)	113.87	78.63	81.93	213.70
Central Highlands (74)	−293.37	−75.83	−43.90	−257.52
Southeast (181)	−12.87	166.34	89.71	100.30
Mekong Delta (511)	147.51	−110.17	37.73	115.12
CIMPS				
North mountains and midlands (606)	238.67 **	−157.26*	36.15	122.43
Red River Delta (738)	−262.11 **	−138.68	−174.64 ***	−307.49 ***
North Central Coast (530)	363.87 ***	42.10	198.75 ***	296.98 ***
South Central Coast (303)	−90.14	103.16	−6.96	8.70
Central Highlands (74)	327.81	199.16	277.89	73.65
Southeast (181)	6.92	−103.54	−48.01	14.72
Mekong Delta (511)	328.96	−1,494.9	283.14	553.77

Note: * $p < 0.1$; ** $p < 0.05$; and *** $p < 0.01$.

In the regressions for the impact of the COMPS on caloric intake, there were hardly any significant coefficients except for a negative coefficient in 1997/98 for the Red River Delta, see Table 8. Market participation has increased for most regions, as shown in Table 8, but it did not affect the development of caloric intake. As the CCPS indicator for the Southeast region was positive, the COMPS indicator for the Southeast region did not show any significant positive impact, although the CCPS measures the value of cash-crop output and COMPS measures the total value of output sold at markets, i.e., CCPS is thus a part of the COMPS. Therefore, the crop-farming households that are involved in cash crops were more market-oriented and show increased caloric intake, while those not involved in cash-crop farming did not show any significant change in caloric intake.

For the regressions on caloric intake, there are only two significant coefficients for the CIMPS, see Table 8. In the North Central Coast region, participation in the input market has had a positive impact on the caloric intake of crop-farming households. In the Red River Delta region, this impact was negative. Both regions showed a similar increase in both the CIMPS and caloric intake, see Table 2, but the impact of the CIMPS on caloric intake was negative. Apparently, the dynamics of the impacts within both regions are different. In the north mountains and midlands regions, the impact varied across the two time periods, positive in 1992/93 and negative in 1997/98. In the other regions, the CIMPS had no significant impact.

The different commercialization indicators appeared to hardly affect the caloric intake indicator. The results in Table 8 show that in most cases, no significant impact of commercialization indicators on caloric intake was found. In the Red River Delta, there were negative impacts from both COMPS and CIMPS. In the North Central Coast, there was a positive impact from the CIMPS, but not from the COMPS, and in the Southeast region, CCPS had a positive impact on caloric intake. In the North mountains and midlands region, the impact of the CIMPS was significant but the direction varied over time. The results indicate that the impact of improved commercialization does not automatically imply increased caloric intake.

4.2. Dietary Diversity (FVS)

The cash-crop indicators showed mixed results across the regions of the Central Highlands and the Southeast (Table 9). In the Central Highlands, the impact of CCPS on dietary diversity was negative but insignificant, while the CCPS had a positive and significant impact on the dietary diversity in the Southeast region. These impacts are similar to the impacts of CCPS on caloric intake, although the impact of CCPS on FVS for the Southeast region was more significant than the impacts on caloric intake.

Table 9. Regression results of cash crop participation share (CCPS), crop output market participation share (COMPS), and crop input market participation share (CIMPS) on Food Variety Score (FVS).

Region (Number of observations)	OLS 1992/93	OLS 1997/98	Pooled	Difference-in-Difference
CCPS				
Central Highlands (74)	−3.16	−1.36	−2.59 **	−0.363
Southeast (181)	0.56	5.43 ***	3.09 ***	3.05 ***
COMPS				
North mountains and midlands (606)	−0.61	1.53 **	0.69	0.50
Red River Delta (738)	−0.94	−3.24 ***	−2.39 ***	−1.79 **
North Central Coast (530)	−0.87	−1.36	−1.16 *	−2.29 **
South Central Coast (303)	1.66	−0.40	0.04	−0.65
Central Highlands (74)	0.46	−0.27	0.15	1.25
Southeast (181)	1.49	1.15	1.28 *	0.20
Mekong Delta (511)	3.20 ***	0.33	1.78 ***	2.68 ***
CIMPS				
North mountains and midlands (606)	−0.62	−1.11	−0.80	−1.58 **
Red River Delta (738)	−4.19 ***	0.11	−1.74 ***	−1.49 *
North Central Coast (530)	−0.23	2.95 ***	1.25 *	2.66 ***
South Central Coast (303)	−0.78	0.0003	−0.32	0.37
Central Highlands (74)	5.01 *	−11.08 ***	0.12	−1.42
Southeast (181)	−2.05	−1.44	−1.88 *	−2.57 *
Mekong Delta (511)	−1.14	−12.70 *	−1.77	−2.32

Note: * $p < 0.1$; ** $p < 0.05$; and *** $p < 0.01$.

The COMPS coefficient in the FVS specification for 1992/93 was significantly positive in the Mekong Delta, while the coefficient was positive but insignificant for 1997/98. In the north mountains and midlands, the COMPS coefficient in 1997/98 was significant and positive; although, there was no significant coefficient in the difference estimator specification. In the Red River Delta, the COMPS coefficient for FVS was insignificant in 1992/93 but significant and negative in 1997/98. In the other regions, no significant coefficients were observed. Therefore, in the Southeast region, the CCPS had a significant impact on FVS while the COMPS did not.

The CIMPS coefficient in the FVS regression for the Red River Delta was negative for 1992/93 and insignificant for 1997/98, while the opposite was true for the Mekong Delta. In the Central Highland, the impact was positive for 1992/93 but negative for 1997/98 and insignificant for the difference estimator specification. In the Southeast region, the CIMPS indicator had a negative impact; although, this was only observed in the difference estimator specification. In the North Central Coast region, the CIMPS coefficient in the FVS regression was negative and insignificant in 1992/93, but positive and significant in 1997/98.

There was a greater number of significant coefficients for the COMPS indicator on dietary diversity than there was on caloric intake, while the signs of the coefficients were mixed. For the CCPS indicator, the impacts on dietary diversity were similar to the impacts on caloric intake. This was also true for the CIMPS indicator for the Red River Delta and the North Central Coast region. In the case of dietary diversity, the CIMPS indicator had a greater number of coefficients that were significant (than any other indicators).

To summarize, there is no consistent evidence for an overall impact of commercialization of crop-growing farmers on food security, either in terms of caloric intake or food diversity in the different regions in Vietnam. The impacts differ in sign, magnitude, and significance. However, in general, the impacts appear to be more positive for the southern regions (Central Highlands, the Southeast, and the Mekong Delta) compared to the northern regions.

5. Discussion and Conclusions

Agricultural commercialization is traditionally measured by the involvement of farmers in cash-crop production because cash crops are produced to be sold or exported. However, agricultural commercialization from a food-system perspective needs to also include market participation at other sectors of the food system, such as input markets. Therefore, we explored three commercialization indicators, namely the cash crop production share (CCPS), the crop output market participation share (COMPS), and the input market participation share (CIMPS).

The results showed that the commercialization of crop-farming households has increased over time after the political and economic regime changes in Vietnam. All three commercialization indicators increased unanimously but the magnitude of the increase differed widely. Therefore, it can be concluded that market participation was not limited to the agricultural output market or the involvement in cash-crop production but also higher participation in input markets. We only considered fertilizer inputs, as no reliable data on other inputs was available from the survey. Furthermore, dietary diversity also increased while caloric intake remained rather constant. Although this is the general trend for Vietnam, there are differences across regions, as observed in earlier studies [10]. Hence, we focused our attention on the impact of commercialization on food security at the regional level.

In general, there was limited variation in the caloric intake indicator, and it remains constant between 1992/93 and 1997/98, while the commercialization indicators exhibited a positive trend. The combination of these observed trends made it less likely to find a positive relationship between commercialization indicators and caloric intake. For dietary diversity, however, we observed an increasing trend.

If we had only looked at the cash crop participation share (CCPS), we would not have been able to analyze commercialization in five of the seven regions in Vietnam because cash-crop production hardly existed in these regions. Only the Central Highlands and the Southeast regions had a significant share of the farmers involved in cash-crop farming in the 1990s. The results for the Southeast region indicate that commercialization had a significant positive effect on dietary diversity. For the Central Highlands, caloric intake dropped in the considered period while commercialization remained relatively constant, noting that the share of cash-crop participation was already high in this region.

However, when considering the output side of commercialization, where households have different shares of crop marketing (different values for COMPS), COMPS had a significant positive effect on the FVS and a negative effect on caloric intake. It is important to note that the significant positive effect of COMPS on the FVS only seems to be the case in the south of Vietnam. In the northern regions, there seems to be no significant effect. In earlier studies, there were no significant coefficients found for the impact of the COMPS indicator on food security [9].

The difference in results between the north and the south is as expected. The benefits of the reforms differed across the northern and southern regions. Before the reforms, rice and cash-crop production were mainly concentrated in the south, and the south also suffered most from the export quota that was in place at that time. Therefore, when this export quota was removed, the southern regions benefited more than the northern regions [7]. Moreover, farming households in the south already had crop specialization, rice, and cash crops. Therefore, as a result of the liberalizations, households in the south were able to more easily adjust their production and the amount sold according to changes in market conditions, taking advantage of a price increase in rice for instance [7,8].

When looking at the commercialization of the input side of the food system, CIMPS showed no significant relationships with the food security indicators. This means that there is no clear significant effect of the CIMPS on food security across regions. However, similar to the impact of the COMPS indicator, we observed some differences between the north and south. The north experienced larger increases in their CIMPS compared to the south. From the regression results, the number of significant negative effects of increased CIMPS on both food security indicators in northern regions was larger than in southern regions. During the liberalizations, fertilizer supply constraints were largely removed [8], and the prices of fertilizer dropped [7]. As a result, the amount of fertilizer use increased. Given the fact that the share of fertilizer purchased at the market was high in the southern regions, the total costs of fertilizer increased over time, which has had a negative impact on income. [8]. This might explain the negative impact of commercialization on food security in the southern regions [8].

The results of our analysis show that the relationship between agricultural commercialization and food security is very complex. The impact depends on both the indicators and the region. Moreover, obvious trends in certain regions do not directly translate into obvious increases or decreases in the impact of commercialization on food security. The promotion of agricultural commercialization in one region, or of one specific type of commercialization, may thus lead to negative impacts, while for another region, or another type of commercialization, it may lead to positive impacts.

One of the shortcomings of our approach is that we did not take into account other aspects or developments, such as hired labor for instance. The general trend that can be concluded for developing countries is that people move away from the agricultural sector. The data showed that there was an 8.4% decrease in crop-producing households between 1992/93 and 1997/98 [10]. For agricultural households, however, we found that the total average income increased mainly due to increases in on-farm income (i.e., higher quantities and returns from agricultural production). This indicates that commercialization of the agricultural activities of households might have been more important than, for instance, the commercialization of labor [10].

Our analysis was limited to the impact of economic and political changes within a five-year period. It is likely that full adjustments to these economic changes would occur over a longer period than that which data were available for. Additionally, if the analysis examined the impact over a longer period, the results could be less ambiguous and more robust. However, with longer periods, more dynamics at the farm household level could be introduced, such as changes in households' composition, changing households' head, and entering or exiting of agricultural activities, which would have affected the regression results.

To analyze the impact of commercialization on food security, it is important to look at the change in expenditures. Earlier studies indicated that households spend a larger share of their increased incomes on non-food items than on food items [10,20]. Our results showed that the FVS, on average, increased with commercialization while caloric intake did not, and this points to an increase in demand for more diverse diets, which was observed in earlier studies [18].

Our research could be extended in multiple ways. The CIMPS indicator measures another element of the food system that can be commercialized. The main drawback of using the CIMPS indicator is that it is only based on one of the input factors of production, namely fertilizer, although there are many other inputs to consider such as hired labor and pesticide use, amongst others. Pesticides, for instance, can only be purchased at the market. The inclusion of pesticides in the CIMPS indicator would have boiled down to pesticide use or not.

Furthermore, we have explored the impacts of the commercialization indicators on food security in separate specifications in our analyses so that we would avoid any possible multi-collinearity issues. From a food-system perspective, it would be interesting to create indicators that combine market participation in input and output markets for instance.

Finally, the impact of the commercialization of agriculture can also be observed outside agriculture, such as farmers exiting the industry, labor moving to other sectors outside the food system, or the entrance of new actors in agriculture such as foreign companies or investors. Additional analyses are needed to explore the impacts of economic and political changes on these factors.

To conclude, one of the reasons to analyze the impact of commercialization in agriculture on food security in the regions in Vietnam was to explore the possibilities to transfer these relationships to other cases like Myanmar. However, it will be hard to find the right circumstances to transfer the relationships given the large variations in results across regions in our study.

Supplementary Materials: The following are available online at http://www.mdpi.com/2071-1050/11/5/1263/s1. The document with the regression results underlying the results of Tables 8 and 9 are available.

Author Contributions: V.L, V.J., and T.A. worked on the concepts and methodologies for this paper. V.L. wrote the paper with contributions of V.J., V.J. analyzed the data with contributions from V.L., T.A. reviewed the paper.

Funding: This research received was funded by the Dutch Ministry of Agriculture, Nature and Food Quality, grant number KB-22.

Acknowledgments: The authors thank Maggie Skirtun and anonymous reviewers for their comments that contributed to improve the quality of this paper. Any remaining errors are ours.

Conflicts of Interest: The authors declare no conflicts of interest. The funders had no role in the design of the study; in the collection, analyses, or interpretation of data; in the writing of the manuscript, and in the decision to publish the results.

Appendix A

Table A1. Calorie conversion rate (calories per kg) for food items [32].

Food Items	Calories per kg	Food Items	Calories per kg
Ordinary rice	3530	Beans	3142
Glutinous rice	3550	Water morning glory	210
Corn/maize	3640	Kohlrabi	300
Cassava	1560	Cabbage	370
Potatoes	1088	Tomatoes	370
Barley, Malt, Millet, Kaoling *	3320	Other vegetables	-
Bread wheat, flour	3015	Oranges	430
(pho) Noodle and instant rice soup	3580	Bananas	830
Rice noodle	3400	Mangoes	290
Vermicelli	1285	Other fruits *	170

Table A1. *Cont.*

Food Items	Calories per kg	Food Items	Calories per kg
Pork	3956	Fish sauce and dipping sauce	332
Beef and buffalo meat	1233	Salt	-
Chicken	1759	Sugar, molasses *	3870
Duck and other poultry meat	1260	Cakes, jams, sweets	4026
Other meat *	2630	Fresh milk	868
Processed meat	3259	Alcohol & beer	470
Fat and oil	9270	Coffee *	560
Fresh fish, shrimp	900	Tea	-
Dried/processed fish and shrimp	2409	Beverages (industrial methods)	470
Other seafood (crab, snails etc.) *	660	Food and drink away from home *	410
Chicken or duck eggs (per one) **	1482	Others *	1700
Tofu	980		
Peanuts, sesame seeds	5445		

Source: Nguyen & Winters [32], adjusted from Vietnam's National Institute of Nutrition (NIN) and General Statistics Office (GSO) of Vietnam. Note: * The conversion rate was not available, so we used the Food and Agriculture Organization (FAO) conversion rates [36]. ** Multiplied with the average weight of a chicken egg in Vietnam [37].

Table A2. FAO adjustment factors for calculating the Adult Male Equivalents (AME) [38].

Age Categories	Males	Females
<1	0.27	0.27
1–3	0.45	0.45
4–6	0.61	0.61
7–9	0.73	0.73
10–12	0.86	0.78
13–15	0.96	0.83
16–19	1.02	0.77
≥20	1	0.73

Table A3. Adult male equivalents per household and household size in the sample for 1992/93 and 1997/98.

	1992/93		1997/98		Change (%)	
Regions	AME	Size	AME	Size	AME	Size
North mountains and midlands	4.14	5.34	4.19	5.18	1.14	−3.09
Red River Delta	3.43	4.94	3.35	4.10	−2.27	−17.00
North Central Coast	3.84	4.94	3.81	4.77	−0.88	−3.51
South Central Coast	4.22	5.33	4.02	5.00	−4.77	−6.25
Central Highlands	4.74	6.14	4.76	6.05	0.36	−1.32
Southeast	4.51	5.61	4.41	5.36	−2.15	−4.43
Mekong Delta	4.71	5.91	4.36	5.34	−7.38	−9.70
Total	4.05	5.16	3.95	4.88	−2.53	−5.43

References

1. Von Braun, J. Agricultural commercialization: Impacts on income and nutrition and implications for policy. *Food Policy* **1995**, *20*, 187–202. [CrossRef]
2. Ingram, J. A food systems approach to researching food security and its interactions with global environmental change. *Food Secur.* **2011**, *3*, 417–431. [CrossRef]
3. Glewwe, P.; Dang, H.-A.H. Was Vietnam's economic growth in the 1990s pro-poor? An analysis of panel data from Vietnam. *Econ. Dev. Cult. Chang.* **2011**, *59*, 583–608. [CrossRef]
4. Rozelle, S.; Swinnen, J.F. Success and failure of reform: Insights from the transition of agriculture. *J. Econ. Lit.* **2004**, *42*, 404–456. [CrossRef]
5. Nguyen, H.H. *Regional Welfare Disparities and Regional Economic Growth in Vietnam*; Wageningen Universiteit: Wageningen, The Netherlands, 2009.
6. World Bank. *Vietnam: Attacking Poverty*; World Bank: Washington, DC, USA, 1999.
7. Benjamin, D.; Brandt, L. Agriculture and income distribution in rural Vietnam under economic reforms: A tale of two regions. In *Economic Growth, Poverty and Household Welfare in Vietnam*; World Bank: Washington, DC, USA, 2004; pp. 133–185.
8. Coello, B. *Agriculture and Trade Liberalization in Vietnam*; Paris School of Economics; Paris School of Economics: Paris, France, 2007; p. 25.
9. Carletto, C.; Corral, P.; Guelfi, A. Agricultural commercialization and nutrition revisited: Empirical evidence from three African countries. *Food Policy* **2016**, *67*, 106–118. [CrossRef] [PubMed]
10. Janssen, V. *The Effect of Agricultural Commercialization on Food Security the Case of Crop Producing Households in Post-Reform Vietnam*; Wageningen UR: Wageningen, The Netherlands, 2018; p. 66.
11. World Bank. *World Development Report 2008: Agriculture for Development*; World Bank: Washington, DC, USA; London, UK, 2007; ISBN 978-0-8213-6809-1.
12. Gebremedhin, B.; Jaleta, M. Commercialization of smallholders: Is market participation enough. In Proceedings of the Contributed Paper Presented at the Joint 3rd African Association of Agricultural Economists (AAAE) and 48th Agricultural Economists Association of South Africa (AEASA) Conference, Cape Town, South Africa, 19–23 September 2010.
13. Fleuret, P.; Fleuret, A. Nutrition, consumption, and agricultural change. *Hum. Organ.* **1980**, *39*, 250–260. [CrossRef]
14. Immink, M.D.; Alarcon, J.A. Household income, food availability, and commercial crop production by smallholder farmers in the western highlands of Guatemala. *Econ. Dev. Cult. Chang.* **1993**, *41*, 319–342.
15. Jaleta, M.; Gebremedhin, B.; Hoekstra, D. *Smallholder Commercialization: Processes, Determinants and Impact*; International Livestock Research Institute: Addis Ababa, Ethiopia, 2009.
16. Kanter, R.; Walls, H.L.; Tak, M.; Roberts, F.; Waage, J. A conceptual framework for understanding the impacts of agriculture and food system policies on nutrition and health. *Food Secur.* **2015**, *7*, 767–777. [CrossRef]
17. DeWalt, K.M. Nutrition and the Commercialization of Agriculture: Ten Years Later. *Soc. Sci. Med.* **1993**, *36*, 1407–1416. [CrossRef]
18. Abdulai, A.; Aubert, D. A cross-section analysis of household demand for food and nutrients in Tanzania. *Agric. Econ.* **2004**, *31*, 67–79. [CrossRef]
19. Gibson, R.S.; Hotz, C. Dietary diversification/modification strategies to enhance micronutrient content and bioavailability of diets in developing countries. *Br. J. Nutr.* **2001**, *85*, S159. [CrossRef] [PubMed]
20. Banerjee, A.; Duflo, E. *Poor Economics: A Radical Rethinking of the Way to Fight Global Poverty*; Penguin Books: London, UK, 2012; ISBN 978-0-7181-9366-9.
21. Bouis, H.E.; Haddad, L.J. *Effects of Agricultural Commercialization on Land Tenure, Household Resource Allocation, and Nutrition in the Philippines*; Research Report; International Food Policy Research Institute; International Food Policy Research Institute in collaboration with the Research Institute for Mindanao Culture: Washington, DC, USA, 1990; ISBN 978-0-89629-081-5.
22. Kennedy, E.; Cogill, B. The commercialization of agriculture and household-level food security: The case of Southwestern Kenya. *World Dev.* **1988**, *16*, 1075–1081. [CrossRef]
23. Achterbosch, T.; van Berkum, S.; Meijerink, G.; Asbreuk, H.; Oudendag, D. *Cash Crops and Food Security: Contributions to Income, Livelihood Risk and Agricultural Innovation*; LEI Wageningen UR: The Hague, The Netherlands, 2014; ISBN 978-90-8615-673-3.

24. Babatunde, R.O.; Qaim, M. Impact of off-farm income on food security and nutrition in Nigeria. *Food Policy* **2010**, *35*, 303–311. [CrossRef]
25. Timmer, C.P. Farmers and Markets: The Political Economy of New Paradigms. *Am. J. Agric. Econ.* **1997**, *79*, 621. [CrossRef]
26. Wood, B.; Nelson, C.; Kilic, T.; Murray, S. Up in Smoke? Agricultural Commercialization, Rising Food Prices and Stunting in Malawi. World Bank Policy Research Working Paper (No. 6650). Available online: https://www.researchgate.net/profile/Talip_Kilic/publication/254387650_Up_in_Smoke_Agricultural_CommercializationRising_Food_Prices_and_Stunting_in_Malawi/links/00b495384e2d34e7b6000000/Up-in-Smoke-Agricultural-Commercialization-Rising-Food-Prices-and-Stunting-in-Malawi.pdf?origin=publication_detail (accessed on 26 February 2019).
27. Niemeijer, R.; Hoorweg, J. 17 Commercialization of Rice and Nutrition: A Case from West Kenya. Available online: http://microdata.worldbank.org/index.php/ddibrowser/1910/export/?format=pdf&generate=yes (accessed on 26 February 2019).
28. Von Braun, J.; de Haen, H.; Blanken, J. *Commercialization of Agriculture under Population Pressure: Effects on Production, Consumption, and Nutrition in Rwanda*; IFPRI: Washington, DC, USA, 1991; ISBN 978-0-89629-087-7.
29. Carletto, C.; Zezza, A.; Banerjee, R. Towards better measurement of household food security: Harmonizing indicators and the role of household surveys. *Glob. Food Secur.* **2013**, *2*, 30–40. [CrossRef]
30. Barrett, C.B. Smallholder market participation: Concepts and evidence from eastern and southern Africa. *Food Policy* **2008**, *33*, 299–317. [CrossRef]
31. General Statistical Office (GSO). State Planning Committee (SPC) Vietnam—Living Standards Survey 1992–1993. Available online: http://microdata.worldbank.org/index.php/ddibrowser/1910/export/?format=pdf&generate=yes (accessed on 26 February 2019).
32. Nguyen, M.C.; Winters, P. The impact of migration on food consumption patterns: The case of Vietnam. *Food Policy* **2011**, *36*, 71–87. [CrossRef]
33. Molini, V. *Food Security in Vietnam During the 1990s: The Empirical Evidence*; UNU WIDER: Helsinki, Finland, 2006; ISBN 978-92-9190-841-7.
34. Coello, B. Exports liberalization and specialization in cash crop: Gains for Vietnamese households? *Econ. Int.* **2009**, 73–99.
35. Markussen, T.; Tarp, F.; Van Den Broeck, K. The Forgotten Property Rights: Evidence on Land Use Rights in Vietnam. *World Dev.* **2011**, *39*, 839–850. [CrossRef]
36. Mishra, V.; Ray, R. Dietary Diversity, Food Security and Undernourishment: The Vietnamese Evidence. *Asian Econ. J.* **2009**, *23*, 225–247. [CrossRef]
37. Duc, N.V.; Long, T. *Poultry Production Systems in Vietnam*; FAO: Rome, Italy, 2008; p. 22.
38. Fiedler, J.L.; Smitz, M.-F.; Dupriez, O.; Friedman, J. Household Income and Expenditure Surveys: A Tool for Accelerating the Development of Evidence-Based Fortification Programs. *Food Nutr. Bull.* **2008**, *29*, 306–319. [CrossRef] [PubMed]

sustainability

MDPI

Article

Healthy Diets and Reduced Land Pressure: Towards a Double Gain for Future Food Systems in Nigeria

Zuzana Smeets-Kristkova [1,2,*], Thom Achterbosch [1] and Marijke Kuiper [1]

1 Wageningen Economic Research, Wageningen University & Research, Prinses Beatrixlaan 582, 2595 BM The Hague, The Netherlands; thom.achterbosch@wur.nl (T.A.); marijke.kuiper@wur.nl (M.K.)

2 Czech University of Life Sciences, Faculty of Economics and Management, Kamycka 129, 165 21 Prague, Czech Republic

* Correspondence: zuzana.kristkova@wur.nl; Tel.: +31-645-748743

Received: 31 December 2018; Accepted: 1 February 2019; Published: 6 February 2019

check for updates

Abstract: Nigeria is one of the most dynamic economies in Africa. Strong GDP and population growth coupled with urbanization trends place tremendous pressures on natural resources and the food systems that are dependent on them. Understanding the impact of these "mega trends" is important to identify key leverage points for navigating towards improved nutrition and food security in Nigeria. This paper contributes to the Foresight Project of the Food Systems for Healthier Diets which aims to analyse how the food system in Nigeria is expected to transform in the next decades, and to identify the leverage points for making sure that the transformation contributes to balanced consumer diets. For the food systems foresight, a well-established global economy-wide model, MAGNET, is applied that enables to capture the interlinkages among different food industry players in one consistent framework. By linking MAGNET to the GENUS nutritional database, it is further possible to relate the developments occurring on a macro-level with detailed macro and micronutrient consumption. Model projections suggest that a process of intensification of agriculture in combination with land substitution appears critical for the evolution of food and nutrition security, and for shifts towards healthy diets for the population. Intensification results in greater diversity of the production systems, which in turn cascades into positive effects on the diversity in the food supply and better food security outcomes.

Keywords: food security; CGE model; nutrition; diet diversity; land substitution; agricultural intensification; baseline projections

1. Introduction

Nigeria is one of the most dynamic economies in Africa. Strong GDP growth and high fertility rates [1] suggest an unfinished demographic revolution, placing tremendous pressures on natural resources and the food systems that are dependent on them. In particular, arable land for expansion is becoming increasingly scarce [2]. Understanding the impact of these "mega trends" on food and nutrition security in the country is highly relevant. Malnutrition in all its forms remains a key concern in Nigeria. While chronic malnutrition remains widespread (in 2013, 37% of children under 5 years of age were stunted [3]), the need to curb the rising prevalence of overweight and obesity among adults and children is emerging as a priority for food security and health policy. Economic development and changing market conditions are associated with shifts in consumption patterns that simultaneously move towards and away from healthy diets [4]. Healthy diets typically have four characteristics—related to quantity, quality, diversity and safety of the diet—and translate into principles for adult consumption [5]: sufficient consumption of fruit, vegetables, pulses; moderate consumption of animal source food, with limited intake of processed meat and sugar-sweetened foods and beverages; avoid transfat and replace consumption of saturated fats with vegetable oils or other sources of unsaturated fat. This paper seeks to contribute with macro-level foresight to the identification of leverage points in the food system of Nigeria for promoting healthy diets in these respects with the exception of safety. Of particular interest is the question how Nigeria's national food system, given its strong bias towards staple production, could transform towards meeting these diverse nutritional needs.

There is increasing recognition that the interplay between market decisions and contextual drivers at multiple levels is important for understanding dietary quality and nutritional outcomes of food systems [5]. The definition of a food system of the High Level Panel of Experts is also used in this paper: "a *food system* gathers all the elements (environment, people, inputs, processes, infrastructures, institutions, etc.) and activities that relate to the production, processing, distribution, preparation and consumption of food, and the output of these activities, including socio-economic and environmental outcomes" [6]. This concept of a food system takes into account the complexity of external driving forces that shape food and nutrition security and dietary patterns across different levels. These drivers appear as the key factors that determine the changes in the food system. Such drivers are commonly grouped into 5 categories: innovation, technology and infrastructure; political and economic; socio-cultural; and demographic drivers [7]. In the expanding literature on drivers of global and national food systems, there is often little attention for changes that originate "from within" the food system [8]. The response of entrepreneurs, consumers and regulators to the external drivers, and the interplay of the behaviour of different actors is what makes the food system complex to understand [5,7,9].

A useful distinction can be made between the indirect and the direct drivers in the food system. This classification follows a governance perspective on drivers: it separates those that can be influenced by food system actors (direct drivers) from those that cannot be influenced (indirect drivers) [9]. Three types of feedback loops in food systems are proposed by this framework: (1) adjustments in the market and external effects that follow from interaction between producer and consumer decisions under the influence of drivers of change; (2) adjustments in the policy framework on the basis of market outcomes and external effects that are inconsistent with prevailing visions in public policy; (3) ramifications in the global food system, e.g., in world commodity markets, biophysical or geopolitical balances, etc.

Although there is a recognition of the need for understanding the relations between drivers of food systems and their impact on diets and nutrition, the empirical evidence from the existing foresight studies is rather scarce, particularly regarding the macro-economic perspective of food systems impact on nutrition. One review identified a gap in foresight research regarding how alternative uses of agricultural land impact food security considering both poor people's access to productive resources and income-earning opportunities as well as their access to food and the prevalence of hunger [10]. These recommendations are fully reflected in this paper where land is identified as the key factor that determines the future development of food systems and food security impacts.

This study thus contributes to the gap in the literature with the objective to provide national-level projections of the demand for food, and to explore how various direct and indirect drivers in the food system (farm input and output prices, demand for on-farm and off-farm labour, consumer preferences, etc.) interact with the general drivers of the food system, and to evaluate ex-ante how these dynamics affect diet outcomes at population level.

The interplay of indirect and direct drivers within the food system are explored in this paper using a computable general equilibrium (CGE) model for Nigeria and the rest of the world. The model is presented in Section 2, along with a scenario framework for exploring future food systems and diets in Nigeria. The results of the impact of drivers and interconnections on food systems outcomes and diets are presented in Section 4, followed by a discussion of the limitations and implications of the paper.

2. Materials and Methods

2.1. Scenario Framework for Exploring Future Food Systems in Nigeria

This paper employs scenario methods to understand how the indirect drivers of the food system interact at multiple scales simultaneously—supra-national level, national and sub-national level for Nigeria and household level—and addresses feedback loops between producers and consumers in Nigeria (feedback loop 1) and between the national and supranational level (feedback loop 3). An advanced quantitative modelling framework is used for this purpose, and will be introduced in the this section. A possible policy response (feedback loop 2) is explored in the discussion. The indirect drivers of change and specific characteristics of the Nigeria food system and its drivers are incorporated in the analysis as follows:

- *Supra-national level*: Included are the unequal levels of wealth and rates of economic growth among countries and regions, the sectoral composition and human capital dimension of growth including changes in population growth. Excluded is the variation across countries in the quality of basic services that are available to the population. For example, Nigeria has the largest population of out-of-school youth in the world [11,12]. The Nigerian population is increasingly youthful and urbanized, with the urban population expected to outstrip the rural population by 2025 [13]. Interactions with global markets for commodities and merchandise, and integration with the global economy through international trade and adjustments of real exchange rates are included; currency fluctuations and stabilization policies are excluded. The differentiated effect of climate change is included at global level. Section 2.3 dwells largely on these drivers.
- *National and sub-national level*: Nigerian food systems are the most important sector in the country, representing 41% of Nigeria's value added in 2011 (see Section 3.3). Included in the analysis are the interaction between agriculture, food supply chains and non-agriculture sectors (energy supply, manufacturing, services), and the interplay between technology change and changes in the factors of production at the national level. Assumptions on the expected rate of technological progress are included in the analysis; excluded are the process of diffusion and adoption. The connection between public research, education and extension services is generally considered weak [14,15], which hampers adoption processes. Nigeria's core infrastructure stock was low (about 20–25% of GDP) until recently, yet investment in infrastructure has greatly increased in the last 4–5 years. The impact of such investments on infrastructural developments such as roads, railways, storage facilities, as well as energy and ICT are excluded from the analysis. Being almost completely rain-fed, cropping systems and the national food system in Nigeria are sensitive to climatic conditions. Included in the scope of the analysis is the impact of global climate change on agricultural yields. Excluded from the analysis are the feedback mechanisms from more extreme weather, degraded land and water resources, and climate change adaptation on the resilience of the primary production systems [16]. Human-induced crises such as forced displacement of populations due to armed conflicts, insurgency, forced evictions and herdsmen-farmers clashes are affecting food systems in Nigeria, e.g., [17,18]. The impact of conflict and political instability on the food system is excluded from the analysis. Political drivers of food systems including leadership and governance and conflicts/humanitarian crises are largely excluded from the analysis, except for limited discussions of land tenure systems (Section 2.3).
- *Household level*: Included are the access to productive assets such as land or fishing grounds, capital and infrastructure for livelihood activities of households, with distinctions between skills levels in the labour market; excluded from the analysis are community-level or other in-country inequalities in access to these assets, as well in access to care, hygiene environment and opportunities for schooling. While demographic change is included in terms of population growth rates, it is not included how the food system must increasingly cater for young and urban consumers, with their nutritional needs and aspirations. Urban patterns of food choice are noticeably different from rural patterns across Africa, and include less consumption of traditional staples, more animal

and dairy products, more processed food, and a much greater proportion of food consumed outside of home [19]. Both the possible globalization of food culture and change of traditional food systems in Nigeria related to tribal culture have been scarcely documented, see e.g., [20]. Included in the scope of the present study are the variation in food access and composition of consumed diets across households depending on their regional location. Excluded from the scope are urban and rural differences in the decision-making regarding consumption in the specific demographic, socioeconomic and cultural context; the unequal socioeconomic conditions and institutional environment across households; the unequal distribution of food, money and power of decision-making between members of the household.

2.2. Economy-Wide Modelling Framework

This study applies the MAGNET (Modular Applied General Equilibrium Tool) model, a well-established CGE model used for global projections on agriculture, bio-based economy, climate, food security and nutrition as well as country-specific assessments (see for instance [21–24]). As an economy-wide model, MAGNET is well placed to examine the costs and benefits of policy scenarios via changes in input and output prices and allocation of competing (agricultural and non-agricultural) uses of primary factors and intermediate inputs [25]. From a food systems perspective, the key strength of MAGNET lies in exploring food systems dynamics, by capturing the interlinkages among different food industry players (farmers, processors, suppliers, traders and consumers) in one consistent framework (see the circular flow of MAGNET in Figure 1). Scenario analysis using MAGNET contributes to an ex-ante identification of challenges and pathways for innovation taking into account trade-offs and synergies between various objectives.

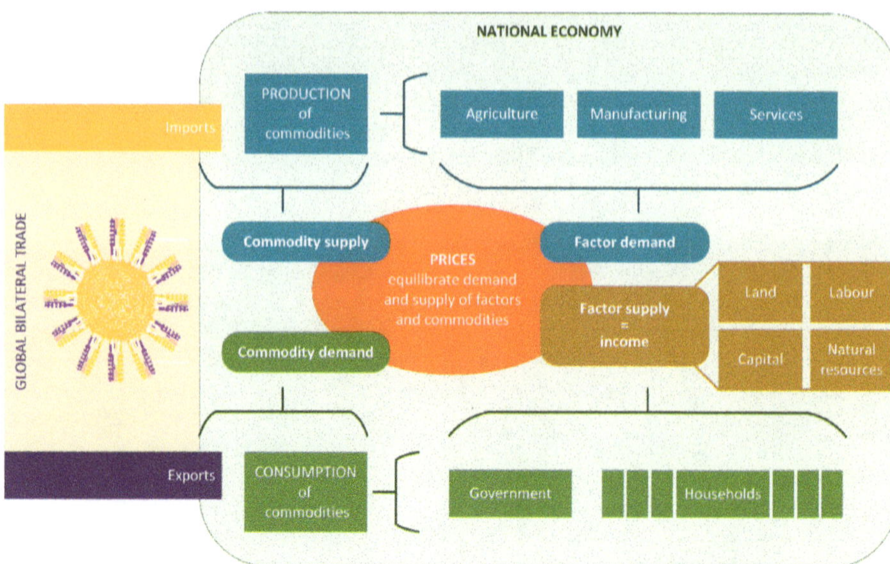

Figure 1. MAGNET—a complete economic model of nations in the global economy.

2.2.1. MAGNET Model Database

MAGNET is a neoclassical recursive dynamic, multi-regional, multi-commodity computable general equilibrium (CGE) model. At its core is the well-known Global Trade Analysis project (GTAP) model and the associated GTAP database. The core of the MAGNET database is the 2011 reference year of GTAP database version 9.2, distinguishing 140 regions, including Nigeria, 8 production factors and 57 sectors [26]. To enhance MAGNET's ability to assess food system and health implications of diet changes, FAOSTAT data is used to split standard GTAP sectors, adding more detail in terms of types of meat (beef, other ruminants, pork, poultry) both in terms of primary products and processed meat products. FISHSTAT data is used to split the GTAP fish sector into aquaculture, wild catch and a fish processing sector producing processed fish for consumers and fishmeal for use as animal feed. To capture the scope for livestock intensification and its potential links to aquaculture, an animal feed sector is defined which uses fishmeal among other crop-based inputs. For crops a fertilizer sector is separated from the broader chemical sector, which can be used as a substitute for land. This allows changes in intensification based on land rent and agricultural prices.

Running the model with the full MAGNET database of 140 regions, 83 sectors and 8 factors is computationally infeasible. Given the purpose of this study, a food-focused sector aggregation was used while the world was aggregated in 11 regions, keeping Nigeria as an individual country (see Table 1). Table 2 lists the food sectors, indicating the amount of food system detail the model is able to capture at macro level. The complete list of sectors included in MAGNET is in Appendix B, Table A1.

Table 1. Regional aggregation in MAGNET.

Code	Description	GTAP Regions Included
NGA	Nigeria	nga
ETH	Ethiopia	eth
SSA	Sub Saharan Africa	ben, bfa, cmr, civ, gha, gin, sen, tgo, xwf, xcf, xac, ken, mdg, mwi, mus, moz, rwa, tza, uga, zmb, zwe, xec, bwa, nam, zaf, xsc
VNM	Vietnam	vnm
BGD	Bangladesh	bgd
APTA	Asia-Pacific trade agreement	chn, hkg, kor, mng, twn, lao, ind, lka
EAS	East and South East Asia	xea, brn, khm, idn, mys, phl, sgp, tha, xse
EU	European Union	aut, bel, cze, dk, est, fin, fra, deu, grc, hun, irl, ita, lva, ltu, lux, mlt, nld, pol, prt, svk, svn, esp, swe, gbr, hr, rou
NAFTA	North American FTA	can usa, mex, xna
LAM	Latin America	arg, bol, bra, chl, col, ecu, pry, per, ury, ven, xsm, cri, gtm, hnd, nic, pan, slv, xca, dom, jam, pri, tto, xcb
ROW	Rest of the World	all remaining regions

Note: GTAP regions are described at: https://www.gtap.agecon.purdue.edu/databases/regions.asp?Version=9.211.

Table 2. Sectoral aggregation in MAGNET (MAGNET codes in brackets).

Category	MAGNET Food System Sectors
Arable and horticulture	paddy rice (pdr), wheat (wht); other grains (grain); oilseeds (oils); raw sugar (sug); vegetables, fruits and nuts (hort); other crops (crops); plant fibres (oagr)
Livestock and meat	beef cattle (cattle); other grazing animals such as sheep,goats,horses (othctl), wool (wol); pigs (pigpls), poultry (pltry); raw milk (milk); cattle meat (bfmt); meat from other grazing animals (othcmt), pork meat (othmt), poultry meat (pulmt); dairy (dairy)
Other food and beverages	sugar processing (sugar); rice processing (pcr); vegetable oils and fats, including crude vegetable oil (vol)[1]; fishing (fish); aqua culture (aqcu), fish processing (fishp), other food and beverages (ofd)
Supplying food system industries	fertiliser (fert), crude vegetable oil by-product oilcake (oilcake); fish meal (fishm), animal feed (feed), chemicals, rubbers and plastics—pesticides (othcrp); 1st generation bioethanol by-product distillers dried grains and solubles (ddgs)
Other sectors	Various industry (including biofuels), transport and service sectors

2.2.2. MAGNET Production Structure Assumptions

Assumptions on the extent to which inputs can be substituted with each other are a key driver of price and thus food system developments. Production in any of the MAGNET sectors listed in Table 2 are modelled using flexible, multilevel nested Constant Elasticity of Substitution (CES) production functions allowing for substitution of different primary production factors (land, labour, capital and natural resources) and intermediate production factors like energy, fertilisers and animal feed components. In primary agriculture, the production tree is more complicated than in the rest of the economy to be able to capture agricultural intensification processes.

The production structure for arable and horticultural sectors comprises four composite levels or nests (Appendix A, Figure A1). In the upper nest, intermediate consumption and value added are combined in fixed proportions (the corresponding elasticity of substitution is zero (the substitution elasticity values are based on GTAP)). This captures the idea that certain intermediate inputs like seeds are always needed and cannot be substituted by for example labour. One level down the value-added composite consists of a land-fertilizer bundle and the remaining production factors bundle, with substitution elasticity close to zero (0.1) (the choice for very low substitution elasticity is based on simulations made in the past where we observed that low substitution elasticities produce more plausible simulation results [23]), suggesting that that there is limited substitution between the inputs. The substitution between land and fertilizer is further defined on the third level, with the elasticity at 0.75, respecting the assumption that it is easier to substitute for inputs that are on a lower level of production structure where inputs are more similar. Because the substitution elasticity is less than 1, the factors behave as complements—an increasing demand for land will tend to increase demand for fertilizer but since the elasticity is not zero land can be substituted for fertilizer. This limited scope for substitution captures the fact that the chemical fertilizers in the MAGNET fertilizer sector can reduce but not fully replace the use of land.

The production tree for the livestock sector is similar to crops (Appendix A, Figure A2) but combines land with feed, which is further composed of concentrated feed and feed from different crop sectors. Again assuming (imperfect) complements rather than substitutes (elasticity is <1) allows limited scope for intensifying livestock production by substituting land with feed. Finally, the production structure in food processing sectors as well as the remaining industry and services consists of only 2 levels (Appendix A, Figure A3). The production tree has only one nest (following GTAP), assuming that all production factors have the same substitution elasticity which is higher than 1, suggesting that the factors behave as substitutes while the bundle of factors behaves as a perfect complement to all intermediate inputs (elasticity of 0 in the top nest).

2.2.3. MAGNET Labour and Land Availability

Assumptions of production functions determine the food production responses to input price changes. Availability of factors, notably labour and land, affect the extent to which sectors need to compete with each other which is reflected by changing wages and land prices.

Modelling of labour markets in MAGNET reflects the presence of rent and wage differentials between agricultural and non-agricultural sectors [27]. This study adopts the assumption that unskilled labour cannot move freely between agriculture and the rest of the economy. However, within the agricultural sector, skilled and unskilled labour behave as perfect substitutes. The market for skilled labour is not segmented and skilled labour is free to move in all sectors in the economy. This modelling assumption is adopted in the light of a projected increase of skills endowment in Nigeria. Allowing more educated labour to be employed elsewhere increases the absorption capacity of the projected boost of skills in the economy, where at the moment 70% of skilled labour is employed in services.

Availability and thus price of land in MAGNET is determined by the change in total agricultural land and the ease with which land can move between agricultural sectors. Total agricultural land supply is not fixed in MAGNET but a function of the real land price. If prices increase, more land is taken into cultivation, but the closer to the upper limit of land potentially suitable for agriculture the more costly land becomes. This land supply module captures that accessible and fertile lands are taken into cultivation first and the further one expands agricultural land, the more costly the conversion becomes. The price elasticity of land supply for all countries in MAGNET has recently been estimated [28]. In Nigeria, the elasticity is set at 0.07, which is lower than for instance in Ethiopia (0.22), and which reflects the rigidity of the land market as well as the limited possibility of further expansion of land in Nigeria.

The land allocation module in MAGNET then allocates total agricultural land as a heterogeneous production factor (e.g., having different biophysical characteristics) depending on the commodity produced by a specific sector. This means that different land types cannot be perfectly substituted and that adjustment costs are involved when land moves from one sector to the other. This is modelled by using a constant elasticity of transformation (CET) function (for a schematic representation of the CET land function, see Appendix A, Figure A4). Effectively, the land allocation module assumes that it is easier to reallocate land within the group of cereals, oilseed and protein crops (COP) activities (NEST 3), while greater adjustment costs are assumed to enable land to move out of COP production into, say, horticultural activities.

2.3. Extension of MAGNET for Modelling Nutrients Supply using the GENUS Database

A key strength of MAGNET is its modular structure which allows the user to easily activate those modules of most relevance to the study at hand. In this case, the study makes use of the MAGNET extension to incorporate the Global Expanded Nutrients Supply (GENUS) database [29]. The GENUS database provides macro and micronutrient data for 225 products in 175 countries, including Nigeria. GENUS allows to disentangle the nutritional aspects of food supply—both in terms of the numerous nutritional indicators, and in terms of a much more detailed food composition (see Appendix B, Table A2 for the nutritional indicators in Nigeria in the Base year). The GENUS database combines the FAO food balance sheets (FBS) with trade data and food compositional tables to construct a global and historical food and nutrient supply database. From the estimates of the domestic food supply in the FBS, the edible food supply is obtained after taking into account slaughtering, peeling, etc. Using region-specific composition tables, nutrient supply is derived and provided at a 95% confidence interval (median, low and upper bounds). See Appendix B, Table A2 for the estimated range of the nutrient supply for Nigeria. In the MAGNET GENUS extension, the individual GENUS food items are mapped to MAGNET commodities. Changes in the growth of quantities demanded by household as modelled by MAGNET are used to update the nutritional indicators in the GENUS data, resulting in consistent assessments of food and nutrient availability for the representative household in MAGNET [21]. In the model version applied in this paper, a single representative household is used.

It is important to make a cross-validation of the nutritional data with other sources in the literature (Table 3). Brouwer et al. (2018) explore food and nutrient intake at the household level based on the General Household Survey (GHS) for Nigeria [30]. The GHS is a survey in the format of the World Bank's Living Standard Measurement Study—Integrated Surveys on Agriculture (LSMS-ISA). It collects data on agricultural practices, socio-economic characteristics of households and communities in a nationally representative sample of 5000 households [31]. Brouwer et al. report average daily caloric consumption from GHS for households in adult female equivalents, which is approximately 20% lower than the usual average adult equivalent. The study of Akerele (2015) computes per capita adult equivalents per rural and urban population, based on the most recent national food consumption and nutrition survey (FCNS) in 2003–04 [32]. In Aromolaran (2010), the caloric data are based on the author's survey of 480 households from semirural areas of south-western Nigeria [33]. When comparing across these sources, it is apparent that all agree on a high share of carbohydrates in the diets (above 60%). Cassava is one of the most important sources of carbohydrates in the Nigerian diet. The GENUS database estimates the average intake of cassava of about 280 g of per day, which is similar to the data collected in a local survey [34]. It is also noted that in Aromolaran, where data comes from a specific region, the share of roots and tubers is even much higher than in the national surveys (56% of total consumption).

Table 3. Structure of caloric consumption in Nigeria by food group, across various resources.

	Database (Year), Source			
Database	*GENUS (2011)*	*General household survey (2015–16)*	*National FCNS (2003–04)*	*Rural South-west Nigeria*
Source	*Smith et al. (2016)*	*Brouwer et al. (2018)*	*Akerele (2015)*	*Aromolaran (2010)*
Cereals	41.7%	39.5%	42.0%	17.0%
Roots and tubers	26.8%	27.9%	24.0%	56.0%
Cereals, roots and tubers	**68.5%**	**67.4%**	**66.0%**	**73.0%**
Sugar	3.4%	1.9%	0.5%	0.6%
Legumes	2.8%	6.9%	10.0%	7.5%
Seeds and nuts	4.8%	1.5%	1.7%	x
Vegetables	1.8%	1.2%	2.5%	2.90%
Fruits	3.5%	2.0%	0.3%	1.35%
Fruits and vegetables	**5.3%**	**3.2%**	**2.8%**	**4.3%**
Tea coffee	0.0%	0.0%	0.1%	0.63%
Spices	0.4%	0.2%	x	x
Alcohol	2.1%	0.2%	0.1%	x
Meat	1.4%	2.1%	1.6%	0.8%
Milk	0.2%	0.6%	0.5%	0.4%
Fish	0.7%	1.9%	5%	1.2%
Eggs	0.4%	0.2%	0.1%	0.4%
Eggs, meat and milk	**2.7%**	**4.7%**	**7.2%**	**2.8%**
Fats and oils	**9.9%**	**13.8%**	**9.0%**	**10.0%**
Total	*100%*	*100%*	*97%*	*99%*
Caloric consumption	2969	2346 (2815) *	2440	1980 **

Note: Categories not reported in the respective sources are marked with x.* GHS data converted to male equivalent. ** Calculated from the statistical regression tables provided by the author. Aggregated food groups are in bold.

It is not straightforward to compare the total caloric consumption across the sources due to differences in the definitions and measurement. The total caloric consumption expressed in female adult equivalent is 2346 Kcal. When converted to male adult equivalent (2815 Kcal), it is close to the caloric consumption reported in the GENUS database (2969 Kcal per capita). However, it is important to note that the GENUS database measures available caloric supply or availability derived from food balances, whereas the nutritional surveys measure the direct caloric intake by households and individuals. To reflect this in our caloric projections, a correction factor to the GENUS caloric data is applied to downscale the caloric availability to intake by average female equivalent reported in the GHS analysis, which is considered the most representative, given that the data from Akerele is based nutritional surveys from 2003–2004 and Aromolaran provides nutritional data from a single region.

In addition to the caloric composition, it is also insightful to compare the nutritional adequacy of GENUS with other sources. A ratio of nutrient intake to the recommended average intake is calculated and compared with the GENUS nutritional data converted to the female equivalent intake. Table 4 shows that the nutrient adequacy is very similar in both data sources and fat, riboflavin, iron and calcium are the most deficient nutrients in Nigeria. On the other hand, the intake of other micronutrients such as vitamin A, B6 and vitamin C is sufficient. However, it is important to note that even when the average intake exceeds the recommended dose, due to variation in incomes and diet patterns, there are households that do not meet the requirements.

Table 4. Micro and macronutrient deficiency calculated as ratio of a nutrient intake to average recommended intake (expressed per adult female equivalent).

		GENUS (2011)	General Household Survey (2015–2016)
Nutrients meeting recommendations	Thiamine	2.9	3.4
	Vitamin A	1.8	3.2
	Vitamin B6	1.8	2.0
	Folate	1.5	1.7
	Protein	1.4	1.5
	Niacin	1.2	1.2
	Vitamin C	2.2	1.1
	Zinc	1.0	1.0
Nutrients not meeting recommendations	Fat	0.7	0.8
	Riboflavin	0.7	0.7
	Iron	0.7	0.7
	Calcium	0.4	0.6

2.4. *Quantified Drivers of Food Systems Change in Nigeria*

To build the foresight on national food systems, the paper follows the scheme presented in the previous section. From the modelling perspective, this means adopting various choices on how the key national drivers of the food systems will evolve in the future. After specifying this, it will be able to assess how the impact of these drivers will affect future food systems and determine the important limitations and opportunities for Nigeria.

The first considered driver is the GDP growth. After 2000, Nigeria enjoyed a decade of favourable economic growth with average annual rates around 8%. Right before the recent crisis, the economy was growing about 5% p.a. [1]. Such high rates, which result in doubling total GDP in only about 10 years, are in line with Solow's theory of economic growth that expects that countries with an initially low level of capital stock grow faster to accumulate new capital. If Nigeria sustained such growth rates into the future, it could easily step up to a higher income level category. However, the recent economic crisis has also revealed some of the bottlenecks of the economy, which are a high dependence on the oil sector, an overvaluation of the exchange rate and armed conflict [35]. This means that the favourable projections of GDP growth are conditional on resolving some of the weaknesses.

To translate the expected economic performance into our MAGNET model, there are several established international macroeconomic forecasts, such as the World Bank [36] and IMF Economic Outlooks [37]. However, these forecasts extend only into the closest future (not far beyond 2020) and therefore they are not suitable for long-term projections beyond 2020. For instance, the extrapolation of the IMF forecast beyond 2022 would be biased downwards due to the effect of the recent crisis (see Figure 1). The shared socioeconomic pathways (SSP) projections [38] are commonly used in foresight modelling exercises because of their long-term span and underlying future storylines. However, the projected GDP growth rates in the SSP2 Middle of the Road scenario (Figure 2) seem to be very optimistic, expecting that Nigeria could reach up to 12,000 GDP per capita by 2050, which is at the edge of an upper middle income economy (comparable to Russian Federation or Turkey) [39]. Optimistic, but not that extreme are the projections of the PWC outlook to 2050, which assume that Nigeria would reach about 4500 GDP per capita by 2050, belonging to the group of upper-middle income economies, comparable for instance to Albania [40]. For this MAGNET analysis, the PWC scenario is chosen, which reckons with optimistic but moderate GDP growth (4–5% annually until 2050). This choice is in line with other studies; a recent study uses an assumption of 5% growth for the periods until 2030 [16]. Adopting the PWC scenario therefore assumes a transition of Nigeria from lower-middle income to an upper-middle-income economy.

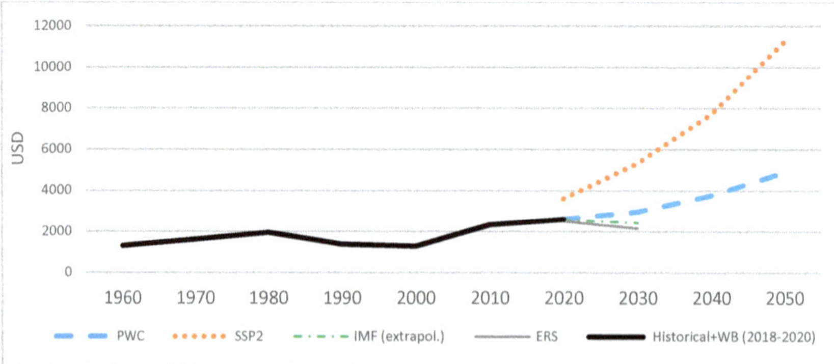

Figure 2. Historical and projected GDP per capita in Nigeria (compilation of sources).

Another key driver for the long-term modelling is the population growth. In this case, the SSP2 projections are based on the UN population growth projections (the medium variant). Figure 3 shows that due to the growth rate exceeding 2% p.a., population will double in Nigeria by 2050. High population growth is also expected in other Sub-Saharan Africa regions, whereas in the EU and other high income countries, the growth is limited.

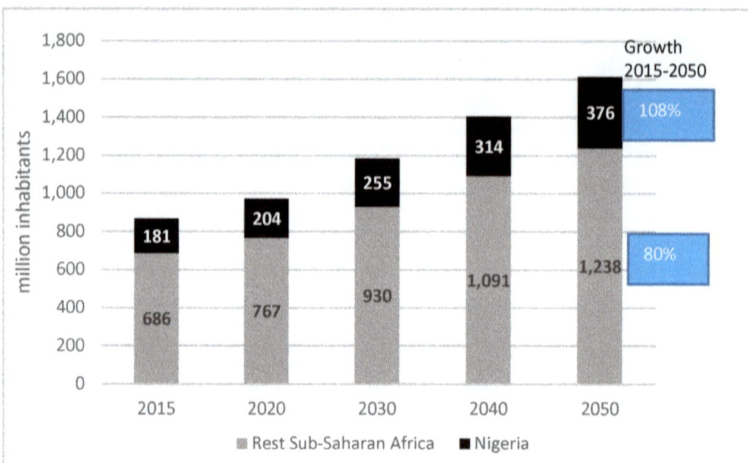

Figure 3. Projected population growth in Nigeria and the rest of Sub-Saharan Africa.

High population growth brings pressure on resources but it can also be an opportunity as a country can benefit from the abundance of labour force. The ratio of skilled and unskilled population will be important for cashing from the demographic dividend—whether the expected mix of skills will find a place in the labour market. In order to translate this into MAGNET, the Wittgenstein labour projections [41] are used. Figure 4 shows that at present, the number of unskilled labour slightly exceeds the number of skilled people in the economy. By 2050, it is expected that the share of skilled labour will reach 70%, up from 40% now, which is comparable with the EU (77%).

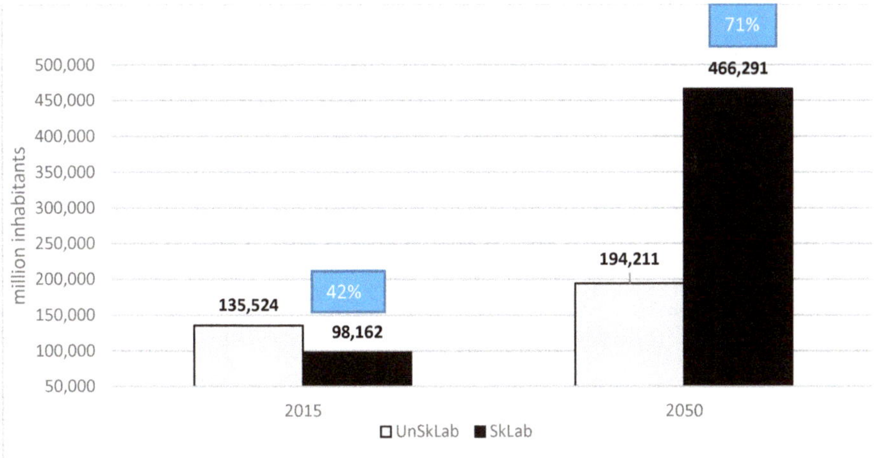

Figure 4. Projections of skilled and unskilled labour force in Nigeria.

The dynamic economic and population growth that is projected into the future will put tremendous pressures on natural resources in order to feed the existing population. Therefore, it is important to take into account the availability of agricultural land for further expansion. As pointed out in the PWC Report on Nigeria's agricultural value chains [42], most of the agricultural growth in the past has been driven by area expansion, with limited contribution of yield growth. This suggests that there is only a limited proportion of land that could yet be brought to cultivation. The estimations from the IMAGE model provide an overview of actual land available that can be used for commercial purposes (e.g., crop land and pasture land versus parks) across the world [2]. Figure 5 shows that already, 93% of available agricultural land is occupied. This is because out of the total 79 million ha of available agricultural land, 70 million ha are being cultivated, from which about 40 million ha are arable land and the rest are pastures and other agricultural land (Table 5). This implies that the only way how to increase land use comes from a conversion of the extensively used pasture land to arable land. Indeed, various literature sources claim that about 40% of agricultural land can still be put in cultivation [42–44]. However, it must be noted that these claims are only feasible by transforming the existing land use, not by adding more land into cultivation. In addition to this, there are also economic barriers to land access. Administrative procedures for acquiring new agricultural land in Nigeria may be cumbersome and have been put forward as a major bottleneck in the food supply system [42,45].

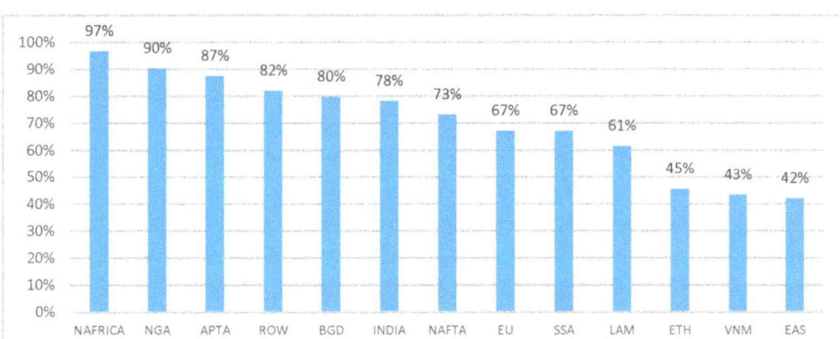

Figure 5. Land pressure in MAGNET countries (% of occupied land in total agricultural land) in the base year.

Table 5. Land use and land potential in Nigeria (2011).

Land Type	Land Used for Agriculture (Million ha)	Land Suitable for Agriculture (Million ha)
Arable land	41	-
Pastures & other	30	-
Total agricultural land	71	-
Land suitable for agriculture (based on IMAGE)	-	79
Available land for expansion	-	8
Land pressure (% of land used or left)	90%	10%

Source: land use (MAGNET database), land potential (IMAGE model).

2.5. *Definition of Alternative Baselines*

It is becoming apparent that Nigeria will soon approach food production limits if non-land inputs are not used more intensively or if there is not a significant change in R&D policy that would boost the crop and animal sector yields. According to FAO, the potential cassava yield is 40 tonnes per hectare, whereas the achieved yield in Nigeria is only 13 tonnes per hectare [46]. Although the SSP2 projections of yields towards 2050 taken from the IMAGE database [47] assume an annual yield growth of about 1%, which is well above the high income countries, it is not enough to make a significant difference in closing the yield gap with the high income countries. At the moment, Nigeria is one of the countries with the lowest use of fertilizer input to land use (Figure 6). Nigerian farmers utilise on average about 10 kg of fertilizer per hectare, which is very little compared to high income countries such as Netherlands, where the consumption is well above 200 kg [48]. In Nigeria, poor infrastructure increases transportation costs that make fertilizer prices unaffordable [49]. Moreover, in the interaction with poor quality of seeds, the productivity impact of fertilizer has its limits [50].

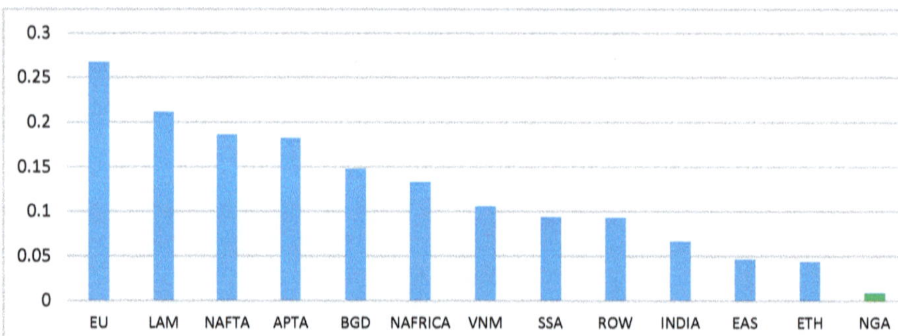

Figure 6. Ratio of fertilizer to land input use in nominal terms (USD) in MAGNET database.

In order to reflect the need for the input intensification in our modelling approach, two foresight scenarios are designed. The first scenario *Land_Fixed* is the "status quo" scenario that counts with high rigidity in the land market where an increasing demand for food driven by GDP and population growth will not be able to respond either by higher land expansion or by more intensive use of other factors. The alternative scenario *Land_Subs* incorporates features of institutional change where increasingly higher land scarcity is adjusted by substituting land for non-land inputs such that they are used more intensively in the production process (Table 6). This is operationalized by increasing the substitution elasticity between land and other inputs both in crops and land-using livestock sectors (corresponding to nest 1 in the Figures A1 and A2). In addition, a higher substitution elasticity between land and feed in the livestock sector and land and fertilizer in the crops sector (nest 3 in Figures A1 and A2) is set. The final choice of parameters is in Table 4. Both scenario versions can be considered as extreme, where under the substitution elasticity of 0.1, which is the default option in MAGNET, there is almost no room for substituting land for other inputs. With the elasticity of 1.2, the inputs behave as substitutes and an increasingly scarcer land can be easily substituted for labour, capital and other inputs. We have performed a sensitivity analysis to assess the response to various levels of substitution elasticity ranging from 0.1 (MAGNET default) to 1.2, which are the upper and lower bounds. The land pressure is notably decreased with substitution elasticities above 0.4.

Table 6. Overview of the alternative baseline scenarios.

Key Assumptions	LAND_FIXED	LAND_SUBS
Technical change	Exogenous—calibrated to GDP growth based on 4–5% p.a.	
GDP growth	Endogenous	
Population Growth	SSP2 (UN Medium Variant)	
Exogenous Yield growth	1% p.a. based on SSP2	
Feed efficiency improvements in livestock sectors	1–2% p.a. based on IMAGE	
CES Substitution elasticity of NEST 1 (land bundle -other factors)	0.1 (MAGNET)	1.2
CES Substitution elasticity of NEST 3 (land-fertilizer or land-feed)	0.75 land-fertilizer, 0.5 land-feed	1.2 both land-fertilizer and land-feed
Land allocation elasticity	−0.2 (NEST1), −0.4 (NEST2), −0.6 (NEST3)	1 (Non-nested)

Another way to mobilize the land market is to allow more flexible conversion between crop land and pastures on the existing land, which makes it easier to increase the share of arable land at the expense of pastures. To operationalize this in MAGNET, a flat land allocation tree is imposed where all types of lands can be perfectly substituted and set the CET elasticity to 1. From the institutional point of view, both measures mean that there is a better mobility of land both for acquiring and for getting rid of.

3. Results

3.1. Agricultural Inputs and Factor Markets

The analysis starts by looking at the agricultural input and factor markets because they directly reveal the pressures or abundance of resources driven by the combination of economic and population growth. First, it is interesting to see what happens with the land prices under both scenarios. Figure 7 shows that if there is no possibility to substitute land for other inputs, land prices will escalate after 2030 due to increasing demand for food. Particularly in the last period, land price growth is enormous, suggesting a real difficulty to meet the demands for food with limited resources. This adverse development could be almost fully avoided if other inputs are used more intensively (Land_Subs scenario). In this case, land prices remain on the same level as in the base year.

Table A3 (Appendix C) shows the comparison of annual growth of factor prices for all production factors. As defined in the scenario framework, in the Land_Fixed scenario land is the key constraining factor in agricultural production, with annual growth of prices reaching up to 13%, particularly in the last two decades. For the other production factors, prices would go down. The difficulties of substituting land for other factors creates a situation where other resources in agriculture are under-utilized and their returns are lower. Particularly, the wages of unskilled labour would go down in primary agriculture. This is related to the fact that the agricultural sector is the most labour intensive sector in the Nigerian economy and the largest employer of unskilled labour (over 70% of unskilled labour works in agriculture). Any decline of agricultural production would be reflected in a decline of agricultural wages.

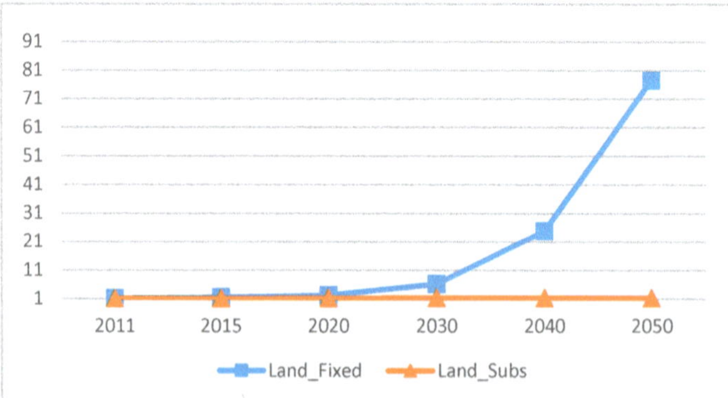

Figure 7. Growth of land price index per scenario. (2011 = 1).

On the other hand, in the Land_Subs scenario, the relative prices of non-land factors decline less notably or even increase (unskilled labour in the rest of the economy). Compared to the Land_Fixed scenario, in 2050, in the primary agriculture, wages of both skilled and unskilled labour would be 60% higher and land prices about 100% lower. This results in a very different composition of value added in both scenarios. In 2011, food system has the largest share in the real value added (41%), followed by industry and services (25%). By 2050, the contribution of food system to the real value added is projected to decline in favour of services and industry, from 41% to 23–28% depending on the scenario (Appendix C, Figure A5). If land remains fixed, the share of food system in total value-added declines more rapidly than in the Land_Subs scenario. On the other hand, it is apparent that the Land_Fixed scenario supports more industrialization of the economy as industry benefits from the absorption of resources from agriculture. The increase of the oil & gas sector's share in the land substitution scenario is driven by increased foreign demand, provoked by more competitive oil and gas export prices in Nigeria compared to the rest of the world.

It is also instructive to analyse the impact of the land market scenarios on the changes in economy-wide factor demand. Table 7 compares the endowment volumes between the two scenarios. It is apparent that with more land substitution, the agricultural sector utilizes more inputs, including the land itself (the total primary agriculture production goes up and therefore also land). This is also transmitted to the sector of food processing where all endowments increase compared to the Land_Fixed scenario. On the other hand, locking land in agriculture releases labour and capital to be employed in industry and therefore there is more value added created in industry in the Land_Fixed scenario. This is especially visible in case of skilled labour. The resulting impact on the demand for labour and capital in the individual food systems sectors is displayed in Appendix C, Table A4. The conclusion that stems from this analysis is that better management of land markets could potentially be a strong leverage point for inclusive growth in food systems activities, yet with a trade-off in terms of industrial development. This trade-off appears in classical theories of rural development, and has been a subject of increasing criticism [49].

Table 7. Endowment volumes growth in Land_Subs in 2050 (% difference from Land_Fixed scenario) by sector in the economy.

Sector Description	Code	Land	Unskilled Labour	Skilled Labour	Capital
Primary agriculture	AGRI_PRIM	1.3	2.5	85.4	17.3
Food processing	AGRI_proc		4.1	0.5	36.4
Industry (non-food)	INDUSTRY		−38.2	−42.2	−11.6
Services and utilities	SERV&UTIL		7.0	1.2	8.9

3.2. Price Transmission in the Food Supply Chain

The developments in the factor markets are transmitted into markets of goods and services. For this analysis, it is interesting to see what the expected price trends in the food supply chain are towards 2050 and how they are affected by the rigidity of land markets. Figure 8 shows the development of prices in the primary agriculture sector. The land constraint is a key factor in determining whether primary agricultural prices will grow or decline. The tipping point is the period after 2020 when there will be no available land to cultivate. Because of this, prices in primary agriculture would be 5 times higher by 2050. Releasing pressure on land would make a significant difference in the production costs of primary agriculture. In this case, producer prices would decline by 2050.

Figure 8. Index of primary agriculture prices in the two scenarios.

Figure 9 shows how the development on land market is transmitted to other sectors and to final consumers. Clearly, in the Land_Fixed scenario, the growth of land prices is so dominant, that all connecting industries face higher production costs, resulting in an increase of food prices for consumers. On the other hand, with a higher land substitutability, producer prices would go down as well as consumer prices of food. The fact that consumer food prices copy more closely the development of primary rather than processed food suggests that the proportion of consumer spending coming from primary agriculture is higher than from processed industry (about 50% of all food expenditures come from land-using sectors such as horticulture and grain). It also tells that there is a low share of imports in food consumption that could potentially moderate the food price inflation (the share of food imports in total food expenditures is 8% and is stable over time and across the scenario).

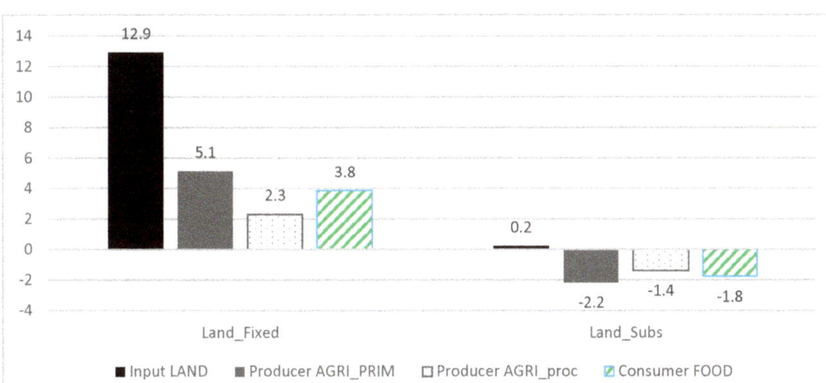

Figure 9. Annual growth of prices between 2020–2050.

It is also interesting to see how the developments in the agri-food markets affect prices in the other industries (Appendix C, Figure A6). Whereas agri-food prices would be significantly higher in the Land_Fixed scenario, producer prices in the other production sectors would decline more than in the Land_Subs scenario. If there is no substitution of labour for land, the surplus of labour from the agricultural sector pushes the wages down. Particularly the sector of services benefits from this, which is the largest employer of skilled labour. In industry, the stronger decline of prices in the Land_Fixed scenario is driven by capital prices. There is again a higher surplus of capital from the food processing sector that is allocated in the rest of the economy.

3.3. Domestic Production, Trade and Value Added

This section looks more closely into the performance of individual agri-food production systems. Figure 10 displays annual growth of production volume of the commodities that represent 99% of total production in 2020 (both in volume and value). The level of production in 2020 is also presented to understand the importance of each sector in the total agri-food complex. It is observed that in the Land_Fixed scenario, meat processing sectors such as poultry meat, pork meat and other meat enjoy unusual production volume growth reaching above 5% per year. On the other hand, in the Land_Subs scenario, land-using sectors such as crops, other agriculture and horticulture flourish. Because the Land_Subs scenario favours those agricultural sectors that have traditionally strong position in the agri-food chain, the total agri-food production volume is higher in the Land_Subs scenario than in Land_Fixed (note that of course, in value terms, it is the other way around).

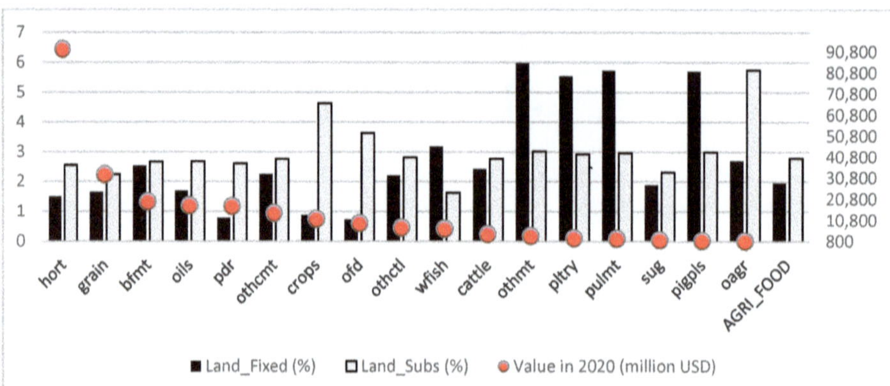

Figure 10. Annual growth of production volume (2020–2050) and volume of production of the major commodities (production volume of sugar, milk, dairy, wheat and aquaculture are too small to be reported).

The divergent development of the agri-food sectors is explained by the growth of production costs and prices. Figure 11 shows that due to an excessive growth of land prices in the Land_Fixed scenario, production costs in land-using sectors such as grain or cattle rise significantly (above 6% p.a.), whereas the non-land agricultural sectors such as other meat, poultry and pigs face a decline of prices. Concretely, prices in pigs and poultry sector reduce 5 times, whereas prices of cattle go up 10 times. Due to these price developments, meat industry enjoys an increased competitiveness compared to the crops sectors. Under the Land_Subs scenario, on the other hand, production costs in primary agricultural sector are lower than in processing industry and the crops sectors expand.

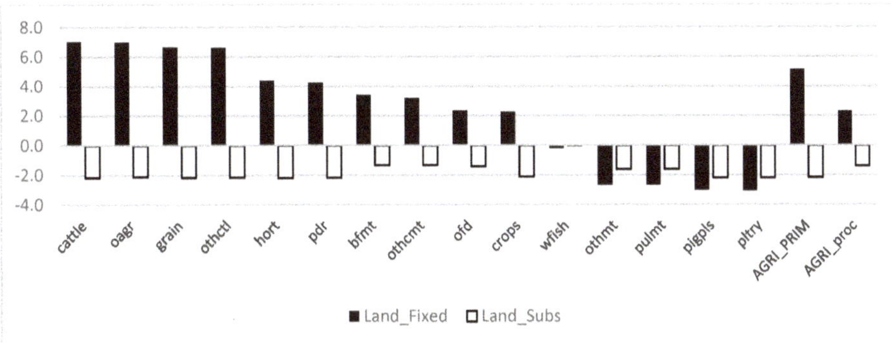

Figure 11. Annual growth of producer market prices of major agro-food commodities (2020–2050).

The projected changes in prices, endowments and productivity result in a very different composition of value added in both scenarios. Food supply chain value in MAGNET is defined as a sum of production value of primary agriculture, food processing, wild fish sector, aquaculture, fish processing and fish meal, fertilizer, feed and pesticides. If land is substituted for other factors, the value share of food processing and supplying industries increases, because producer prices and costs in primary agriculture are comparatively lower than in food processing in this scenario. Under the Land_Fixed scenario, due to the excessive growth of land prices, the value of food supply chain is dominated even more strongly by primary agriculture.

The developments in the food systems sectors are also projected to have a strong impact on the external position of Nigeria and the domestic supply and self-sufficiency. Figure 12 shows that the land substitutability plays a significant role in the competitiveness of Nigeria on foreign agri-food markets. Until 2020, trade balance remains stable and negative at around 20 billion USD. After 2020, the trends diverge notably where under the Land_Fixed scenario, trade balance would deteriorate significantly up to a negative 140 billion USD, whilst in Land_Subs scenario, it would improve to a negative 11 billion USD.

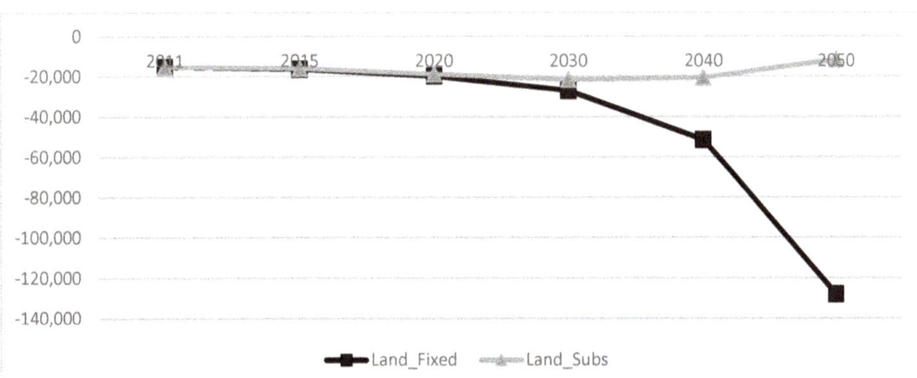

Figure 12. Agri-food trade balance expressed in world prices in million USD.

Aggregate agri-food trade balance is driven by the competitiveness of individual agri-food sectors. In 2020, there are only few sectors which have a positive net trade with abroad, namely the sectors of horticulture, grain and other crops, oils and wild fish (similar in both scenarios). By 2050, in the Land-Fixed scenario, the trade balance in all these traditionally trade-oriented sectors would turn negative (except for wild fish). There is also a very strong deterioration of trade balance in other processed food ("*ofd*"), due to an increased household consumption of processed food from abroad. On the other hand, there would be a new development on the meat markets, previewing a large increase in net exports of poultry and other meat (driven by the relative decline of producer prices of these sectors).

In the Land_Subs scenario, the competitiveness of crops sectors is improved and the deficit of Nigeria's trade balance is only moderate. The sectors that have a positive balance in 2020 remain with surplus and other sectors such as beef meat and paddy rice newly gain competitiveness on external markets.

3.4. Food Environment and Consumption

In this section, the attention shifts to the consumer side of the economy to assess the impact of the projected changes in food production on households' living standards and food security. Table 8 shows that agri-food consumer prices are expected to increase by 4% in the Land_Fixed scenario, whereas in the Land_Subs scenario, they decline by 1.8%. As a result of that, the quantity of food consumed is higher in the second scenario. Due to excessive growth of food prices, private expenditures on food grow quite significantly in the Land_Fixed scenario and they are also reflected in the growth of total household expenditures. The lower panel in the table analyses food accessibility as the compounding impact of food prices and household earnings, using a cereal price index divided by wage of skilled and unskilled labour as an indicator. In the Land_Fixed scenario, the accessibility of staple food such as cereals declines as the cereal price index is up to 7 times higher compared to the wage of unskilled labour in agriculture, and 4 times higher compared to unskilled labour wage in other industry. The relative accessibility of food is, on the other hand, increasing in the Land-Subs scenario, where cereal prices are below the wages. Similar developments are recorded for the skilled labour, which shows that food-security problems would be threatening both skilled and unskilled labour households in the Land_Fixed scenario.

Table 8. Aggregate consumer prices and expenditures of Nigerian households.

(% annual growth, 2020–2050)	**Land_Fixed**	**Land_Subs**
Price index agri-food consumption	4.2	−1.8
Agri-food consumption quantity	2.0	2.4
Agri-food expenditures	6.3	0.6
Total Household Expenditures	5.6	2.6
(Ratio of cereal consumer price index to unskilled labour wages, by sector of employment, 2050)	**Land_Fixed**	**Land_Subs**
Household employed in primary agriculture (2011 = 1)	6.89	0.52
Household employed in the rest of the economy (2011 = 1)	3.69	0.28

Given that agri-food expenditures grow significantly, it is interesting to see if the share of food expenditures in total household expenditures increases as well (Appendix C, Figure A7). In the base year (2011), the share of food expenditures in total expenditures in Nigeria is relatively high, reaching almost 70%. This is in line with Akerele (2015), who highlights that expenditure on food claimed more than 60% of household income in 2012. In the Land_Subs scenario, the share of food expenditures after 2020 declines to 30%, whereas in the Land_Fixed scenario, due to excessive growth of prices, the share of food expenditures is expected to exceed 80%. Basically, most household income would be spent on food in this case, which is alarming.

In the Land_Fixed scenario, the food groups that contribute most to total food expenditures are horticulture and other crops, cereals, red meat and other food (Figure 13). Contrary to that, in the Land_Subs scenario, the expenditures growth remains very moderate, except for milk & dairy, sugar and processed food, which are food groups with lower share in total expenditures.

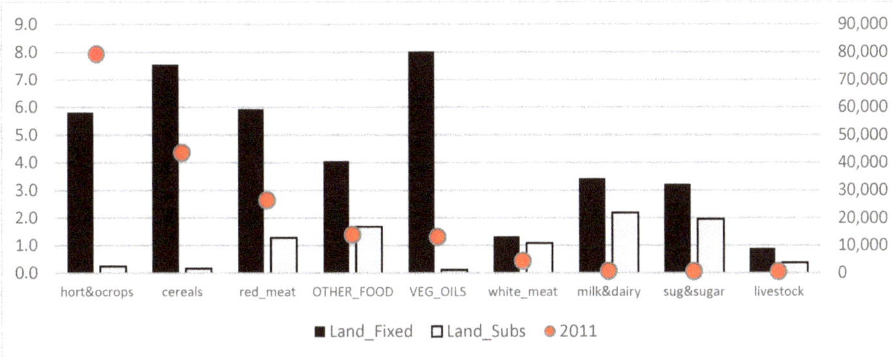

Figure 13. Annual growth of household expenditures on individual food groups (%, 2020–2050, left-hand scale) and value of expenditures in 2011 (million USD, right-hand scale).

3.5. Nutrient Availability

In this section, the impact of the food system developments on household nutrition is analysed. Using the GENUS database, the calories linked to the main food groups in MAGNET are traced. A correction factor is applied to express the total caloric supply in terms of caloric intake of adult female equivalent. Figure 14 shows that in the Land_Fixed scenario, the projected caloric consumption would be in the range of 2600–2700 Kcal, which is about 100 Kcal more than in case of the Land_Subs scenario (4% difference). This shows that improved food accessibility is not necessarily accompanied by higher caloric consumption.

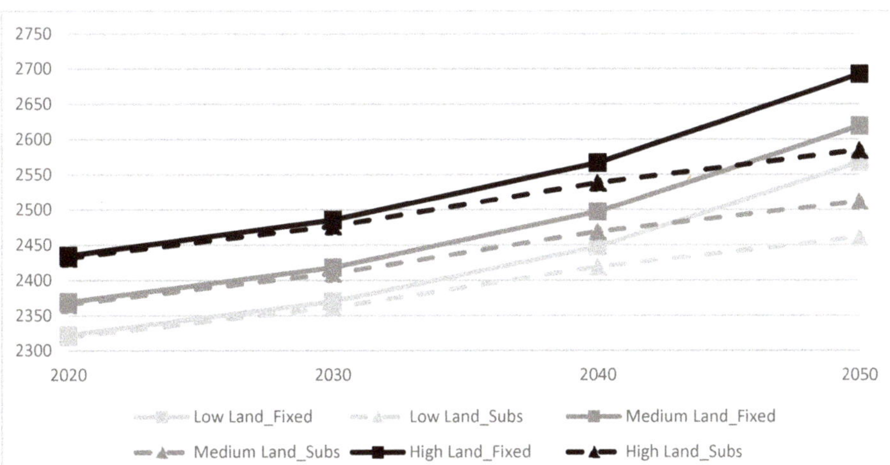

Figure 14. Projections of household caloric consumption (kcal per day) in the low, medium and high variant (converted to adult female equivalent).

Figure 15 shows the composition of caloric consumption across main food groups. As mentioned before, roots and tubers, grain and grain flour are the key sources of energy—they provide about 2000 Kcal out of 2900 Kcal of daily consumption. By 2050, these foods will remain the most important caloric sources but there will be some differences depending on the scenario. In the Land_Fixed scenario, due to a bigger role of food processing, the caloric consumption of more processed foods such as grain flour, vegetable oils and sugar would be higher compared to Land_Subs. Given that these foods have high caloric content, this explains why the caloric consumption is slightly higher in the Land_Fixed scenario.

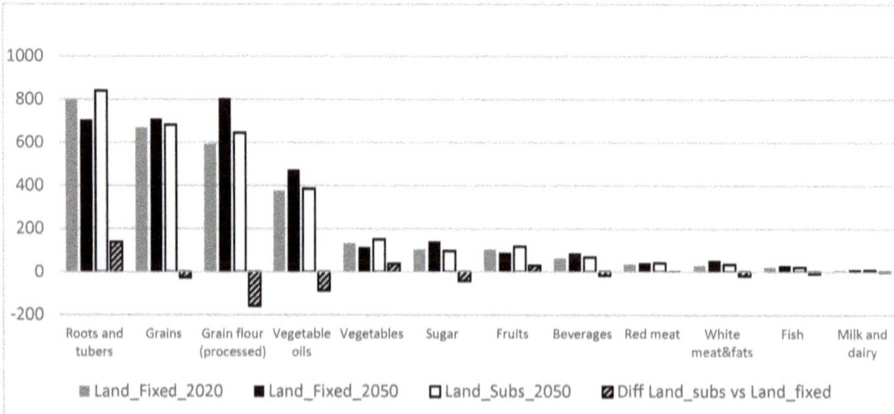

Figure 15. Projections of caloric consumption (kcal per capita day) per food group.

It is also important to assess whether the energetic consumption corresponds to a higher intake of micronutrients. For instance, although in Land_Subs scenario the total caloric consumption is slightly lower, the caloric intake from fruits and vegetables is higher, which is important from the health perspective.

Figure 16 shows that nutritional intake is projected to increase for all deficient nutrients, except for calcium, where projections diverge per scenario. In the Land_Fixed scenario, calcium intake would decline by 1% compared to 2020, whereas in the Land_Subs scenario, nutrient intake increases to 8% compared to 2020. Fifty per cent of calcium in Nigeria is obtained from cassava, yams, citruses and okra. Because the Land_Subs scenario favours the horticulture and crops food systems, it also leads to a higher intake of calcium. On the other hand, in the Land_Fixed scenario the intake of carbohydrates and fats would be up to 23% higher than in 2020.

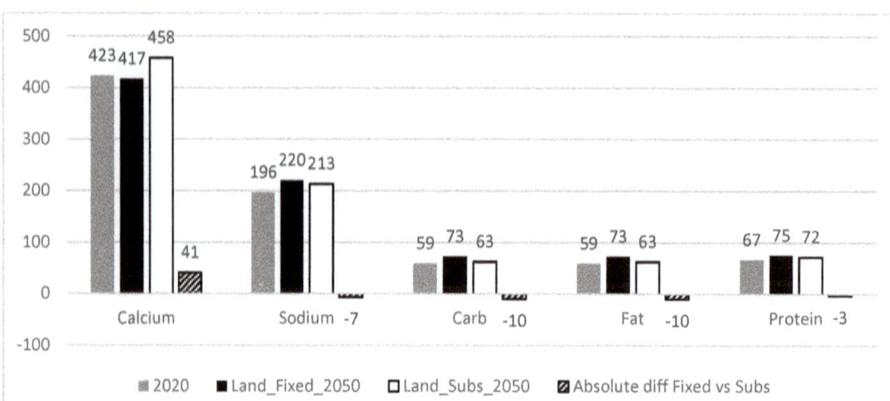

Figure 16. Overview of micronutrient intake in different scenarios (mg/capita/day or g/capita/day).

It is also important to look at how the nutrient gap evolves in time per scenario (Figure 17). Clearly, the projected changes in nutrient intake are too small to make a significant improvement in the nutrient gap. The exception is the intake of vitamin A, where the ratio could increase to 1.8 in the

Land_Fixed scenario. This is driven by a push in palm oil consumption which represents about 70% of all vitamin A intake in Nigeria. On the other hand, the Land_Subs scenario would favour more the intake of thiamine, vitamin C, vitamin B6 and folate. However, the calcium, fat, riboflavin and iron would remain highly deficient in the Nigerian nutrition by 2050.

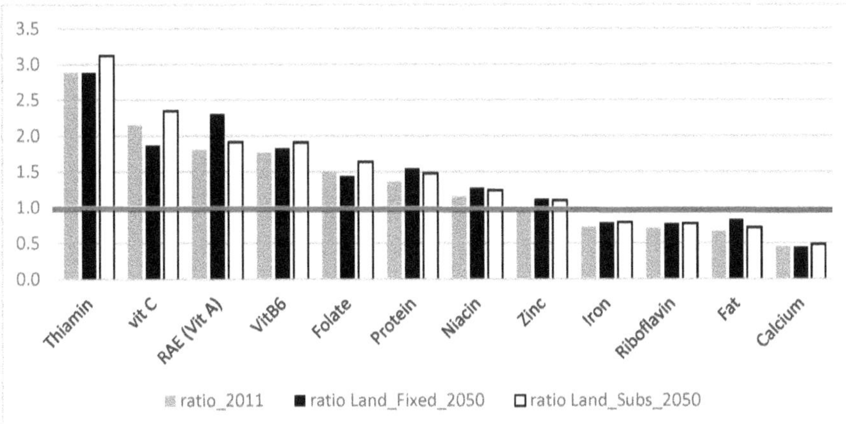

Figure 17. Overview of micronutrient gap in 2011 and 2050. Note: ratio is calculated as a nutritional intake divided by recommended dose. GENUS data is translated to female adult equivalent to correspond to the recommended dose in LSMS-ISA.

4. Discussion

It was shown by means of a modelling exercise that structural change in agriculture and transformation in the food system are important elements in diet change in Nigeria. The Nigerian (agri-fish-) food system is undergoing substantial change under the influence of global and domestic drivers, and model analysis gives insight into the processes of adjustment. Nigeria is currently the country with the lowest level of input use in agriculture (in value terms) in the global database that underpins the MAGNET model, and with a reserve of just 10% of agricultural land that can be brought into production. Model projections suggest that a process of intensification of agriculture in combination with land substitution appears critical for the evolution of food and nutrition security, and for shifts towards healthy diets for the population. The strength of the analytical framework employed in this study is its capability to account for economy-wide adjustments of producer and consumer decisions under the influence of global drivers of change and the drivers related to the rigidity in the land market. This no.1 feedback effect, as defined in the introduction, appears to be particularly strong in relation to adjustments in factor markets for labour in response to the land rigidity scenarios combined with economic growth and expansion of the population. A major assumption underpinning the improved food accessibility and shifts towards greater diversity and quality in the diet under the Land_Subs scenario stems from adjustments in the labour market. In particular, the model projects that unskilled agricultural employment grows by over 2%, and skilled labour by over 80%, suggesting that skilled work becomes firmly established in the agriculture sector. This is a substantial departure from today's realities, in which farming is predominantly a low-input activity. More detailed assessments will need to be done to assess confidence whether the Nigerian labour market would support such a transformational shift.

At least two implications of these scenario analyses can be assessed. First, without a significant governmental policy directed to R&D, it is not plausible to expect significant boosts of yields in the future. Coupled with the limited possibilities for the expansion of agricultural land, to prevent the collapse of the system under the dynamic growth of population and incomes, an increase in input intensification will be necessary. Western Africa has one of the lowest shares of agricultural R&D spending as a proportion of agricultural GDP in the world (0.5% vs 5% in high income countries). In earlier research it was estimated that under a 0.5% share of R&D spending in agricultural GDP, land productivity can grow up to 1% annually (Figure 18) [51]. In order to close the yield gap with high income countries, yields would need to grow by 3.5% annually, which requires a much larger share of R&D spending than is the current spending.

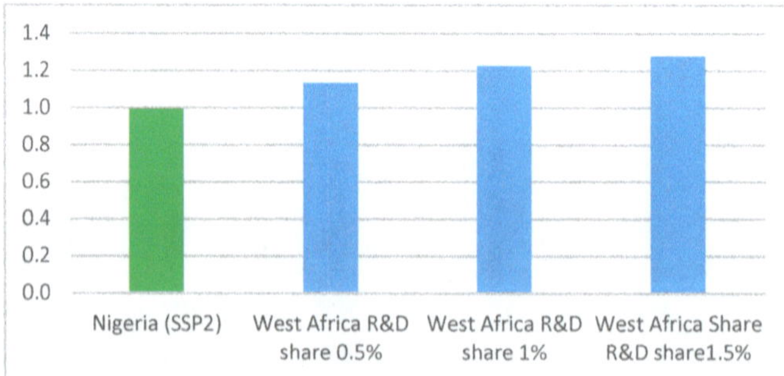

Figure 18. Expected annual growth of land productivity (2011–2050) in the SSP2 scenario and under alternative intensities for R&D. *Sources*: for Nigeria, the IMAGE model [47]; for West-Africa under different R&D intensities, the MAGNET model [51].

The second implication is related to the nature of the land tenure system in Nigeria that has been critiqued to be expensive, cumbersome, time-consuming, and risky, creating a major bottleneck for improving the food supply system [42,45]. If rigidity on the land market is project to remain the status quo, and under contextual changes towards 2050 (as presented by assumptions on buoyant economic growth, population growth and urbanization and technology change), demand growth from the rising middle classes are projected to take the form of more processed food and higher consumption of animal source foods. The model suggests strong import competition in Nigeria for these sectors. Meat and dairy would be produced only in part from domestic livestock systems and with limited backward linkages into domestic cropping systems because of the scarcity and low productivity on crop land. As a result, a significant portion of demand is projected to be sourced from imports. A speculative interpretation of the scenario results for expansion in the meat and dairy sector with a higher input from skilled labour and capital is that "footloose" livestock production systems may thrive in the vicinity of urban centres. These are capital-intensive production systems, aimed at maximum technical efficiency, yet with possible trade-offs in terms of benefits to ecological and human livelihood systems [52]. While these systems may provide clear benefits in terms of an effective and efficient food provision, if these systems depend on global imports of feed or intermediate products (frozen meat, milk powder), the general equilibrium effects on food security and diet quality at population level may be less beneficial for society than under an alternative scenario that promotes a more diverse domestic supply. With land substitution demand, pull may benefit the rural development and nutrition agenda for the rural population. A striking result is that intensification in the analysis results in greater diversity of the production systems, which in turn cascades into positive effects on the diversity in the food supply. This suggests that intensification in Nigeria would lead to the availability of foods with higher density in micronutrients than without intensification.

The results of our study can be put into context with other empirical evidence, although, as argued in the introduction section, the macroeconomic perspective on the linkage between food systems and diets is not yet sufficiently covered in the existing literature. One recent study applied a CGE methodology as well, to carry out an economic assessment of climate change impacts in Nigeria [16]. Whereas the authors warn against possible growth of food prices and higher food dependency on foreign imports due to climate change, they do not analyse the nutritional aspects of climate change impacts in Nigeria. Another study examined the agricultural transformation and its relation to hunger and poverty eradication in Nigeria [43]. The study argued that a rural transition is a condition for the alleviation of hunger and poverty. First, a structural change in labour markets to commercialize agriculture and develop agro-based industries was considered to be a major agricultural development pathway in Nigeria. This is in line with our study that clearly shows that in the absence of land substitution, labour currently employed in agriculture will have to seek employment outside of the primary sector, accelerating the process of urbanization. However, it can also be argued that in case of higher intensification, the agricultural labour could be used more productively within the agricultural sector and contribute thus to growth of rural wages. Second, the potential importance of increased capitalisation of the agricultural sector by improving access to credit was considered as a condition for the commercialisation pathway to materialise [43]. The findings from this study corroborate this; under assumptions that land is better substitutable for other inputs, the food-industry can absorb as much as 40% more capital, which points out to the need to improve access to credit to stimulate the intensification process of Nigerian agri-food system.

Various global initiatives employ food systems foresight on the global scale. A recent review identified a paucity of studies that apply comparable methodologies at regional and national levels [8]. It is instructive, therefore, to reflect on this study on the food systems foresight on Nigeria from the perspective of an earlier global approach that addressed similar themes using comparable methods as ours [53]. The baseline scenario in this study identified similar challenges in achieving the goals of achieving food and nutrition security in the absence of agricultural intensification. These are upward pressure on land prices, high food prices threating economic growth, insufficient agricultural productivity growth and a prevalence of micronutrient deficiency. In view of this, the results of this study, while focussed on Nigeria, can provide lessons for other developing countries facing similar issues of rapid transformation coupled with the triple burden of malnutrition. Results of this study indicate that different trajectories of the food system affect average diet developments towards more calories, carbohydrates and fats or alternatively towards an increased importance of fruit and vegetables. While showing that food system developments matter for the undernutrition, micronutrient deficiency and obesity challenges of fast changing countries like Nigeria, complementary micro-level analyses are needed to assess the food system impacts on the nutrient transition of vulnerable population groups.

There are several limitations of this study that should be mentioned. Because of the widespread presence of informal arrangements in Nigerian markets, it is difficult to represent them in a broad-based modelling framework such as the global computable equilibrium model used here. A further limitation is that the behavioural decisions of consumers are modelled for a single representative household. No distinction is made regarding the livelihood system or geography of the household, even though these conditions will obviously drive both production and consumption decisions as well as dietary outcomes. Analyses of household consumption point to large differences across households in relation to socioeconomic, geographic and cultural variables [30,32,33]. In particular, the regional and rural/urban dimensions of nutrition warrant a deeper analysis if it is made useful for policy-making in Nigeria.

With respect to the areas of future research, a further analysis of the heterogeneity of household response to food systems drivers is considered as key issue. As consumer diets should be seen as outcomes as well as drivers of the performance of national food systems, maintaining a link with the macro-level framework as presented here is considered to be important while exploring response at the micro-level in greater depth. The question raised before, whether Nigeria's national food system has the potential to nourish its population with a healthy diet, can therefore be answered only in part, with these limitations in mind.

5. Conclusions

This paper provides a perspective on the future of food systems in Nigeria, taking into account an interplay of various macroeconomic and biophysical drivers. Because of its highly dynamic economic growth and demographic boom, the country represents an interesting case of studying the impact of these drivers on the food systems.

Given the historical increase in land expansion and low yield growth, land availability was identified as the key constraining factor determining the future of the food systems and food security.

The projections obtained in this study distinguish two future worlds for Nigeria and it is the land rigidity and extensification that determines which of the two worlds would become reality. In the first world, where land would be used extensively without inputs substitution, increasing pressures on the land market would result in excessive growth of producer prices. Food systems production value would increase about six times to the benefit of some agri-food players, mainly (white) meat processing sectors that could turn net exporters for the first time. For the economy as a whole, it would bring a positive structural transformation towards industrialization, higher role of services, less agriculture and less oil & gas. This is because industry and services would benefit from the release of non-land resources from agriculture. On the other hand, the world would become less favourable for consumers due to rising food prices, declining wages and increasing wage disparity between agriculture and the rest of the economy. Most of the income of the consumers would be spent on food and the access to food as one of the dimensions of food security would worsen not only for the unskilled but also for the skilled labour endowed households. From the nutritional point of view, the households would consume more calories, but these would come from more processed foods such as flour and palm oil, as well as from white meat.

In the alternative world, the non-land inputs, particularly labour, would substitute increasingly scarcer land. Higher land productivity due to the use of more inputs would rise wages in agriculture resulting in a more pro-poor growth. The traditionally trade-oriented sectors would restore competitiveness and, eventually, agri-food trade would enjoy a positive trade balance in many commodities. The structural transformation from agriculture to industry would be also expected, but with a higher share of food-economy. Because of decreasing food prices and increasing wages, food security would improve and the share of food expenditures in total expenditures would be comparable to a middle-income economy. Although the caloric consumption would come from more from primary agriculture, the share of fruits and vegetables would be higher, with more positive health impacts.

Main policy recommendations for fostering food systems development with positive nutritional impacts in Nigeria are directed to increased investments in agricultural R&D to alleviate the land pressure, reducing the rigidity of land markets to stimulate entrance and exit from the land market and to support intensification by improved access to capital markets and by replacing land for agricultural labour to stimulate agricultural wages.

As argued in the discussion, for multiple reasons the evolution of impact of these food systems changes on diet and nutrient gaps warrants further analysis and interpretation in a combined micro-level and macro-level framework. Concretely, the absorption capacity of skilled labour inside agriculture should be further assessed. An important area of future research is the analysis of heterogeneity of household response to food systems by linking the macro-level framework to simulating behaviour of individual households. This can enable tracing how the macro-drivers of food system effect individuals' nutritional outcomes and provide more insights into the nutrition inequality.

Author Contributions: Conceptualization, Z.S.K. and T.A.; methodology, Z.S.K. and M.K.; investigation, Z.S.K. and T.A.; data curation, Z.S.K. and M.K.; writing—original draft preparation, Z.S.K. and T.A; writing—review and editing, Z.S.K., T.A. and M.K.; visualization, Z.S.K.; project administration and funding acquisition, T.A.

Funding: This research was funded by CGIAR program Agriculture for Nutrition and Health (A4NH), grant no. P329 "Key leverage points in the food system", Dutch Ministry of Agriculture, Nature and Food Quality under grant no. WUR-KB22-001-003, and the European Union under Grant Agreement no. 633692 (SUSFANS).

Acknowledgments: We thank Jonathan Doelman, PBL Netherlands Environmental Assessment Agency, for sharing data on land use in Nigeria, and for his comments on the scenario design. We thank participants at the A4NH technical stakeholder workshop on food systems diagnosis and foresight (Abuja, 2–3 July 2018), SDG conference (Wageningen, 30 August 2018) and Tropentag (Ghent, September 2018) for comments and discussions.

Conflicts of Interest: The authors declare no conflict of interest. The funders had no role in the design of the study; in the collection, analyses, or interpretation of data; in the writing of the manuscript, or in the decision to publish the results.

Appendix A The Nested Structure of CES and CET Functions in MAGNET

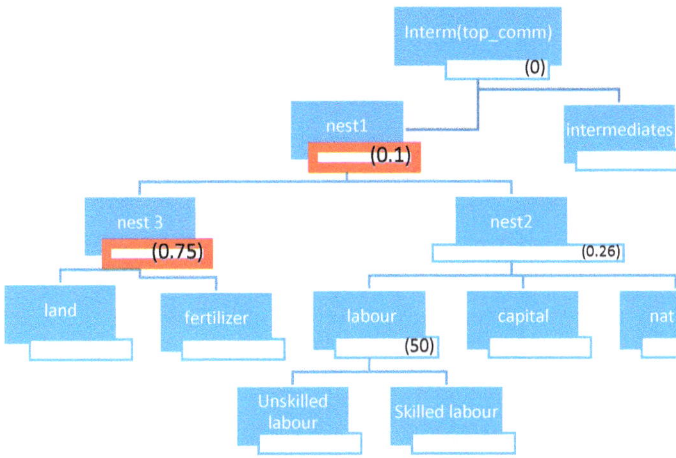

Figure A1. The nested production structure in the crop production sectors (substitution elasticity is in the brackets).

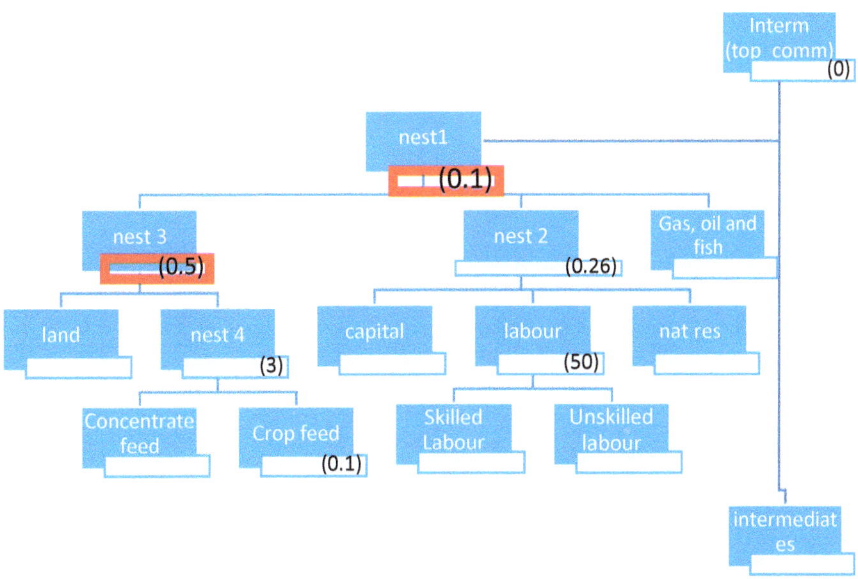

Figure A2. The nested production structure in the livestock production sectors (substitution elasticity is in the brackets).

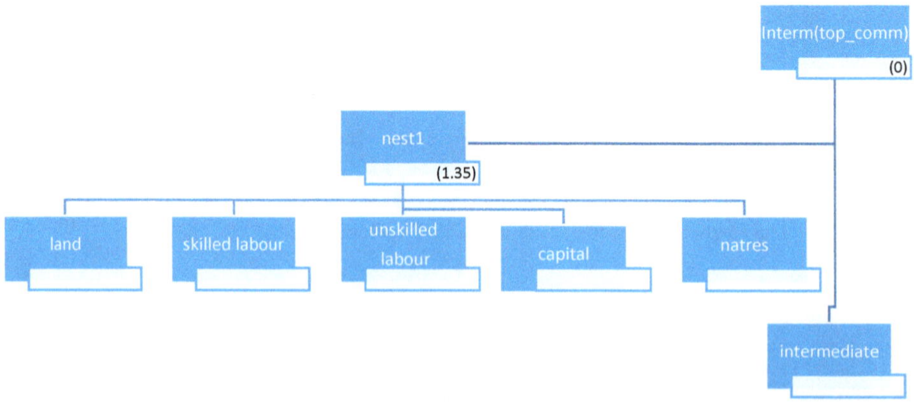

Figure A3. The nested production structure in the food processing industry, other industry and services production sectors (substitution elasticity is in the brackets).

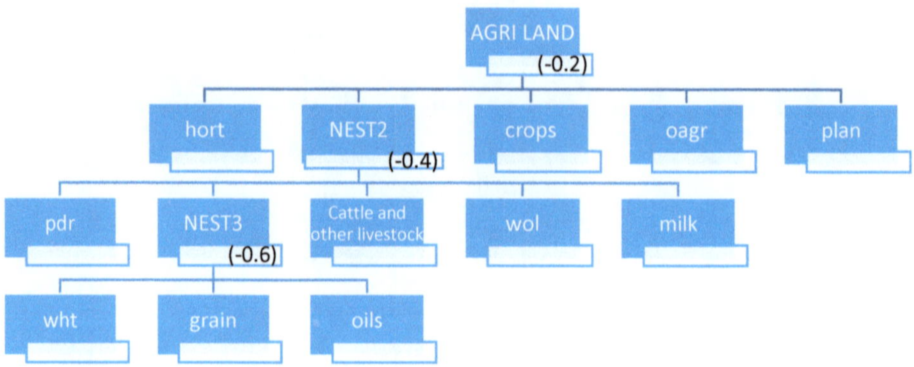

Figure A4. The structure of land allocation in the CET function (substitution elasticity is in the brackets).

Appendix B MAGNET and GENUS Database

Table A1. List of MAGNET sectors and mapping to GTAP.

GTAP Code	Description	MAGNET Code	Description
pdr	Paddy rice	pdr	Paddy and processed rice
wht	Wheat	wht	Wheat
gro	Cereal grains nec	grain	Cereal grains nec
osd	Oil seeds	oils	Oil seeds
c_b	Sugar cane, sugar beet	sug	Sugar cane, sugar beet
v_f	Vegetables, fruit, nuts	hort	Vegetables, fruit, nuts
ocr	Crops nec	crops	Crops nec
pfb	Plant-based fibers	oagr	Other agriculture
Bfctl *	beef	cattle	cattle sector
ctl	Cattle, sheep, goats, horses	othctl	sheep, goats, horses
Pltry *	poultry live animals	pltry	poultry sector
wol	Wool, silk-worm cocoons	wol	Wool, silk-worm cocoons
oap	Animal products nec	pigpls	Pig and other animal product
rmk	Raw milk	milk	Raw milk
BFCMT *	beaf meat	bfmt	beef meat
cmt	Meat: cattle, sheep, goats, horse	othcmt	Meat: other cattle, sheep, goats, horse
Poum *	Poultry meat	pulmt	poultry meat
omt	Meat products nec	othmt	Other meat product nec
mil	Dairy products	dairy	Dairy products

Table A1. *Cont.*

GTAP Code	Description	MAGNET Code	Description
sgr	Sugar and molase	sugar	Sugar and molasses
vol	Vegetable oils and fats	vol	Vegetable oils and fats
pcr	Processed rice	pcr	Processed rice
ofd and b_t	Food products nec, Beverages and tobacco	ofd	Processed food
Feed *	Animal feed	feed	Animal feed
Fsh *	Fishing	wfish	Wild fish
Aqcltr *	Diadromis fish	aqcltr	Aquaculture
Fishp *	Fish processing	fishp	Fish processing
Fishm *	Fish meal	fishm	fish meal

Note: GTAP codes with * refer to sectors that are newly disaggregated in MAGNET. Original GTAP sectors are found at: https://www.gtap.agecon.purdue.edu/databases/v9/v9_sectors.asp.

Table A2. GENUS macro and micronutrient data for Nigeria (2011).

	1 Median	2 Low	3 High
1 EdFd	1549	1549	1549
2 Calorie	2969	2911	3052
3 Protein	66	62	71
4 Fat	58	56	65
5 Carb	58	56	65
6 VitC	218	203	274
7 VitA	1118	151	1299
8 Folate	473	463	516
9 Calcium	423	365	508
10 Iron	23	19	26
11 Zinc	13	12	15
12 Potas	5379	5100	5613
13 Fiber	49	47	55
14 Copper	3	2	3
15 Sodium	195	183	210
16 Phosph	1452	1366	1768
17 Thiamin	2	2	3
18 Ribofl	1	1	1
19 Niacin	16	14	27
20 B6	3	3	3
21 Magnsm	623	594	803
22 SatFat	22	21	25
23 MonoUSF	20	18	22
24 PolyUSF	14	12	15

Appendix C Detailed MAGNET Results

Table A3. Annual growth of factor prices (2020–50) and % difference of prices in 2050 (Land_Subs vs Land_Fixed scenario).

Factor	Sector	Land_Fixed	Land_Subs	% Diff. 2050
Land	AGRI_PRIM	12.8	0.2	−98
UnSkLab	AGRI_PRIM	−1.5	0.0	62
	OTHER_SECTORS	0.5	2.0	55
SkLab	AGRI_PRIM	−1.5	0.0	60
	OTHER_SECTORS	−1.4	0.0	58
Capital	AGRI_PRIM	−2.3	−1.3	37
	OTHER_SECTORS	−2.2	−1.2	37

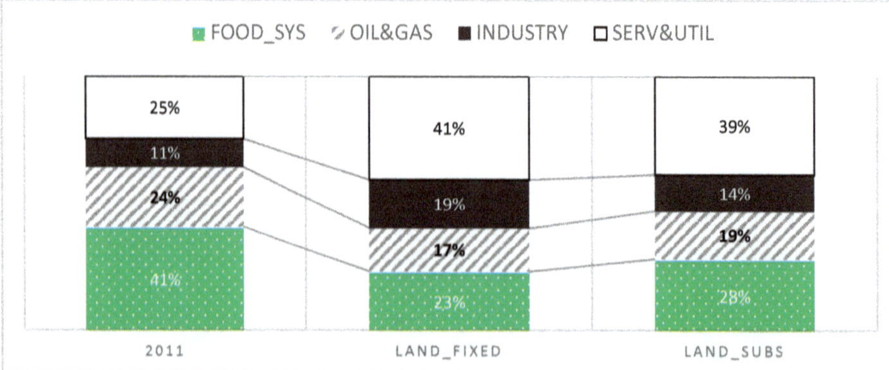

Figure A5. The structure of real value added.

Table A4. Absolute difference of sector endowment volumes in Land_subs vs. Land_Fixed scenario in 2050 (million USD in constant prices of 2011).

	UnSkLab	SkLab	Capital
AGRI_PRIM	**2659**	**16,281**	**9151**
hort	912	7307	4387
crops	6741	2806	4094
pdr	2172	1809	1821
grain	−3514	1799	−333
oils	−553	1435	585
othctl	−924	484	−82
wht	1046	343	583
oagr	448	234	298
cattle	−804	215	−185
sug	−199	64	−40
pigpls	−936	−80	−696
pltry	−1747	−142	−1291
AGRI_proc	**57**	**134**	**5087**
ofd	157	3003	2499
othcmt	35	409	1695
sugar	12	206	54
pcr	8	149	405
vol	2	38	35
bfmt	−13	−673	1708
pulmt	−53	−1096	−492
othmt	−98	−2034	−940
FISH SECTORS	**−3472**	**−89**	**−3373**
wfish	−3462	−24	−3360
aqcltr	−3	0	−4
fishp	−7	−65	−9
AGRI_FOOD	**−760**	**16,325**	**10,863**
INDUSTRY	**−609**	**−20,568**	**−24,790**
SERV&UTIL	**4791**	**4765**	**5587**

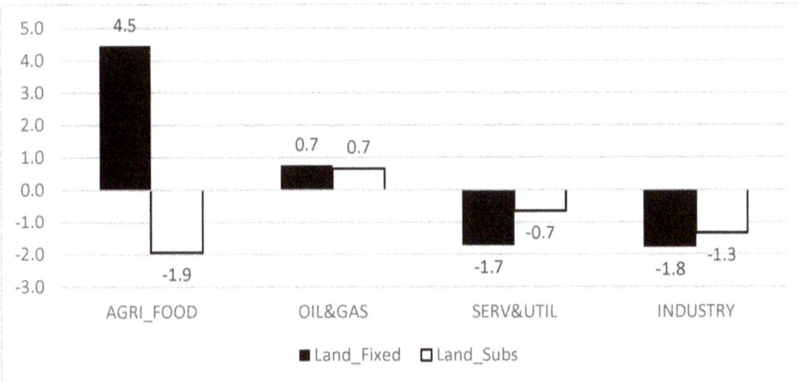

Figure A6. Annual growth of producer prices in Nigerian economy between 2020–2050.

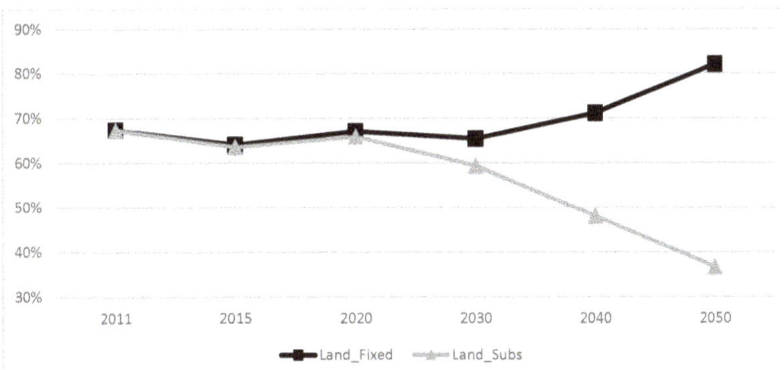

Figure A7. Share of food expenditures in total household expenditures.

References

1. World Bank. *World Development Indicators*; World Bank: Washington, DC, USA, 2018.
2. Mandryk, M.; Doelman, J.; Stehfest, E. Assessment of Global Land Availability: Land Supply for Agriculture. FOODSECURE Technical Papers; LEI Wageningen UR. 2015. Available online: https://www.pbl.nl/en/publicaties/assessment-of-global-land-availability-land-supply-for-agriculture (accessed on 5 September 2018).
3. National Population Commission Nigeria. *I. C. F. International Nigeria Demographic and Health Survey, 2013*; National Population Commission Nigeria: Abuja, Nigeria, 2014.
4. Imamura, F.; Micha, R.; Khatibzadeh, S.; Fahimi, S.; Shi, P.; Powles, J.; Mozaffarian, D. Dietary quality among men and women in 187 countries in 1990 and 2010: A systematic assessment. *Lancet Glob. Health* **2015**, *3*, e132–e142. [CrossRef]
5. HLPE. *Nutrition and Food Systems*; The High Level Panel of Experts on Food Security and Nutrition of the Committee on World Food Security: Rome, Italy, 2017.
6. HLPE. *Food Losses and Waste in the Context of Sustainable Food Systems*; The High Level Panel of Experts on Food Security and Nutrition of the Committee on World Food Security: Rome, Italy, 2014.
7. Ingram, J. A food systems approach to researching food security and its interactions with global environmental change. *Food Secur.* **2011**, *3*, 417–431. [CrossRef]
8. Van Dijk, M.; Meijerink, G. *A Review of Global Food Security Scenario and Assessment Studies: Results, Gaps and Research Priorities*; FOODSECURE Working Papers; LEI: Wageningen, The Netherlands, 2014.
9. Zurek, M.; Hebinck, A.; Leip, A.; Vervoort, J.; Kuiper, M.; Garrone, M.; Havlík, P.; Heckelei, T.; Hornborg, S.; Ingram, J.; et al. Assessing Sustainable Food and Nutrition Security of the EU Food System—An Integrated Approach. *Sustainability* **2018**, *10*, 4271. [CrossRef]

10. De Haen, H.; Réquillart, V. Linkages between sustainable consumption and sustainable production: Some suggestions for foresight work. *Food Secur.* **2014**, *6*, 87–100. [CrossRef]
11. Abdullahi, D.; John, M.S. The Political will and Quality Basic Education in Nigeria. *J. Power Polit. Gov.* **2014**, *2*.
12. WENR. *Education in Nigeria*; World Education News & Reviews: New York, NY, USA, 2017.
13. UNDESA. *World Urbanization Prospects—2018 Revision*; United Nations Population Division: Geneva, Switzerland, 2018.
14. CTA. *A Case Study Report on Nigeria's Agricultural Extension and Advisory System*; Technical Center for Agricultural and Rural Cooperation: Wageningen, The Netherlands, 2011.
15. DLEC Project. *Nigeria: In-Depth Assessment of Extension and Advisory Services*; Developing Local Extension Capacity Project; Care, Digital Green, IFPRI and USAID; United States Agency for International Development: Washington, DC, USA, 2017.
16. Bosello, F.; Campagnolo, L.; Cervigni, R.; Eboli, F. Climate Change and Adaptation: The Case of Nigerian Agriculture. *Environ. Resour. Econ.* **2018**, *69*, 787–810. [CrossRef]
17. Ayuba, I.G.U.; Ayuba, L.I. The Effect of Internal Displacement and Variation in Crop Production in Barkin Ladi Local Government Area of Plateau State. *J. Environ. Earth Sci.* **2016**, *6*, 35–40.
18. Tari, V.A.; Kibikiwa, M.G.; Umar, K. The effects of Boko Haram insurgency on food security status of some selected local government areas in Adamawa State, Nigeria. *Sky J. Food Sci.* **2016**, *5*, 12–18.
19. Van Berkum, S.; Achterbosch, T.; Linderhof, V.; Godeschalk, F.; Vroege, W. *Dynamics of Food Systems in Sub-Saharan Africa: Implications for Consumption Patterns and Farmers' Position in Food Supply Chains*; Wageningen Economic Research: The Hague, The Netherlands, 2017.
20. Okeke, E.C.; Eneobong, H.N.; Uzuegbunam, A.O.; Ozioko, A.O.; Kuhnlein, H. Igbo Traditional Food System: Documentation, Uses and Research Needs. *Pak. J. Nutr.* **2008**, *7*, 365–376. [CrossRef]
21. Kuiper, M.; Kalter, J.; Mertens, E.; Latka, C.; Heckelei, T.; Leip, A.; Kanellopoulos, A.; Havlik, P.; Achterbosch, T. *The SUSFANS Toolbox for Assessing EU Sustainable Food and Nutrition Security*; SUSFANS: The Hague, The Netherlands, 2018.
22. Van Meijl, H.; Tsiropoulos, I.; Bartelings, H.; Hoefnagels, R.; Smeets, E.; Tabeau, A.; Faaij, A. On the macro-economic impact of bioenergy and biochemicals—Introducing advanced bioeconomy sectors into an economic modelling framework with a case study for the Netherlands. *Biomass Bioenergy* **2018**, *108*, 381–397. [CrossRef]
23. Kuiper, M.; Shutes, L.; Verma, M.; Tabeau, A.; van Meijl, H. *Exploring the Impact of Alternative Population Projections on Prices, Growth and Poverty Developments*; Background Paper to the UNCTAD-FAO Commodities and Development Report 2017 Commodity Markets, Economic Growth and Development; FAO: Rome, Italy, 2018; ISBN 978-92-5-130096-1.
24. Helming, J.; Tabeau, A. The economic, environmental and agricultural land use effects in the European Union of agricultural labour subsidies under the Common Agricultural Policy. *Reg. Environ. Chang.* **2018**, *18*, 763–773. [CrossRef]
25. Woltjer, G.; Kuiper, M.; Kavallari, A.; van Meijl, H.; Powell, J.; Rutten, M.; Shutes, L.; Tabeau, A. *The MAGNET Model: Module Description*; LEI Wageningen UR: The Hague, The Netherlands, 2012.
26. Aguiar, A.; Narayanan, B.; McDougall, R. An Overview of the GTAP 9 Data Base. *J. Glob. Econ. Anal.* **2016**, *1*, 181–208. [CrossRef]
27. Keeney, R.; Hertel, T. *GTAP-AGR: A Framework for Assessing the Implications of Multilateral Changes in Agricultural Policies*; GTAP Technical Paper No. 24; Purdue University: Purdue, IN, USA, 2005.
28. Tabeau, A.; Helming, J.; Philippidis, G. *Land Supply Elasticity: Overview of Available Estimates and Recommended Values for Magnet*; EU: Brussels, Belgium, 2017; ISBN 978-92-79-69102-7.
29. Smith, M.R.; Micha, R.; Golden, C.D.; Mozaffarian, D.; Myers, S.S. Global Expanded Nutrient Supply (GENuS) Model: A New Method for Estimating the Global Dietary Supply of Nutrients. *PLoS ONE* **2016**, *11*, e0146976. [CrossRef] [PubMed]
30. Brouwer, I.D.; Nijhuis, A.; Talsma, E.F.; Trijsburg, L.; Pastoria, G.; Mekonnen, D.A.; Morley, T.; Linderhof, V.; Achterbosch, T. The state of the dietary and nutrient gaps in Nigeria. In *Proceedings of the A4NH Stakeholder Workshop "Food Systems for Healthier Diets in Nigeria"*; IFPRI: Washington, DC, USA, 2018.
31. NBS. *Nigeria—General Household Survey, Panel 2015–2016, Wave 3*; NBS: Abuja, Nigeria, 2017.
32. Akerele, D. Household Food Expenditure Patterns, Food Nutrient Consumption and Nutritional Vulnerability in Nigeria: Implications for Policy. *Ecol. Food Nutr.* **2015**, *54*, 546–571. [CrossRef] [PubMed]

33. Aromolaran, A.B. Does increase in women's income relative to men's income increase food calorie intake in poor households? Evidence from Nigeria. *Agric. Econ.* **2010**, *41*, 239–249. [CrossRef]
34. Onyemauwa, C.S. Analysis of Household Consumption of Cassava Products in Ohaozara, Ebonyi State, Southeast Nigeria. *Researcher* **2010**, *2*, 1–6.
35. Arndt, C. *Nigeria's Macroeconomic Crisis Explained*; NSSP Working Paper; International Food Policy Research Institute (IFPRI): Abuja, Nigeria, 2018.
36. World Bank. *Macro Poverty Outlook for Sub-Saharan Africa: Nigeria*; World Bank: Washington, DC, USA, 2018; pp. 266–267.
37. International Monetary Fund World Economic Outlook Database April 2018. Available online: https://www.imf.org/external/pubs/ft/weo/2018/01/weodata/index.aspx (accessed on 15 November 2018).
38. Riahi, K.; van Vuuren, D.P.; Kriegler, E.; Edmonds, J.; O'Neill, B.C.; Fujimori, S.; Bauer, N.; Calvin, K.; Dellink, R.; Fricko, O.; et al. The Shared Socioeconomic Pathways and their energy, land use, and greenhouse gas emissions implications: An overview. *Glob. Environ. Chang.* **2017**, *42*, 153–168. [CrossRef]
39. Fricko, O.; Havlik, P.; Rogelj, J.; Klimont, Z.; Gusti, M.; Johnson, N.; Kolp, P.; Strubegger, M.; Valin, H.; Amann, M.; et al. The marker quantification of the Shared Socioeconomic Pathway 2: A middle-of-the-road scenario for the 21st century. *Glob. Environ. Chang.* **2017**, *42*, 251–267. [CrossRef]
40. PWC. *The World in 2050*; Pricewaterhouse Coopers: London, UK, 2017.
41. Lutz, Butz, and Kc Wittgenstein Centre Data Explorer. Available online: http://dataexplorer.wittgensteincentre.org/shiny/wic/ (accessed on 15 November 2018).
42. PWC. *Transforming Nigeria's Agricultural Value Chain*; PricewaterhouseCoopers: London, UK, 2017.
43. Nwankpa, N. Sustainable Agricultural Development in Nigeria: A Way Out of Hunger and Poverty. *Eur. J. Sustain. Dev.* **2017**, *6*, 175–184. [CrossRef]
44. Laborde, D.; Lallemant, T.; McDougal, K.; Smaller, C.; Fousseini, T. *Transforming Agriculture in Africa & Asia: What Are the Policy Priorities?* ISSD and IFPRI: Washington, DC, USA, 2018.
45. UNDP. *The Roles and Opportunities for the Private Sector in Africa's Agro-Food Industry*; United Nations Development Programme (UNDP): Johannesburg, South Africa, 2012.
46. FAO Nigeria at a Glance. Available online: http://www.fao.org/nigeria/fao-in-nigeria/nigeria-at-a-glance/en/ (accessed on 15 November 2018).
47. Stehfest, E.; van Vuuren, D.; Bouwman, L.; Kram, T. *Integrated Assessment of Global Environmental Change with IMAGE 3.0: Model Description and Policy Applications*; Netherlands Environmental Assessment Agency (PBL): Bilthoven, The Netherlands, 2014; ISBN 94-91506-71-4.
48. World Bank. *World Development Indicators—Fertilizer Consumption*; World Bank: Washington, DC, USA, 2018.
49. World Bank. *World Development Report 2008: Agriculture for Development*; The World Bank: Washington, DC, USA, 2008.
50. Liverpool-Tasie, L.S.; Barrett, C.B.; Sheahan, M.B. Understanding Fertilizer Use and Profitability of Rice Production across Nigeria's Diverse Agroecological Conditions. In Proceedings of the Annual Bank Conference on Africa 2014. Available online: http://catalog.ihsn.org/index.php/citations/61607 (accessed on 5 September 2018).
51. Kristkova, Z.S.; van Dijk, M.; van Meijl, H. Assessing the Impact of Agricultural R&D Investments on Long-Term Projections of Food Security. In *World Agricultural Resources and Food Security*; Frontiers of Economics and Globalization; Emerald Publishing Limited: Bingley, UK, 2017; Volume 17, pp. 1–17.
52. Baltussen, W.; Achterbosch, T.; Arets, E.; de Blaeij, A.; Erlenborn, N.; Fobelets, V.; Galgani, P.; de Groot Ruiz, A.; Hardwicke, R.; Hiemstra, S.J.; et al. *Valuation of Livestock Eco-Agri-Food Systems: Poultry, Beef and Dairy*; Wageningen Economic Research Report; 2017-039; Wageningen Economic Research: Wageningen, The Netherlands, 2017; ISBN 978-90-8615-734-1.
53. Dobermann, A.; Nelson, R.; Beever, D.; Bergvinson, D.; Crowley, E.; Denning, G.; Giller, K.; d'Arros Hughes, J.; Jahn, M.; Lynam, J. Solutions for Sustainable Agriculture and Food Systems. U. N. Sustain. Dev. Solut. Netw. Paris Fr. 2013. Available online: http://unsdsn.org/what-we-do/thematic-networks/sustainable-agriculture-and-food-systems/ (accessed on 5 September 2018).

sustainability

MDPI

Article

Social Capital and Adoption of Alternative Conservation Agricultural Practices in South-Western Nigeria

Seyi Olalekan Olawuyi * and Abbyssinia Mushunje

Department of Agricultural Economics and Extension, University of Fort Hare, P/Bag X1314, Alice 5700, Eastern Cape Province, South Africa; amushunje@ufh.ac.za

* Correspondence: seyidolapo1704@gmail.com

Received: 29 September 2018; Accepted: 26 November 2018; Published: 30 January 2019

check for updates

Abstract: The major concern of most African countries, including Nigeria, in recent times is how to increase food production because of food insecurity issues, which by extension, is a major contributing factor to the prevalence of poverty. Therefore, adoption of conservation agricultural practices is regarded as a pathway to drive the achievement of food and nutrition security, as well as the needed optimal performance in the agri-food sector. Reportedly, scaling up of the limited adoption of these practices could be facilitated through kinship ties, peer influence, and social networks that govern mutual interactions among individuals; therefore, this motivated the study. Using cross-sectional data obtained from 350 sample units selected from South-Western Nigeria through a multistage sampling technique, this study applied descriptive statistical tools and cross-tabulation techniques to profile the sampled subjects while count outcome models were used to investigate the factors driving counts of conservative agriculture (CA) adoption. Similarly, a marginal treatment effects (MTEs) model (parametric approach) using local IV estimator was applied to examine the effects of CA adoption on the outcome (*log* of farmers' farm income). Additionally, appropriate measures of fit tests statistics were used to test the reliabilities of the fitted models. Findings revealed that farmers' years of farming experience ($p < 0.1$), frequency of extension visits ($p < 0.05$), and social capital viz-a-viz density of social group memberships ($p < 0.05$) significantly determined the count of CA practices adopted with varying degrees by smallholder farmers. Although, social capital expressed in terms of membership of occupational group and diversity of social group members also had a positive influence on the count of CA practices adopted but not significant owing largely to the "information gaps" about agricultural technologies in the study area. However, the statistical tests of the MTEs indicated that the treatment effects differed significantly across the covariates and it also varied significantly with unobserved heterogeneity. The policy relevant treatment effect estimates also revealed that different policy scenarios could increase or decrease CA adoption, depending on which individuals it induces to attract the expected spread and exposure.

Keywords: adoption; conservation agriculture; social capital; count outcome models; pca; marginal treatment effects; Nigeria

1. Background Information

Sustainable economic growth and development in a developing economy like Nigeria is achievable through the agricultural sector and its sub-sectors which are concentrated in rural areas, home to the majority (about 75%) of the households practicing farming for family sustenance and/or earning income from the sales of agricultural products [1]. In addition to the persistent use of traditional farming practices, these rural farming households cultivate crop varieties that are low-yielding on small and scattered farmland holdings (smallholder farmers). This act depletes the soil organic matter with devastating consequences on production output, income generation as well as the ecosystem. Similarly, non-access to agricultural credit and limited technical know-how are part of the challenges facing the development of farming activities in sub-Saharan Africa, including Nigeria [2]. These challenges call for holistic interventions that are sustainable, promote a safe environment, and ultimately increase production output. Thus, a practice with zero environmental and human hazards which have literatures converging [3–11] on its capability to use renewable local farm resources for sustainable and increased production output is called conservative agriculture (CA).

Generally, CA is regarded as a resource saving agricultural practice that can help farmers simultaneously harvest high yield and conserve the environment [12]. Besides, the water retention characteristic of CA makes it suitable in water deficient farming areas. The basic CA principles include the following practices: minimum soil disturbance, the use of crop biomass for permanent soil cover, and sequential rotation practice for different unrelated crops; all these can potentially strengthen farmers' resilience to climate change and enhance the sustainability of agro-ecosystems [13–16]. The diagrammatic view of these three CA packages required for full adoption, according to these authors is shown in Figure 1.

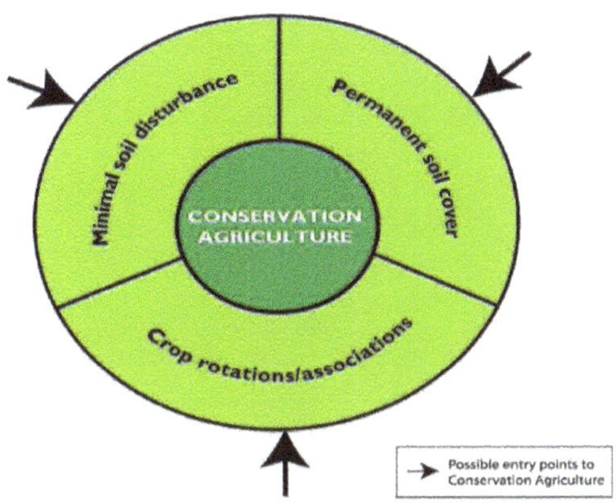

Figure 1. Basic principles of conservative agriculture (CA) practices. Source: Calegari and Ashburner [13] as cited in Ndah, Schuler, Uthes, and Zander [17].

Equally, the major concern of most African countries (including Nigeria) in recent times is how to increase food production [18]. Meanwhile, rural food insecurity is a major contributing factor to widespread poverty in Africa, and Nigeria is no exception, where most farmers are peasants. Therefore, CA is regarded as a panacea to achieving food security and the needed optimal performance in agricultural production, as it is now being promoted, without any negative consequences on the environment. However, the tendency of CA in preserving the environment (erosion inclusive) and improving soil properties cannot be under estimated [19]. This is because its success is reportedly premised on the production environment and readiness of smallholder farmers to accept, adopt, and continue to use this innovative method for sustainable management agricultural systems. The potential of these practices to mitigate adverse effects of climate change and extreme weather events was also emphasized by De Lucas et al. [20] and Deligios et al. [21]. Expectedly, farmers' decisions to accept CA innovation according to Silici [4] could be facilitated through social capital (SC); that is ties, kinship, peer influence, and social groups (formal or/and informal) vis-a-viz social networks that govern the interactions among social group members. Hence, the motivations to factor in the social aspect of farmers' economic behavior in a bid to thoroughly understand the process of CA uptake and adoption. The main focus point of agricultural research and scientific debates from different fora for several decades and up till now is centered on agricultural sustainability and how to gain proper understanding about the push and pull factors driving producers' decision on agricultural technology adoption [10,20]. Several past studies on adoption of new agricultural innovation majorly pointed to human and physical capital among other factors as predictive determinants of technology adoption [22–29], using a standard utility model at the individual adopter's level. Similarly, Pino et al. [30] citing Kirton [31] and Rogers [32] emphasized farmers' innovativeness—an individual's characteristics as a driver of technologies adoption in a study conducted in Italy. The majority of these studies tend to ignore that individual decisions are not just made, rather such are entrenched in a more complex and organized system of communities whose individual decisions are products of shared common interests, collective participation, and concerns based on mutual trust [4,9,33]. Collectively, all these attributes are put together as "social capital".

According to Lollo [34], the first mention of social capital concept was in 1916 by Lyda Judson Hanifan in his seminar paper titled "The Rural School Community Center" published in the United States. The paper discussed community involvement and how neighbors could possibly work together to foster the performance and success of the schools. Suffice it to say that Hanifan [35] invoked the idea of social capital by referring to it as:

> "those tangible assets or substances that count for most in the daily lives of people, namely: goodwill, fellowship, mutual sympathy, and social intercourse among the individuals and families who make up a social unit. This further suggests that individual is helpless socially, if left to himself. But, if he interacts with his neighbour, with chain of interconnectivity, there will be an accumulation of social capital, which may immediately satisfy his social needs and bear a social potentiality sufficient enough for the improvement in living conditions of individuals. The community as a whole in turn will benefit by this cooperation (collective participation), while individual will eventually find in his associations the advantages of the help, the sympathy, and the fellowship of his neighbours." ([35], p.130)

In lieu of this position, the concept of social capital vis-a-viz a social network framework has been advocated for as a crucial factor to understand the interconnectivity existing between people, and foster the aims and objectives of community development experts and stakeholders towards achieving equitable and sustainable agricultural growth and development [36]. Therefore, social capital can succinctly be conceptualized as features (i.e., reciprocity, norms, and trust) existing between people of the same or diverse cultural background which facilitates cooperation among individuals for their mutual and societal benefits [37–39].

Importantly, these features encourage collective action/participation towards achieving bonding social networks and the much needed sustainable development [40]. Collective action/participation is recognized as a crucial component of rural and economic development as well as local-level institutions management [41] through which efficient flow of important information can be achieved among the resource-poor farmers [42]. In a similar manner, Woolcock [40] and Aker [43] also affirmed that, social capital can be facilitated through participation in formal and informal networks, registered social organizations or community-based organizations as well as social movements. Hence, investment in collective action/participation activities based on social capital-trust, with the expectation of reciprocity and through mutual cooperation and co-existence, sharing of useful information among members can definitely be helpful in pushing for uptake and adoption of improved agricultural technologies towards achieving increased production output, better income and welfare, as well as the attainment of Sustainable Development Goal two (SDG 2) [44].

Consequent on the above arguments, this study investigated the pathways through which social networks can possibly drive adoption and adoption-count of alternative CA practices as well as the possible effects and impacts of CA adoption on farmers' farm income in South-Western Nigeria.

2. Materials and Methods

2.1. The Study Area

This research work was carried out in South-Western Nigeria which consists of six states, namely: Ekiti, Lagos, Ogun, Ondo, Osun, and Oyo states. But, for the purpose of this research work, Oyo, Osun, and Ondo states were used. The choice of these states was premised on the fact that adoption of improved agricultural technologies (such as improved maize seeds, improved rice varieties and cassava vitamin A fortified cassava varieties) had earlier been reported in these states of South-Western Nigeria [45–48]. Moreover, the majority of the rural households in these states are into farming and farming related activities. Importantly, the overview of the study area is presented in Figure 2.

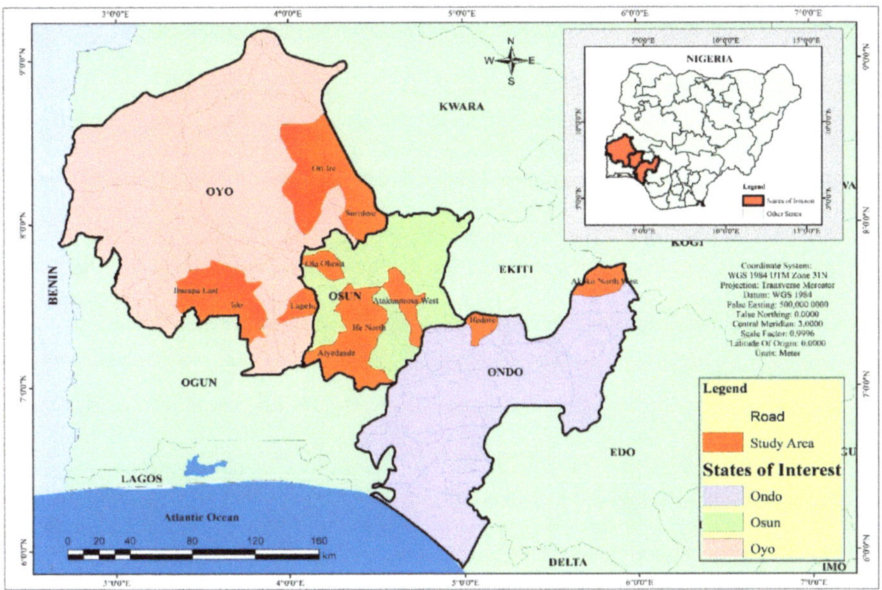

Figure 2. Map of South-Western Nigeria showing the states and Local Government Areas (LGAs) of interest. Source: National Space Research and Development Agency of Nigeria (NASRDA) [49].

2.2. Sampling Technique and Data Collection

Multistage sampling technique was used to select the representative sample of 350 smallholder farmers and responses were elicited with the aid of a carefully prepared questionnaire which is in line with the guidelines provided in "Qualitative expert Assessment Tools for assessing the adoption of CA in Africa (QAToCA)" taking into consideration the "regional factor" caution [50]. Hence, smallholder farmers represent the entity under study (that is, the unit of analysis).

South-Western Nigerian states are stratified into agro-ecological zones which have been pre-determined by the Ministry of Agriculture, Natural Resources, and Rural Development in each of the states. Therefore, Oyo, Osun, and Ondo states are stratified into four, three, and two Agricultural Development Programme (ADP) zones, respectively, based on rurality. First, a simple random sampling technique was used to select 50% of the ADP zones in each of the three states to arrive at 2 ADPs from Oyo State, 2 ADPs from Osun State, and 1 ADP from Ondo State, respectively. Equally, the second stage made use of simple random sampling technique to select one-third (1/3) of the Local Government Areas (LGAs) from each of the ADPs selected in the chosen states. The third stage also involved simple random sampling to choose three villages from each of the LGAs selected in the second stage while the fourth stage involved the use of a proportionate to size sampling technique to select 350 registered smallholder farmers used as sample size for this study.

The proportionality factor applied for a bias-free sample size selection was:

$$N_i = n_i/N \times 350 \quad (1)$$

where:

N_i = number of respondents/instruments selected in each of the ith state (i = 1, 2, and 3);

n_i = the population of all registered farmers in ith states selected;

N = total population of all registered farmers in all the three states selected;

350 = total number of respondents sampled across the selected states.

Importantly, this research observed the following ethical considerations in the study area: anonymity, informed consent, privacy, confidentiality, as well as professionalism.

2.3. *Data Analytical Techniques*

The analytical tools used include: descriptive statistics such as frequency counts, percentages, and mean and standard deviation. Similarly, inferential statistics applied include: binary probit regression model, count outcome models (Poisson and Negative Binomial regression models), marginal treatment effects model, as well as principal components analysis (PCA) to generate index of social capital benefits. More so, measures of fit statistics tests were applied to ascertain and affirm the reliabilities of the fitted models. However, cautions were taken in the estimated models to avoid what is known as "forbidden regression" ([51], pp. 265–268). This is a situation where the models' results produce consistent estimates only under very restrictive assumptions which rarely hold in practice.

2.3.1. Model Specification

Binary Probit Regression Model

Binary probit regression is usually applied to model dichotomous outcome variable [52]. According to Sebopetji and Belete [53], the probit model assumes that while 0 and 1 values are only observed for the response variable Y, there is a latent and unobserved continuous variable Y^* that determines the value of the response variable Y. Therefore, Y^* can be expressed as:

$$Y^* = X^1\beta + \varepsilon_i \quad (2)$$

such that:

$$Y = 1\ (Y^* > 0). \text{ That is, } Y = 1 \text{ if } Y^* > 0 \text{ i.e., } (\varepsilon < X^1\beta), 0, \text{ otherwise.}$$

where:

Y = vector of the response variable (CA adoption = 1, 0, otherwise);

X = vector of explanatory variables, β = probit coefficients, ε_i = random error term.

Count Models

In estimating the Poisson model, according to Williams [54], let y be a random variable representing the number of occurrences of an event during an interval of time; such that: y has a Poisson distribution with parameter $\mu > 0$ iff:

$$\frac{\Pr(y|\mu) = \exp(-\mu)\mu\mu^y}{y!} for\ y = 0, 1, 2, 3, \ldots\ldots\ldots\ldots n \quad (3)$$

Equally, borrowing from Bruin [55], the negative binomial distribution model is expressed as:

$$\Pr(Y = y|\lambda, \alpha) = \frac{\Gamma(y + \alpha^{-1})}{y!\Gamma(\alpha^{-1})}\left(\frac{\alpha^{-1}}{\alpha^{-1} + \lambda}\right)^{\alpha^{-1}}\left(\frac{\lambda}{\alpha^{-1} + \lambda}\right)^{y} \tag{4}$$

Here, the negative binomial distribution has two parameters namely: λ and α, where:

λ = the mean or expected value of the distribution; and α = the over dispersion parameter.

However, the likelihood function for the negative binomial model according to Bruin [54] is given by:

$$L(\beta|y, \mathbf{X}) = \prod_{i=1}^{N}\Pr(y_i|x_i) = \prod_{i=1}^{N}\frac{\Gamma(y + \alpha^{-1})}{y!\Gamma(\alpha^{-1})}\left(\frac{\alpha^{-1}}{\alpha^{-1} + \mu_i}\right)^{\alpha^{-1}}\left(\frac{\mu_i}{\alpha^{-1} + \mu_i}\right)^{y_i} \tag{5}$$

Therefore, the relationship between the count of CA practices adopted by farmers and the specified covariates is expressed as:

$$Y_i = f\ (FC, HC, IS, SC, Expt) \tag{6}$$

where:

Y_i = count of alternative CA practices adopted by ith farmer; FC = farmers and farm-based attributes; HC = human capital; IS = institutional supports; SC = social capital and networks components; $Expt$ = exposure time period.

The explanatory variables are explicitly defined as follow:

X_1 = gender (male = 1, 0, otherwise); X_2 = age (years); X_3 = years of formal education (years);

X_4 = land acquisition (inheritance = 1, 0, otherwise); X_5 = CA farm size (plot/ha-continuous);

X_6 = total years of experience in farming (years); X_7 = frequency of extension visits (actual number-continuous); X_8 = occupational group membership (yes = 1, 0, otherwise);

X_9 = participation in collective action/initiatives (yes = 1, 0, otherwise); X_{10} = density of social groups membership (actual number-continuous); X_{11} = diversity of social group members (heterogeneity index) (%); X_{12} = participation in decision making (decision making index) (%);

* years of experience in CA practices (a proxy for exposure period) (years).

Marginal Treatment Effects Model

The marginal treatment effects model (MTE) using local IV is usually applied to capture heterogeneity in the treatment effects alongside the unobserved dimension otherwise known as resistance to treatment. According to Andresen [56] as well as Abadie and Imbens [57], MTEs generate selection on unobserved gains. This suggests that individuals who choose treatment because of their low-resistance capacity are likely to have different gains compared to individuals with high-resistance capacity. According to Andresen [56], MTEs model specification is based on the generalized Roy model. This is specified as:

$$Y_j = \mu_j(X) + U_j \quad \text{for } j = 0, 1 \tag{7}$$

$$Y = DY_1 + (1 - D)\ Y_0 \tag{8}$$

$$D = \mathbf{I}\ \{\mu_D\ (Z) > V\} \text{ where } \quad Z = (X, Z_{-}) \tag{9}$$

Y_1 and Y_0 are the potential outcomes in the treated and untreated state; that is, log of farmers' income with and without the treatment (CA adoption) which are modeled as functions of observables covariates. This of course may have the possibility of fixed effects. Equation (9) represents the selection equation, which contains the latent index of **I** as an indicator function. This also presents selection modeling into treatment equation in an implicit form conditioned on the observables covariates and instruments Z_{-} which does not influence potential outcomes but the probability of treatment. More importantly, identification of the MTEs model requires the following assumptions:

- Conditional independence: $(U_0, U_1, V) \perp Z - | X$

- Separability: $E(U_j \mid V, X) = E(U_j \mid V)$

3. Results and Discussion

3.1. Probit Regression Estimates

The results in Table 1 reveal the estimates of the marginal effects at the means (MEMs) obtained from the binary probit model. Findings from the estimation indicated that, for farmers with average values of being a male gender (0.69), age (52.13), years of formal education (6.88), years of exposure to CA farming system (12.97), and frequency of farmers' contact with extension agents (1.92), the predicted probability of adopting CA farming practices was approximately 0.07 points more compared to female counterparts. In terms of age, the predicted probability of CA adoption was 0.005 points more for older farmers than younger ones. However, the predicted probability of CA adoption was0.09 points more for farmers who had regular contact with extension agents than those with few contacts. Conversely, the predicted probability of CA adoption was 0.004 point less for farmers with many years of experience and exposure to CA system than the new entrants. Importantly, the findings revealed that the gender of the farmers ($p < 0.1$), age ($p < 0.1$), years of formal education (a proxy for human capital) ($p < 0.1$), years of exposure to CA system ($p < 0.1$), and frequency of farmers' contact with extension agents ($p < 0.01$) significantly predicted adoption of conservation agriculture in the study area.

Table 1. Marginal effects (at the means) estimates of the binary probit model.

		Delta-Method		
Adoption of CA	*dy/dx*	*std. err.*	*z*	*p* > | *z* |
1.gender	0.0670	0.0395	1.70 ***	0.089
Age	0.0052	0.0027	1.92 ***	0.054
years of formal education	0.0085	0.0044	1.95 ***	0.051
years of CA farming experience	−0.0042	0.0024	−1.73 ***	0.083
farm size under CA cultivation	0.0102	0.0180	0.57	0.570
log of output	0.0341	0.0238	1.43	0.153
duration of residency	0.0027	0.0022	1.21	0.225
labor contribution	0.0005	0.0013	0.38	0.703
risk attitude	0.1136	0.0955	1.19	0.235
1.access to extension service	−0.2377	0.1680	−1.41	0.157
frequency of extension visit	0.0896	0.0357	2.51 *	0.012
regional characteristics				
region 2	−0.0237	0.0583	−0.41	0.685
region 3	0.0122	0.0969	0.13	0.900

Note: dy/dx for factor levels is the discrete change from the base level. * $p < 0.01$; *** $p < 0.1$ probability levels respectively. Source: Data analysis, 2018.

Furthermore, to validate the model's goodness-of-fit, the study applied Hosmer, Lemeshow, and Sturdivant [58] fit-test procedure. The findings from this test evidently revealed that the model fits reasonably well (see Table A1).

3.2. Econometrics Results: Effects of Social Capital on CA Adoption

3.2.1. Poisson and Negative Binomial Distribution Models: Empirical Results

The estimation of Poisson distribution regression model (PRM) and the associated goodness-of-fit tests indicated that the Poisson estimation suffers from over-dispersion problem as expected. Evidently, the Pearson's goodness-of-fit test result shows that the distribution of CA practices adoption counts significantly differs for a Poisson distribution. Consequently, the unacceptably large value obtained and recorded for chi-square in the post estimation (likelihood ratio test) is an indication that the Poisson distribution model is not a suitable option because over-dispersion is suspected. This estimation is consistent with the guidelines provided by Baum [59]. In lieu of this, it is clearly impossible to make any meaningful inference from the Poisson regression model estimates to avoid a misleading conclusion. Given the distribution of data, the negative binomial distribution model was considered an appropriate option over the Poisson model to address the over-dispersion issue. More so, the incident rate ratio (IRR) of the negative binomial regression model was computed and reported as suggested by Piza [60] to show the impact of explanatory variables in terms of a percentage change in the observed response variable (in this case, counts of CA practices adopted). In essence, "the IRR represents the change in the response variable in terms of a percentage change, with the precise percentage determined by the amount the IRR is either above or below 1" [60]. Equally, it is important to stress that, count regression techniques model the *log* of incident counts [54].

The findings indicated in Table 2 report the fitted negative binomial regression model. Similarly, the statistical significance ($p < 0.01$) of alpha coefficient, and the likelihood ratio test of alpha also attest to the non-appropriateness of the Poisson regression model. Therefore, this permits a strong rejection of the null hypothesis that the errors do not exhibit an over-dispersion problem. Hence, the negative binomial model is deemed fit for describing the influencing dynamics governing smallholder farmers' adoption count of alternative CA practices in the study area. These procedures and findings are in tandem with Pedzisa [8] whose study investigated the intensity of adoption of CA by smallholder farmers in Zimbabwe. The result from Table 2 revealed that, for every one unit increase in the male gender compared to the female counterpart, the *log* count of CA practices adopted by female gender is expected to increase by approximately 0.76; with an estimated statistical significance (p-value) of 0.099 (that is, $p < 0.1$). A viable explanation for this is that, increase in the count of CA practices adopted by male gender serves as a positive motivating factor for the female counterpart to increase the count of CA practices adopted by them in a bid to also achieve maximum benefits accrued from CA adoption. Similarly, for every unit increase in the number of social groups to which farmers belong, the *log* count of CA practices adopted is expected to decrease by approximately 0.20. This suggests that membership in many social groups significantly ($p < 0.01$) influences the *log* count of CA practices adopted in the study area, though with inverse relationship. This result reinforces earlier findings that there is a persistent information gap among members of various social groups; rather much focus is placed on the social events than sharing useful information about improved and beneficial agricultural techniques such as CA.

Table 2. Negative binomial regression model estimates.

Count of CA Practices	Coefficient	IRR	*z-Statistics*	$p > \lvert z \rvert$
1.gender	−0.2421	0.7850	−1.65 ***	0.099
Age	0.0121	1.0122	1.46	0.145
years of formal education	0.0042	1.0042	0.28	0.777
1.land acquisition	0.0639	1.0660	0.40	0.691
farm size cultivated under CA	−0.0125	0.9876	−0.20	0.841
total years of farming experience	0.0134	1.0135	1.90 ***	0.057
frequency of extension visits	0.1345	1.1439	2.03 **	0.042
1.occupational group membership	0.1483	1.1598	0.92	0.357
1.participation in collective action	−0.0753	0.9274	−0.51	0.613
density-social groups membership	−0.1956	0.8224	−2.53 *	0.011
diversity of social group members	0.2797	1.3227	0.43	0.664
involvement in decision-making	−0.7197	0.4869	−1.18	0.239
constant	−0.8022	0.4483	−1.00	0.320
Ln (years of CA farming experience)	1	1		
Lnalpha	0.2140	0.0914		
Alpha	1.2386	0.1132		

Likelihood-ratio test of alpha = 0: chibar^2 (01) = 1028.23, Prob >= chibar^2 = 0.000. Number of observations = 350, Log likelihood = −948.64879, Dispersion = mean. Prob > chi^2 = 0.0005, Pseudo R^2 = 0.0180, LR chi^2 (12) = 34.86. * $p < 0.01$; ** $p < 0.05$; *** $p < 0.1$ level respectively; IRR = incident rate ratio. *Wald test of lnalpha*: [lnalpha] _cons = 1; chi^2 (1) = 73.91; prob > chi^2 = 0.0000. Source: Data analysis, 2018.

On the other hand, the results also indicated that, for every one unit increase in human capital designate-total years of farming experience, the *log* count of CA practices adopted is expected to increase by approximately 0.01; suggesting that a unit increase in the years of farming experience significantly ($p < 0.1$) increases the *log* count of CA practices adopted by the smallholder farmers in the study area. This result is in line with *a-priori* expectations. Expectedly, frequency of contact with extension agents was found to have a direct and significant ($p < 0.05$) influence on the *log* count of CA practices adopted. This implies that, for every one unit increase in the frequency of extension visits in the study area, the *log* count of CA practices adopted is expected to increase by approximately 0.14. By implication, such visit is expected to induce positive adoption behavior among the smallholder farmers. In the same vein, the likelihood ratio test shown in the negative binomial model output is a test of the over-dispersion parameter alpha. The results of the Wald test revealed that, alpha parameter is significantly different from zero which of course reinforces the earlier submission that the Poisson regression model is not appropriate for the distribution of the count data under consideration.

According to Piza [60], the interpretation of the results is more or less similar with all the count regression models. This implies that model parameters tend to communicate the same information in both Poisson and negative binomial regression models. The author further noted that reporting IRR can communicate clearly and precisely the influence of explanatory variable influence on the outcome variable than the model regression coefficient. Hence, it is more tenable to report the incidence rate ratio of the negative binomial regression model in estimating the influence or effect of the explanatory variables on the response variable than reporting regression coefficients arising from Poisson or negative binomial distribution models. This position was also upheld by Cameron and Trivedi [61] as well as Long and Freese [52]. However, the IRR estimates in Table 2 revealed that, CA adoption count is expected to decrease by a factor of 0.80 or approximately 20% with every unit increase in male gender, given that other explanatory variables in the model are held constant. This suggests that male gender compared to female counterparts is expected to have a rate of 0.80 points less for count of CA practices adopted. In the same vein, holding all other covariates in the model constant, the IRR value of 0.82 for density of members in social groups suggests a factor of 0.82 or an approximately 18% decrease in the count of CA practices adopted. This is also an indication that diffusion of information about relevant agricultural technologies is a "missing gap" among the social groups in the study area. Conversely, as expected, if farmers' years of farming experience were to increase by one unit, count of CA practices adopted is expected to increase by a factor of 1.01 or approximately 1%, while holding other explanatory variables in the model constant. Furthermore, the findings also indicated that, all things being equal, CA adoption count is expected to increase by a factor of 1.14 or approximately 14% with every point/unit increase in the frequency of visits by extension agents, given that all other explanatory variables in the model are held constant. Conclusively, gender of the farmer ($p < 0.1$), farmers' years of farming experience ($p < 0.1$), frequency of visits by the extension agents ($p < 0.05$), and density of social group membership ($p < 0.01$) significantly drive the count of CA practices adopted or rate ratio for CA adoption by smallholder farmers in the study area. Importantly, the basic CA practices adopted by farmers to preserve the ecosystem services in preferential order are: sequential rotation practice for different unrelated crops, the use of crop biomass for permanent soil cover, as well as minimum soil tillage. These findings partly agree with Abebe and Sewnet [62] who investigated determinants of soil conservation practices adoption in North-West Ethioia. Findings from their study indicated the influence of farmers' and plot-level features, human capital, trainings and institutional support as the main drivers of adoption but never considered the role of social capital in adoption process which our study emphasized on. The importance of social capital in agricultural technologies adoption was also noted in the studies conducted by Hunecke et al. [10] and Husen et al. [9].

Similarly, the computed average marginal effects estimates in Table 3 revealed that, after controlling for other variables, on the average, farmers with appreciable years of farming experience used about 0.089 (8.9% points) of CA practices more than those with fewer years of experience in farming, and on average, farmers who were constantly in touch with extension officers adopted 0.896 (89.6% points) of CA practices more compared to those with less contact. Conversely, on the average, farmers who belong to many social groups adopted 1.304 points of CA practices less than those who belong to fewer social groups. The implication of this is that activities of social groups in the study area tend to tilt towards social engagement alone other than sharing useful and beneficial information about agricultural technologies. This result also reinforced the earlier submission made about the social groups in the study areas. Meanwhile, as indicated in Table A2, the evaluation of information measures (that is, Akaike's and Bayesian Information Criterion—AIC and BIC) clearly revealed that negative binomial regression model fits better, owing to a smaller AIC and BIC statistics values. This is in line with Williams [54,63].

Table 3. Average marginal effects estimates of the negative binomial model.

Count of CA Practices	*dy/dx*	*z-Statistics*	*p* > |*z*|
1.gender	−1.6770	−1.56	0.118
Age	0.0810	1.43	0.153
years of formal education	0.0281	0.28	0.778
1.land acquisition	0.4187	0.40	0.687
farm size cultivated under CA	−0.0834	−0.20	0.841
total years of farming experience	0.0896	1.85 ***	0.064
frequency of extension visit	0.8966	2.00 **	0.045
1.occupational group membership	0.9605	0.94	0.347
1.participation in collective action	−0.4935	−0.51	0.608
density-social groups membership	−1.3041	−2.41 **	0.016
diversity of social group members	1.8650	0.43	0.664
involvement in decision-making	−4.7995	−1.17	0.243

* $p < 0.01$, ** $p < 0.05$, and *** $p < 0.1$, respectively. Note: dy/dx for factor levels is the discrete change from the base level. Source: Data analysis, 2018.

3.2.2. Goodness-of-Fit Test/Fit-Test Statistics

Evidently, it is clear from the result presented in Table A3 that both the negative binomial model and zero-inflated negative binomial model consistently fit better than either of the Poisson model or zero-inflated Poisson model. Importantly, BIC favors the negative binomial regression model while AIC favors the zero-inflated negative binomial model. This finding also provides the necessary and sufficient condition that the Poisson regression model is unfit for the estimation in question because it suffers from an over-dispersion problem. Hence, the justification for the use of the negative binomial model to examine the effects of social capital viz-a-viz social networks on CA adoption counts in South-Western Nigeria.

3.2.3. Marginal Treatment Effects Estimates: Empirical Results

The MTE model estimation was fitted through local IV and separate approach estimators with reference to parametric assumptions. However, the local IV was favored due to the model performance. The output from this estimation as shown in Table 4 highlights the impact evaluation of the specified covariates on the outcomes as measured by farmers' farm income. Likewise, the differences in the average outcomes across the fitted covariates could be inferred directly from the first panel of the output as indicated by β_0. In this instance, the coefficient for years of farming experience in the first panel of the output table indicates that one more year of farming experience translates into approximately 1.83% higher income, albeit with a non-linear effect. Arising from this, it is difficult to confidently infer that it is the actual effects of extra years of farming experience that drives the higher income if we fail to observe a strong exogeneity assumption as required on the fitted covariates. Equally, the coefficient of farm size under CA system from the first panel of the output table also suggests that an extra hectarage of farm size leads to about 25.22% decrease in farmers' income. However, without accounting for strong exogeneity assumption on this factor, this reason alone cannot substantiate the farmers' inability to produce within the production possibility frontier, given the economies of scale in terms of farm size increase.

Table 4. Parametric marginal treatment effects estimates.

Log of Farmers' Farm Income	Coefficient	t	$p > \|t\|$
β_0			
1.gender	−0.3044	−0.92	0.359
Age	−0.0176	−0.78	0.437
years of formal education	−0.0099	−0.28	0.782
1.marital status	0.0224	0.08	0.934
total years of farming experience	0.0183	2.37	0.018 **
farm size cultivated under CA	−0.2523	−1.73	0.084 ***
total available farm size	0.0894	1.43	0.154
1.credit access	0.0197	0.12	0.902
1.information acquisition	−0.0125	−0.07	0.942
index of social capital benefits	0.0825	0.82	0.415
1.access to extension	0.3504	0.65	0.518
frequency of extension visit	−0.3312	−0.34	0.218
regional factor			
2	0.3914	1.42	0.155
3	−0.1718	−0.34	0.733
Constant	11.74	11.97	0.000 *
$\beta_1 - \beta_0$			
1.gender	6.12	3.08	0.002 *
Age	0.34	2.53	0.012 *
years of formal education	0.41	2.02	0.045 **
1.marital status	2.19	1.38	0.167
total years of farming experience	−0.11	−2.63	0.009 *
farm size cultivated under CA	2.93	3.46	0.001 *
total available farm size	−1.02	−2.73	0.007 *
1.credit access	0.68	0.76	0.446
1.information acquisition	1.40	1.50	0.134
index of social capital benefits	−1.27	−2.19	0.029 **
1.access to extension	−8.02	−2.37	0.019 **
frequency of extension visit	5.34	3.14	0.002 *
regional factor			
2	−4.29	−2.76	0.006 *
3	6.61	2.22	0.027 **
Constant	−68.36	−2.81	0.005 *
K			
Mills	−30.91	−2.43	0.016 **
Effects			
parametric normal MTE model (Local IV)			
Ate	−38.03	−2.97	0.003 *
Att	6.84	1.11	0.270
Atut	−47.52	−2.87	0.004 *
Late	8.16	1.82	0.069 ***
$mprte_1$	−8.53	−3.05	0.003 *
$mprte_2$	−7.28	−2.50	0.013 *
$mprte_3$	−12.48	−3.57	0.000 *
parametric polynomial MTE model (Separate approach)			
Ate	−1.70	−0.33	0.741
Att	−2.29	−0.74	0.460
Atut	−1.58	−0.25	0.799
Late	−0.09	−0.04	0.967
$mprte_1$	−3.74	−1.61	0.107
$mprte_2$	−3.87	−1.55	0.122
$mprte_3$	−4.30	−1.85	0.065
Test of observable heterogeneity, p-value			0.0129 *
Test of essential heterogeneity, p-value			0.0157 *

* $p < 0.01$; ** $p < 0.05$; *** $p < 0.1$ level respectively. Note: mprtes indicate stylized marginal policy relevant treatment effects. Source: Data analysis, 2018.

In a similar manner, the second panel of the output with $\beta_1 - \beta_0$ in Table 4 explains the observed differences in treatment effects across covariate values, which also indicates treatment status and covariate interactions. Thus, the coefficient for gender indicates that a male farmer has 6.12 points higher advantage in terms of income generated as a result of CA adoption. The coefficient for age of the farmers suggests that an increase in age translates to about a 34.2% increase in farmers' income, while an extra year of formal education suggests a farmer has about a 40.2% increase in income. However, the estimated coefficient for years of farming experience suggests than an increase in this farmers' characteristics translates to approximately 10.86% decrease in the farmers' income, while an extra increase in farm size under the CA system suggests about a 2.93-point increase in farmers' income ceteris paribus. Similarly, an increase in the farmers' total farm size indicates an approximately 1.02-point decrease in these farmers' income which is somewhat erroneous and contrary to expectation; given the economies of scale in terms of farm size increase and all else equal, an increase in total farm size is expected to drive increased farm output and by extension, increased farmers' income. The results also indicated that social capital is a significant factor towards CA adoption, but the benefits of social interaction is not maximally explored based on the direction of movement of this variable; that is, an increase in social capital benefits was found to drive an approximately 1.27-point decrease in farmers' revenue. More so, the coefficients of extension delivery services (i.e., access and frequency of access) translated to about an 8.02-point decrease and a 5.34-point increase in farmers' income, respectively, suggesting that the performance of an extension delivery system in the study area was not optimal. Importantly, for regional factor influence, a region (that is, Oyo State region) was arbitrarily set to be the basis of comparison since few research institutes (such as the International Institute of Tropical Agriculture (IITA)) are domiciled in this region. Therefore, compared to the counterpart farmers in region1 (Oyo State), the coefficients of region2 and region3 (Osun and Ondo states) suggest that an increase in adoption of CA by farmers in these regions will induce about a 4.29-point decrease and a 6.61-point increase in farmers' income, respectively, all else equal. However, drawing conclusions on the treatment by relying on these findings alone without accounting for the possible non-linear effects may be erroneous and misleading for a valid, tenable, and causal inference about these findings.

To this effect, the third panel in the output table addressed this concern where under different treatment effects parameters and policy changes. The full distribution of marginal treatment effects parameters presented include: average treatment effects (ATEs), average treatment effects on the treated (ATT), average treatment effects on the untreated (ATUT—the spill-over effects), as well as the policy relevant treatment effects (MPRTEs—which points at the average effects of making marginal shifts to the propensity scores for both the treated and untreated individuals). This is also necessary to fully understand the treatment effects heterogeneity in relation to the framework guiding MTEs potential from a hypothetical policy that shifts the propensity to choose treatment which is the CA adoption. More importantly, as noted by Zhou and Xie [64], this approach preserves all of the treatment effects heterogeneity that is consequential for selection bias. In lieu of this, the output from the third panel highlighting the average difference in the outcome between the treated and untreated groups revealed that ATT > ATE > ATUT > LATE ≈ 0; such that, income is higher among the farmers who adopted the CA system than the counterparts who did not adopt CA for whom average income is virtually zero. More so, these treatment effects parameters are statistically significant at various levels; but an exception is made of ATT which is not significant at any level. However, MPRTEs estimated under the stylized policy changes represented by MPRTE_1, MPRTE_2, and MPRTE_3 respectively indicate a substantial marginal income among these farmers (treated group). It is important to note that the exact magnitude of MPRTE depends heavily on the form of the policy change, especially under the normal parametric model which this study considered. For instance, under the first policy change where the policy changes increase everyone's probability of adopting CA by the same amount, the parametric estimate of MPRTE is -8.527, suggesting that an extra effort to adopt CA would translate to about an 8.5-point decrease in farmers' income among the marginal entrants on CA adoption. Equally, under the second policy change where this change favors farmers who appear more likely to adopt CA, the marginal income is approximately a 7.3-point decrease if there is a change in policy that permits and increases everyone's probability of adopting CA proportionally. Besides, this scenario can even go as high as about a 12.5-point decrease in income under the third policy change where the change favors those farmers who appear less likely to adopt CA. However, the same pattern of results is observed under the polynomial MTEs model. The implication of this is that a different policy experiment could increase or decrease CA adoption, depending on which individuals it induces to gain and attract the expected spread and exposure.

In addition, considering the p-values for the two statistical tests shown in Table 4, the first one represents a joint test for the second panel of the output $\beta_1 - \beta_0$, which is also a test of whether the treatment effect differs across the covariates. The second one indicates a test for essential heterogeneity, which is also a joint test of all coefficients in k(u). From all indications, the first test revealed that the treatment effects differ significantly across the covariates in the second panel of output while the second test indicated that the treatment effects vary significantly with unobserved heterogeneity in the sample. Evidently, there are significant differences in the treatment effects across the sample. Therefore, this finding suggests that different policy scenarios or situations could increase or decrease CA adoption, depending on which individuals it induces to attract the expected spread and exposure. However, for parametric joint normal assumption using local IV, Figures 3 and 4 depict the density distribution of propensity scores, MTE curve plot, as well as the associated confidence intervals for the treated and untreated farmers. This will permit to make necessary inferences about the common support. In this case, downward sloping of the estimated MTE plot is observed, with relatively high treatment effects at the beginning of the U_D distribution (addressing propensity not to be treated), which eventually declines to negative effects at the right end of the distribution. This pattern of slope (downward) is in tandem with Roy model which predicts a positive selection on unobservable benefits.

For robust estimation, this study further applied parametric polynomial MTE model and the separate estimation approach by relaxing the joint normal distribution assumption as well as plotting MTE curves for both normal and polynomial functions of the MTE models as indicated in Figures 5 and 6, respectively. Here, the MTE plot for normal is downward sloping with negative treatment effects, which is consistent with the first estimate while MTE plot for the polynomial is relatively flat at the start of the U_D distribution. This eventually slopes upward above zero towards the tail end of the U_D distribution. Similarly, treatment parameter weights were estimated and the resultant plots are shown in Figure 7. In this case, the MTE curve at the average of the covariate and the MTE curve for adopters are evidently convex upward; that is, the plots slope consistently upward without overlapping from the start to the end of U_D distribution. This suggests that farmers are motivated to adopt CA because of the instrumented participation in collective action (social capital) have different values of covariate. Therefore, this influences the treatment but not the outcome. However, the weight distribution indicated that the adopters have a much lower probability to have unobserved resistance towards the mid-point of the distribution. This further suggests that the farmers have MTEs slightly above the average. Hence, the farmers (adopters) who are influenced by the instrument are the ones with slightly above average increase in farm income. Similarly, separate estimation procedure was carried out in fitting the polynomial model by plotting the resulting potential outcomes to investigate if the observed MTE downward plot trend is generated by upward slopping of Y_1, and downward sloping of Y_0, or a combination of the two scenarios. Recall that the difference between outcome for the treated Y_1 and outcome for the untreated Y_0 represents MTE. Therefore, the plot as shown in Figure 8 indicated that though these farmers are relatively similar, the farmers who have high resistance to treatment perform poorly in terms of income realized from farm output than their low resistance counterparts who are also adopters. Hence, it can be inferred that, all else equal, there is a substantial effects and impacts of the treatment (that is, adoption of CA practices) on the farmers' farm income.

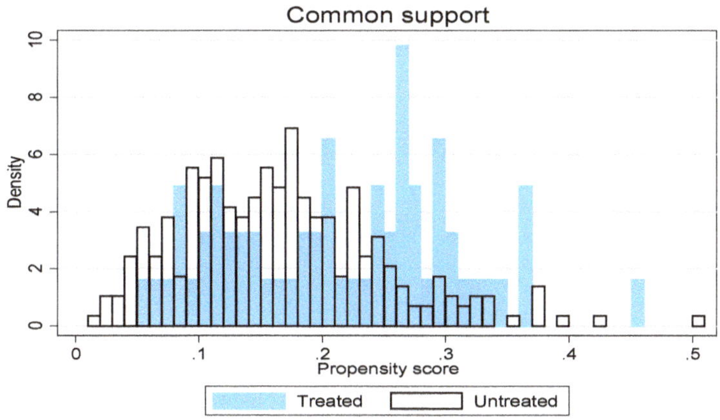

Figure 3. Common support for joint normal assumption. Source: Data analysis, 2018.

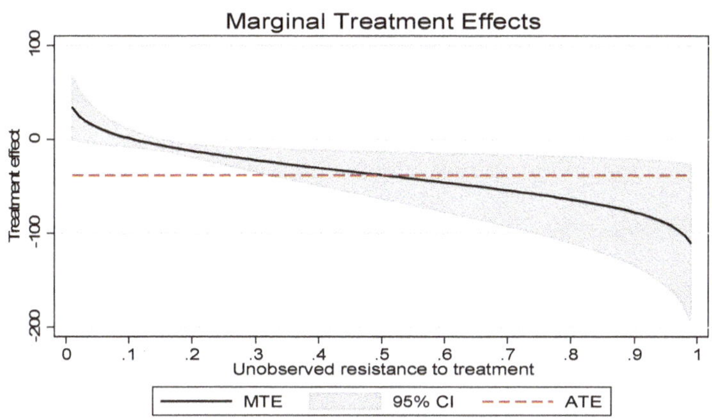

Figure 4. Marginal treatment effects plot for joint normal assumption. Source: Data analysis, 2018.

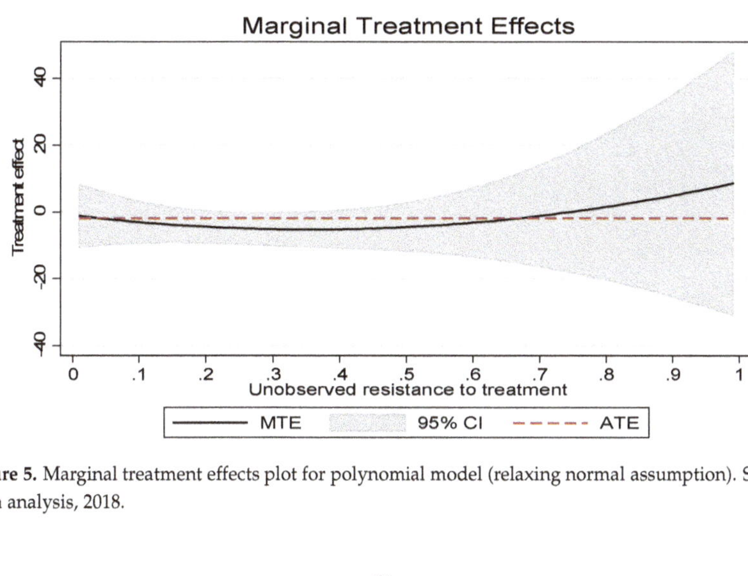

Figure 5. Marginal treatment effects plot for polynomial model (relaxing normal assumption). Source: Data analysis, 2018.

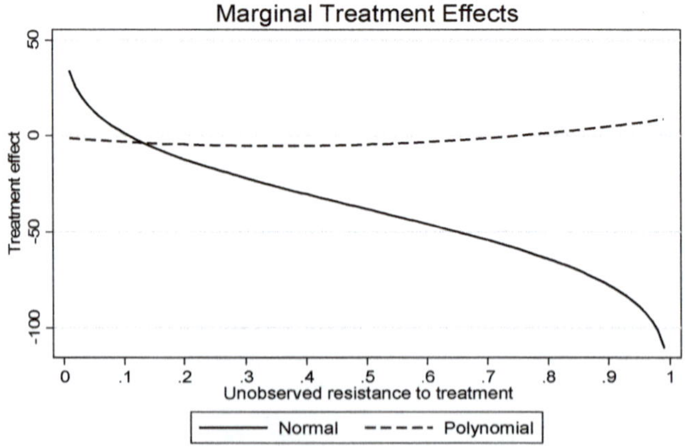

Figure 6. Marginal treatment effects plot for normal and polynomial models. Source: Data analysis, 2018.

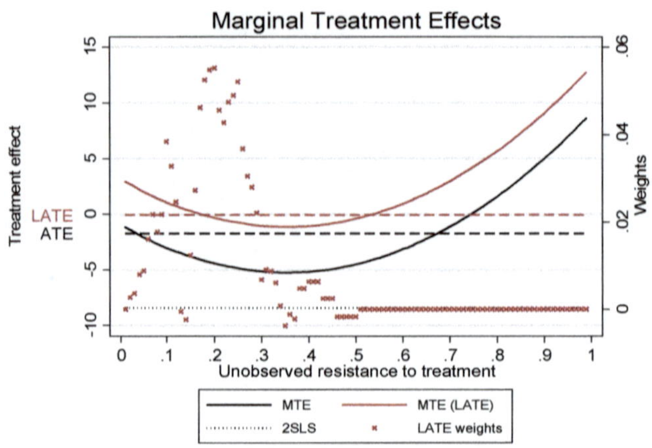

Figure 7. Marginal treatment effects plot polynomial (using late memory). Source: Data analysis, 2018.

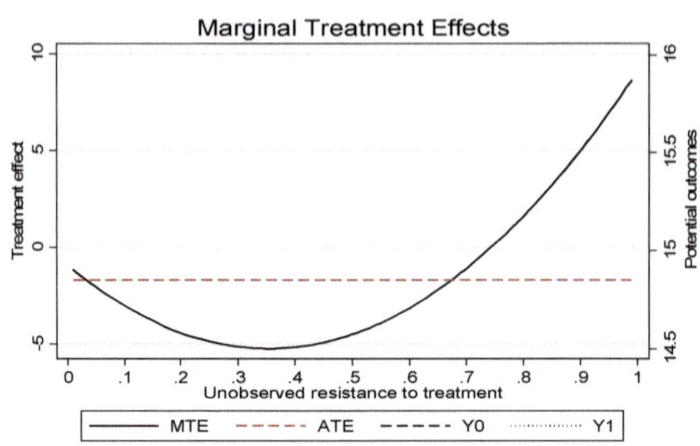

Figure 8. Marginal treatment effects plot (separate approach other than local IV). Source: Data analysis, 2018.

4. Concluding Remarks and Policy Statements

Conclusively, the study found that farmers' years of farming experience ($p < 0.1$), frequency of visits by the extension agents ($p < 0.05$), and social capital viz-a-viz density of social groups membership ($p < 0.05$) significantly determined the count of CA practices adopted with varying degrees by smallholder farmers in the study area. Although social capital expressed in terms of membership of occupational group and diversity of social group members also had positive influence on the count of CA practices adopted, but these features were not significant owing largely to the "information gaps" about the improved agricultural technologies. Suffice it to say that, there is the possibility of apathy among the farmers within the social structure to acquire more information about the improved agricultural technology because of the long-term benefits associated with adoption of CA alternative practices; hence, activities of various social groups, importantly, farmers' occupational group largely center on social engagements.

Therefore, from the findings, the study highlighted the relevance of gender in lieu of the count of CA technologies adopted. Equally, the skewed pattern of CA adoption towards male gender as a significant predictor of adoption was also revealed. Therefore, there is a need to address the core issue of women marginalization in farming activities and farming related policies, most especially the bias towards women in land tenure arrangement. Importantly, there is need for a greater re-visitation of extension delivery systems associated with diffusion of information about CA practices in Nigeria through continuing and ongoing supports of extension services using farmer-led extension approaches facilitated by public extension agencies and NGOs saddled with outsourced extension services. On a general note, findings from count model mirror the significant importance and positive impact of social capital accumulation viz-a-viz social networks in the adoption process. The underlying aim is to understand peer group influence within a social structure impact diffusion of information among networks members and how to constantly explore these links to promote effective dissemination and flow of information on improved agricultural technologies towards sustained adoption of CA in Nigeria. Similarly, since policy relevant treatment effects indicated that different policy scenarios or situations could increase or decrease CA adoption, depending on which individuals it induces to attract the expected spread and exposure, there is a need to intensify the effort and policies to change the reality of farming especially among smallholder farmers in Africa and Nigeria in particular, from the traditional, inappropriate and unproductive tillage-based farming systems to a more and highly-productive, profitable, sustainable, and environmentally sound conservation agriculture system.

Author Contributions: S.O.O. conceptualized the study, conducted the research, and wrote the draft manuscript while A.M. reviewed, edited, and validated the manuscript. All authors therefore approved the final manuscript.

Funding: This research received no external funding.

Conflicts of Interest: The authors declare no conflict of interest.

Appendix A

Table A1. Quantiles of estimated probabilities (*Goodness-of-fit test*).

Group	*Prob*	*Obs_1*	*Exp_1*	*Obs_0*	*Exp_0*	*Total*
1	0.0677	5	1.7	30	33.3	35
2	0.0896	1	2.8	34	32.2	35
3	0.1158	2	3.6	33	31.4	35
4	0.1336	6	4.4	29	30.6	35
5	0.1603	3	5.2	32	29.8	35
6	0.1829	4	6.0	31	29.0	35
7	0.2089	4	6.8	31	28.2	35
8	0.2431	12	7.9	23	27.1	35
9	0.3086	10	9.4	25	25.6	35
10	0.5211	14	13.2	21	21.8	35

Number of observations = 350, number of groups = 10. Hosmer–Lemeshow chi^2 (8) = 15.47, $prob > chi^2 = 0.0507$.
Source: Data analysis, 2018.

Table A2. Akaike's information criterion and Bayesian information criterion.

Model	*Obs.*	*ll(null)*	*ll(model)*	*df*	*AIC*	*BIC*
Poisson	350	−1548.42	−1462.76	13	2951.53	3001.68
Negative Binomial	350	−966.08	−948.65	14	1925.30	1979.31

Source: Data analysis, 2018.

Table A3. Tests and Fit Statistics.

PRM	BIC = 376.870	AIC = 6.769	Prefer	Over	Evidence
vs. NBRM	BIC = −171.005	diff = 547.875	NBRM	PRM	Very strong
	AIC = 5.193	diff = 1.576	NBRM	PRM	
	LRX^2 = 553.733	prob = 0.000	NBRM	PRM	$p = 0.000$
vs. ZIP	BIC = 121.142	diff = 255.728	ZIP	PRM	Very strong
	AIC = 6.006	diff = 0.764	ZIP	PRM	
	Vuong = 5.241	prob = 0.000	ZIP	PRM	$p = 0.000$
vs. ZINB	BIC = −160.147	diff = 537.017	ZINB	PRM	Very strong
	AIC = 5.191	diff = 1.578	ZINB	PRM	
NBRM	BIC = −171.005	AIC = 5.193	Prefer	Over	Evidence
vs. ZIP	BIC = 121.142	diff = −292.147	NBRM	ZIP	Very strong
	AIC = 6.006	diff = −0.813	NBRM	ZIP	
vs. ZINB	BIC = −160.147	diff = −10.858	NBRM	ZINB	Very strong
	AIC = 5.191	diff = 0.002	ZINB	NBRM	
	Vuong = 1.323	prob = 0.093	ZINB	NBRM	$p = 0.093$
ZIP	BIC = 121.142	AIC = 6.006	Prefer	Over	Evidence
vs. ZINB	BIC = −160.147	diff = 281.289	ZINB	ZIP	Very strong
	AIC = 5.191	diff = 0.815	ZINB	ZIP	
	LRX^2 = 287.147	prob = 0.000	ZINB	ZIP	$p = 0.000$

Source: Data analysis, 2018. Note that: PRM = Poisson regression model; NBRM = Negative binomial regression model; ZIP = Zero inflated poisson model; ZINB = Zero inflated negative binomial regression model.

References

1. Oyakhilomen, O.; Zibah, R.G. Agricultural Production and Economic Growth in Nigeria: Implication for Rural Poverty Alleviation. *Q. J. Int. Agric.* **2014**, *53*, 207–223.
2. Kassie, M.; Pender, J.; Yesuf, M.; Köhlin, G.; Bluffstone, R.; Mulugeta, E. Estimating returns to soil conservation adoption in the Northern Ethiopian Highlands. *Agric. Econ.* **2008**, *38*, 213–232. [CrossRef]
3. IIRR; ACT. Conservation Agriculture. A manual for farmers and extension workers in Africa. In *International Institute of Rural Reconstruction, Nairobi*; African Conservation Tillage Network: Nairobi, Kenya, 2005.
4. Silici, L. The Role of Social Capital in the Adoption and the Performance of Conservation Agriculture: The Practice of Likoti in Lesotho. Ph.D. Thesis, Departimento di Economia, Universita degli Studi Roma Tre, Rome, Italy, 2009.
5. Kassam, A.; Friedrich, T.; Derpsch, R.; Lahmar, R.; Mrabet, R.; Basch, G.; González-Sánchez, E.J.; Serraj, R. Conservation Agriculture in the dry Mediterranean climate. *Field Crops Res.* **2012**, *132*, 7–17. [CrossRef]
6. Ngwira, A.R.; Johnsen, F.H.; Aune, J.B.; Mekuria, M.; Thierfelder, C. Adoption and extent of conservation agriculture practices among smallholder farmers in Malawi. *J. Soil Water Conserv.* **2014**, *69*, 107–119. [CrossRef]
7. Kassam, A.; Friedrich, T.; Derpsch, R.; Kienzle, J. Overview of the worldwide spread of Conservation Agriculture. *Field Actions Sci. Rep.* **2015**, *8*, 1–12.
8. Pedzisa, T.; Rugube, L.; Winter-Nelson, A.; Baylis, K.; Mazvimavi, K. The Intensity of adoption of Conservation agriculture by smallholder farmers in Zimbabwe. *Agrekon* **2015**, *54*, 1–22. [CrossRef]
9. Husen, N.A.; Loos, T.M.; Siddig, K.H. Social Capital and Agricultural technology Adoption among Ethiopian Farmers. *Am. J. Rural Dev.* **2017**, *5*, 65–72. [CrossRef]

10. Hunecke, C.; Enger, A.; Jara-Rojas, R.; Poortvliet, P.M. Understand the Role of Social Capital in Adoption Decisions: An Application to Irrigation Technology. *Agric. Syst.* **2017**, *153*, 221–231. [CrossRef]
11. D'Souza, A.; Mishra, A.K. Adoption and Abandonment of Partial Conservation Technologies in Developing Economies: The Case of South Asia. *Land Use Policy* **2018**, *70*, 212–223. [CrossRef]
12. Silici, L.; Ndabe, P.; Friedrich, T.; Kassam, A. Harnessing sustainability, resilience and productivity through conservation agriculture: The case of likoti in Lesotho. *Int. J. Agric. Sustain.* **2011**, *9*, 1–8. [CrossRef]
13. Calegari, A.; Ashburner, J. Further Experiences with Conservation Agriculture in Africa. In Proceedings of the 18th World Congress of Soil Science, Philadephia, PA, USA, 9–15 July 2006.
14. Montemurro, F.; Fiore, A.; Campanelli, G.; Tittarelli, F.; Ledda, L.; Canali, S. Organic fertilization, green manure, and vetch mulch to improve organic zucchini yield and quality. *HortScience* **2013**, *48*, 1027–1033.
15. EIP-AGRI Focus Group. Soil Organic Matter in Mediterranean Regions. 2015. Available online: http://www.ec.europa.eu/eip/agriculture/sites/agri-eip (accessed on 20 October 2018).
16. Deligios, P.A.; Tiloca, M.T.; Sulas, L.; Buffa, M.; Caraffini, S.; Doro, L.; Sanna, G.; Spanu, E.; Spissu, E.; Urracci, G.R.; et al. Stable nutrient flows in sustainable cropping systems of globe artichoke. *Agron. Sustain. Dev.* **2017**, *37*, 1–12. [CrossRef]
17. Ndah, H.T.; Schuler, J.; Uthes, S.; Zander, P. Adoption Decision Theories and Conceptual models of Innovations Systems. In Proceedings of the CA2 Africa Inception Workshop, Leibniz-Centre for Agricultural Landscape Research (ZALF), Nairobi, Kenya, 2–4 March 2010.
18. Todaro, M.P.; Smith, S.C. *Economic Development*, 10th ed.; Pearson Addison Wesley: Boston, MA, USA, 2009.
19. Hobbs, P.R. Conservation agriculture: What is it and why is it important for future sustainable food production. *J. Agric. Sci.* **2007**, *145*, 127–137. [CrossRef]
20. De Lucas, A.I.; Molari, G.; Seddaiu, G.; Toscano, A.; Bombino, G.; Ledda, L.; Milani, M.; Vittuari, M. Multidisciplinary and Innovative Methodologies for Sustainable Management in Agricultural Systems. *Environ. Eng. Manag. J.* **2015**, *14*, 1571–1581.
21. Deligios, P.A.; Chergia, P.A.; Sanna, G.; Solinas, S.; Todde, G.; Narvarte, L.; Ledda, L. Climate change adaptation and water saving by innovation irrigation management applied on open field globe artichoke. *Sci. Total. Environ.* **2019**, *649*, 461–472. [CrossRef] [PubMed]
22. Foster, A.D.; Rosenzweig, M.R. Microeconomics of Technology Adoption. *Annu. Rev. Econ.* **2010**, *2*, 395–424. [CrossRef] [PubMed]
23. Abdulai, A.; Owusu, V.; Bakang, J.E. Adoption of safer irrigation technologies and cropping patterns: Evidence from Southern Ghana. *Ecol. Econ.* **2011**, *70*, 1415–1423. [CrossRef]
24. Awotide, B.A.; Diagne, A.; Awoyemi, T.T. Agricultural Technology Adoption, Market Participation and Rural Farming Households' Welfare in Nigeria. In Proceedings of the 4th International Conference of the African Association of Agricultural Economists, Hammamet, Tunisia, 22–25 September 2013.
25. Abdulai, A.; Huffman, W. The Adoption and Impact of Soil and Water Conservation Technology: An Endogenous Switching Regression Application. *Land Econ.* **2014**, *90*, 26–43. [CrossRef]
26. Ademola, A.O.; Olujide, M.G. Soil Conservation Practices of Arable Crop Farmers in Atisbo Local Government Area of Oyo State, Nigeria. *Adv. Res.* **2014**, *2*, 878–888. [CrossRef]
27. Afolami, C.A.; Obayelu, A.E.; Vaughan, I. Welfare impact of adoption of improved cassava varieties by rural households in South Western Nigeria. *Agric. Food Econ.* **2015**, *3*, 1–17. [CrossRef]
28. Obisesan, A.A. Causal Effect of Off-Farm Activity and Technology Adoption on Food Security in Nigeria. *AGRIS On-Line Pap. Econ. Inform.* **2015**, *7*, 3–11.
29. Wossen, T.; Bergen, T.; Di-Falco, S. Social capital, risk preference and adoption of improved farm land management practices in Ethiopia. *Agric. Econ.* **2015**, *46*, 81–97. [CrossRef]
30. Pino, G.; Toma, P.; Rizzo, C.; Miglietta, P.; Peluso, A.; Guido, G. Determinants of farmers' intention to adopt water saving measures: Evidence from Italy. *Sustainability* **2017**, *9*, 77. [CrossRef]
31. Kirton, M. Adaptors and innovators: A description and measure. *J. Appl. Psychol.* **1976**, *61*, 622–629. [CrossRef]
32. Rogers, E.M. What are innovators like? *Theory Pract.* **1963**, *2*, 252–256. [CrossRef]
33. Oreszczyn, S.; Lane, A.; Carr, S. The role of networks of practice and webs of influencers on farmers' engagement with and learning about agricultural innovations. *Agric. Syst.* **2010**, *26*, 404–417. [CrossRef]
34. Lollo, E. Toward a theory of social capital definition: Its dimensions and resulting social capital types. In Proceedings of the 14th World Congress of Social Economics, Glasgow, UK, 20–22 June 2012.

35. Hanifan, L.J. The Rural School Community Center. *Ann. Am. Acad. Political Soc. Sci.* **1916**, *67*, 130–138. [CrossRef]
36. Nath, T.K.; Inoue, M.; Pretty, J. Formation and Function of Social Capital for Forest Resource Management and the Improved Livelihoods of Indigenous People in Bangladesh. *J. Rural Community Dev.* **2010**, *5*, 104–122.
37. Putnam, R. Social capital: Measurement and consequences. *Canad. J. Policy Res.* **2001**, *2*, 41–51.
38. Bowles, S.; Gintis, H. Social capital and community governance. *Econ. J.* **2002**, *112*, F419–F436. [CrossRef]
39. Grootaert, C.; Bastelaer, T.V. *Understanding and Measuring Social Capital: A Synthesis of Findings and Recommendations from the Social Capital Initiatives*; Social Capital Initiative Working Paper No. 24; The World Bank: Washington, DC, USA, 2001.
40. Woolcock, M. The place of social capital in understanding social and economic outcomes. *Can. J. Policy Res.* **2001**, *2*, 11–17.
41. McCarthy, N.; Dutilly-Diane, C.; Drabo, B. Cooperation, collective action and natural resources management in Burkina Faso. *Agric. Syst.* **2004**, *82*, 233–255. [CrossRef]
42. Jagger, P.; Luckert, M.K. Investments and returns from cooperative and household managed woodlots in Zimbabwe: Implications for rural afforestation policy. *Land Use Policy* **2008**, *25*, 139–152. [CrossRef]
43. Aker, J.C. *Social Networks and Household Welfare in Tanzania: Working Together to Get out of Poverty*; University of California-Berkely: Berkeley, CA, USA, 2007.
44. United Nations. *Sustainable Development Knowledge Platform*; Department of Economic and Social Affairs (UNDESA): New York, NY, USA, 2015; Available online: https://sustainabledevelopment.un.org/sdgs (accessed on 9 August 2018).
45. Nguezet, P.M.D.; Diagne, A.; Okoruwa, V.O.; Ojehomon, V.E. Impact of Improved Rice Technology Adoption (NERICA varieties) on Income and Poverty among Rice Farming Households in Nigeria: A Local Average Treatment Effect (LATE) Approach. *Q. J. Int. Agric.* **2011**, *50*, 267–291.
46. Oparinde, A.; Banerji, A.; Birol, E.; Ilona, P. *Consumer Willingness to Pay for Biofortified Yellow Cassava: Evidence from Experimental Auctions in Nigeria*; HarvestPlus Working Paper 13; International Food Policy Research Institute: Washington, DC, USA, 2014; pp. 1–2.
47. Obisesan, A.A.; Amos, T.T.; Akinlade, R.J. Causal Effect of Credit and Technology Adoption on Farm Output and Income: The Case of Cassava Farmers in South-West Nigeria. In Proceedings of the 5th International Conference of the African Association of Agricultural Economists, Addis Ababa, Ethiopia, 23–26 September 2016.
48. Awotide, B.A.; Karimov, A.A.; Diagne, A. Agricultural technology adoption, commercialization and smallholder rice farmers' welfare in rural Nigeria. *Agric. Food Econ.* **2016**, *4*, 1–24. [CrossRef]
49. National Space Research and Development Agency of Nigeria (NASRDA). *Map of Nigeria Showing South-Western Region of Nigeria*; NASRDA: Abuja, Nigeria, 2018.
50. Corbeels, M.; Apina, T.; Koala, S.; Schuler, J.; Triomphe, B.; El Mourid, M.; Tittonell, P. *Impact and Adoption of Conservation Agriculture in Africa: A Multi-Scale and Multi-Stakeholder Analysis*; Centre de Coopération Internationale en Recherche Agronomique pour le Développement: Montpellier, France, 2010.
51. Wooldridge, J.M. *Econometric Analysis of Cross Section and Panel Data*; MIT Press: Cambridge, MA, USA, 2010; pp. 265–268.
52. Long, J.S.; Freese, J. *Regression Models for Categorical Dependent Variables using STATA*, 3rd ed.; Stata Press: College Station, TX, USA, 2014.
53. Sebopetji, T.O.; Belete, A. An application of Probit analysis to factors affecting small-scale farmers' decision to take credit: A case study of the Greater Letaba Local Municipality in South Africa. *Afr. J. Agric. Res.* **2009**, *4*, 718–723.
54. Williams, R. Models for Count Outcomes. University of Notre Dame. 2015. Available online: https://www3.nd.edu/~{}rwilliam/ (accessed on 7 August 2017).
55. Bruin, J. Newtest: Command to Compute New Test. UCLA: Statistical Consulting Group, 2006. Available online: https://stats.idre.ucla.edu/stata/ado/analysis/ (accessed on 2 April 2018).
56. Andresen, M.E. Exploring marginal treatment effects: Estimation using STATA. *Stata J.* **2018**, *18*, 118–158. [CrossRef]
57. Abadie, A.; Imbens, G.W. Matching on the estimated propensity score. *Econometrica* **2016**, *84*, 781–807. [CrossRef]

58. Hosmer, D.W., Jr.; Lemeshow, S.A.; Sturdivant, R.X. *Applied Logistic Regression*, 3rd ed.; Wiley: Hoboken, NJ, USA, 2013.
59. Baum, C.F. *Models for Count Data and Categorical Response Data*; Boston College and DIW Berlin, University of Adelaide: Adelaide, Australia, 2010; Available online: www.bc.edu/ec-c/s2013/327/S5CountCategorical0511.slides.pdf/ (accessed on 2 April 2018).
60. Piza, E.L. *Using Poisson and Negative Binomial Regression Models to Measure the Influence of Risk on Crime Incident Counts*; Rutgers Center on Public Security: Newark, NJ, USA, 2012.
61. Cameron, A.C.; Trivedi, P.K. *Regression Analysis of Count Data*, 2nd ed.; Cambridge University Press: New York, NY, USA, 2013.
62. Abebe, Z.D.; Sewnet, M.A. Adoption of soil conservation practices in North Achefer District, Northwest Ethiopia. *Chin. J. Popul. Resour. Environ.* **2014**, *12*, 261–268. [CrossRef]
63. Williams, R. *Scalar Measures of Fit: Pseudo R^2 and Information Measures (AIC and BIC)*; University of Notre Dame: Notre Dame, IN, USA, 2018; Available online: https://www3.nd.edu/~{}rwilliam/ (accessed on 7 August 2018).
64. Zhou, X.; Xie, Y. Heterogeneous Treatment Effects in the Presence of Self-Selection: A Prospensity Score Perspective. 23 January 2018. Available online: https://scholar.harvard.edu/files/xzhou/ (accessed on 22 August 2018).

sustainability

MDPI

Article

Approach for Designing Context-Specific, Locally Owned Interventions to Reduce Postharvest Losses: Case Study on Tomato Value Chains in Nigeria

Christine Plaisier [1,*], Milindi Sibomana [2], Johannes van der Waal [3], Luud Clercx [3], Coen P. A. van Wagenberg [1] and Youri Dijkxhoorn [1]

1 Wageningen Economic Research, 2595 BM Den Haag, The Netherlands; coen.vanwagenberg@wur.nl (C.P.A.v.W.); youri.dijkxhoorn@wur.nl (Y.D.)

2 M.AGRO-SPEC Ltd., P.O. Box 6734 Kigali, Rwanda; sylversi@yahoo.co.uk

3 AgroFair Projects B.V., 2991 LN Barendrecht, The Netherlands; hans-willem.van.der.waal@AGROFAIR.NL (J.v.d.W.); Luud.clercx@agrofair.nl (L.C.)

* Correspondence: christine.plaisier@wur.nl; Tel.: +31-70-3358330

Received: 29 November 2018; Accepted: 31 December 2018; Published: 7 January 2019

Abstract: Development projects on interventions to reduce postharvest losses (PHL) are often implemented largely independently of the specific context and without sufficient adaptation to the needs of people who are supposed to use them. An approach is needed for the design and implementation of specific, locally owned interventions in development projects. Our approach is based on Participatory Development and includes Living Lab and World Cafés. We applied the approach in a case study on reducing PHL in tomato value chains in Nigeria. The approach consists of nine steps. After scoping the sector, selected value chain stakeholders (case: farmers, transporters, traders, retailers) were gathered in Living Lab workshops. In the workshop, participants analyzed the product, information, and monetary flows in their own value chain, identified causes for PHL, and selected potential interventions to reduce these (case: plastic crates instead of raffia baskets to transport tomatoes). Selected interventions were implemented, tested, and monitored in pilot projects with the workshop participants. This was followed by an evaluation workshop. At the end of the case study, 89% of participants bought crates to keep using them in their value chain. Our approach resulted in context-specific, locally owned interventions to reduce PHL in the case study on tomato value chains in Nigeria. Its application in other countries, commodities, or interventions is needed to determine the effectiveness of the approach in a broader scope.

Keywords: value chain development; participatory approach; context-specific interventions; behavioural change; postharvest losses; tomato; Nigeria; supply chain; raffia basket; plastic crate

---

1. Introduction

Feeding Africa's urban population is a task that is becoming ever more challenging. Currently, urban areas in Africa comprise of 472 million people. That number is expected to double over the next 25 years as more migrants are pushed into the cities from the countryside, with annual growth rates of up to 4% for the largest cities [1]. Good health and wellbeing, sustainable cities and communities, and responsible consumption and production are all relevant sustainable development goals (SDG) in this respect [2]. To ensure these SDGs are met, the provision of good-quality food in adequate quantities is of crucial importance. Much attention has been paid to the supply side of food by improving the yield and productivity of agricultural production. Less attention has been paid to the importance of optimal supply networks, which are the links between agricultural production and (urban) consumers. In such supply networks, a large amount of the food produced for human consumption is lost or wasted as a result of damage, rotting, pests, and diseases [3,4]. Reducing such so-called postharvest losses (PHL) is a key pathway to food and nutrition security in sub-Saharan Africa [5].

Many initiatives have been taken to reduce PHL in sub-Saharan Africa [5,6]. Development agencies, governments, non-governmental organizations, and private companies have been keen to invest in installations and equipment to improve supply networks, such as by putting in place cold stores and collective market structures. However, Ika [7] found that 50 to 64% of value chain development projects in Africa fail because of inadequate beneficiary needs analysis, poor stakeholder management, and overemphasis on financial and technical feasibility at the expense of social, cultural, environmental, and political feasibility. In other words, such projects suffer from a lack of ownership and insufficient adaptation of interventions to the needs of the people who were supposed to use them. Such projects have become "white elephants"—beautiful temporary gifts, but useless [8]. One cause for this is the fact that innovations to reduce PHL in sub-Saharan Africa were developed and tested without sufficient participation of local stakeholders [5]. Existing guides for value chain development are designed to implement interventions largely independently of the specific context, and insufficiently incorporate co-creation, co-testing, and co-analyzing of interventions with local stakeholders [9]. As a consequence, there is still a significant lack of adoption of the innovations presented, resulting in ever-present high postharvest losses observed across various agricultural value chains in Nigeria. Ideally, all actors, from producers to transporters and traders, are involved in problem identification, solution generation, pilot testing, and intervention calibration. This would improve the adoption potential of generated solutions. An approach is needed for the design of such context-specific and locally owned interventions to reduce PHL. This study aims to develop and test such an approach.

Our study follows the thinking of Participatory Development, which advocates the active participation of stakeholders in the decision-making process [10,11]. Participatory methods can enhance the uptake and sustainable use of new (technological) solutions [5]. In our approach for designing context-specific, locally owned interventions to reduce PHL, we included the Living Lab as a participatory process to co-create, co-test, and co-analyze the interventions with relevant value chain stakeholders. The Living Lab is a user-centered development concept with two essential elements, namely a real-life test and experimentation environment [12]. The Living Lab provides for a real-world setting, involving multiple stakeholders from multiple organizations, stages, or backgrounds, and their interaction. Application of the Living Lab results in users who are aware that they are co-involved in and co-owners of the innovation process [12]. The Living Lab helps create trust and commitment, which are prerequisites for sustainable and effective cooperation in supply chains [13].

We applied the approach to reducing PHL in a case study on reducing PHL in tomato value chains in Nigeria. Nigeria is the most populous country of sub-Saharan Africa, with an estimated population of about 190 million people. Nigeria is one of the leading producers of tomatoes in Africa. According to the statistics from the Food and Agricultural Organization of the United Nations, in 2016, Nigeria ranked number one in Africa in areas planted with tomatoes and number four in the world, and number 14 in production volume in the world. Tomatoes are an important vegetable in the local Nigerian cuisine, because they are used daily [14]. According to Adeoye et al. [15], over 90% of studied consumers in the city of Ibadan in Nigeria purchase tomatoes in the urban market. Of these consumers, over 90% purchase fresh tomatoes. Compared to other developing countries in Africa, Nigeria lags behind in agricultural productivity development due to long periods of underinvestment in public infrastructure, such as roads, energy generation, and clean water supply [16]. Tomato supply chains are affected by a lack of investments in storage, packaging, transportation and marketing infrastructure, and are highly fragmented. Prior research has highlighted numerous problems in the tomato supply network in Nigeria, especially around PHL. Pre-consumer PHL range from an estimated 25% [17] to as high as 50% [18,19]. Some research suggested improvements to reduce PHL [6,20,21]. However, the results of these studies were often fragmented in time, space, and focus, and did not address the question on how to effectively test and embed suggested improvements in the value chain. Potential solutions need a strong support base across the value chain as Nigerian tomato value chains are often informal and fragmented, and lead firms or value chain captains are lacking.

This study's research aim was to develop an approach with which to design context-specific, locally owned interventions to reduce PHL. The remainder of this paper is structured as follows. In Section 2, we present the approach for designing and testing context-specific interventions to reduce PHL that are locally owned by all value chain actors. In Section 3, we present the results of the approach applied to the PHL in tomato value chains in Nigeria. Section 4 provides the discussion, and Section 5 the conclusions.

2. Materials and Methods

The approach for designing and testing context-specific interventions to reduce PHL, which are locally owned by all value chain actors (Figure 1) was developed based on consultations with an international and local research team and actors in the tomato value chain in Nigeria. The approach is based on Participatory Development, which actively includes stakeholders in the decision-making process [10,11]. In the approach, we applied Living Lab workshops [12] with a "World Café" setting [22]. The research team consisted of three Dutch, two Nigerian, and one Rwanda expert from development organizations with extensive experience in implementing solutions in developing countries, with three experts from Wageningen Economic Research with significant experience in value chain development and impact evaluation of such solutions, and three experts from Wageningen Food & Biobased Research on PHL. The research team was supported by local enumerators for translation and monitoring the pilot projects. Members of the research team were involved in developing all steps of the approach. The actors in the tomato value chain were involved in steps 4 to 9 of the approach. The approach consists of nine steps, which are described below.

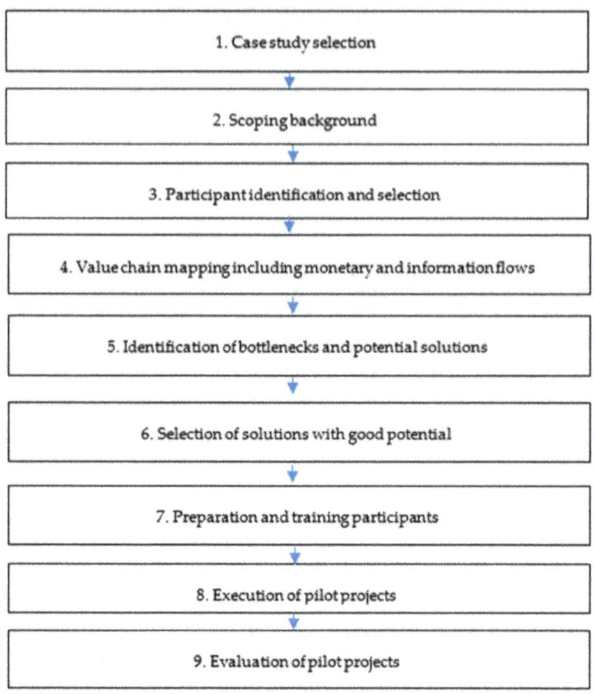

Figure 1. Approach for designing context-specific solutions to reduce postharvest losses, which are locally owned by all value chain actors.

2.1. Case Study Selection

In Step 1, the case studies to be analyzed are defined. The product(s) and geographical region(s) of production and consumption are selected for which improvements in PHL are potentially viable. To this end, a literature review and or expert knowledge can be used.

2.2. Scoping Background

Step 2 of the approach consists of gathering in-depth background information on the PHL in the selected case studies from key value chain actors (such as producers, transporters, processors, traders, and retailers), and key informants from the government and other relevant organizations, e.g., development organizations. For example, this can be done through a scoping survey and interviews among the stakeholders. It addresses the general characteristics of the value chain actors (gender, age, education, and production), the value chain and marketing characteristics (purchase and sale points, payment moment, bargaining power, customer relationship) and the potential causes for PHL related to shelf life, tomato yield, and transport efficiency. The appropriate selection strategy of participants depends on the context and envisioned scope. Ideally, a random sampling strategy is applied with representation of the diversity participants of various value chain stages, production tiers, i.e., small to commercial farmers, single-vehicle to fleet hauliers, and wholesalers to retailers to determine the status quo in the different areas.

2.3. Participant Identification and Selection

In Step 3 of the approach, a limited number of value chain actors are selected for participation in Living Lab workshops and pilot projects. Potential participants represent all actors active in the value chain. Again, the appropriate selection strategy of participants depends on the context, envisioned scope, and the project period and timing, but the selection should be based on three criteria:

(1) Participants are active actors actually working in the value chain in at least one of the activities, such as production, transportation, processing, trading, or retailing;
(2) Participants are already trading with at least one other participant in the value chain;
(3) Participants are willing to implement potential innovations to reduce PHL and are committed to participate for the entire project trial.

2.4. Value Chain Mapping Including Monetary and Information Flows

In Step 4, the participants map their own value chain, including monetary and information flows. To this end, the participants are brought together in a Living Lab workshop. Living Lab workshops are especially suited when network partners, together with end-users need to develop innovation processes to address specific challenges [23]. Involvement of value chain actors in the problem contextualization and innovation development process is critical for the development of sustainable innovation solutions [24,25]. The Living Lab workshop also offers opportunities for informal contacts and thus helps to improve the bonds between the value chain actors. [26]. With Higgins and Klein [26], we believe that inviting the respective parties to engage in the Living Lab's real-world experiment is a promising option because they can be more willing to overcome established attitudes and obstacles, as long as it is "only" in an experimental setting. As such, it may enable the establishment of forums and supportive environments for innovators which can otherwise be stuck in existing adversarial relations, hierarchies, and traditional practices. The experimental setting also encourages a critical attitude and searching for creative solutions. Moreover, the Living Lab itself can give a symbolic meaning to the process of facilitating broader collective action. The Lab can signal commitment, momentum of change, and the opportunity to act and take charge of developments that are critical for the development of the participating parties.

In the Living Lab workshop, participants are grouped in round tables, with each table containing actors from each part of the chain, effectively forming a platform to engage on the different aspects of the value chain. This Living Lab workshop follows a structured approach in three stages. Stage 1 of the Living Lab workshop is a feedback and validation session discussing the outcomes of the scoping survey. In Stage 2, the participants map the connections in, and product flows through their own supply chain. This involves free mapping to show connections in the supply chain as perceived by the stakeholders and identification of points of tomato losses (hotspots) in the supply chain scheme. Participants are asked to brainstorm and identify all possible stakeholders and relations. Some work out a linear map, others a network map, with less or more detail. This exercise is useful to visualize and identify the players and relations in the chain. The final map is visualized on paper.

After mapping the supply chain, a depiction of monetary and information flows in the supply chain is created (Stage 3). Participants are asked to identify different inflows and outflows (or processes) of money, information, and tomato operations at the levels of farmers, hauliers, traders, and retailers, as the illustration or precision of the relations between the stakeholders of the chain.

In Stage 3, the participants together identify the main bottlenecks in their chains, as well as possible solutions (based on the maps and information flows depicted in the previous stages). The participants start with a broad approach in which every actor identifies the main problems in the chain. The problems can vary from lack of quality inputs, to the state of the road, to the institutional environment.

2.5. Identification of Bottlenecks and Potential Solutions

In Step 5, the main bottlenecks causing PHL and potential solutions for these bottlenecks are identified. Each group of participants identifies bottlenecks in their own value chain, considering the value chain map, and monetary and information flows identified in the previous step. Then, each group identifies possible solutions for each identified bottleneck.

The bottlenecks and potential solutions are presented in tabular form, per value chain actor and per link of the value chain. This is followed by a plenary session in which each group presents their results to all other groups. After the plenary session, a so-called "World Café" [22] is organized, in which participants from each group, except a reporter, move to another group to observe and discuss the bottlenecks and solutions of the other group. The café's ambience allows for a more relaxed and open conversation to take place. This type of conversation is a creative process for leading collaborative dialogue, sharing knowledge, and creating possibilities for action in groups of all sizes [27]. The environment is set up like a café, with paper-covered round tables. Participants are provided with pens, paper, and stickers and are encouraged to draw and record their conversations on the paper tablecloths or other materials to capture free-flowing ideas as they emerge. Participants discuss the issue at hand around their table, and they move to a new table at regular intervals. One participant (the table host) remains and summarizes the previous conversation to the newly arrived participants. By moving participants around the room, the conversations at each table are cross-fertilized with ideas from other tables. At the end of the process, the main ideas are summarized in a plenary session, and follow-up possibilities are discussed [22,27,28]. Members of the groups can indicate their preferences on the worksheets with material like stickers, pencils, symbols (for example, applying green dots for preferred solutions). The other groups review the worksheet and can also indicate their preference (e.g., with a blue dot). Participants then reconvene at their original group to revise their own bottlenecks and solutions. This results in a final list of bottlenecks and potential solutions for each group.

2.6. Selection of Solutions with Good Potential

In Step 6, participants identify the most promising solutions for application in pilot projects. Out of all the potential solutions portrayed, each group identifies the two most important bottlenecks in their own value chain and the two most viable potential solutions for these bottlenecks, which they could commit to testing in a pilot project. The potential solutions should be guided by a set of SMART criteria—i.e., the solutions should be Specific (what is included/excluded in the activities),

Measurable (what is the envisaged result: how to measure), Achievable/Attractive (achievable within the scope of the project, in potential self-sustainable, is the solution attractive, what is the role division), Realistic (can we do it given the constraints), and Time-Specific (within the available timeframe and with milestones set).

Next in the second "World Café" set-up, each group receives the selected bottleneck-intervention combinations of another group. Each group votes for which of these bottleneck-intervention combinations is the most feasible. At the end of this activity, all votes for each bottleneck-intervention combination are added. The combinations with the most votes from all the groups are then selected to be implemented in pilot projects. The number of bottleneck-intervention combinations selected for pilot projects cannot be too large, because the participants will have to implement these and too many pilots implemented simultaneously could negatively affect the implementation.

2.7. Preparation and Training Participants

In Step 7 of the approach, all preparatory actions for the pilot projects are performed. This includes logistics, permissions, and any other organization needed before implementation, as well as training of the participants of the workshop on the implementation of the selected interventions and training of enumerators for evaluation of the impact of the solutions. When all logistics, hardware, and permissions are ready, an official kick-off workshop of approximately two days is organized to prepare the pilot projects with the value chain actors and the research team. In this workshop, all risks and potential challenges in the implementation of the pilot projects are discussed and anticipated for as much as possible. The participants design the appropriate starting and ending dates, the planning, timelines, and requirements, and again, all commit fully to the implementation. It is important that all agree on the set-up and planning and that the process is highly participatory. All should be given the opportunity to express their ideas and doubts, which enables a constructive discussion and customized design before the official take-off.

2.8. Execution of Pilot Projects

In Step 8, the value chain actors put the proposed solutions in practice and the local research team monitor the results of the intervention. To overcome the initial errors, doubts, and hiccups encountered along the way, it is important to have a local representative of the team available to coach all participants during the implementation period. This person must be able to reach all participants, be aware of the planning of each value chain, and able to monitor the processes. To monitor the extent to which the intervention solves the bottlenecks, an intervention-specific monitoring tool must be developed, and a local monitoring team should be established.

2.9. Evaluation of Pilot Projects

In Step 9, the research team, together with all participants, evaluates each pilot project and identified bottlenecks for continuation and upscaling. A workshop is organized to discuss and validate the results of the monitoring tool and the participants' personal experiences. Besides, participants are also challenged to outline the basic features for continuation of the project that would benefit all the value chain actors.

3. Results

3.1. Case Study Selection

In this study, we selected tomato value chains delivering from two regions in Nigeria with a tomato cultivation tradition to consumers in the urban areas of Lagos and Ibadan. The first group of value chains runs from farms in the South-West (Oyo, Osun, Ondo, and Katsina states) to the Mile 12 market in Lagos or Sasa market in Ibadan, with distances from 95 to 305 km from farm to retailer. The second group of value chains runs from farms in the North (Kano and Kaduna states) to retailers in Lagos (Okomaiko, Agege, Iyana Ipaja markets), with distances from 1085 to 1316 km from farm to retailer. With regard to this study, the sites and value chains of interest were chosen mainly because:

- The North-South value chain represents the highest volume of movement of tomatoes in the country, so it was important to understand dynamics there with regard to tomato losses and identify mitigation strategies;
- The South-West value chain also represents a secondarily high-volume tomato value chain in the country and with proximity to the highest consumption area, therefore this study would potentially uncover different tomato loss drivers, which may also result in different loss mitigation strategies which could be just as effective.

3.2. Scoping Background

Scoping surveys were held with actors from the supply chains from both regions. In the South-West, actor-specific scoping surveys were conducted among 48 farmers, 44 transporters, and 48 traders/retailers. The survey for traders/retailers was performed on the largest markets in the area, Sasa and Mile 12, which were the major end markets for the supply chain of tomatoes. In the North, the scoping surveys were conducted among 151 farmers, 89 transporters, and 109 traders/retailers.

Survey participants were randomly selected after the researchers had "followed the chain", which means that in an initial visit (prior to survey execution), the researchers had interviews with various actors in all parts of the supply chain in the areas of interest, e.g., the market (traders, retailers, wholesalers), farmer groups, and individual farmers who often supplied these markets (or supply other market actors), as well as hauliers who transported fresh produce (including tomatoes). These initial meetings enabled identification of the chain structure and networks, which enabled the research team to have access to participants who were later selected for the survey.

Simple random sampling and a structured questionnaire was employed as per [20,29]. The main criteria for participants in the survey was that they had to be involved commercially in the tomato chain regardless of level—in other words, small, medium, or large-scale growers, traders, or hauliers had to be involved directly in growing, transporting, or selling the product.

Limitations in the sampling methodology are acknowledged, and this technique was chosen primarily to give the researchers insight into the status quo of the tomato value chain in Nigeria, which would be compared to the outcomes of similar, more in-depth studies, e.g., by the Growth and Employment in States- Wholesale and Retail sector project (GEMS4) [20,29]. Similar to these studies, the data was analyzed using descriptive statistics, such as frequency counts and percentages. The software IBM SPSS Statistics version 17 (International Business Machines Corporation, New York, USA) was used for data analysis. Furthermore, validation of these results were to be done in a workshop setting where in-depth questions and discussions would allow the researchers more perspective into the different value chains. The workshop participants, as described in the next section, were not limited to scoping survey participants only.

Table 1 provides the general characteristics of the survey respondents, Table 2 the chain and marketing characteristics of the tomato value chains of the respondents, and Table 3 the potential causes for PHL in these tomato value chains mentioned by the respondents. The majority of respondents were male, a member of an association (either producer, transporter, or trader), and used raffia baskets as packaging material. This is similar to the results reported by [20,29]. The majority of transactions took place at the farm gate, collection centers, or markets. Poor infrastructure conditions, unsuitable tomato varieties, and poor postharvest handling were identified as critical drivers of PHL for farmers, and transporters were more concerned with poor infrastructure conditions, roadblocks, and poor postharvest handling. In both regions, limited postharvest infrastructure and unsuitable tomato varieties were identified as critical drivers of PHL for farmers, while traders were more concerned with poor transportation conditions. The challenges highlighted from these activities are consistent with earlier findings published by [19]. This served as an indication that the sampling method was effective enough to broadly capture the state of the tomato value chain in Nigeria, which was the aim of this step.

Table 1. General characteristics of the respondents to a survey in tomato value chains from the South-West and North of Nigeria to the urban centers of Lagos and Ibadan held in 2017.

Characteristic	Option	South-West			North		
		Producer (n = 48)	Transporter (n = 44)	Trader/Retailer (n = 48)	Producer (n = 151)	Transporter (n = 89)	Trader/Retailer (n = 109)
Gender (%)	Male	85	100	55	100	100	99
Education (%)	Primary school (in)complete	23	46	52	32	28	25
	Secondary school (in)complete	46	48	43	34	44	24
	Arabic school		2		72	45	74
	Higher education	32	5	7	4	10	11
Age (%)	25–35 years	28	23	27	40	36	37
	36–45 years	24	57	64	39	60	38
	>45 years	48	43	9	21	3	24
Member of association (%)	Yes	92	85	72	64	60	77
Main function of association (%) [1]	Information and training	73		42			
	Seeds and other inputs	56					
	Sales	46					
	Government liaison		45	31			
	Transport regulations		41				
	(Off)loading coordination		41				
	Collect Levies			23			
	Price setting			19			
Packaging material (%) [1]	Raffia basket	63	100	90	102	97	100
	Sack	2	100		2	3	
	Crate	13	8		1	3	
	Mixed/other	6	2	10		4	
Tomato Growing Experience (%)	1–3 years	35			13		
	3–10 years	28			53		
	>10 years	37			34		
Tomato land size (ha)	Wet season	2			3		
	Dry season	1.2			2.7		
Productivity (kg/ha)	Wet season	3523			10,455		
	Dry season	4838			14,698		
Tomato variety used (%)	Hybrid-Padma F1	43					
	Carrot-Open-pollinated variety	30					
	UC82B-Open-pollinated variety				72		
	Roma vf-Open-pollinated variety				21		
	Rio-Open-pollinated variety				7		
	Other	27					

Table 1. *Cont.*

Characteristic	Option	South-West			North		
		Producer (n = 48)	Transporter (n = 44)	Trader/Retailer (n = 48)	Producer (n = 151)	Transporter (n = 89)	Trader/Retailer (n = 109)
Truck/car ownership (%) [1]	Owns the truck		41			24	
	Driver of individual/company		85			76	
Truck/car backhaul (%)	Empty		95			17	
	Loaded		5			83	

[1] Multiple answers were possible, so the sum can exceed 100%.

Table 2. Value chain and marketing characteristics of tomato value chains from the South-West and North of Nigeria to the urban centers of Lagos and Ibadan, mentioned by respondents in response to a survey held in 2017.

Characteristic	Option	South-West			North		
		Producer (n = 48)	Transporter (n = 44)	Trader/Retailer (n = 48)	Producer (n = 151)	Transporter (n = 89)	Trader/Retailer (n = 109)
Farmer sale point, transporter/trader/retailer collection point (%) [1]	Market	56			42		
	Collection Centre	44	70	50	58	92	51
	Farm gate	25	50		27	43	
	Trader			17			14
	Other			10			3
Bargaining power (%)	Yes	72			52		
Knowledge of in-transit losses (%)	Yes	36			43		
Market information knowledge (%)	Yes	72			87		
Moment buyer pays farmer (%) [1]	In advance	2			9		
	At sale (farm gate)	88			56		
	Afterwards	6			15		
	Combination	11			20		
Trader customer (%)	Wholesale			71			56
	Retail			29			44
Trader/retailer customer base (%) [1]	Fixed			60			54
	Changing			40			92

[1] Multiple answers were possible, so the sum can exceed 100%.

Table 3. Potential causes for postharvest losses in tomato value chains from the South-West and North of Nigeria to the urban centers of Lagos and Ibadan, mentioned by respondents in response to a survey held in 2017.

Characteristic	Option	South-West		North	
		Producer (n = 48)	Transprter (n = 44)	Producer (n = 151)	Transporter (n=89)
Shelf life limiting factors (%) [1]	Lack of/poor infrastructure	33	58	44	49
	Unsuitable/poor variety	23	38	38	52
	Poor (post-)harvest handling	17	44	61	26
	Poor packaging	19	29	15	29
	Low level/lack of technology and skills	23		33	
	Poor market facilities		33		35
	Lack of market incentives	17		35	
	Harvest at very ripe stage	17		15	
	Other		29	13	8
Factors affecting tomato yield (%) [1]	Pests and diseases	75		70	
	Excess rainfall	44		17	
	Drought/lack of water	23		29	
	Bad seed	21		29	
	Low level/lack of technology and skills	17		23	
	Lack of fertilizer	13		43	
	High temperature			26	
Factors limiting efficient transport (%) [1]	Road blocks/delays		84		81
	Poor/lack of infrastructure		80		88
	Road congestion		59		29
	Vehicle breakdown		43		35
	Delays in (off)loading		11		16
	Poor market infrastructure		9		10
	Other	14	14		11

[1] Multiple answers were possible, so the sum can exceed 100%.

3.3. Selection of Participants

Participants were selected from the value chain actors that participated in the scoping survey (for the simple random sampling strategy of the survey respondents, see Section 2.2). The following criteria were applied to select and invite the participants of the workshop (see also Section 2.3): (i) Participants are active actors actually working in the value chain in at least one of the activities, such as production, transportation, processing, trading, or retailing; (ii) participants are already trading with at least one other participant in the value chain; and (iii) participants are willing to implement potential innovations to reduce postharvest losses, and commit to participate for the entire project period.

In the North, 27 value chain actors (15 farmers, 4 transporters, 5 traders, 3 retailers) were selected, and in the South-West, 24 were selected (8 farmers, 4 transporters, 6 traders, 6 retailers). Budgetary constraints limited the number of participants and number of workshops.

3.4. Value Chain Mapping Including Monetary and Information Flows

Two Living Lab workshops were held, one in Ibadan (South-West) and one in Kano (North). Each workshop lasted for two days. The Living Lab workshops started with a plenary session, discussing and validating the outcomes of the scoping surveys. In both workshops, five groups of participants were established. Each group represented a value chain and contained at least one farmer, one transporter, one trader, and one retailer. Local enumerators were present to guide participants through the assignments and to assist and translate where needed.

In the South-West, two groups indicated that product flow was from producers to the traders via transporters, while one group specified that producers were directly connected with retailers and traders without transporters. Farmers in the South-West have more interaction with the end markets, and take their own produce to the market or bear the cost of transportation to the market. This could be related to the shorter distance between production and the market in the South-West (100–200 km) compared to the North (800–1,200 km). In contrast to the South-West, aggregation markets were present in the North.

Monetary and information flows were generally similar in the South-West and North. A difference is that farmers and traders in the South-West indicated to have more expenses compared to those in the North. This could be an actual reality, or it may be that farmers and traders in the South-West were more aware of their outgoing expenditures. Bribes to police or security officials was noted more frequently in the South-West value chain compared to the Northern value chain. With regard to information flows, it was evident that actors in the North were more active in sharing information on supply–demand variations in the market, i.e., glut-scarcity periods.

3.5. Identification of Bottlenecks and Potential Solutions

The main bottlenecks causing PHL were the occurrence of pests and diseases, low access to (quality) inputs, poor road infrastructure, inappropriate harvest and postharvest handling practices, and the seasonality of the production system. Access to good quality and unadulterated insecticides and quality control of this by the government were identified as solutions to prevent losses due to pests and diseases. Access to good quality seed from seed companies at an affordable cost and from an easy to access site to their communities were identified as a solution to prevent losses due to low access to (quality) inputs. As a specific problem, participants from the South-West mentioned the challenge caused by herdsmen moving their animals across tomato fields. In the North, a suggested solution for delivery delays was reduction of checkpoints on the road. However, the same actors also acknowledged insecurity as an issue and recommended increased police presence on the roads. In the North, participants identified bad harvesting practices as an important factor contributing to losses, with a recommendation for training on harvesting and handling of tomatoes on the farm to reduce losses. In the South-West, participants mentioned the hours of exposure which harvested tomatoes had to high temperatures while waiting for the transporter, as an important possible cause of PHL. Significant reduction in quality occurred during transportation, which attributed to the use of raffia baskets and which were often overloaded and squeezed during stowage, resulting in mechanical damage. A proposed solution to this was the use of plastic crates. The seasonal production with a glut period was identified in both regions as an issue, although on-site engagement with farmers showed that it was a more dominant challenge in the North. Participants in the North indicated that managing product flows during the glut and scarcity periods could be an opportunity for improving efficiency and reducing tomato loss in this chain.

3.6. Selection of Solutions with Good Potential

In each region, the participants mentioned and agreed upon two solutions to be tested in a pilot project by the participants. In this paper, we will only present the solution selected in both regions, which is sufficient to show the functioning of the approach. The solution agreed upon was the transport of tomatoes in returnable plastic crates instead of in the usual raffia baskets to reduce losses.

The solution not only had a 100% support base among the participants, but it also met the criteria of being specific, measurable, achievable (and affordable), realistic, and possible within the time boundaries of the project.

3.7. Preparation of Pilot Projects

The project team purchased the necessary materials and equipment, such as 600 plastic crates, and 15 analogue and 5 digital scales for weighing. A tool to measure and monitor the impact of the plastic crates on PHL both in weight and in quality was developed. Local enumerators were recruited for applying the measurement tool and monitoring the pilot projects during execution.

In each region, all participants and local enumerators were gathered in a two-day workshop to design and implement the plan for each solution, and to train the participants on the use of the solutions and on the tool to measure the impact of the solutions. In the South-West, 24 value chain actors and five local enumerators participated, and in the North, 27 value chain actors and four local enumerators took part. Per value chain and region, the participants designed an implementation plan which included a start and end date, responsibilities, tasks, and roles of each value chain actor, and the organization and logistics of the pilot project adapted to the local situation. Many issues and concerns were raised, such as the amount of tomatoes which were to be in a crate compared to the raffia baskets (as tomato amounts are measured in baskets), price-setting, the returning of crates, and payment for the transporter (payment per item carried). During the workshop, participants and local enumerators could familiarize themselves with the plastic crates, and they were provided with tomatoes, crates, baskets, and scales to understand how they should be comparing them to the raffia baskets in terms of handling, packing, weighing, and pricing. A draft version of the measurement tool was discussed with all participants and local enumerators and customized to ensure it was applicable to the local situation. For example, the quality grades in the measurement tool were aligned with the actual quality grades used in the value chain. In the South-West, three quality grades, A, B, and C, were used, whereas in the North, an additional grade, D was used. Points of weight measurement in the value chain were at the farm directly after harvest, at the farm just before transport, at arrival at the wholesale market, at the wholesale market just before leaving, and at arrival on the retail market. The tomatoes were only graded on quality at the first and last point of measurement to minimize disturbance. Batches of tomatoes were divided over crates and baskets and moved in the same transport vehicle to ensure similar conditions. The crates and baskets were marked at the farm, and these were followed in person throughout the value chain by the local enumerators. During the workshop, all participants and local enumerators received detailed training on the final measurement tool.

3.8. Execution of Pilot Projects

The pilot project on plastic crates in the South-West was conducted in December 2017 and in the North from February–March 2018. The plastic crates were tested compared to the conventional raffia baskets as a control. Tomatoes were collected from five farms in the South-West and three farms in the North on two separate days within a two-week period. Handling of the tomatoes was done as closely as possible to the normal situation in all links of the value chain. At farm level, tomatoes were sorted as usual by the farm workers during harvesting. Rotten or heavily damaged tomatoes were left on the field. After the enumerators weighed and graded the harvest, part of the harvested tomatoes were stacked in crates and the rest in baskets. It was ensured that quality grades were divided equally over crates and baskets. Any further preparation for the market until pickup by a transporter was done as usual. In some cases, the tomatoes were waiting for various hours. Before the actual loading took place, the tomatoes were weighed again to identify possible losses. The tomatoes were transported to the wholesale markets, in cars, small vans, and buses in the South-West and in large lorries in the North. At the market, the crates and raffia baskets were weighed again. In most cases, the tomatoes arrived around midnight at the market and were bought by the retailers in the early morning. Just before the retailers collected the tomatoes, they were weighed again to identify any possible weight loss. At arrival on the retail market, the tomatoes in both crates and baskets were weighed and graded using the same grades as used on the farm level.

3.9. Evaluation of Pilot Projects

Members of the project team analyzed the data collected with the measurement tool [30]. Crates were found to outperform baskets in both regions. Weight loss was between 5 and 12% lower with crates than with baskets. Similarly, the loss in best-quality A-grade tomatoes was between 16 and 20% lower with crates than with baskets. Here, we do not present further details on the PHL results that were measured in the pilot project, because the aim of this paper is to present the approach for designing context-specific interventions owned by local value chain actors. The details on the PHL results can be found in [30]. The results were discussed and presented in two-day evaluation workshops with the participating actors, one in the South-West and one in the North. All participants were invited to evaluate the pilot projects, and were guided with group assignments to facilitate the evaluation of their experiences and to ensure that all participants could express themselves. Value chain actors were also challenged to outline the basic components of a business plan that would benefit all the value chain actors.

All participants were happy to find the reduction in losses by using crates instead of baskets. Each value chain actor mentioned the benefits of crate use at each specific stage in the chain. Overall, retailers and traders stated a preference for tomatoes transported in crates compared to raffia baskets, because of the increased volume of quality products. Transporters appreciated the ease associated with loading and offloading crates, and it also meant that the wooden planks and grass thatch used to separate two layers of raffia baskets during transport were not needed anymore. Farmers appreciated the ease of postharvest handling and the stacking possibilities of crates. All participants mentioned that the introduction of plastic crates could also contribute to the standardization of measuring units and the introduction of scales—each crate had the same size and could contain around 23 kg of tomatoes. At the time, several types and sizes of baskets were being used, and weighing scales were not used. A large majority of participants (87%) preferred plastic crates to raffia baskets. Only 9% of participants preferred raffia baskets over plastic crates, and this was specifically for the greater level of convenience in returning from the retailer to the farm and stocking in a car. The remaining 4% of participants did not have a preference. After the evaluation pilot, the participants were provided the opportunity to purchase the plastic crates against half the price of a new crate. 67% of the participants bought the plastic crates they were using during the pilot, and 22% bought not only the crates used in the pilot but also some additional ones. This indicates a high adoption rate of the implemented intervention with plastic crates, and shows that it was a good fit for the local context.

In the final oral evaluation session, all participants expressed their appreciation of the approach applied. According to the participants, the design of the project helped them to overcome their initial skepticism, and to build trust in the innovation itself and in their co-value chain members. They were proud to be part of a community of change by acting and taking charge of an important developmental change.

However, the evaluation workshops also revealed several challenges and hiccups. The participants identified bottlenecks, which needed attention before upscaling could take place. Because of the current structure and organization of the value chain, not all actors could benefit equally. Retailers and traders had most of the financial gains. Most farmers did not benefit from the increased value of the tomatoes transported in crates, because pricing between farmer and trader occurred at the farm before transportation in the majority of cases. Transporters could even have a lower income, because they were paid per item and the transporter carried less items when transporting the larger and rigid crates compared to the smaller and more flexible baskets. Another challenge mentioned was the returning of the empty crates. In addition, no common agreement was reached on which actor(s) should purchase and own the crates and how this ownership could be protected. Crates are more expensive than baskets, but should last up to five years instead of one year. A final challenge was found in setting prices, because at the time, pricing was based on the size of a raffia basket and not on the weight of the content. The benefit of reduced losses was not totally reflected in a higher price when scales were not used, and price was set on the size of packing material.

4. Discussion

This paper presents an approach for designing context-specific interventions owned by local value chain actors. The approach was applied to develop an intervention to improve food security by reducing PHL in the tomato value chain in Nigeria. The approach considers contextual and cultural factors by actively involving value chain stakeholders working together to design an intervention.

This solved potential problems which can occur in top-down initiatives following a technical–rational project management approach, as put forward by Ika [7] and Robinson and Torvik [8]. Although many guides for value chain development include some value chain actor participation to design interventions, most guides result in interventions implemented largely independently of the specific context, due to insufficient co-creation, co-testing, and co-analyzing of potential interventions with value chain stakeholders [9]. Especially in a value chain such as the tomato value chain in Nigeria, with a low level of governance and fragmented, undifferentiated local markets, active participation is needed to reach sustainable interventions owned by value chain actors. Potential solutions are taken off the drawing board and trialed in practice, testing both the technical feasibility of potential interventions and how the interventions are embedded in the business models of existing value networks. As such, our approach ensures the support base, commitment, and ownership of participants, which are crucial conditions for sustainable and effective development interventions aiming at a certain change. The Living Lab and World Café which were set up also stimulated trust among and between the participants, i.e., trust in the intervention itself but also in their peers, which were the other actors in their value chain.

However, this approach also has its limitations; it is relatively resource-intensive in terms of financial and human resources, and the Living Lab and World Café only allowed for a limited number of participants. The whole approach is also time-consuming.

The results of the pilot projects in the case study in Nigeria showed that the PHL were lower for tomatoes transported in plastic crates compared to those transported in raffia baskets, and that the percentage of good-quality tomatoes was higher when transported in plastic crates than in baskets. In an experiment which simulated transport, [31] also found that the mechanical damage to tomatoes resulting from impact and vibration was lower when using plastic crates compared to when using raffia baskets. The majority of the pilot project participants were convinced of the benefits of using crates instead of baskets, and experienced actual financial gains when using crates. This indicates that the application of our approach to reduce PHL in tomato value chains in Nigeria resulted in context-specific interventions that were owned by the pilot project participants.

Our approach resulted in a context-specific intervention owned by the local value chain actors who participated in the pilot projects. Ownership of an intervention in a pilot project is a first step for improvement of the efficiency and effectiveness in all tomato value chains in Nigeria. The pilot project participants are only a small group of all actors active in the tomato sector in Nigeria, and tomato consumers were also not incorporated. Several challenges for a broader implementation of the plastic crates in tomato value chains in Nigeria were identified, alongside the following issues: (i) Investment in and ownership of the crates; (ii) an effective crate-returning system; (iii) distribution of the financial benefits between value chain actors; (iv) pricing of transactions in the value chain; and (v) lack of using scales and factual measures to define the price. When upscaling the pilot project in our case study to national Nigerian level, the increase in tomato availability and increase in quality requires a food system approach to analyze the national impact. This analysis should include the different roles and reactions to these increases in the different value chain actors, consumers of all income levels, and other stakeholders. At national level, new governance mechanisms might be needed in tomato value chains, such as producer, trader, or transporter networks or associations, chain leadership development, or third parties with sufficient financial and organizational strength for the acquisition and leasing of crates and weighing scales.

We developed the approach for context-specific, locally owned interventions to reduce PHL and applied it in a case study on reducing PHL in tomato value chains in Nigeria by introducing plastic crates to transport the tomatoes. In this case study, the purchase of the crates by the value chain participants in the study after the pilot projects showed that these stakeholders intended to use the crates after the project had finished. Application of the approach to reduce PHL in other countries or commodities or in other interventions to improve food and nutrition security is required to also determine its effectiveness in a broader scope of development projects.

5. Conclusions

This study provided an approach for designing, implementing, and testing context-specific interventions to reduce PHL with local value chain actors. In a Living Lab workshop with World Cafés, value chain actors together analyzed their own value chain and identified bottlenecks and potential interventions to solve each bottleneck. The most promising bottleneck intervention combination was tested in pilot projects, i.e., replacing the raffia baskets with plastic crates during tomato transport. The vast majority of pilot project participants in the case study of PHL in the tomato value chain in Nigeria indicated a preference of plastic crates to raffia baskets and bought crates after the project finished. This shows that the approach was effective in designing a context-specific, locally owned intervention. Application in other countries, commodities, or interventions is recommended to determine the effectiveness of the approach in a broader scope.

Author Contributions: C.P. facilitator living lab workshops and field work coordination. J.v.d.W. inception phase and contextual study. L.C. and C.P.A.v.W.: Project Management. M.S.: facilitator living lab workshops and data analyst and reporter survey. Y.D.: postharvest losses protocol design and testing, training local researchers, field measurement.

Funding: The authors wish to thank the Multi-Donor Trust Fund for Sustainable Logistics (MDTD-SL) administrated by the World Bank, CGIAR-A4NH Flagship Food Systems for Healthier Diets, and the KB-22 theme "Global food and nutrition security" of Wageningen UR funded by the Dutch Ministry of Agriculture, Nature and Food Quality (grant KB-22-002-010) for funding this study.

Acknowledgments: Thanks also goes to Frits Blessing, Peace Eruore Quadt (N-N-Solutions), Olalekan Emmanuel Akinwekomi and Thompson Ogunsanmi, both from the International Fertilizer Development Centre, for local support in Nigeria in design and execution of this project.

Conflicts of Interest: The authors declare no conflict of interest. The founding sponsors had no role in the design of the study, in the collection, analyses, or interpretation of data, in the writing of the manuscript, and in the decision to publish the results.

References

1. Lall, S.V.; Henderson, J.V.; Venables, A.J. *Africa Cities. Opening Doors to the World*; International Bank for Reconstruction and Development, the World Bank: Washington, DC, USA, 2017.
2. United Nations. *Transforming our World: The 2030 Agenda for Sustainable Development Resolution A/RES/70/1*; General Assembly of the United Nations: New York, NY, USA, 2015.
3. FAO. *Global Food Losses and Food Waste-Extent, Causes and Prevention*; Food and Agricultural Organization of the United Nations: Rome, Italy, 2011.
4. Sheahan, M.; Barrett, C.B. Review: Food loss and waste in sub-saharan Africa. *Food Policy* **2017**, *70*, 1–12. [CrossRef] [PubMed]
5. Affognon, H.; Mutungi, C.; Sanginga, P.; Borgemeister, C. Unpacking postharvest losses in sub-saharan Africa: A meta-analysis. *World Dev.* **2015**, *66*, 49–68. [CrossRef]
6. Sibomana, M.S.; Workneh, T.S.; Audain, K. A review of postharvest handling and losses in the fresh tomato supply chain: A focus on sub-saharan Africa. *Food Secur.* **2016**, *8*, 389–404. [CrossRef]
7. Ika, L.A. Project management for development in Africa: Why projects are failing and what can be done about it. *Proj. Manag. J.* **2012**, *43*, 27–41. [CrossRef]
8. Robinson, J.A.; Torvik, R. White elephants. *J. Public Econ.* **2005**, *89*, 197–210. [CrossRef]
9. Donovan, J.; Franzel, S.; Cunha, M.; Gyau, A.; Mithöfer, D. Guides for value chain development: A comparative review. *J. Agribus. Dev. Emerg. Econ.* **2015**, *5*, 2–23. [CrossRef]
10. Mohan, G. Participatory development: From epistemological reversals to active citizenship. *Geogr. Compass* **2007**, *1*, 779–796. [CrossRef]
11. Cornwall, A. *Beneficiary, Consumer, Citizen: Perspectives on Participation for Poverty Reduction*; Swedish International Development Cooperation Agency: Stockholm, Sweden, 2002.
12. Dell'Era, C.; Landoni, P. Living lab: A methodology between user-centred design and participatory design. *Creat. Innov. Manag.* **2014**, *23*, 137–154. [CrossRef]
13. Morgan, R.M.H.; Hunt, S.D. The commitment-trust theory of relationship marketing. *J. Mark.* **1994**, *58*, 20–38.

14. Grubben, G.; Klaver, W.; Nono-Womdim, R.; Everaarts, A.; Fondio, L.; Nugteren, J.A.; Corrado, M. Vegetables to combat the hidden hunger in Africa. *Chron. Horticult.* **2014**, *54*, 24–32.
15. Adeoye, I.B.; Adegbite, O.O.; Fashogbon, A.E.; Layade, A.A. Consumer purchasing behavior for tomatoes. *Int. J.Veg. Sci.* **2016**, *22*, 259–265. [CrossRef]
16. Leke, A.; Fiorini, R.; Dobbs, R.; Thompson, F.; Suleiman, A.; Wright, D. *Nigeria's Renewal: Delivering Inclusive Growth in Africa's largest Economy*; McKinsey Global Institute, McKinsey & Company: Michigan, NY, USA, 2014.
17. Bolarin, F.M.; Bosa, S.O. Post-harvest losses: A dilemma in ensuring food security in Nigeria. *J. Nat. Sci. Res.* **2015**, *5*, 151–154.
18. Ibeawuchi, I.I.; Okoli, N.A.; Alagba, R.A.; Ofor, M.O.; Emma-Okafor, L.C.; Peter-Onoh, C.A.; Obiefuna, J.C. Fruit and vegetable crop production in Nigeria: The gains, challenges and the way forward. *J. Biol. Agric. Healthc.* **2015**, *5*, 194–202.
19. Muhammad, R.H.; Hionu, G.C.; Olayemi, F.F. Assessment of postharvest knowledge of fruit and vegetable farmers in Garun Mallam L.G.A. of Kano, Nigeria. *Int. J. Dev. Sustain.* **2012**, *1*, 510–515.
20. Adepoju, A.O. Post-harvest losses and welfare of tomato farmers in Ogbomosho, Osun State, Nigeria. *J. Stored Prod. Postharvest Res.* **2014**, *5*, 8–13.
21. Ugonna, C.U.; Jolaoso, M.A.; Onwualu, A.P. Tomato value chain in Nigeria: Issues, challenges and strategies. *J. Sci. Res. Rep.* **2015**, *7*, 501–515. [CrossRef] [PubMed]
22. Brown, J.; Isaacs, D.; The World Café Community. *The World Café: Shaping Our Futures through Conversations that Matter*; Barrett-Koehler Publishers, Inc.: San Francisco, CA, USA, 2005.
23. Katzy, B.R.; Pawar, K.; Thoben, K.D. Editorial: A living lab research agenda. *Int. J. Prod. Dev.* **2012**, *17*, 1–7.
24. Eriksson, M.; Niitamo, V.P.; Kulkki, S.; Hribernik, K.A. Living labs as a multi-contextual R&D methodology. In Proceedings of the 12th International Conference on Concurrent Enterprising: Innovative Products and Services through Collaborative Networks, ICE 2006, Milan, Italy, 26–28 June 2006.
25. Følstad, A. Living labs for innovation and development of information and communication technology: A literature review. *Electr. J. Virtual Org. Netw.* **2008**, *10*, 99–131.
26. Tan, Y.H.; Bjørn-Andersen, N.; Klein, S.; Rukanova, B. *Accelerating Global Supply Chains with It-innovation: Itaide Tools and Methods*; Springer: Berlin/Heidelberg, Germany, 2011.
27. CGIAR, F.; KM4Dev; UNICEF; UNDP. The Knowledge Sharing Toolkit. Available online: http://www.kstoolkit.org/The+World+Cafe (accessed on 6 March 2017).
28. WorldCafe. Guide to the World Cafe Method. Available online: http://www.click4it.org/images/a/a5/World.pdf (accessed on 8 June 2017).
29. Adeoye, I.B.; Odeleye, O.M.O.; Babalola, S.O.; Afolayan, S.O. Economic analysis of tomato losses in Ibadan metropolis, Oyo State, Nigeria. *African J. of Basic Appl. Sci.* **2009**, *1*, 87–92.
30. Kok, M.G.; Dastoum, S.; Groot, J.J.; Plaisier, C.; Dijkxhoorn, Y.; van Wagenberg, C.P.A. *White Paper on Measurement Protocol on Food Loss (Forthcoming)*; Wageningen Food & Biobased Research: Wageningen, The Netherlands, 2018.
31. Idah, P.A.; Yisa, M.G.; Chukwu, O.; Morenikeji, O.O. Simulated transport damage study on fresh tomato (lycopersicon esculentum) fruits. *Agric. Eng. Int. CIGR J.* **2012**, *14*, 119–126.

sustainability

MDPI

Article

A Systems Approach to Food Loss and Solutions: Understanding Practices, Causes, and Indicators

Monika Verma *, Christine Plaisier †, Coen P. A. van Wagenberg † and Thom Achterbosch

Wageningen Economic Research, Wageningen University and Research, 2595 BM Den Haag, The Netherlands; christine.plaisier@wur.nl (C.P.); coen.vanwagenberg@wur.nl (C.P.A.v.W.); thom.achterbosch@wur.nl (T.A.)

* Correspondence: monika.verma@wur.nl; Tel.: +31-070-335-8277

† These authors contributed equally to this work.

Received: 30 November 2018; Accepted: 18 January 2019; Published: 22 January 2019

check for updates

Abstract: Reasons behind food loss can be very specific for each product and supply chain stage but it is also affected by factors independent of the product and stage. This work focuses on such generic factors and develops a framework to analyze food loss as a systemic outcome. The framework highlights the interconnected nature of problem across supply chain stages and therefore emphasizes the need to look at the whole system instead of specific stages, when proposing solutions. Practices and underlying causes contributing to food loss are identified for each stage of the supply chain using a literature search. Deductive logic is used to fill the gaps where literature was found to be scarce, and to derive socio-economic indicators that signal the presence of identified causes. Using this framework, we propose a non-exhaustive list of 30 socio-economic indicators, which can signal the presence of the 22 practices and 60 causes associated with food loss in supply chains. This list can serve as a starting list for practitioners and policymakers to build on when analyzing food losses in supply chains in their region. We evaluate the framework using a field-study of a tomato supply chain in Nigeria, and conclude that it can be a useful tool to identify practices, causes, and indicators of food loss.

Keywords: systems approach; conceptual framework; food loss practices; food loss causes; food loss solutions; supply-chain stages; literature; socio-economic indicators; tomato; Nigeria

---

1. Introduction

Identifying reasons behind food loss is necessary for proposing solutions to combat the problem. These reasons differ according to the nature of the product and the stage of food supply chain, and are therefore more likely to be explored for specific combinations of products and supply chain stages. Looking at narrowly defined stand-alone product-stage combinations can give a comprehensive look into very specific causes but a broad general understanding of the issue in a macro sense can remain elusive. This work aims to promote a systemic rather than the stand-alone view of supply chains. This is done by showing how reasons for food loss are linked across stages of a single or more supply chains, and therefore proposed solutions should account for such inter-stage linkages. While the need for such holistic approaches accounting for dynamics of the whole supply chain instead of considering specific points in isolation is becoming apparent [1], this is precisely the kind of synergy that is often overlooked by taking a product specific approach to looking at food loss. The purpose of this study is therefore to develop a conceptual framework identifying root/structural causes of food loss with emphasis on across stage and across chains interrelations.

Inspiration for our framework comes from three concepts—micro, meso, and macro causes of food loss—as proposed by the High Level Panel of Experts report (HLPE) [2]. HLPE defines micro-level causes as actions or inaction of individual actors occurring at the same stage of supply chain where food is discarded, e.g., consumers not checking their refrigerators regularly to keep stock of what is available and discarding products that go bad. Meso-level causes refer to the way different actors are organized within or across different stages, e.g., lack of better transport infrastructure hinders how quickly and efficiently the suppliers and processors/retailers can conduct business. Macro-level causes are structural in nature, e.g., lack of institutional and legal capacity that could help coordinate actors to take actions and move towards better outcomes.

The use of terms micro-, meso-, and macro-level are difficult to explain to field actors, therefore we propose to use more comprehensible concepts of practices, causes, and indicators. In addition, HLPE does not give a clear link between the micro-, meso-, and macro-level causes. Building on HLPE, this work attempts to establish a more structured link between the practices, causes, and indicators. We start by identifying practices across supply chain stages, then identify causes underlying such practices, and finally we arrive at socio-economic indicators that can signal the presence of the identified causes in an economy.

While built as progressing from practices to indicators, given the strong links at each stage, this framework can as easily be used to look in the opposite direction: starting broad and filtering down to specifics. The information on socio-economic indicators is usually more readily available (from local governments and international bodies). Such information can be used to identify infrastructure categories that need more attention. The framework can then be used to list possible problems (practices and causes) associated with those infrastructural categories.

The benefit of using such a framework comes from being able to identify broad factors across supply chain stages applicable to most agricultural products. The link between indicators and food loss at more than a single supply chain stage implies, that addressing the causes associated with these indicators should, and could yield multi-stage benefits. Also, while important on their own, structural reasons of food loss can further be responsible for determining the presence and magnitude of loss at micro-level [2].

For practitioners, the framework emphasizes the need to be aware of the fact that their specific interventions might not yield the full potential effect because of the interrelations across supply chain stages and causes. For policy makers looking to make an impact on food loss, the proposed easier to observe socio-economic indicators can be used as red flags regarding the existence and severity of food loss in a region. Identifying indicators and seeing how many practices they can influence, made possible by viewing food loss as a result of the whole system, can help target efforts towards the practices and causes with links to multiple stages of supply chain. This work covers four supply chain stages: production and pre-harvest, harvest and initial on-farm handling, transportation and storage, and processing.

The article is structured as follows. Section 2 outlines the conceptual framework by means of a stylized example and identifies the components of the framework. Based on the details of Section 2, a synthesis of identified practices, causes, indicators, and their influence across stages is provided in Section 3. Section 4 looks at a specific supply chain—tomato supply chain in Nigeria—to assess whether the field data and observations support or refute the causes, practices, and indicators proposed under the systems framework approach. Finally, Section 5 gives the discussion and conclusions.

2. Materials and Methods

Section 2.1 lays out the conceptual framework. Section 2.2 identifies the most direct and clear links between indicators and causes.

2.1. Conceptual Framework: A Systems Approach

Figure 1a shows a stylized graphical representation of the conceptual framework. We list *practices* leading to food loss at each of the four above mentioned supply chain stages by asking: *"what practices exist?"* in the field, that contributes to food loss. Practices can be seen as actions or inactions on part of supply-chain participants. The practices are further explored in order to identify underlying *causes* that can help explain the existence of these practices by asking: *"why do these practices exist?"* For example, inadequate credit markets can help explain both, lack of proper storage as well as lack of proper harvesting techniques. Both questions above, are answered relying heavily, but not exclusively, on findings of the agronomy literature (see Section 2.2). For identifying indicators, given the absence of literature linking characteristics of supply chain to those of economy, deductive logic is used to see what kind of socio-economic characteristics of the economy can reflect the existence of the practices and causes. If they can help explain the existence of causes and practices behind food loss, efforts made towards improving these indicators should also contribute towards reduction in food loss. Therefore, the question asked to identify the indicators is: "*how to reduce or minimize the loss resulting from these causes?*", essentially asking what factors explain the existence of these causes and can therefore help reduce loss resulting from these causes.

Figure 1b represents the complexity of the relationship between practices, causes, and indicators proposed in Figure 1a. For clarity of depiction, only two stages of the supply chain are shown, though the concept applies to all stages. The rectangular boxes called stage A and stage B in the figure depict the different supply chains stages. The red circles represent causes with ties to observed practices. Note that a single cause can contribute to losses in more than a one stage. For example, the lack of availability of credit can lead to the use of sub-quality seeds at the production stage, as well as to use of poor harvest equipment at the harvest and on-farm handling stage. Finally, the existence of causes is linked to indicators which can be grouped in four broad categories of infrastructure: knowledge, physical, financial, and institutional. These categories are chosen as most causes of food loss are believed to result from financial, knowledge (managerial, technical, organizational), institutional, and physical infrastructure bottlenecks [3]. We define the categories as follows. Knowledge infrastructure includes actors and process that determine how knowledge is created, shared, and changed/updated. In this context, knowledge infrastructure covers knowledge institutions, and extension and information networks. Physical infrastructure includes basic physical structures required for an economy to function and survive, such as transportation networks, power grid, sewage and waste disposal, etc. Institutions (both formal and informal), and institutional arrangements influencing rules and processes regarding how economy operates, form institutional infrastructure. State of the financial sector and its operations, and ease of credit access for all agents, forms the financial infrastructure.

The stylized representation helps to understand the food loss problem as a systems approach problem and therefore helps brings forth insights that otherwise escape scrutiny. Below are some examples of types of insights that can emerge from our stylized representation:

- *What kind of infrastructure is relevant for what stage?* Different categories of infrastructure play a primary/substantial role in determining the state of food loss at different stages. For example, food loss at stage A is affected by causes (1) and (2), which are linked to knowledge, financial and institutional features of economy. Stage B practices are affected by causes (2), (3), and (4), and are thereby linked to physical and institutional infrastructure. As only the institutional category is shared across the two stages, addressing institutional infrastructure issues could help reduce food loss across both stages.
- *Some practices might be easier to target than others* In terms of the stylized figure above, reducing losses resulting from practice A1 calls for working on both knowledge and financial infrastructure issues. For A2, in addition to the aforementioned, we also need to pay attention to institutional factors. This increases the complexity and number of factors that an effective solution looking to reduce food loss at A2, would need to take into account, and it might therefore be harder to address. Only addressing knowledge and financial issues might not work because of institutional bottlenecks. As an example, cold storage facilities (physical) built without addressing electric grid and distribution issues (institutional and physical).
- *Tackling certain causes may affect more practices and possibly yield higher reduction in food loss* Addressing cause (1) affects both practices in Stage A and might therefore be more effective in reducing food loss in Stage A. Note that focusing on cause (1) to address losses in stage A, does not necessarily mean higher impacts in terms of food loss reduction: the impact also depends on magnitude

issues, the initial states of food loss and different infrastructures in a region, and other country specific features.

➢ *Sustainability/acceptance – need to involve all actors* Given the entangled nature of the problem and chains, the implementation success and propagation of any proposed solution hinges highly on its acceptability for all involved actors.

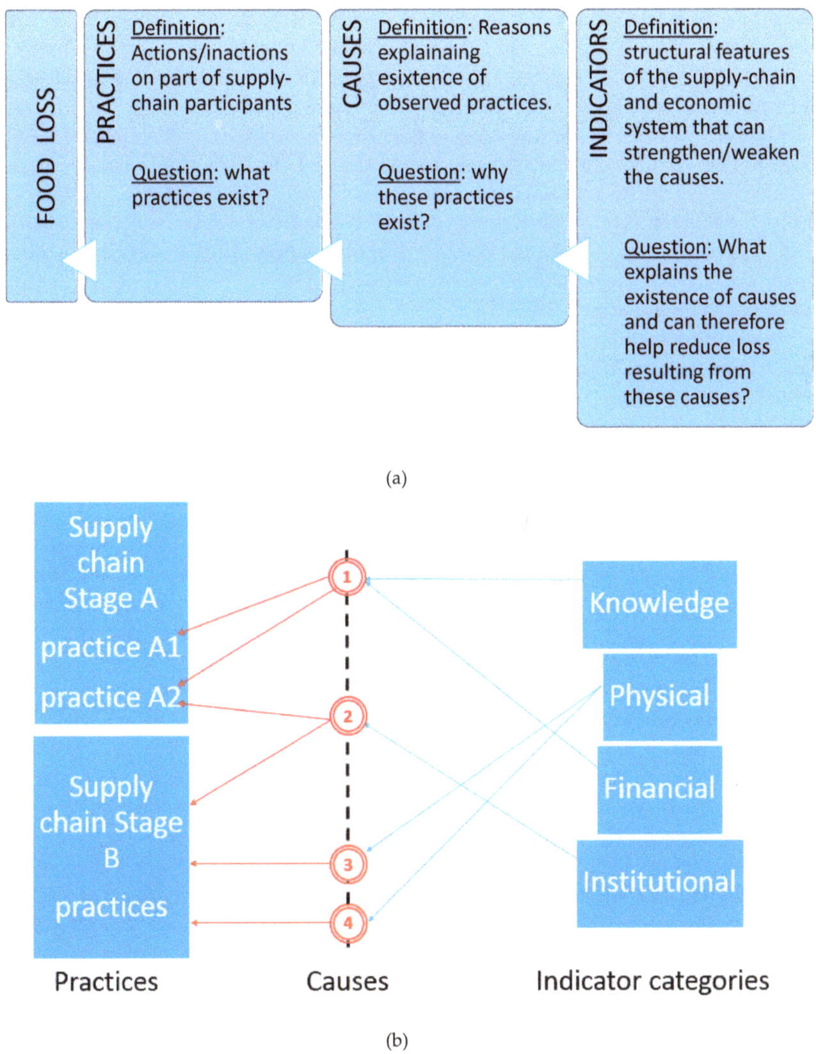

Figure 1. Stylized figure showing conceptual framework for the proposed systems approach. (**a**): link between the practices contributing to food loss, underlying causes, and identifying indicators. (**b**): complex nature of relationship between practices, causes, and indicators. A1 and A2 are examples of two different practices specific to the supply chain's stage A. The concentric red circles represent causes of food loss, not necessarily specific to any given stage. The figure depicts four such different causes. Knowledge, physical, financial, and institutional are the four types of infrastructure categories, to which the presence of the stage-generic causes can be linked.

2.2. Operationalizing the Framework: Identifying Practices and Causes, and Proposing Indicators

2.2.1. Production and Pre-Harvest Stage

Production can be compromised in terms of quantity as well as quality even before it is harvested. While the focus on food loss is often restricted to post-harvest stages of supply chain, pre-harvest and harvest stages can and do influence the extent of post-harvest loss [4,5]. For example, wheat exposed to showers at a late stage of maturity is likely to have a shorter storage life irrespective of quality of storage facilities. Similarly, at least 5–10% of rice crop in Asia is lost annually to rodents [6] even before it is harvested, amounting to an equivalent of 11 kg/capita; which might become worse with more frequent outbreaks expected with climate change [7]. Also, if for any reason the product does not conform to certain standards for size, color, or shape, it might be rejected at a later stage [8]. For example, if a particular crop of mangoes fails to develop a required bright red hue, it is either rejected or fetches a much lower price. Note that this rejection can happen much later at the retail stage even though the cause—failure to develop the desired characteristics—occurred at the production/pre-harvest stage.

This is the stage of supply chain along with harvesting that is often overlooked in analysis of food loss [9] by economists; however, it is well explored by agronomists and agricultural engineers. Furthermore, in face of changing climate, we should expect the extent of such losses to rise [10,11] and therefore the need for attention to this stage.

➢ Practices contributing to pre-harvest loss

Pre-harvest losses could occur due to the presence or absence of practices ranging from choice of sub-optimal crop varieties and seeds [12,13] for local conditions; sub-optimal planting schedule [14,15]; to inefficient farm management practices regarding use of soil [16], water [17], nutrient [17,18]; and pest control [19].

Yet other causes that can explain losses at the production stage but do not have a bearing on agents' agricultural practices can be external factors like a bad weather spell [20–22]. While these losses are counted, the underlying causes are not a part of the food system. However, such causes can and do often lead to loss-averting behavior by agents, such as planting more to hedge against such risk of loss. Such causes should therefore be considered because their effect on agent behavior can be modified by means of policies and coordination, for example, crop insurance [23].

➢ Causes underlying practices

Reasons that could influence the choice of crop variety and lead to planting of varieties unsuitable for local conditions could be many, such as lack of adequate information [24] and unavailability of the right seed varieties [25–27], either physically or economically.

Suboptimal farm management practices are also often seen as a result of lack of adequate information [28,29], unavailability of sufficient credit [26] to make changes towards better practices. The importance of such information in making critical and sustainable farming decisions is well understood in agronomy [30]. The scale of operation [25] and absence of clear regulations regarding farm management [27] can also influence how much time and effort a farmer spends on such activities.

Unforeseen consequences of government policies is another possible cause that can promote bad farm management practices (for example, using too much fertilizer) or the use of varieties unsuitable for local conditions. For example, providing free electricity for irrigation in India has promoted paddy cultivation supported by injudicious ground-water pumping in parts of the country, which would normally not grow the crop in absence of such a policy [31]. This is an example where the higher price for locally unsuitable variety along with government support for irrigation makes the adoption of the crop possible. Such crops are more susceptible to loss in event of failure/delay of the policy support they get. It is therefore important to pay attention to unintended consequence of a government support programs.

➢ Indicators of causes of production and pre-harvest loss

Lack of adequate information and physical unavailability of varieties suitable to local conditions can arise from absence of regional agricultural research institutes and extension services or lack of their active participation in extension work. These, in turn, can result from neglecting agricultural sector and focusing more on industry in an attempt to grow faster. This is often the case with underdeveloped and developing economies.

Lack of economic access to suitable crop varieties could be a manifestation of lack of credit availability and immature state of financial infrastructure. The same two reasons explain equally well the use of suboptimal farm management practices. The small scale of operation (as is often the case in developing world) in absence of co-operatives and associations could result in a lack of incentives for investing in better farm practices. Such lack of organized efforts is also reflected in the inability of farmers to negotiate better contracts and prices for their produce, leading them to alternative hedging practices like planting more.

The above-mentioned indicators are often found to be poorly performing in developing countries, which are also the regions believed to be suffering more from problem of food loss in comparison to the developed world, which usually performs better on these indicators.

2.2.2. Harvest and Initial On-Farm Handling Stage

The next important stage in the food supply chain is harvesting including on farm sorting, threshing, and initial handling, described broadly as the "agricultural production" category in Gustavsson et al. [32]. Harvesting losses cause loss of output for not only current crops but may also have implications for quality and therefore buyers' acceptability of future crops, as suggested in Gulden et al. [33]. In terms of products, roots and tubers, and fruits and vegetables seem to be more susceptible to on farm losses than crops like cereals and oil seeds [32].

Unlike production losses (Section 2.2.1), cutting and threshing losses (not including losses from other initial handling processes like drying) seem not to significantly differ across traditional and mechanized supply chains (Figure 1, [34]). This seems to indicate that large-scale mechanization (as seen in agriculture in developed world) is not necessarily better when it comes to preventing harvesting losses.

➢ Practices contributing to harvest loss

Actions at harvest stage can broadly be grouped into poor timing of harvest [35,36], poor methods and equipment choice for harvesting and initial handling [37], and inability to harvest or decision not to harvest the crop [38].

The time of harvest can affect the loss of agricultural produce in multiple ways. Apart from determining yield and quality of produce [39], harvest time also determines the moisture content of crop [40] and thereby its susceptibility to infestation. Similarly, maturity at the time of harvest can affect the extent of mold, insect, and aflatoxin contamination for grains [41,42]. Using contaminated grains as feed is also not feasible as it can lead to contaminated animal products [43]. Sometimes harvesting is delayed because the crop is left to dry in a field before it is harvested and often for longer than the recommended duration [41,44]. The timing of harvest of horticultural crops determines levels of ethylene, which affects its post-harvest shelf life (Chapter 2, [45]). Not only the maturity of horticultural products, but even the time of day chosen for harvest, can determine their post-harvest chemistry, and handling needs and in absence of the proper handling the likelihood of spoilage [4].

Timing, harvest method, and initial handling procedures can all affect the nutrient content (quality loss) of horticulture crops [46]. Improper harvest methods and initial handling can result in cuts, bruising, and surface abrasion in roots, tubers, fruits, and vegetables while harvesting leading to loss of water and nutrients (Chapter 2, [45]). Some examples of such practices are: use of mechanical combine harvester without specialized headers [47], particularly when crop is fallen or lodged; losses like spillage and heat injuries associated with mechanical harvesting [48]; failing to sort infected crop during harvesting often leading to contamination spreading to good harvest [49–51]; contamination during harvesting particularly from use of unhygienic equipment and unhygienic handling of dairy and slaughtered animals [52,53], often resulting in entire batches being discarded. Similarly, multiple or rough handling of horticultural produce can result in avoidable loss during harvest and/or transportation [48]. Note that while such factors are applied to both mechanical and manual harvesting, they are more relevant for mechanical ones as proper management of procedures is required (Chapter 2, [45]).

Inability to harvest or the decision to leave crops in field also contribute to harvest loss. Data from the U.S. shows that, on average, about 7% of all planted crops are not harvested, the same figure for fruits and vegetables stand at 6%, and the numbers can be as high as 50% for some particularly bad years [8]. Moreover, one crop not harvested and left on field can provide food to rodent populations that can harm other standing crops [6].

➢ Causes underlying practices

The reasons for sub-optimal harvest timing could be many: economic hardship [54] and need of cash [55], lack of adequate infrastructure and transport [55,56] (Chapter 13) for timely delivery of product to markets contributing to early harvest, labor shortage [57,58] contributing to late or no harvest, lack of information on best practices [59], and credit constraints (can delay or push forward the time of harvest).

Mechanical inefficiency [60], often in combination with plant spacing, can also contribute to some amount of harvest lost. According to some estimates the harvest loss should be about 2–4% but is often as high as 10% or more (Figure 2, [2]), even in developed countries.

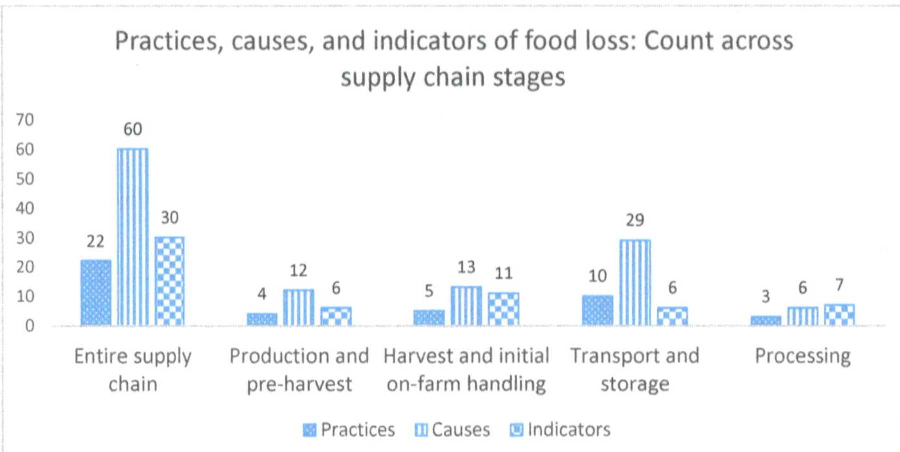

Figure 2. Practices, causes, and indicators' count by supply-chain stage.

Often times all or part of horticulture crop is not harvested or not sent forward into the supply chain owing to failed aesthetic standards [61], bad weather [62] or a plunge in market price for crop [63,64].

➢ Indicators of causes of harvest and initial on-farm handling loss

From the analysis so far in Sections 2.2.1 and 2.2.2, it can be argued that within on-farm losses that are often clumped together in food loss analysis (Figure 2, [55]), the real difference between developed and developing regions comes not so much from harvesting losses but from initial handling and production practices. This seems to suggest that prevalence of traditional (less mechanized) harvesting cannot be taken as an indicator for the presence of large-scale harvesting loss.

Lack of means for dealing with economic hardship and need for cash leading to premature harvesting would be reflected in the absence of financial support system for farmers, which can in turn be measured by the absence of formal (agricultural development banks) and/or informal (co-operatives) channels of credit. Similarly, harvesting early to reach markets in time can be caught by measuring distance of markets from farms and the condition and extent of transport network. The need to reach distant markets could indicate a lack of local markets but also a shift towards a supermarket model of produce distribution. Agricultural labor shortage is again often a result of urbanization and migration of population towards urban areas. Early harvest due to lack of information on the best time to harvest is often due to lack of education and extension services. For example, education of farmer—a measure of ability of farmer to either already have the necessary information or be able to seek it—was the only variable that was found to be significant at 1% level for reducing losses for both wheat and rice in India in 2004 [65].

A decision not to harvest because of produce lacking in appearance can indicate a lack of alternative channels of disposal for fresh produce than the established supermarket chains, a lack of consumer awareness, or a lack of processing facilities to sell produce as canned or preserved. If crop is left on farm and not harvested because the farm price cannot cover production costs, it can indicate a lack of farm collectives or support programs and of storage and processing facilities.

2.2.3. Transport and Storage Stage

Transport and storage are important features of a modern food supply chain. Urbanization and the associated "supermarket revolution" [66] leads to changing nature of the food supply chain. This often results in increased distance and time between production and consumption [67], thereby increasing the demand and importance of transport and storage in food supply chain [68]. However, the improvements in efficiency come with a delay and products spend longer time in conditions not quite suitable for maintaining quality and/or quantity. Aulakh et al. [69] claim the loss during storage to be the highest in the supply chain, and as much as 50% to 60% of cereals stored can be lost due to technical inefficiencies [70]. Given the lack of clear chronological order between transport and storage, the two are combined for the purpose of this analysis.

In regions of world where transport and storage facilities are better, such losses are rare and come from equipment malfunctioning. Not surprisingly, the extent of losses are higher for developing regions of the world and, across products, losses are higher for fruits and vegetables owing to their highly perishable nature [32].

➢ Practices contributing to transport and storage loss

Spoilage can occur during transportation itself owing to lack of temperature control, but also because of rough and multiple handling during loading and off-loading [4,71], and lack of proper storage [72] at the docks. Other factors that contribute to transport losses are theft [73], spillage, stress or heat injuries [45,71,74] in fresh produce, as well as in livestock due to improper securing and packaging of cargo and longer time spent in transit [71]. As much as 16% of expected income from cattle is lost in Ghana and 45% of cattle in Ethiopia is affected during transport [2]. Loss of fresh produce is estimated to be between 30–50% in developing countries in the transport and storage stages [75].

Sometimes the produce is stored on-farm instead of being transported to market or is transported to an off-farm storage facility instead of a market. Grains stored at home openly or in traditional sacks [76,77]; tubers not sorted, cured, and treated before storage [78,79]; ineffective fumigation of grain silos [80,81]; and not using cold storage for horticultural crops all result in avoidable storage losses [75,82]. Traditional storage methods and equipment usually do not provide sufficient protection against temperature fluctuations and humidity. Better practices and technologies can reduce storage losses by almost 98% of those seen with traditional methods [70], irrespective of length of storage period. For example, Purdue improved crop storage (PICS) has been shown to reduce grain damage in terms of weight loss due to insects from over 40% to under 1% [83]. Unsanitary and unhygienic handling [22] during storage and transportation further adds to these losses.

Despite the use of the best storage facilities and practices, the losses can be considerable depending on product quality at time of harvest. This again points to the importance of link between product loss across stages and the need to tackle the problem in a system-wide context.

➢ Causes underlying practices

One reason behind transport losses is the high cost of transportation. Transport costs in Africa can be five times as high as those in Asia [84]. High transport costs render decent transport facilities outside the reach of small to medium scale farms [8], often making temperature-controlled transportation inaccessible to such farms. Cheaper but unreliable transport alternatives like donkeys, bicycles, or bad quality vehicles often result in delays in addition to those caused by bad road infrastructure and road blocks [82]. Both contribute toward a loss of quantity and quality. Transport losses are compounded because of improper cargo handling (unhygienic handling, piling, sitting on produce, etc.) often due to lack of knowledge about best practices [82]. Inefficient logistic planning further complicates matters (Chapter 2, [71]). Delays in reaching off-farm destinations could be caused not only for domestic produce but also for exported and imported products at the port of entry or exit. The important role that a cheaper and faster transport plays in spurring international trade is well-established [85].

Most storage loss, whether because of insects, pests, rodents, or temperature and moisture, can be attributed to poor storage conditions or equipment. For farmers in developing countries, cold storage for fresh produce is often not available or is a very expensive technology [82]. Given the fact that most such countries are geographically concentrated in tropical zones [75], high and variable temperatures accelerate the spoilage. Even when cold storage is available, mixing and piling together all different products at different stages of maturity results in losses [82] that can be easily avoided if this knowledge is to be made available to farmers. Furthermore, many developing countries experience an erratic, inconsistent power supply which renders any existing cold storage facilities ineffective [86–88]. While grains and tubers do not require cold storage, traditional storage methods often result in high losses mostly on account of lack of knowledge on best practices [22] such as drying and packaging.

➢ Indicators of causes of transport and storage loss

Only purchasing the best equipment available is often not enough to reduce losses seen at the storage and transport for a host of factors as pointed out by [75]; using forklift/pallet trucks to avoid multiple handling also requires better surfaces for these machines to operate on, which cannot be achieved without planning infrastructural investments. Similarly, building cold storage without ensuring a regular and reliable power supply does not help. Therefore, developing low-cost technologies suited to local conditions [82] while working towards improving the local conditions should be a preferred approach. This requires promoting local research and extension efforts and building on the existing indigenous knowledge infrastructure alongside physical infrastructure. Both are severely lacking in the developing world.

Also, a lack of credit for investment and general apathy to agriculture can be seen in the inadequate (less than 5% of agricultural research funds worldwide) investment funds devoted to combat such losses (Chapter 1, [45]). Individual farmers do not have enough resources to make such investments on their own, while the absence of cooperatives rules out any collective organized effort.

Lack of credit, state of physical and knowledge infrastructure, and absence of collectives in presence of large share of small-scale farms are therefore good telltale signs of high transport and storage loss.

2.2.4. Processing

Processing can increase the shelf life of products by transforming it into canned and preserved varieties. In this way, this processing can be seen as one form of storage. In developing countries, edible food often ends up being lost due to the absence of processing facilities.

Food surplus in periods of excess supply or food that is slightly bruised/aesthetically unappealing can be processed for preservation and consumption at a later time.

An undesirable result of processing is often found to be a loss of vitamins and trace minerals [89], indicating that processing should not be a primary focus as a solution. Fortification can address this criticism but places a double burden on small-scale, capital-poor agents. Also, while processing serves the increasing demand for convenience food (pre-cut and ready to use produce), minimally processed produce does not last as long as intact produce (Chapter 2, [71]). This results in increased food waste at retail and consumer level. It therefore seems, that processing avoids food loss but adds to food waste and the reduction in food loss often comes at price of increased plastic/packaging waste [2].

➢ Practices contributing to processing loss

Processing facilities in developing countries are mostly traditional for example fermentation, pickling (Chapter 3, [71]) and can often only handle small volumes [32]. Also, the produce preserved in traditional methods would often have limited local demand due to tastes and often fails to comply with food safety standards and other requirements such as labelling [87]. In addition, unsanitary and unhygienic handling of produce and dairy during the process can lead to easily avoidable losses [52,53,87].

In the developed world, defects in packaging, such as wrong size, shape, and appearance make certain batches of processed product redundant [75]. Similarly, increased focus on achieving conformity results in excessive trimming of otherwise edible product. European Commission [90] estimated the processing sector in EU to be the second highest waste generator (19%) in 2012 after consumers (53%). Some estimates indicate that only 50% of the potato at the processing stage comes out in the form of a final output [91].

➢ Causes underlying practices

As indicated in the previous section, losses in developing countries are often attributable to absent, inadequate, or limited capacity processing facilities [2]. The small scale of operation of individual farmers and the absence of coordinated collectives makes processing costs for individual farmers too high [87]. Processing facilities also requires investments in technical know-how capacity beyond the reach of small and oftentimes uneducated farmers. Also, in most developing countries, governments usually emphasize increased agricultural production rather than integrated efforts towards production and post-harvest management (Chapter 1, [71]).

The reasons for a high processing loss in the developed world on the other hand seems more to do with consumer attitudes towards acceptability [92] than technical or institutional limitations affecting the processing sector. Technical malfunctioning [2] can also result in processing loss.

➢ Indicators of causes of processing loss

Capacity limits or non-existence of processing facilities, absence of organized community efforts, along with small scale farm operations and biased agricultural policy in favor of increased production are some indicators that can indicate the possible presence of big processing loss.

In the face of given consumer attitudes, absence of food-collectives, and possibility of legal obligation on food donations by processors can be indicative of the presence of processing losses.

While this section identifies only the most direct and clear links, a full matrix of the links, both direct and apparent, is available to interested readers upon request from the authors.

3. Results

This section builds on Section 2.2. and provides the information in tabular and graphical form. We present only the most direct and clear links. A full matrix of links is available upon request from the authors.

Figure 3 shows the number of indicators identified for the entire supply chain and for each of the individual four stages. In total, there are 30 indicators: 6 each relevant for production and pre-harvest, and transport and storage stages, 7 for processing stage, and 11 for harvest and initial on-farm handling stage. One important insight emerging from Figure 3 is that there is overemphasis on technological (knowledge and physical facilities) and financial bottlenecks in proposing solutions against food loss, but the often overlooked institutional infrastructure seems just as important. Collective efforts can yield benefits for small farmers as well as for the entire supply chain [93,94].

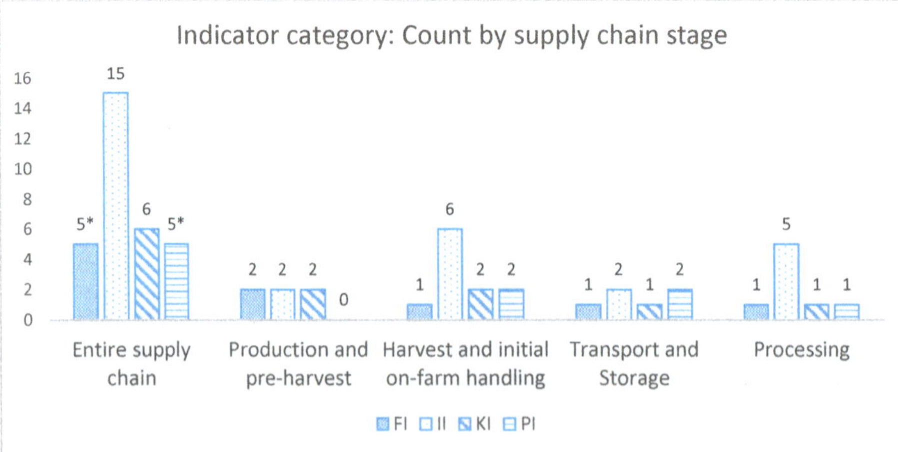

Figure 3. Indicator count by category and supply-chain stage. FI: financial infrastructure; II: institutional infrastructure; KI: Knowledge infrastructure; PI: physical infrastructure. (*)FI and PI categories have one common indicator counted in both categories.

A list of the 30 indicators identified across the four stages is presented in Table 1.

Table 1. Infrastructural indicators proposed to assess the state of food loss in a region.

Supply-chain Stage [1]	Proposed Indicator	Infrastructure Category of Proposed Indicator [2]
PH	Lack of general credit availability	FI
	Immature state of general financial infrastructure	FI
	Ag policy: neglecting agricultural and focusing more on industry/Injudicious ag support	II
	Absence of co-operatives and farmer associations	II
	Absence of regional agricultural research institutes and extension services	KI
	Lack of active participation of regional agricultural research institutes in extension work	KI
HH	Absence of financial support system for farmers	FI
	Distance of markets/seed centers from farms	II
	Lack of local markets to sell product	II
	Supermarket model of produce distribution	II
	Urbanization and migration of population	II
	Lack of alternative demand channels of disposal of produce (processing, storage, others)	II
	Lack of farm collectives or support programs	II
	Lack of education and extension services	KI
	Lack of consumer awareness	KI
	Transport network	PI
	Lack of processing and storage facilities	PI
TS	Lack of credit for investment	FI
	General apathy to agriculture	II
	Absence of cooperatives	II
	Lack of research and extension efforts	KI
	Lack of technologies/facilities suited to local conditions	PI
	Lack of infrastructural investments	PI
P	Absence of organized community efforts	II
	Small-scale or non-diversified farm operations	II
	Biased agricultural policy in favor of increased production	II
	Absence of food-collectives	II
	Legal obligation on food donations	II
	Consumer attitudes	KI
	Capacity limits or non-existence of processing facilities	PI/FI

[1] PH: production and pre-harvest; HH: harvest and initial on-farm handling; TS: transport and storage; P: processing.
[2] FI: financial infrastructure; II: institutional infrastructure; KI: knowledge infrastructure; PI: physical infrastructure.

Some of the proposed indicators are specific to farming and agriculture, while others reflect the state of the wider economy. Some are connected to more than one stage (column 1, Table 1), and therefore seem to reappear in the list with slight differences. For example, community/co-operative efforts appear in all four stages, but it is considered a different indicator as community effort can take the form of different practices across different stages. Finally, the categorical classification (FI: financial infrastructure; II: institutional infrastructure; KI: knowledge infrastructure; PI: physical infrastructure) is also reported in column 3.

As the performance of a region on each indicator is seen as a signal regarding presence and severity of the food loss problem, the better the performance of a region in all categories, the less severe food loss is likely to be an issue for that region.

Figure 2 provides similar information for the practices and causes, as Figure 3 does for indicators. In all, there are 22 broad practice groups, 60 causes underlying the existence of such practices, and 30 indicators which can reflect the presence of these practices and causes in the supply chain.

Table 2 provides a list of all practices and causes by each stage. Existence of more causes (60) than practices further iterates the fact that each practice can be liked to multiple causes, and addressing a single cause might not yield desired reduction in food loss attributable to that practice. With multiple practices responsible for food loss, the complexity further increases.

Table 2. Practices and causes identified in literature by each supply chain stage.

Supply Chain Stage	Practices	Causes
Production and pre-harvest	Use of sub-optimal crop varieties and seeds	Lack of information on varieties, physical unavailability of right seeds and varieties, economic unavailability of right seeds and varieties, wrong market(price) signals
	Sub-optimal planting schedule	Lack of information about planting; multiple or early planting due to economic hardship
	Inefficient farm management	Lack of information about farm management, credit unavailability, scale of farm operation, absence of clear farm management regulations
	External factors	Bad weather (often leading to planting more to hedge), bad general economic conditions (often leading to planting more to hedge)
Harvest and initial on-farm handling	Poor timing of harvest	Economic hardship and need for cash, inadequate infrastructure and transport to ensure timely delivery, labor shortage, lack of information about harvesting
	Poor harvest methods and equipment	Lack of information about harvesting, mechanical inefficiency, credit unavailability
	Poor on-farm initial handling, during and post-harvest	Lack of information
	Inability to harvest	Labor shortage; bad weather
	Decision to not harvest	Low market price, aesthetic product standard requirements, bad weather
Transport and storage	Lack of temperature control	Tropical climate, expensive storage technologies, erratic and unreliable power supply
	Rough and multiple handling	Lack of knowledge about best practices, inefficient logistic planning, unreliable transport alternatives
	Improper storage at docks	Expensive storage technologies, tropical climate, lack of knowledge about best practices
	Theft and spillage	Unreliable transport alternatives; inadequate road and rail infrastructure, road blocks; lack of knowledge about best practices; inefficient logistic planning
	Stress and heat injuries	Unreliable transport alternatives; inadequate road and rail infrastructure, road blocks; lack of knowledge about best practices; inefficient logistic planning; tropical climate; erratic and unreliable power supply
	Long transit time	Inadequate road and rail infrastructure, road blocks; unreliable transport alternatives; inefficient logistic planning
	Storing unsorted, untreated, uncured crop	Lack of knowledge about best practices
	Insanitary and unhygienic storing	Lack of knowledge about best practices
	Ineffective fumigation	Lack of knowledge about best practices, lack of technical capacity and availability of treatment
	Ineffective storage sacks/silos	Tropical climate, expensive storage technologies, lack of knowledge about best practices
Processing	Traditional processing	Lack of or limited processing facilities, high costs of modern processing equipment
	Unsanitary and unhygienic handling	Lack of information on best practices, lack of government efforts to improve postharvest practices
	Discards due to product aesthetic standards	Consumer attitudes towards product acceptability, technical malfunctioning

4. Framework Assessment: Comparison with Tomato Supply Chains in Nigeria

Does the framework proposed in Section 2 help explain practices and causes seen in tomato supply chain in Nigeria? Data on tomato supply chain in Nigeria is gathered from two sources: a) responses of supply chain participants [95], and b) a non-exhaustive literature search on tomato losses in Nigeria. The subsections list the practices and causes of losses identified by literature and by field workshops for each stage.

4.1. Practices, Causes, and Indicators Identified in Literature

4.1.1. Production and Pre-Harvest

Tomatoes in Nigeria are grown outside and farmers do not control many variables that affect the quantity and quality of harvest, such as temperature, relative humidity, solar radiation and growing media [96], and soil moisture and presence of pathogens [97]. Farmers generally lack access to improved varieties and quality seeds [98].

Reliance on rainwater because of lack of proper irrigation [99] causes tomato production to be concentrated in the wet season. This causes seasonal production peak and low prices [99]. Heavy rainfall in wet season promotes growth of fungi [100,101], causing leaf diseases, such as wilt [102], blight [103], and defoliation and yellowing [100] of field crop, but also losses at later stages [101]. Recommended pesticides are not used because of high costs and lack of the necessary expertise to ensure their proper application [104]. Credit facilities needed to address such problems are often not available to farmers [101].

4.1.2. Harvest and Initial On-Farm Handling

Tomatoes are harvested half or fully ripe [105] including those intended for distant markets. Sometimes harvesting is done during the hotter moments of the day, which can result in heat stress to tomatoes [105]. Farmers use woven palm leaf collection baskets with hard and sharp edges for harvest collection and load these as fully as possible [99]. Baskets and tomatoes are not disinfected [99], and rotten fruits are mixed with healthy ones in baskets and in storage facilities, causing rapid spread of pathogens [99,106]. Most farmers store harvested tomatoes under tree shade without any further protection [99,105]. Without adequate storage facilities, pathogens can develop quickly [98]. Adequate storage facilities and post-harvest technologies are not used because farmers do not know about these options, have inadequate technological knowledge, have insufficient contact with extension workers, or the technologies are not available [98].

4.1.3. Transport and Storage

Cold storage facilities are often inadequate due to lack of electricity supply for farmers [107]. Farmers do not use appropriate post-harvest crop handling techniques, appropriate storage facilities are too expensive or not available, appropriate transport modes are not available, road conditions are bad, and market information and access are insufficient [106]. Fungicides are often too expensive to use during on-farm storage even if they were available [108], and farmers do not have access to credit facilities [101].

The raffia baskets are often used to move and store tomatoes without being disinfected between batches. A fungal spore left by one batch can easily infect subsequent batches [101]. Appropriate vehicles are often not available [98] and rail system cannot be used due to unusual delays [101]. Due to lacking adequate storage facilities at the farm [106], farmers need to move tomatoes quickly after harvest; therefore, all kinds of vehicles are used for transporting tomatoes to markets, many of them old and unfit [101]. Ropes used to secure baskets to such vehicles result in excessive local pressure and damage to tomatoes [109]. During transportation when baskets are overfilled or transporters use only leaves to separate the baskets [101], tomatoes are crushed [109]. In addition, flexible baskets forced into inadequate spaces in a vehicle, in conjunction with vibrations and impacts due to bad road surfaces, result in the compression and damage of the tomatoes [98,109].

Knowledge on the correct handling, storing, and transporting of tomatoes is lacking due to insufficient assistance and extension services, prevalence and perseverance of existing suboptimal practices, and a lack of interest from policy-makers [109]. Available training and research in the agricultural sector is inadequate [101]. Packaging containers such as plastic crates are not readily available, not available in sizes similar to the familiar raffia baskets, and are too expensive [110]. Additionally, many tomato fields are remote and are either not connected by good roads or the roads are in deplorable condition [99].

4.1.4. Processing

Sun-drying being simple and cheap is often used as a preservation technique though it reduces the ascorbic acid in tomatoes by almost 70% [101]. Knowledge on correct preservation and processing is lacking, because available training and research in the agricultural sector is inadequate [101].

4.2. Field Observations on Practices and Causes

Supply chain participants (farmers, transporters, traders, and retailers) in the tomato supply chain in Nigeria were gathered in workshops with the aim to identify practices and causes for postharvest losses in their supply chain [95]. This section summarizes the results of these workshops.

4.2.1. Production and Pre-Harvest

Workshop participants identified seed quality, pest and disease occurrence, and weather fluctuations as main reasons for on-farm loss of tomatoes. Few farmers have irrigation facilities. The participants also reported high seasonality of tomato production leading to oversupply, low prices, and high losses in the peak season.

Farmers confirmed that high-quality seeds are either not available or not accessible due to lack of access to credit facilities. Lack of knowledge required for pest and disease control and unavailability of quality pesticide were also mentioned. The general lack of knowledge on good agricultural practices, both at farmer and extension worker level, was evident. Absence of farm records made the use of farm data for decision making and planning at the farm impossible. There is a general lack of producer cooperatives or collective action. Periods of over- and under-supply also usually result in produce not being harvested as costs often exceeding revenue. According to all participants, the focus of government policies is on other sectors (i.e., oil). No investments are made in agricultural research, nor in development and adoption of good agricultural practices.

4.2.2. Harvest and Initial On-Farm Handling

Workshop participants indicated that tomatoes are harvested at a late stage of maturity because the price is often set at the farm gate and the more mature the tomatoes the higher the price the farmer receives. In addition, participants mentioned a shortage of labor and that the available labor is often unskilled and unaware of proper harvesting techniques. Mechanization is not practical, as it requires high investments, which farmers cannot afford.

Use of raffia baskets for the tomato collection is widespread. According to the farmers, using plastic crates for tomatoes would not be possible because of the large scale of production would require a huge number of crates and it was unclear who should own the crates. They reported that investment needed towards introducing the plastic crates would be too high to bear for a single actor. The long geographic spread of the chain also poses a challenge of returning empty crates to their owners.

Beneath these practices is a lack of awareness and knowledge on good harvesting and handling techniques and on the subsequent consequences for the tomato quality. Besides, the lack of access to credit facilities hampers investments in better materials and equipment.

4.2.3. Transport and Storage

All participants perceived transportation as a very serious cause of tomato loss. In addition, participants mentioned that the infrastructure is very poor, and many official and unofficial road controls lead to huge delays.

Transport fees are paid for each basket transported. The per basket mode of payment provides incentive for the transporters to take as many baskets as possible in a single trip. Given the flexibility of traditional raffia baskets this leads to tomato loss during transportation. Loading and off-loading is done in a very poor way and without much care. Furthermore, transporters are not held accountable for the condition and delivery time of tomatoes at their destination. This leaves no incentive for transporters to change the manner of their operations.

Most existing markets are without any shed, equipment, hygiene, or covering facilities for storing produce. For underlying causes, the participants point at general lack of knowledge on proper handling, lack of investments in improving the infrastructure, corruption and lack of security and control, and actors not being held accountable.

4.2.4. Processing

According to workshop participants, tomatoes are hardly ever processed to make tomato paste. There are a few companies willing to work with farmers producing tomatoes in open fields instead of green houses. Lack of cooperatives and collective action and the absence of crop scheduling hampers a stable supply of tomatoes to potential processing companies. Besides, there are doubts about the quality of tomatoes as a result of the low level of knowledge and application of good agricultural practices. For the drying practices, farmers lack the appropriate facilities and equipment as well as the knowledge and awareness on the appropriate drying techniques and hygiene practices. Participants also perceived the business environment to be marked by corruption and lack of transparency and suspected that it results in lack of interest from external investors to establish processing facilities in the country.

As can be seen, the conceptual framework lists many of the factors identified using field observations and associated literature of a specific case study without having to dig deeper into the specific case in question.

5. Discussion and Conclusions

The current work provides a conceptual framework to identify macro/structural factors responsible for food loss at various supply chain stages. According to the findings, poor institutional infrastructure, appears to be as prevalent a reason behind food loss as bad physical and financial infrastructures. This suggests that while technology-based solutions to food loss are important, the role of institutions deserves more attention. The framework further brings to the fore, the complex nature of the inter-connected reasons underlying food loss in supply chains, and emphasizes the need to see food loss as a systemic outcome.

The main advantages of this approach are:

- It can quickly help to identify problem areas in the supply-chain without the need to gather data on the whole supply chain of a product in any region. Once identified, the points of possible intervention should be explored further to devise the right solution.
- This manner of looking for solutions using a systems view lowers the risk of ineffective solutions and unintended negative consequences of proposed intervention.
- While food loss is hard to measure, data on indicators listed above are more easily available and comparable across regions.
- A solution towards bettering a given indicator affects not one but multiple causes associated with that indicator whether in the same or in different stages of the supply-chain; therefore, this approach can be used to identify solutions with most potential. Similarly, a given cause can be

perpetuated due to bad performance on multiple indicators. Therefore, a single intervention aiming a single indicator/category would often fail to achieve its full potential in the presence of other bottlenecks in the system.

This work is not a substitute for quantifying the impact of various causes on extent of food loss. While an attempt is made to be thorough with the posed framework, the list of possible practices, causes, and indicators should not be treated as exhaustive or complete but as a starting base to build on. Also, while we could identify that Nigerian tomato supply chain seems to comply with the proposed framework, it should be tested for more products and regions. Furthermore, while the indicators are suggested because they are expected to be correlated to food losses, this remains to be confirmed. With limited existing work on evaluating the impact of improving infrastructure on food loss and waste [111], as a next step, more effort should be made to estimate the magnitude of effect on food loss from efforts towards improvement in these indicators.

We conclude that our framework can be a useful tool to identify socio-economic indicators that can signal the presence of food loss in supply chain stages. Furthermore, it can be used for linking practices in a given supply chain stage with their underlying causes that appear across multiple supply chain stages.

Author Contributions: Conceptualization and methodology, M.V.; validation, C.P. and C.P.A.v.W.; formal analysis, M.V., C.P., and C.P.A.v.W.; writing, M.V., C.P., C.P.A.v.W.; visualization, M.V., C.P.; project administration, M.V. and T.A.; funding acquisition, M.V. and T.A.; resources, T.A.

Funding: This research was funded by Dutch Ministry of Economic Affairs, grant number KB22-002-005.

Acknowledgments: The authors acknowledge use of Nigerian case-study results from projects funded by CGIAR-A4NH Flagship Food Systems for Healthier Diets for workshops in Nigeria, and from Multi-Donor Trust Fund for Sustainable Logistics (MDTD-SL) administrated by the World Bank. They also wish to thank the editor, the editorial staff, and the anonymous reviewers, for helping bring the article to its current state.

Conflicts of Interest: The authors declare no conflict of interest. The funders had no role in the design of the study; in the collection, analyses, or interpretation of data; in the writing of the manuscript, or in the decision to publish the results.

References

1. Affognon, H.; Mutungi, C.; Sanginga, P.; Borgemeister, C. Unpacking Postharvest Losses in Sub-Saharan Africa: A Meta-Analysis. *World Dev.* **2015**, *66*, 49–68. [CrossRef]
2. HLPE Food Losses and Waste in the Context of Sustainable Food Systems. *The High Level Panel of Experts on Food Security and Nutrition*; FAO: Rome, Italy, 2014; p. 117.
3. Kader, A.A. Increasing food availability by reducing postharvest losses of fresh produce. In *Acta Horticulturae*; International Society for Horticultural Science (ISHS): Leuven, Belgium, 2005; pp. 2169–2176.
4. Kader, A.A. Postharvest biology and technology: An overview. In *Postharvest Technology of Horticultural Crops*; University of California Agriculture and Natural Resources: Oakland, CA, USA, 2002; ISBN 978-1-879906-51-8.
5. Ippolito, A.; Nigro, F. Impact of preharvest application of biological control agents on postharvest diseases of fresh fruits and vegetables. *Crop Prot.* **2000**, *19*, 715–723. [CrossRef]
6. Singleton, G. Impacts of Rodents on Rice Production in Asia 2003. Available online: http://books.irri.org/971220183X_content.pdf (accessed on 18 January 2019).
7. John, A. Rodent outbreaks and rice pre-harvest losses in Southeast Asia. *Food Secur.* **2014**, *6*, 249–260. [CrossRef]
8. Gunders, D. *Wasted: How America Is Losing Up to 40 Percent of Its Food from Farm to Fork to Landfill*; Natural Resources Defense Council: New York, NY, USA, 2012.
9. IFPRI Toward a sustainable food system: Reducing food loss and waste. In *Global Food Policy Report*; IFPRI: Washington, DC, USA, 2016; p. 10.
10. Gourdji, S.; Läderach, P.; Valle, A.M.; Martinez, C.Z.; Lobell, D.B. Historical climate trends, deforestation, and maize and bean yields in Nicaragua. *Agric. For. Meteorol.* **2015**, *200*, 270–281. [CrossRef]
11. Lesk, C.; Rowhani, P.; Ramankutty, N. Influence of extreme weather disasters on global crop production. *Nature* **2016**, *529*, 84–87. [CrossRef] [PubMed]

12. Bandyopadhyay, R.; Kumar, M.; Leslie, J.F. Relative severity of aflatoxin contamination of cereal crops in West Africa. *Food Addit. Contam.* **2007**, *24*, 1109–1114. [CrossRef] [PubMed]
13. Öztürk, F.P.; Kaçal, E.; Sarısu, H.C.; Karamürsel, D.; Emre, M. Economic evaluation of preharvest and harvest losses in "0900 ziraat" sweet cherry cultivar. In *Acta Hortic*; International Society for Horticultural Science (ISHS): Leuven, Belgium, 2010; pp. 261–267.
14. Scott, R.K.; English, S.D.; Wood, D.W.; Unsworth, M.H. The yield of sugar beet in relation to weather and length of growing season. *J. Agric. Sci.* **1973**, *81*, 339–347. [CrossRef]
15. Shivrain, V.K.; Burgos, N.R.; Gealy, D.R.; Smith, K.L.; Scott, R.C.; Mauromoustakos, A.; Black, H. Red Rice (*Oryza sativa*) Emergence Characteristics and Influence on Rice Yield at Different Planting Dates. *Weed Sci.* **2009**, *57*, 94–102. [CrossRef]
16. George, T.; Magbanua, R.; Garrity, D.P.; Tubana, B.; Quiton, J. Rapid Yield Loss of Rice Cropped Successively in Aerobic Soil. *Agron. J.* **2002**, *94*, 981–989. [CrossRef]
17. Jatav, K.S.; Agarwal, R.; Singh, R.; Shrivastava, M. Growth and Yield Responses of Wheat [*Triticum aestivum* L.] to Suboptimal Water Supply and Different Potassium Doses. *J. Funct. Environ. Bot.* **2012**, *2*, 39. [CrossRef]
18. Dangler, J.M.; Locascio, S.J. External and Internal Blotchy Ripening and Fruit Elemental Content of Trickle-irrigated Tomatoes as Affected by N and K Application Time. *J. Am. Soc. Hortic. Sci.* **1990**, *115*, 547–549.
19. Oerke, E.-C. Crop losses to pests. *J. Agric. Sci.* **2006**, *144*, 31–43. [CrossRef]
20. Thompson, L.M. Weather Variability, Climatic Change, and Grain Production. *Science* **1975**, *188*, 535–541. [CrossRef] [PubMed]
21. Lusk, J.L.; Tack, J.; Hendricks, N.P. Heterogeneous Yield Impacts from Adoption of Genetically Engineered Corn and the Importance of Controlling for Weather. *Agric. Product. Prod. Behav.* **2018**.
22. Abass, A.B.; Ndunguru, G.; Mamiro, P.; Alenkhe, B.; Mlingi, N.; Bekunda, M. Post-harvest food losses in a maize-based farming system of semi-arid savannah area of Tanzania. *J. Stored Prod. Res.* **2014**, *57*, 49–57. [CrossRef]
23. Hill, R.V.; Viceisza, A. A field experiment on the impact of weather shocks and insurance on risky investment. *Exp. Econ.* **2012**, *15*, 341–371. [CrossRef]
24. Gardas, B.B.; Raut, R.D.; Narkhede, B. Modeling causal factors of post-harvesting losses in vegetable and fruit supply chain: An Indian perspective. *Renew. Sustain. Energy Rev.* **2017**, *80*, 1355–1371. [CrossRef]
25. Ghimire, R.; Wen-chi, H.; Shrestha, R.B. Factors Affecting Adoption of Improved Rice Varieties among Rural Farm Households in Central Nepal. *Rice Sci.* **2015**, *22*, 35–43. [CrossRef]
26. Shiferaw, B.; Kebede, T.; Kassie, M.; Fisher, M. Market imperfections, access to information and technology adoption in Uganda: Challenges of overcoming multiple constraints. *Agric. Econ.* **2015**, *46*, 475–488. [CrossRef]
27. Floro IV, V.O.; Labarta, R.A.; López-Lavalle, L.A.B.; Martinez, J.M.; Ovalle, T.M. Household Determinants of the Adoption of Improved Cassava Varieties using DNA Fingerprinting to Identify Varieties in Farmer Fields: A Case Study in Colombia. *J. Agric. Econ.* **2017**, *69*, 518–536. [CrossRef]
28. Bozoğlu, M.; Ceyhan, V. Measuring the technical efficiency and exploring the inefficiency determinants of vegetable farms in Samsun province, Turkey. *Agric. Syst.* **2007**, *94*, 649–656. [CrossRef]
29. Wilson, P.; Hadley, D.; Asby, C. The influence of management characteristics on the technical efficiency of wheat farmers in eastern England. *Agric. Econ.* **2001**, *24*, 329–338. [CrossRef]
30. Tanaka, D.L.; Krupinsky, J.M.; Liebig, M.A.; Merrill, S.D.; Ries, R.E.; Hendrickson, J.R.; Johnson, H.A.; Hanson, J.D. Dynamic cropping systems: An adaptable approach to crop production in the Great Plains. *Agron. J.* **2002**, *94*, 957–961. [CrossRef]
31. *Planning Commission Report of the High Level High Level Expert Group on Water Logging in Punjab*; Government of India: New Delhi, India, 2013; p. 95.
32. Gustavsson, J.; Cederberg, C.; Sonesson, U. *Global Food Losses and Food Waste: Extent, Causes and Prevention; Study Conducted for the International Congress Save Food! at Interpack 2011, [16–17 May], Düsseldorf, Germany*; Food and Agriculture Organization of the United Nations: Rome, Italy, 2011.
33. Gulden, R.H.; Shirtliffe, S.J.; Thomas, A.G. Harvest losses of canola (*Brassica napus*) cause large seedbank inputs. *Weed Sci.* **2003**, *51*, 83–86. [CrossRef]
34. Hodges, R.J.; Buzby, J.C.; Bennett, B. Postharvest losses and waste in developed and less developed countries: Opportunities to improve resource use. *J. Agric. Sci.* **2011**, *149*, 37–45. [CrossRef]

35. Bednarz, C.W.; Shurley, W.D.; Anthony, W.S. Losses in Yield, Quality, and Profitability of Cotton from Improper Harvest Timing. *Agron. J.* **2002**, *94*, 1004–1011. [CrossRef]
36. Philbrook, B.D.; Oplinger, E.S. Soybean Field Losses as Influenced by Harvest Delays. *Agron. J.* **1989**, *81*, 251–258. [CrossRef]
37. Price, J.S.; Hobson, R.N.; Neale, M.A.; Bruce, D.M. Seed Losses in Commercial Harvesting of Oilseed Rape. *J. Agric. Eng. Res.* **1996**, *65*, 183–191. [CrossRef]
38. Kantor, L.S.; Lipton, K.; Manchester, A.; Oliveira, V. Estimating and Addressing America's Food Losses. *Food Rev.* **1997**, *20*, 2–12.
39. Dag, A.; Kerem, Z.; Yogev, N.; Zipori, I.; Lavee, S.; Ben-David, E. Influence of time of harvest and maturity index on olive oil yield and quality. *Sci. Hortic.* **2011**, *127*, 358–366. [CrossRef]
40. Buckmaster, D.R.; Heinrichs, A.J. Losses and Quality Changes during Harvest and Storage of Preservative-treated Alfalfa Hay of Varying Moisture Content. *Trans. ASAE* **1993**, *36*, 349–353. [CrossRef]
41. Kaaya, A.N.; Warren, H.L.; Kyamanywa, S.; Kyamuhangire, W. The effect of delayed harvest on moisture content, insect damage, moulds and aflatoxin contamination of maize in Mayuge district of Uganda. *J. Sci. Food Agric.* **2005**, *85*, 2595–2599. [CrossRef]
42. Hell, K.; Fandohan, P.; Ranajit, B.; Kiewnick, S.; Sikora, R.; Cotty, P.J. Pre- and postharvest management of aflatoxin in maize: An African perspective. In *Mycotoxins: Detection Methods, Management, Public Health and Agricultural Trade*; Leslie, J.F., Bandyopadhyay, R., Visconti, A., Eds.; CABI: Wallingford, UK, 2008; pp. 219–229. ISBN 978-1-84593-082-0.
43. Prandini, A.; Tansini, G.; Sigolo, S.; Filippi, L.; Laporta, M.; Piva, G. On the occurrence of aflatoxin M1 in milk and dairy products. *Food Chem. Toxicol.* **2009**, *47*, 984–991. [CrossRef] [PubMed]
44. Alakonya, A.E.; Monda, E.O.; Ajanga, S. Effect of Delayed Harvesting on Maize Ear Rot in Western Kenya. *Am.-Eurasian J. Agric. Environ. Sci.* **2008**, *4*, 372–380.
45. Kader, A.A.; Rolle, R.S. *The role of Post-Harvest Management in Assuring the Quality and Safety of Horticultural Produce*; FAO Agricultural Services Bulletin; Food and Agriculture Organization of the United Nations: Rome, Italy, 2004; ISBN 978-92-5-105137-5.
46. Lee, S.K.; Kader, A.A. Preharvest and postharvest factors influencing vitamin C content of horticultural crops. *Postharvest Biol. Technol.* **2000**, *20*, 207–220. [CrossRef]
47. Hobson, R.N.; Bruce, D.M. PM—Power and Machinery: Seed Loss when Cutting a Standing Crop of Oilseed Rape with Two Types of Combine Harvester Header. *Biosyst. Eng.* **2002**, *81*, 281–286. [CrossRef]
48. Kitinoja, L.; Kader, A.A. *Small-Scale Postharvest Handling Practices: A Manual for Horticultural Crops*, 5th ed.; Postharvest Horticulture Series; Postharvest Technology Research and Information Center, University of California-Davis: Davis, CA, USA, 2015.
49. Mutiga, S.K.; Were, V.; Hoffmann, V.; Harvey, J.W.; Milgroom, M.G.; Nelson, R.J. Extent and Drivers of Mycotoxin Contamination: Inferences from a Survey of Kenyan Maize Mills. *Phytopathology* **2014**, *104*, 1221–1231. [CrossRef] [PubMed]
50. Kabak, B.; Dobson, A.D.W.; Var, I. Strategies to Prevent Mycotoxin Contamination of Food and Animal Feed: A Review. *Crit. Rev. Food Sci. Nutr.* **2006**, *46*, 593–619. [CrossRef] [PubMed]
51. Choudhary, A.K.; Kumari, P. Management of Mycotoxin Contamination in Preharvest and Post Harvest Crops: Present Status and Future Prospects. *J. Phytol.* **2010**, *2*, 37–52.
52. Oliver, S.P.; Jayarao, B.M.; Almeida, R.A. Foodborne Pathogens in Milk and the Dairy Farm Environment: Food Safety and Public Health Implications. *Foodborne Pathog. Dis.* **2005**, *2*, 115–129. [CrossRef] [PubMed]
53. Sofos, J.N. Challenges to meat safety in the 21st century. *Meat Sci.* **2008**, *78*, 3–13. [CrossRef] [PubMed]
54. Sunderlin, W.D.; Resosudarmo, I.A.P.; Rianto, E.; Angelsen, A. The Effect of Indonesia's Economic Crisis on Small Farmers and Natural Forest Cover in the Outer Islands 2000. Available online: http://dlc.dlib.indiana.edu/dlc/bitstream/handle/10535/4358/Sunderlin-the-effect-of-indonesias-economic-crisis-on-small-farmers-and-natural-forest-cover-in-the-outer-islands.pdf?sequence=1&isAllowed=y (accessed on 18 January 2019).
55. Godfray, H.C.J.; Beddington, J.R.; Crute, I.R.; Haddad, L.; Lawrence, D.; Muir, J.F.; Pretty, J.; Robinson, S.; Thomas, S.M.; Toulmin, C. Food Security: The Challenge of Feeding 9 Billion People. *Science* **2010**, *327*, 812–818. [CrossRef] [PubMed]
56. Sudheer, K.P.; Indira, V. *Post Harvest Technology of Horticultural Crops*; New India Publishing: New Delhi, India, 2007; ISBN 978-81-89422-43-1.

57. Richards, T.J.; Patterson, P.M. Hysteresis and the Shortage of Agricultural Labor. *Am. J. Agric. Econ.* **1998**, *80*, 683–695. [CrossRef]
58. Guan, Z.; Wu, F.; Roka, F.; Whidden, A. Agricultural Labor and Immigration Reform. *Choices* **2015**, *30*, 1–9.
59. Aguilar-Gallegos, N.; Muñoz-Rodríguez, M.; Santoyo-Cortés, H.; Aguilar-Ávila, J.; Klerkx, L. Information networks that generate economic value: A study on clusters of adopters of new or improved technologies and practices among oil palm growers in Mexico. *Agric. Syst.* **2015**, *135*, 122–132. [CrossRef]
60. Casamali, B.; Williamson, J.G.; Kovaleski, A.P.; Sargent, S.A.; Darnell, R.L. Mechanical Harvesting and Postharvest Storage of Two Southern Highbush Blueberry Cultivars Grafted onto *Vaccinium arboreum* Rootstocks. *HortScience* **2016**, *51*, 1503–1510. [CrossRef]
61. Stuart, T. *Waste: Uncovering the Global Food Scandal*; W. W. Norton: New York, NY, USA, 2009; ISBN 978-0-393-07735-3.
62. Dawson, A. *Unprecedented Year for Potatoes as Thousands of Acres go Unharvested*; Manit. Co-Oper: Winnipeg, MB, Canada, 2018.
63. Vietnamnews Price Drop Forces Farmers to Leave Vegetables Unharvested. Available online: http://vietnamnews.vn/society/424383/price-drop-forces-farmers-to-leave-vegetables-unharvested.html (accessed on 5 January 2019).
64. Freshplaza Argentina: Apples are Left Unharvested due to Low Prices. Available online: www.freshplaza.com/article/2190490/argentina-apples-are-left-unharvested-due-to-low-prices/ (accessed on 5 January 2019).
65. Basavaraja, H.; Mahajanashetti, S.B.; Udagatti, N.C. Economic Analysis of Post-harvest Losses in Food Grains in India: A Case Study of Karnataka. *Agric. Econ. Res. Rev.* **2007**, *20*, 117–126.
66. Reardon, T.; Gulati, A. The Supermarket Revolution in Developing Countries: Policies for Competitiveness with Inclusiveness. Available online: https://ideas.repec.org/p/fpr/polbrf/2.html (accessed on 22 January 2019).
67. Thyberg, K.L.; Tonjes, D.J. Drivers of food waste and their implications for sustainable policy development. *Resour. Conserv. Recycl.* **2016**, *106*, 110–123. [CrossRef]
68. Seto, K.C.; Ramankutty, N. Hidden linkages between urbanization and food systems. *Science* **2016**, *352*, 943–945. [CrossRef]
69. Aulakh, J.; Regmi, A.; Fulton, J.R.; Alexander, C. *Estimating Post-Harvest Food Losses: Developing a Consistent Global Estimation Framework*; Agricultural and Applied Economics Association: Washington, DC, USA, 2013.
70. Kumar, D.; Kalita, P. Reducing Postharvest Losses during Storage of Grain Crops to Strengthen Food Security in Developing Countries. *Foods* **2017**, *6*, 8. [CrossRef]
71. Rolle, R.S.; Asian Productivity Organization, Food and Agriculture Organization of the United Nations (Eds.) *Postharvest Management of Fruit and Vegetables in the Asia-Pacific Region*; Asian Productivity Organization, Food and Agricultural Organization: Tokyo, Japan; Rome, Italy, 2006; ISBN 978-92-833-2369-3.
72. Salin, V.; Nayga, R.M. A cold chain network for food exports to developing countries. *Int. J. Phys. Distrib. Logist. Manag.* **2003**, *33*, 918–933. [CrossRef]
73. Sheahan, M.; Barrett, C.B. Review: Food loss and waste in Sub-Saharan Africa. *Food Policy* **2017**, *70*, 1–12. [CrossRef] [PubMed]
74. Lara, L.J.; Rostagno, M.H. Impact of Heat Stress on Poultry Production. *Animals* **2013**, *3*, 356–369. [CrossRef] [PubMed]
75. *IMECHE Global Food: Waste Not, Want Not*; Institute of Mechanical Engineers: London, UK, 2013; p. 36.
76. Thamaga-Chitja, J.M.; Hendriks, S.L.; Ortmann, G.F.; Green, M. Impact of maize storage on rural household food security in Northern Kwazulu-Natal. *J. Consum. Sci.* **2004**, *32*. [CrossRef]
77. Tefera, T. Post-harvest losses in African maize in the face of increasing food shortage. *Food Secur.* **2012**, *4*, 267–277. [CrossRef]
78. Parmar, A.; Hensel, O.; Sturm, B. Post-harvest handling practices and associated food losses and limitations in the sweetpotato value chain of southern Ethiopia. *NJAS Wageningen J. Life Sci.* **2017**, *80*, 65–74. [CrossRef]
79. Ray, R.C.; Ravi, V. Post harvest spoilage of sweetpotato in tropics and control measures. *Crit. Rev. Food Sci. Nutr.* **2005**, *45*, 623–644. [CrossRef] [PubMed]
80. Bell, C.H. Fumigation in the 21st century. *Crop Prot.* **2000**, *19*, 563–569. [CrossRef]
81. Flinn, P.W.; Hagstrum, D.W.; Reed, C.; Phillips, T.W. United States department of agriculture–Agricultural Research Service stored-grain areawide Integrated Pest Management program. *Pest Manag. Sci.* **2003**, *59*, 614–618. [CrossRef] [PubMed]

82. Ait-Oubahou, A. Postharvest technologies in sub-saharan Africa: Status, problems and recommendations for improvements. In *Acta Hortic*; International Society for Horticultural Science (ISHS): Leuven, Belgium, 2013; pp. 1273–1282.
83. Ng'ang'a, J.; Mutungi, C.; Imathiu, S.M.; Affognon, H. Low permeability triple-layer plastic bags prevent losses of maize caused by insects in rural on-farm stores. *Food Secur.* **2016**, *8*, 621–633. [CrossRef]
84. Worldbank (Ed.) *Reshaping Economic Geography*; World Development Report; World Bank: Washington, DC, USA, 2009; ISBN 978-0-8213-7607-2.
85. Hummels, D.L.; Schaur, G. Time as a Trade Barrier. *Am. Econ. Rev.* **2013**, *103*, 2935–2959. [CrossRef]
86. Saurav, S.; Potti, R. Cold Chain Logistics in India: A Study. In *Innovative Solutions for Implementing Global Supply Chains in Emerging Markets*; IGI Global: Hershey, PA, USA, 2016; pp. 159–172.
87. Mhazo, N.; Mvumi, B.M.; Nyakudya, E.; Nazare, R.M. The status of the agro-processing industry in Zimbabwe with particular reference to small- and medium-scale enterprises. *Afr. J. Agric. Res.* **2011**, *7*.
88. Walters, S.A.; Groninger, J.W.; Oval Myers, J. Rebuilding Afghanistan's agricultural economy: Vegetable production in Balkh Province. *Outlook Agric.* **2012**, *41*. [CrossRef]
89. Schroeder, H.A. Losses of vitamins and trace minerals resulting from processing and preservation of foods. *Am. J. Clin. Nutr.* **1971**, *24*, 562–573. [CrossRef] [PubMed]
90. *FUSIONS Estimates of European Food Waste Levels*; Food Use for Social Innovation by Optimising Waste Prevention Strategies; European Commission: Brussels, Belgium, 2016.
91. Higgins, K.T. Flash Dryers and Potato Waste. Available online: https://www.foodengineeringmag.com/articles/83689-flash-dryers-and-potato-waste (accessed on 29 November 2018).
92. Neff, R.A.; Spiker, M.L.; Truant, P.L. Wasted Food: U.S. Consumers' reported awareness, attitudes, and behaviors. *PLoS ONE* **2015**, *10*, e0127881. [CrossRef] [PubMed]
93. ABB Ability System 800xA for world's Largest Milk Cooperative Amul in India—ABB Process Control Solutions for Food and Beverage Manufacturing (ABB Process Automation and Control Solutions by Industry). Available online: https://new.abb.com/control-systems/industry-specific-solutions/food-beverage/system-800xa-for-world-s-largest-milk-cooperative-amul-india (accessed on 28 November 2018).
94. Jamal, R. Saving Milk from Spoilage in Rural India. Available online: https://www.theguardian.com/sustainable-business/saving-milk-rural-india-business-sustainable (accessed on 28 November 2018).
95. Plaisier, C.; Sibomana, M.; van der Waal, J.; Clercx, L.; van Wagenberg, C.P.A.; Dijkxhoorn, Y. Approach for Designing Context-Specific, Locally Owned Interventions to Reduce Postharvest Losses: Case Study on Tomato Value Chains in Nigeria. *Sustainability* **2019**, *11*, 247. [CrossRef]
96. Tilahun, S.; Do, S.P.; Mu, H.S.; Cheon, S.J. Review on factors affecting the quality and antioxidant properties of tomatoes. *Afr. J. Biotechnol.* **2017**, *16*, 1678–1687.
97. Okoli, C.A.N.; Erinle, I.D. Factors responsible for market losses of tomato fruits in the Zaria area of Nigeria. *J. Hortic. Sci.* **1989**, *64*, 69–71. [CrossRef]
98. Akangbe, J.A.; Ogundiran, T.J.; Komolafe, S.E.; Ifabiyi, J.O.; Ajibola, B.O. Tomato farmers adoption level of postharvest value addition technology and its constraints in Surulere Area of Oyo State, Nigeria. *J. Agric. Soc. Res.* **2014**, *14*, 91–97.
99. Arah, I.K.; Kumah, E.K.; Anku, E.K.; Amaglo, H. An Overview of Post-Harvest Losses in Tomato Production in Africa: Causes and Possible Prevention Strategies. *J. Biol. Agric. Healthc.* **2015**, *5*, 78–88.
100. Wokoma, E.C.W. Preliminary Report on Diseases of Tomato in Choba, Rivers State. *J. Appl. Sci. Environ. Manag.* **2008**, *12*. [CrossRef]
101. Etebu, E.; Nwauzoma, A.B.; Bawo, D.D.S. Postharvest spoilage of tomato (Lycopersicon esculentum Mill.) and control strategies in Nigeria. *J. Biol.* **2013**, *3*.
102. Ganiyu, S.A.; Popoola, A.R.; Enikuomehin, O.A.; Bodunde, J.G. Influence of grafting on growth and yield performance of two tomato cultivars grown in open field in Nigeria. *J. Plant Pathol.* **2018**, *100*, 43–50. [CrossRef]
103. Ekundayo, E.A.; Ekundayo, F.O.; Osinowo, I.A. Antifungal activities of *Trichoderma viride* and two fungicides in controlling diseases caused by *Sclerotium rolfsii* on tomato plants. *Adv. Appl. Sci. Res.* **2015**, *6*, 12–19.
104. Umeh, V.C.; Onukwu, D. Development of environmentally-friendly tomato insect pest control options under tropical conditions. *J. Veg. Sci.* **2005**, *11*, 73–84. [CrossRef]

105. Olayemi, F.; Adegbola, J.; Bamishaiye, E.; Daura, A. Assessment of post-harvest challenges of small scale farm holders of tomotoes, bell and hot pepper in some local government areas of Kano State, Nigeria. *Bayero J. Pure Appl. Sci.* **2011**, *3*. [CrossRef]
106. Adepoju, A.O. Post-harvest losses and welfare of tomato farmers in Ogbomosho, Osun state, Nigeria. *J. Stored Prod. Postharvest Res.* **2014**, *5*, 8–13.
107. Olosunde, W.A.; Aremu, A.K.; Onwude, D.I. Development of a solar powered evaporative cooling storage system for tropical fruits and vegetables. *J. Food Process. Preserv.* **2016**, *40*, 279–290. [CrossRef]
108. Animashaun, M.O.; Bishop, C.F.H.; Cullum, J. The use of hot water treatment for the control of Alternaria alternata in tomato. In *Proceedings of the Acta Horticulturae*; International Society for Horticultural Science (ISHS): Leuven, Belgium, 2016; pp. 233–240.
109. Raji, O.A.; Oriola, O.K. Packaging and handling methods as sources of mechanical damage in tomatoes in transit. *J. Eng. Appl. Sci.* **2007**, *2*, 1450–1454.
110. Idah, P.A.; Ajisegiri, E.S.A.; Yisa, M.G. Fruits and Vegetables Handling and Transportation in Nigeria. *AU J. Technol.* **2007**, *10*, 175–183.
111. Rosegrant, M.W.; Magalhaes, E.; Valmonte-Santos, R.A.; Mason-D'Croz, D. Returns to Investment in Reducing Postharvest Food Losses and Increasing Agricultural Productivity Growth. Available online: http://www.ifpri.org/publication/returns-investment-reducing-postharvest-food-losses-and-increasing-agricultural (accessed on 22 January 2019).

sustainability

MDPI

Article

Drivers of Vegetable Consumption in Urban Nigeria: Food Choice Motives, Knowledge, and Self-Efficacy

Ireen Raaijmakers [1,*], Harriette Snoek [1], Busie Maziya-Dixon [2] and Thom Achterbosch [1]

[1] Wageningen Economic Research, Wageningen University & Research, P.O. Box 29703, 2502 LS The Hague, The Netherlands; harriette.snoek@wur.nl (H.S.); thom.achterbosch@wur.nl (T.A.)

[2] International Institute of Tropical Agriculture (IITA), Headquarters and West Africa Hub, PMB 5320, Oyo Road, Ibadan 200001, Oyo State, Nigeria; B.Maziya-Dixon@cgiar.org

* Correspondence: ireen.raaijmakers@wur.nl; Tel.: +31-(0)70-3358208

Received: 27 November 2018; Accepted: 11 December 2018; Published: 14 December 2018

Abstract: Objective: This study aimed to provide insights into vegetable consumption behavior of urban Nigerian consumers across different Socio-Economic Classes (SEC), their main food choice motives, and the associations of these motives and other drivers with vegetable consumption. **Methods:** An online survey was conducted in which 1220 women from Lagos (N = 808) and Ibadan (N = 412) metropolis from different SEC participated. **Results:** On average, respondents reported to consume 2.6 portions of vegetables per day. Most vegetables were bought at open and traditional markets, were bought fresh rather than processed, and were consumed cooked. Respondents from the second richest and upper middle SEC consumed most vegetables () and higher SEC consumed a larger variety of vegetables compared to those from lower classes. Respondents who reported to have a higher knowledge of vegetable consumption, had a higher belief in one's own ability to prepare vegetables (self-efficacy), and those that valued the food motive Mood and Health more, reported a higher vegetable intake. **Conclusions:** Vegetable consumption in the studied cities in Nigeria was below recommendations. Increasing knowledge and self-efficacy might be a way to increase consumption, especially in combination with interventions in the food environment and product design focused on the motives Health and Mood, and considering the importance of differences between SEC.

Keywords: vegetable consumption; food choice motives; knowledge; self-efficacy; socio-economic classes; food environment; Nigeria

1. Introduction

Globally more people live in urban than in rural areas and by 2050, 66% of the world's population is projected to be urban. It is expected that more than half of this growth will occur in Africa, whereby Nigeria will stand out [1–3].

Urbanization, in combination with economic and social development, leads to a change in dietary patterns and nutrient intake: This process is called 'nutrition transition' [4], and contributes to increasing health burdens and Non-Communicable Diseases (NCDs) worldwide and especially in developing countries [5]. It is shown that when income rises the consumption of foods associated with a high-quality diet increases (including fruit, vegetables and milk). However, the consumption of products associated with a low quality diet (e.g., fast food, sugar-sweetened beverages) increases even more strongly. When income rises, the budget share of vegetables in total food expenditures declines [6]. With its large and quickly expanding urban population with rapidly accumulating wealth and rapid changes in food habits, Nigeria will face new, multiple and different challenges regarding food security and food systems, health burdens and NCDs [7,8]. While the nutrition transition is still in an early stage in Nigeria [9,10], an increase in the incidence of obesity and related NCDs is already observed in urban and rural areas in Nigeria [10,11].

This study aims to contribute to sustainable healthy eating patterns in urban areas in Low- and Middle-Income Countries (LMICs), in this case urban Nigeria. For this, insights into the underlying determinants of healthy food choices is essential, in the broader context of the food environment. Insights into the motives and barriers that consumer experience, as well as the relation between drivers of behavior, and food purchase and consumption behavior, provides insights into the opportunities and threats for changing the diet.

Within the present study, we focus on vegetable consumption behavior, as vegetable consumption is a commonly recommended element in a balanced and healthy eating pattern. An adequate consumption of vegetables could lead to significant improvements in public health, as it reduces the risk of the development of chronic diseases (e.g., heart diseases, high blood pressure, diabetes and obesity), several cancers and prevents or alleviates several micronutrient deficiencies (e.g., References [12–14]).

Despite its importance the daily consumption of vegetables is insufficient in Nigeria [15–18]. Reliable data on food intake in populations in developing countries (including Nigeria) are scarce and limited, meaning that the mentioned numbers may deviate from actual consumption [6]. In the latest national survey 12.4% of the households reported to consume leafy vegetables, and 16.3% consumed non-leafy vegetables, at least once or twice per week. In urban areas, 11.1% of the households indicated to consume at least once or twice a week leafy vegetables and 16.6% indicated to consume non-leafy vegetables at the same frequency [15].

Several potential barriers to increasing vegetable consumption in urban areas of Nigeria are observed in the literature. Limited year-round availability, affordability, need for convenience, food safety issues and the attraction to the modern or Western lifestyles are mentioned as constraints for healthy food choices by urban middle class consumers in Lagos [19]. Next, cultural beliefs and taboos, and religious beliefs are also found to influence the food choices of consumers [20]. Regarding the vegetable availability, this is region- and season-dependent, and products are mostly eaten fresh, since storage possibilities are few and substantial losses occur due to inadequate preservation and transport. Also at the national level, the availability of vegetables is insufficient to meet the recommended levels of intake [21]. For lower Social Economic Classes (SEC), the affordability of vegetables is problematic due to low purchasing power of households, and necessities to prioritize energy-dense foods which are generally cheaper. Across all urban consumers, including the lower SEC, constraints in the time available for shopping and preparation of food appears to drive consumers towards increased consumption outside the home. Convenient foods are typically high in fat and carbohydrates, and low in vegetables and other nutrient-dense foods. Those seeking to shift to healthier, but convenient alternatives, such as fish, fresh fruits and vegetables, are faced with the increasingly expensive costs of nutritious foods relative to the fast-food alternatives.

Motivation represents the individual's willingness to change behavior [22]. The motivational factors determining an individual's intention are the attitude towards and social norms regarding the behavior [23]. Consumers have different motivations for choosing different types of food products. These so-called food choice motives (FCM) are consumers' motives, reasons or motivations for choosing or eating food products and provide valuable insight into the underlying consumer drivers [24]. They are associated with intake of food products, including vegetable intake [25,26]. Individuals are motivated to behave when they can discern that their self-interest will be served. As such, self-interest is a strong component of motivation [27]. Steptoe and colleagues (1995) developed an instrument to assess the impact of different reasons for making food choices, the Food Choice Questionnaire (FCQ) [24]. This multidimensional scale consists of 36 items, representing both health and non-health related food characteristics, classified into nine different motivational dimensions, measuring the importance of Health, Mood, Sensory appeal, Natural content, Weight control, Familiarity, Price and Ethical concern in food choice. Despite its relevance, the FCQ is mostly applied in high income countries, and to a limited extent in LMIC. The FCQ was applied in one African country, namely

Cape Verde [28]. In this study the identified motives slightly differed with the motives found in the study by Steptoe et al. (1995). The most important motive was Well-being (combination of health and mood), followed by Sensory appeal, Nutritional aspects and diet (combination of nutrition and weight control), Natural content and Price [28].

A well-known model to describe why consumers perform certain behaviors and how these behaviors can be changed is the Motivation—Opportunity—Ability (MOA) model [22]. In this model, people need the motivation and the environmental or contextual opportunity to eat healthily. On top of the motivation and opportunity people need to be able to conduct the intended behavior and therefore ability is the third factor in the MOA model. Ability refers to skills and knowledge to perform behaviors. On the one hand, this refers to more practical skills and knowledge that are needed, such as cooking techniques for preparing vegetables, knowledge on recommended vegetable intake, etc. Subjective knowledge, someone's own perception of his/her level of knowledge has been related to the acceptation and evaluation of products [29]. In LMICs subjective knowledge has been related to food safety [30,31]. Another central concept in the ability literature is self-efficacy or perceived behavior control. This is the belief that someone has the capability to perform a certain behavior [32]. It is specific to a certain behavior, for example someone can be confident about being able to limit his or her intake of sugary drinks, but not to have adequate amounts of fruit intake. Self-efficacy is assumed to reflect true personal abilities and skills and therefore relate to behavior [33], and an important predictor of health behavior change [34]. In the Theory of Planned Behavior for example, perceived behavior control is related to both intention and behavior (e.g., Reference [23]). However, in LMICs some studies applied the self-efficacy scale in domains related to computer use, job search and HIV, but very limited to healthy eating or the consumption of fruits and vegetables.

One of the most relevant socio-demographic variables that influence food choice and consumption is SEC. SEC relates to the persons' position in society and is operationalized in various ways, including income, occupational level, educational level or wealth (assets) [35–37]. Research conducted in high income countries has found that SEC influence food choice and intake. More precisely, it was found that low SEC consumers are more likely to have a less healthy diet and consume less fruit, vegetables and fibers compared to high SEC consumers [38,39]. A study conducted in Uruguay confirmed the influence of income level on the underlying FCM and barriers to the adoption of healthy eating between low and middle SEC. It was found that low SEC respondents described their choices as mainly driven by economic factors and physical needs (e.g., satiety), whereas product-related characteristics (i.e., convenience) were mainly determined for middle SEC respondents [40].

The Present Study

To summarize, current obesity rates and micro-nutrient deficiencies in LMICs underline the need for dietary changes and even more when considering development in urbanization and nutritional transition. Nigeria is one of the countries for which this is particularly true. Motives and the ability to change are important drivers of consumer behavior, but little is known about the importance of these determinants in LMICs in general and in Nigeria in particular.

With this study, we aimed to get more insights regarding the vegetable consumption behavior of urban Nigerians across different SEC, their main FCM, and the associations of these motives and other drivers with vegetable consumption. Specific objectives were to first to describe the local vegetable situation, vegetable intake and purchase behavior; second, to describe the importance of the different FCM for the urban Nigerian consumer; third, to determine the association between motivation and ability (subjective knowledge and self-efficacy) with vegetable intake, and fourth, to investigate differences in vegetable consumption and determinants of consumption across the different SEC. Additionally to the results of this study, implications will be discussed in the food system perspective as consumer and consumer choices cannot be considered separately from the food environment, i.e., the context in which food choices are made. This food environment in turn consists of a large number of chains and actors and is a dynamic system in which influences and trade-offs occur.

2. Materials and Methods

2.1. Study Design and Respondents

Data were collected through an online survey in Lagos and Ibadan, Nigeria. The International research agency IPSOS located in Lagos, Nigeria, collected the data in November 2016. The questionnaire was administered by a trained interviewer using a structured interview reading out loud the questions from the questionnaire on a mobile device, and if applicable supported by show cards. Show cards were developed and used to present the included vegetable answering scales visually to the respondent. The show cards with vegetables were used to have a shared perception of vegetables and the show cards with answering categories were used so that respondents did not have to memorize them. Respondents were recruited across different districts in Lagos and Ibadan, and were only included if they were the key decision makers in the purchase of groceries within their household and when they were one of the persons that bought the groceries. Respondents freely participated and received an incentive after finishing the questionnaire. A pretest of the questionnaire was conducted before the start of the fieldwork.

In total, 1220 female respondents were included in the study. The average age of the sample was 32.4 years (range 18–55). The demographics of the respondents are shown in Table 1.

Table 1. Demographic characteristics of the study sample (N = 1220 females).

		N =	%/Mean
City	Lagos	808	66.2
	Ibadan	412	33.8
Average age (range 18–55)			32.42
Family status	Married/living with partner	896	73.4
	Single	315	25.8
	Divorced	9	0.7
People living in the household	One	41	3.4
	Two	84	6.9
	Three	260	21.3
	Four	385	31.6
	Five	277	22.7
	Six	107	8.8
	Seven or more	66	5.4
Children living in the household	Yes	902	73.9
	No	278	22.8
Income level (monthly net income)	Below N10,000	31	2.5
	N10,001–N20,000	100	8.2
	N20,001–N30,000	117	9.6
	N30,001–N40,000	161	13.2
	N40,001–N50,000	232	19.0
	N50,001–N60,000	145	11.9
	N60,001–N80,000	120	9.8
	N80,001–N100,000	75	6.1
	N100,001–N120,000	35	2.9
	Above N120,001	44	3.6
	Don't know/Refuse	160	13.1
Employment Status	Work full-time	208	17.0
	Work part-time	100	8.2
	Work informally (e.g., seamstress at home)	15	1.2
	Unemployed	144	11.8
	Retired	4	0.3
	Student (not employed)	105	8.6
	Housewife (not employed)	41	3.4
	Self-employed	603	49.4
Ethnicity	Hausa	23	19.9
	Ibo	221	18.1
	Yoruba	919	75.3
	Others	57	4.7

Table 1. *Cont.*

		N =	%/Mean
Religion	Muslim	317	26.0
	Christian	901	73.9
	Others	2	0.1
Socio-Economic Class, based on assets [1]	A—Richest	56	4.6
	B—2nd Richest	105	8.6
	C1—Middle class	129	10.6
	C2—Middle class	246	20.2
	D—2nd Poorest	684	56.1
Key decision maker for grocery shopping	Yes, I am the key decision maker	886	72.6
	Yes, I am one of the key decision makers within our household	334	27.4
Buying groceries	Yes, I am buying groceries for our household	940	77.0
	Yes, I am one of the persons within our household that buys groceries for our household	280	23.0

[1] Respondents were allocated to the different socio-economic classes by their assets, such as ownership of durable, facilities (cooking, water, sanitary), housing and are, educational level, and occupation.

2.2. Measures

2.2.1. Vegetable Buying and Consumption Behavior

To examine the *buying behavior*, questions related to the following topics were included: (i) The form in which vegetables were bought (fresh, dried, canned and frozen), and (ii) buying place (market, street vendor, convenience or small grocery store, and supermarket). The consumption pattern of different types of vegetables (e.g., tomatoes, onions, cucumber, carrots, okra) was explored by asking the consumption frequency with the following categories to choose from: Never, less than once a month, monthly, weekly and daily.

To estimate the respondents' usual *vegetable intake*, the standardized Food Frequency Questionnaire (FFQ), developed and validated by Van Assema et al. (2002) was applied. This FFQ measures usual fruit and vegetable intake. FFQs are considered a suitable tool to rank individuals according to their usual consumption of foods or food categories [41]. As we aimed to identify and rank the SEC on their usual vegetable consumption behavior, the FFQ fitted the best whereas it is less suitable for establishing the level of intake of a population. Respondents indicated their usual consumption frequency (number of days per week) and usual consumption amount of both cooked and raw vegetables (number of servings in spoons). These data were converted in three steps to determine total vegetable intake: Converting intake levels into meaningful data (into portion sizes), multiplying the intake frequency by portion sizes, and adding together the subgroups raw and heated vegetables [42].

2.2.2. Socio-Psychological Determinants

To measure the underlying *food choice motives* the Food Choice Questionnaire (FCQ) developed by Steptoe et al. (1995) was used. The FCQ consists of 36 items, representing both health and non-health related food characteristics. Each item was introduced by the affirmative sentence "It is important to you that the food you eat on a typical day … " followed by each motive, and evaluated by the respondent on a 7-point Likert scale, going from 1 = not important at all to 7 = extremely important. An Exploratory and Confirmatory Factor Analysis (EFA and CFA) were conducted to determine the underlying structure of the questionnaire. The EFA indicated four factors based on the scree plot of the EigenValue, and eight factors based on an EigenValue of above 1.0., with a total explained the variance of 57.4%. These results differ from the nine factors presented by Steptoe et al. (1995). CFA was conducted with nine factors (fixed). The output did not reveal the pattern mix "Rotation failed to converge in 25 iterations (convergence = 0.004)". Iterations of 35 were needed to conduct the CFA with nine factors. This output revealed that the factor "Convenience" would be split into

"Convenience-preparing" and "Convenience-buying". After examining the results, the eight-factor solution was chosen, as the four-factor structure did not provide a clear pattern. All items loaded 0.30 or more on one of the factors. Regarding the sample size of this study, this is enough to have practical significance [43]. Five of the items that loaded more than 0.30 on more than one factor have been deleted. These items were "Looks nice", "Can be bought in shops close to where I live or work", "Is easily available in shops and supermarkets", "Has a pleasant texture", and "Is like the food I ate when I was a child". Next one factor was deleted, as it only included one item "Is high in fibre and roughage". The items factor loading and Cronbach's α of the remaining items are shown in Table 2.

Table 2. Factor loading and Cronbach's α for Food Choice Questionnaire (FCQ).

Factor	Item	Factor Loading	Cronbach's α
Mood	Keeps me awake/alert	0.783	0.83
	Cheers me up	0.713	
	Helps me relax	0.638	
	Helps me to cope with life	0.613	
	Is good for my skin/teeth/hair/nails etc.	0.514	
	Makes me feel good	0.495	
	Helps me to cope with stress	0.486	
	Is packaged in an environmentally friendly way	0.404	
	Smells nice	0.403	
Convenience	Is easy to prepare	0.850	0.87
	Can be cooked very simply	0.825	
	Takes no time to prepare	0.803	
Ethical concern	Comes from countries I approve politically	0.887	N.A [1]
	Has the country of origin clearly marked	0.880	
Natural Content and Weight Control	Is low in fat	−0.735	0.83
	Is low in calories	−0.646	
	Helps me control my weight	−0.637	
	Contains no artificial ingredients	−0.606	
	Contains natural ingredients	−0.575	
	Contains no additives	−0.551	
Price	Is not expensive	0.846	0.73
	Is cheap	0.809	
	Is good value for money	0.597	
Familiar	Is familiar	0.790	N.A [1]
	Is what I usually eat	0.776	
Health	Contains a lot of vitamins and minerals	0.752	0.69
	Keeps me healthy	0.719	
	Is nutritious	0.716	
	Is high in protein	0.512	

[1] As the factor consists of two items it was not possible to calculate the Cronbach's α.

The respondents' own perception of their knowledge about vegetables was measured with the three item-scale *subjective knowledge* scale developed by Aertsens and colleagues (2011) [44]. The scale included the following three items: "In comparison with an average person you know a lot about vegetables", "You know a lot about how to judge the quality of vegetables", and "People who know you, consider you as an expert in the field of vegetables". Answers were given on a 7-point Likert scale ranging from 1 = strongly disagree to 7 = strongly agree. The Cronbach's α in this sample was 0.83.

The respondents' beliefs in their own ability to prepare and increase their vegetable consumption (*self-efficacy*) was measured with the following nine items: "You know how to prepare all vegetables", "You have a cook who prepares the vegetables for you", "You can distinguish vegetables of good quality from vegetables with a low quality", "You like all kind of vegetables", "You lack cooking skills to make all kind of vegetables", "You feel stressed when you have to prepare all kind of vegetables", "A lot of vegetables are difficult to cook", "You are too busy to make meals with vegetables", and "You do not believe that vegetables are health". Answers were given on a 7-point Likert scale ranging from 1 = strongly disagree to 7 = strongly agree. After recoding the negatively formulated items,

EFA indicated two-factor structure, based on the scree plot of the EigenValue, with a total explained the variance of 53.3%. After examining the results, it was decided to use one factor structure, as one factor included all the recorded items. The following items were deleted as this would increase the Cronbach's α: "You have a cook who prepared the vegetables for you" (from α0.56 to α0.71) and "You like all kinds of vegetables" (from α0.71 to α0.72). The final scale consists of seven items, with a Cronbach's α of 0.72.

3. Results

3.1. Vegetable Buying and Consumption Behavior

The vast majority (97.7%) of the respondents indicated to buy their vegetables; only 2.3% of the sample (n = 28) indicated to both buy *and* grow their own vegetables, and no one relied only on self-grown vegetables. 99.8% of the respondents indicated to buy fresh vegetables for their household. Fresh vegetables were most often bought at open markets (58.0%), followed by street vendors (19.6%) and convenience stores/small grocery stores (19.2%). Supermarkets were the least likely outlet for vegetables (3.2%). Similar results were found for canned, dried and frozen vegetables, although frozen and canned vegetables were bought relatively more in supermarkets and convenience stores.

All respondents indicated to consume vegetables with on average 17.8 portions per week, 2.55 portions a day. Looking at the average consumption per week, the respondents indicated to consume 12.9 (SD = 8.0) of cooked and 4.9 (SD = 5.8) of raw vegetables. One portion (serving spoon) equals 50 g. Cooked vegetables were consumed on a daily basis by 44.3% of the respondents whereas for raw vegetables this was 6.5%. Almost all respondents consumed fresh vegetables (99.8%) and a majority consumed canned vegetables (58.9%), whereas dried and especially frozen vegetables were consumed by a smaller percentage of the population (35.8% and 13.3% respectively). Tomatoes, onions, small sweet peppers, hot peppers, carrots and green leafy vegetables were the most frequently consumed types of vegetables (consumed by >90% of the respondents). Also, bell peppers, cucumber, okra, baby corn, cabbage, green beans, and garden egg were consumed by a large majority (>70%) of the respondents. Lettuce was consumed by 43.9% and pumpkin by 33.1% of the respondents. Other vegetables (i.e., broccoli, beet roots, karalla, and zucchini) were consumed only by a minority of the sample (<10%).

Significant differences regarding vegetable consumption were found between the different SEC for both heated ($F(4, 1219) = 3.1, p < 0.01$) and raw vegetables ($F(4, 1219) = 11.9, p < 0.001$). Post-hoc analyses showed that respondents from the second richest and upper middle class (SEC B and C1) consumed more vegetables compared to the poor (SEC C2 and D) which was mostly attributable to the consumption of raw vegetables (see Table 3). The rich and upper middle class (SEC A, B and C1) were also more likely to consume frozen and canned vegetables than the poor (SEC C2 and D), and they consumed a greater variety of vegetables, since they consume more often the less traditional vegetable species.

3.2. Socio-Psychological Determinants

Regarding the FCM, overall the motive Health was considered the most important (M = 6.36). The motives Mood, Natural, Price, Convenience and Familiar all scored high, more specifically between 5.91 and 5.31 on average (see Table 3). Ethical concerns were considered the least important motive. The mean scores for subjective knowledge were M = 5.66, and self-efficacy was M = 5.58.

Significant differences were found between the SEC regarding the FCM Price, Mood, and Familiar. Price was considered less important in the middle and highest SEC compared to the lower SEC. Familiar and Mood were most important for the middle and less important for the highest SEC. Next, small, but significant, differences were found between the SEC groups in perceived knowledge. The middle-class group reported that they had a higher knowledge of vegetable consumption compared to the lower SEC.

Table 3. Vegetable consumption and the different socio-economic classes.

		Total	A	B	C1	C2	D	F(4, 1219)
Vegetable intake [1]								
Total vegetable intake	Mean	17.81	20.21 [ab]	21.15 [a]	21.62 [a]	16.71 [b]	16.78 [b]	9.08 ***
	SD	11.14	13.73	15.6	13.07	10.30	4	
Intake heated vegetables	Mean	12.90	13.45	14.40	14.50	12.21	12.57	3.05 **
	SD	8.00	9.37	9.32	9.35	7.55	7.49	
Intake raw vegetables [1]	Mean	4.91	6.76 [abd]	6.75 [ab]	7.12 [ab]	4.50 [ade]	4.21 [e]	11.89 ***
	SD	5.82	6.623	8.86	6.39	5.58	4.89	
Food Choice Motives [2]								
Health	Mean	6.36	6.35	6.46	6.40	6.30	6.35	2227
	SD	0.50	0.39	0.46	0.48	0.51	0.52	
Mood	Mean	5.86	5.61 [c]	6.01 [a]	5.93 [ab]	5.86 [b]	5.85 [b]	4311 **
	SD	0.63	0.77	0.58	0.64	0.56	0.64	
Natural content and	Mean	5.78	5.76	5.96	5.84	5.77	5.75	1682
Weight control	SD	0.80	0.66	0.75	0.78	0.76	0.84	
Price	Mean	5.69	5.38 [c]	5.69 [ab]	5.52 [bc]	5.64 [ab]	5.77 [a]	4318 **
	SD	0.89	0.84	0.95	0.90	0.90	0.87	
Convenience	Mean	5.63	5.94	5.69	5.77	5.58	5.59	2173
	SD	1.09	0.65	1.16	0.98	1.07	1.14	
Familiar	Mean	5.61	5.32 [d]	5.78 [ab]	5.79 [a]	5.54 [cd]	5.60 [bc]	3.495 **
	SD	0.96	0.99	0.90	0.87	0.98	0.97	
Ethical concern	Mean	4.24	4.04	4.48	4.29	4.33	4.18	1291
	SD	1.59	1.60	1.56	1.70	1.53	1.59	
Subjective knowledge [3]	Mean	5.66	5.91 [ab]	5.76 [ab]	5.87 [a]	5.57 [b]	5.62 [b]	2375 *
	SD	1.19	1.04	1.07	1.16	1.26	1.19	
Self-efficacy [3]	Mean	5.58	5.34	5.72	5.67	5.56	5.57	2195
	SD	0.85	0.85	0.92	0.89	0.90	0.81	

[1]: Number of self-reported vegetable portions per week. [2]: 7-point Likert scale is applied ranging from 1 = not important at all to 7 = extremely important. [3]: 7-point Likert scale is applied ranging from 1 = strongly disagree to 7 = strongly agree. * $p < 0.05$, ** $p < 0.01$, *** $p < 0.001$. [abcde] different letters indicate a significant difference between clusters.

3.3. Determinants of Vegetable Consumption

Respondents who reported to have a higher knowledge of vegetable consumption, who valued the food motive Mood and Health more, and who also had a higher self-efficacy reported a higher vegetable intake. These associations were found after controlling for the positive relation between a higher household size and a higher SEC status with vegetable intake. Age did not have an additional association with vegetable intake. Although significant, the associations however were weak and in total only 15.3% of the variance in food intake was explained by the variables (see Table 4).

Table 4. Results stepwise regression analysis on the drivers of vegetable intake.

	Standardized Beta Coefficients	*t*-Value	*p*-Value	R^2 Change
	First step			
(Constant)		13.181	0.000	
Household size	0.093	3.298	0.001	
SEC A [1]	0.058	2.023	0.043	0.038, F(5, 1214) = 9.50, $p < 0.001$
SEC B [1]	0.110	3.813	0.000	
SEC C1 [1]	0.132	4.546	0.000	
SEC C2 [1]	−0.002	−0.057	0.955	
	Second step [2]			
Mood	0.316	8.608	0.000	
Convenience	0.043	1.349	0.177	
Ethical concern	−0.023	−0.782	0.434	
Natural content and weight control	−0.006	−0.154	0.877	0.114, F(9, 1205) = 17.99, $p < 0.001$
Price	−0.049	−1.573	0.116	
Familiar	0.038	1.305	0.192	
Health	−0.083	−2.480	0.013	
Knowledge	0.085	2.883	0.004	
Self−efficacy	0.077	2.455	0.014	

[1] Socio-economic status (SEC) as a dummy variable with the lowest SEC (D) as the reference. [2] Beta's are reported for the step when the variable was introduced.

4. Discussion

4.1. Summary of the Main Results

This study provided insights into the vegetable consumption behavior of urban Nigerians across different SEC. And it adds insights to the existing literature as it identified the main FCM of urban Nigerians, and the associations of these motives and other drivers with vegetable consumption.

On average, the total consumption of vegetables was 2.55 portions per day whereas it is recommended to eat at least 4 portions (200 g). Vegetables were considered a standard element of meals, but a limited variety of vegetables was commonly consumed, both in terms of types of vegetables eaten, degree of processing (i.e., mostly fresh), and outlets (i.e., mostly traditional open markets). The respondents in the higher SEC consumed a greater variety of vegetables, especially the ones that are considered exotic (e.g., broccoli, cauliflower) and they also ate more raw vegetables. Regarding the drivers of vegetable consumption, we found support for the importance of motives and ability variables. Respondents who reported a higher knowledge of vegetables and who had a higher belief in ones' own ability to prepare vegetables (self-efficacy) reported a higher vegetable intake. Also, those who valued the FCM Mood and Health more, reported a higher vegetable intake. Health was considered the most important FCM by the respondents, followed by Mood, Natural, Price, Convenience and Familiar while Ethical concern was considered least important. Implications of these findings will be considered in detail below.

4.2. Implications of the Main Results

The average vegetable consumption was below recommended levels. This is in accordance with our expectations, as previous research revealed a low average vegetable consumption (e.g., References [15–17]). Reliable information on vegetable consumption in Nigeria is scarce and the available data reveals a large range in the estimated consumption amount of vegetables from 59 g to 170 g [16,17]. This large range might be due to the influence of seasonality or due to different definitions of vegetables in different studies (e.g., green leafy vegetables only versus all vegetables). Only one study explicitly mentioned in its discussion that tomatoes, onions and peppers were excluded, because of their ubiquitous use in the preparation of most of the soups in the Nigerian culture [18]. The results of our pilot study showed that respondents have different interpretations of what they consider as vegetables. For example, tomatoes and onions were considered spices, rather than vegetables, whereas spinach or other leafy green vegetables were considered vegetables. Overall, this result indicates that it is of great importance to define and categorize the term vegetables in surveys. However, it should be taken into account that it is of great importance to tailor the applied questionnaire or instrument as much as possible to the local perceptions and definitions of vegetables. On the other hand, in data collection it is also of great importance that the used definition of vegetables is clearly marked. In our study we tried to overcome this challenge, by showing a clear explanation our definition of vegetables by including pictures of the vegetables that were seen at the open local markets and in supermarkets and other outlets during a previous trip. Regardless of how vegetables are defined in this survey, results indicate that vegetable consumption should be increased across all the SEC.

The limited variety of vegetable intake should be considered in interventions, especially for the low SEC groups and at the same time might provide opportunities, for example in terms of processed vegetables (i.e., dried vegetables). In the dry season, dried vegetables might be a good suggestion as the availability of fresh vegetables is lower and prices are higher [17]. Moreover, future research could focus on the specific motives to buy fresh or processed vegetables, the selection for the more traditional or exotic ones, and the specific motives to purchase vegetables at a specific outlet (e.g., open market, supermarket or small convenience store), this to get more insights into ways to increase variety. More specifically, we found that richer and upper middle-class respondents consumed more vegetables and especially more raw vegetables, a larger variety of vegetables, and more canned vegetables compared those respondents that were in the poorer SEC groups. This indicates that there are opportunities to increase the intake of a more varied vegetable basket and preparing and processing methods. On the other hand, it is unclear what the motives and barriers are behind those SEC differences. Do consumers from lower SEC have different attitudes and beliefs, or does the availability or accessibility differs, or both? These research questions might be of interest in future research.

When looking at the FCM, it was shown that the motives Health, Mood, Natural and Weight control were considered the most important motives in making food choices. Ethical concern was considered least important. The order of importance of the FCM was broadly in line with other studies that used the original FCQ (e.g., Reference [45,46]), with the exception that the Nigerian consumer considered the motive Familiar as more important than the European consumer (M = 5.31 versus M = 2.85) [45]. Food consumption practices in Nigeria are found to be influenced by many social-cultural factors, including cultural traditions, food beliefs or religious circumstances [47]. Future interventions and product design should consider Health and other motives important to consumers. To stimulate vegetable consumption the motives Health and Mood should be integrated into an intervention or product design as they are related to vegetable consumption. For example, the motive Health could be further operationalized in mentioning the health benefits of vegetables.

Next, it is important to realize that the revealed eight-factor structure in this study is not in correspondence with the nine-factor structure presented by Steptoe et al. (1995). This result is in line with other studies in LMICs that applied the FCQ [46]. Therefore, a good comparison of the most and least important FCM between countries and over time is not possible as the results of the CFA and EFA differs between the original FCQ and the ones that are conducted in the developing countries. This due to the fact that some other studies added extra items or conducted a different statistical test. A review by Cunha et al. (2018) showed that several studies have shown the invariance of the FCQ across cultures, while others present the need for adaptations of the FCQ [46]. Also, for Nigeria, the original FCQ might not fit the local context. There is some research conducted on the different motives that Nigerian consumers have. Culture, food safety/risk, healthiness and convenience are considered important motives for selecting a certain food product [19]. Future research should focus on the validity of the FCQ for the Nigerian context and context-specific motives might be useful to further improve the measurement scale. For the other drivers, the results were in line with previous studies that showed an association between self-efficacy and subjective knowledge with food intake. Increasing ability aspect of vegetable intake seems to be a promising way to move forward. Ability should be considered in combination with motivation and opportunity; the so-called MOA model [22]. While abilities are the individual's skills and/or knowledge that enable behavior change [27] and motivation represents the individual's willingness to change behavior; opportunity is the environmental or contextual mechanisms that enable behavior change, and ability. Collective changes in consumer behavior can open pathways to more sustainable food systems that enhance food security and nutrition and health. Therefore, we discuss the implications of the results from a food systems perspective in Section 4.4.

4.3. Study Limitations

This study has limitations. First, and most importantly, vegetable consumption is based on self-reported data. The results should be interpreted carefully, as we lack insights on how reliable self-reported vegetable consumption is. In addition, the FFQ was used which is a valid method to measure vegetable intake at a level where consumers can be ranked, but it is less suitable for establishing more detailed information on intake and quantification of intake [41]. Respondents might have over- or underreported their consumption. In developing countries consumers might be more prone to report socially desirable aspects, rather than real behavior [41]. However, respondents also might have underreported their vegetable consumption as a result of short memory or low educational level. However, in this study, this is less of a problem because we only compare the different SEC [41] and look at associations between drivers and intake while we do not draw conclusions in terms of the actual intake. Another limitation in the interpretation of the study results was the focus on vegetables rather than meals and preparation and the limited geographical position. Vegetables are a crucial part of a healthy diet and often consumption is below recommendation which justifies the focus on vegetables. On the other hand, looking at combinations of vegetables with other food products, and studying preparation methods will help to formulate implications of the outcomes in relation to healthy eating patterns. Similarly, it is not feasible within the scope of the project to include the whole country, at the same time we must be aware that the results might differ in other regions of Nigeria and should try to gain some insights regarding the degree of these differences. Additionally, the timing of the research in the wet season, the relatively high-income level of our sample, and the focus on urban areas with relatively high availability of vegetables throughout the year limits the generalizability of the results.

4.4. Food Systems—Implications and Future Direction

To be able to effectively address current nutrition challenges, research and intervention strategies on consumer healthy eating behavior should not be considered in isolation, but in a broad setting. Dietary behavior related to consumer purchase behavior is shaped in the context of the food environment; food environments, in turn, are shaped by the activities of all actors in the food system [48]. The food system approach considers all the different activities in our food systems from production to consumption (and the relationships between them), as well as the outcomes of these activities on a range of domains, such as food security (including nutrition), socio-economics (income, employment) and the environment/climate (biodiversity, climate) [49]. In this way it provides good insights into particular parts of the food system and insights into opportunities for the development of food system interventions and effective entry points for longer-term policy [6,50]. In urban areas in Africa, food systems rapidly transform in many ways with changes in food supply (food environment) and food demand (consumers) [8]. Regarding the demand side, shifts in preferences, attitudes regarding foods, income and household structures will occur [8]. Consumers are part of the system and developed certain preferences through their knowledge, available time, resources (purchasing power), age, sex, culture, religion, etc. These preferences provide an entry point for the different dimensions of a food system: The food environment can be changed to influence consumer behavior at the level of production (product characteristics such as taste), retail (nudging, logos, prices) or governance (directly through regulations or indirectly through price and availability). In turn, changing preferences will again influence the system and might have side-effects on other parts of the system (e.g., environmental impact). The results of the present study provides insights into consumer behavior that could be used to develop such kind of intervention strategies, in particular the importance of health and convenience for vegetable consumption. The motive *health* is considered one of the most important motives in making food choices. Other research conducted in Nigeria confirms that urban Nigerians have become increasingly concerned about the amount of fat and sugar in their diet and the adverse health effects resulting from this [19]. An example of an intervention that affected the different dimensions within the whole food system is the Mexico sugar-sweetened beverage tax, The tax (enabling environment) specifically targeted the food environment (affordability aspect), and had an impact on the consumption of sugar containing beverages (food supply chain) and changed consumer choices (consumer characteristics) [6,51]. *Convenience* was a main barrier for vegetable intake in our study. Hollinger and Staatz (2015) showed that there is a growing need for convenience foods; there is less time to buy and prepare foods [19]. In the United States research has shown that mobile produce markets emerged as a strategy to improve vegetable access and consumption among lower-income consumers (food supply chain and enabling environment) [52]. The results of the study indicate that also in urban Nigeria such an intervention might possibly increase accessibility and consumption of vegetables.

5. Conclusions

The burden of NCDs is on the rise in Nigeria. One of the major contributors to the risks of the NCDs is poor eating habits. Current vegetable consumption is below recommendations and this study provides insights into drivers to increase consumption in the context of Nigeria's burgeoning city regions of Lagos and Ibadan. The current consumption patterns also show a low variety in terms of vegetables types, outlets, and types of processing of vegetables. Increasing knowledge and the belief in one's own ability to prepare vegetables (self-efficacy) might be a way to increase vegetable consumption, especially in combination with interventions and product design focused at the motives Health and Mood and taking into account the importance of Price and differences between SEC. In the design of an intervention and/or experiment it would be more beneficial to target on specific SEC and consider that these groups differ in their vegetable consumption and purchase behavior, FCM, and subjective knowledge. For example, for the low SEC an intervention could focus on the limited variety of vegetable intake. This intervention should then also integrate the FCM, and other drivers that are relevant for the vegetable intake of low SEC consumers. Another implication of the study is that overall in Nigerian studies vegetables should be further defined as consumers have different definitions in mind regarding vegetables. Next, the importance of FCM in food choices is well known, however, to measure them there is a need for an FCQ that fits the local context.

Author Contributions: All authors worked on the conceptualization of the study. I.R. and H.S. analyzed the data and wrote the paper. All the authors reviewed and edited the manuscript.

Funding: Funding has been generously provided by the Netherlands Ministry of Agriculture, Nature and Food Quality (grant WUR-KB22-003-001). This work and follow-up research is performed in collaboration with the CGIAR research program Agriculture for Nutrition and Health, flagship *Food systems for healthier diets*.

Acknowledgments: We thank Gemma Tacken for her contributions to the design and efforts in the project management during the project. We thank seminar participants at IITA in 2017, and ANH Academy 2018 for their useful feedback.

Conflicts of Interest: The authors declare no conflict of interest. The funders had no role in the design of the study; in the collection, analyses, or interpretation of data; in the writing of the manuscript, and in the decision to publish the results.

References

1. United Nations, Department of Economic and Social Affairs, Population Division. *World Urbanization Prospects: The 2014 Revision, Highlights (ST/ESA/SER.A/352)*; United Nations: San Francisco, CA, USA, 2014.
2. United Nations, Department of Economic and Social Affairs, Population Division. *World Population Prospects: The 2015 Revision, Key Findings and Advance Tables*; Working Paper No. ESA/P/WP.241; United Nations: San Francisco, CA, USA, 2015.
3. Bloch, R.; Monroy, J.; Fox, S.; Ojo, A. *Urbanisation and Urban Expansion in Nigeria*; Urban Research Nigeria (URN) Research Report; ICF International: London, UK, 2015.
4. Popkin, B.M.; Adair, L.S.; Ng, S.W. Global nutrition transition and the pandemic of obesity in developing countries. *Nutr. Rev.* **2012**, *70*, 3–21. [CrossRef] [PubMed]
5. Gayathri, R.; Ruchi, V.; Mohan, V. Impact of nutrition transition and resulting morbidities on economic and human development. *Curr. Diabetes Rev.* **2017**, *13*, 452–460. [CrossRef] [PubMed]
6. Global Panel on Agriculture and Food Systems for Nutrition. *Food Systems and Diets: Facing the Challenges of the 21st Century*; Global Panel on Agriculture and Food Systems for Nutrition: London, UK, 2016.
7. Oyewole, O.E.; Atinmo, T. Nutrition transition and chronic diseases in Nigeria. *Proc. Nutr. Soc.* **2015**, *74*, 460–465. [CrossRef] [PubMed]
8. Seto, K.C.; Ramankutty, N. Hidden linkages between urbanization and food systems. *Science* **2016**, *352*, 943–945. [CrossRef] [PubMed]
9. Abrahams, Z.; Mchiza, Z.; Steyn, N.P. Diet and mortality rates in Sub-Saharan Africa: Stages in the nutrition transition. *BMC Public Health* **2011**, *11*, 801. [CrossRef] [PubMed]
10. Steyn, N.P.; Mchiza, Z.J. Obesity and the nutrition transition in Sub-Saharan Africa. *Ann. N. Y. Acad. Sci.* **2014**, *1311*, 88–101. [CrossRef]
11. Maiyaki, M.B.; Garbati, M.A. The burden of non-communicable diseases in Nigeria; in the context of globalization. *Ann. Afr. Med.* **2014**, *13*, 1–10. [CrossRef]

12. Aune, D.; Giovannucci, E.; Boffetta, P.; Fadnes, L.T.; Keum, N.; Norat, T.; Greenwood, D.C.; Riboli, E.; Vatten, L.J.; Tonstad, S. Fruit and vegetable intake and the risk of cardiovascular disease, total cancer and all-cause mortality—A systematic review and dose-response meta-analysis of prospective studies. *Int. J. Epidemiol.* **2017**, *46*, 1029–1056. [CrossRef]
13. Hung, H.C.; Joshipura, K.J.; Jiang, R.; Hu, F.B.; Hunter, D.; Smith-Warner, S.A.; Colditz, G.A.; Rosner, B.; Spiegelman, D.; Willett, W.C. Fruit and vegetable intake and risk of major chronic disease. *J. Natl. Cancer Inst.* **2004**, *96*, 1577–1584. [CrossRef]
14. Yahia, E.M. The Contribution of Fruit and Vegetable Consumption to Human Health. In *Fruit and Vegetable Phytochemicals*; De La Rosa, L.A., Alvarez-Parrilla, E., González-Aguilar, G.A., Eds.; John Wiley & Sons, Ltd.: Hoboken, NJ, USA, 2017; pp. 3–51.
15. Maziya-Dixon, B. *Nigeria Food Consumption and Nutrition Survey 2001–2003: Summary*; International Institute of Tropical Agriculture: Ibadan, Nigeria, 2004.
16. Chubike, N.E.; Okaka, J.; Okoli, E.C. Evaluation of vegetable consumption in South Eastern Nigeria. *Int. J. Nutr. Metab.* **2013**, *5*, 57–60.
17. Hart, A.; Azubuike, C.; Barimalaa, I.; Achinewhu, S. Vegetable consumption pattern of households in selected areas of the old Rivers State in Nigeria. *Afr. J. Food Agric. Nutr. Dev.* **2005**, *5*, 1–18.
18. Olatona, F.; Sosanya, A.; Sholeye, O.; Obrutu, O.; Nnoaham, K. Knowledge of fruits and vegetables, consumption pattern and associated factors among adults in Lagos State, Nigeria. *Res. J. Health Sci.* **2018**, *6*, 50–62. [CrossRef]
19. Hollinger, F.; Staatz, J.M. *Agricultural Growth in West Africa, Market and Policy Drivers*; FAO: Rome, Italy, 2015.
20. Ijewere, A.A.; Odia, E.O. Cultural Influences on Product Choice of the Nigerian Consumer. *Indian J. Econ. Bus.* **2012**, *11*, 139–155.
21. Food and Agriculture Organization. Balance Sheet Nigeria 2017. Available online: http://www.fao.org/faostat/en/#data/FBS (accessed on 13 December 2018).
22. ThØgersen, J. Understanding of consumer behaviour as a prerequisite for environmental protection. *J. Consum. Policy* **1995**, *18*, 345–385.
23. Ajzen, I.; Fishbein, M. *Understanding Attitudes and Predicting Social Behaviour*; Prentice-Hall: Upper Saddle River, NJ, USA, 1980.
24. Steptoe, A.; Pollard, T.M.; Wardle, J. Development of a measure of the motives underlying the selection of food: The food choice questionnaire. *Appetite* **1995**, *25*, 267–284. [CrossRef] [PubMed]
25. Konttinen, H.; Sarlio-Lähteenkorva, S.; Silventoinen, K.; Männistö, S.; Haukkala, A. Socio-economic disparities in the consumption of vegetables, fruit and energy-dense foods: The role of motive priorities. *Public Health Nutr.* **2013**, *16*, 873–882. [CrossRef]
26. Appleton, K.M.; Hemingway, A.; Saulais, L.; Dinnella, C.; Monteleone, E.; Depezay, L.; Morizet, D.; Armando Perez-Cueto, F.J.; Bevan, A.; Hartwell, H. Increasing vegetable intakes: Rationale and systematic review of published interventions. *Eur. J. Nutr.* **2016**, *55*, 869–896. [CrossRef] [PubMed]
27. Rothschild, M.L. Carrots, sticks, and promises: A conceptual framework for the management of public health and social issue behaviors. *J. Mark.* **1999**, *63*, 24–37. [CrossRef]
28. Cabral, D.; de Almeida, M.D.V.; Cunha, L.M. Food Choice Questionnaire in an African country–Application and validation in Cape Verde. *Food Qual. Prefer.* **2017**, *62*, 90–95. [CrossRef]
29. Flynn, L.R.; Goldsmith, R.E. A short, reliable measure of subjective knowledge. *J. Bus. Res.* **1999**, *46*, 57–66. [CrossRef]
30. Lagerkvist, C.J.; Hess, S.; Okello, J.; Karanja, N. Consumer willingness to pay for safer vegetables in urban markets of a developing country: The case of Kale in Nairobi, Kenya. *J. Dev. Stud.* **2013**, *49*, 365–382. [CrossRef]
31. Ngigi, M.; Okello, J.; Lagerkvist, C.; Karanja, N.; Mburu, J. Urban consumers' willingness to pay for quality of leafy vegetables along the value chain: The case of Nairobi Kale consumers, Kenya. *Int. J. Bus. Soc. Sci.* **2011**, *2*, 208–216.
32. Bandura, A. Self-efficacy: Toward a unifying theory of behavioral change. *Psychol. Rev.* **1977**, *84*, 191–215. [CrossRef] [PubMed]
33. Brug, J. Determinants of healthy eating: Motivation, abilities and environmental opportunities. *Fam. Pract.* **2008**, *25*, i50–i55. [CrossRef] [PubMed]

34. Schwarzer, R. Modeling health behavior change: How to predict and modify the adoption and maintenance of health behaviors. *Appl. Psychol.* **2008**, *57*, 1–29. [CrossRef]
35. Galobardes, B.; Shaw, M.; Lawlor, D.A.; Lynch, J.W.; Smith, G.D. Indicators of socioeconomic position (part 1). *J. Epidemiol. Community Health* **2006**, *60*, 7–12. [CrossRef]
36. Galobardes, B.; Shaw, M.; Lawlor, D.A.; Lynch, J.W.; Smith, G.D. Indicators of socioeconomic position (part 2). *J. Epidemiol. Community Health* **2006**, *60*, 95–101. [CrossRef]
37. Adler, N.E.; Newman, K. Socioeconomic disparities in health: Pathways and policies. *Health Aff.* **2002**, *21*, 60–76. [CrossRef]
38. Chen, E.; Miller, G.E. Socioeconomic status and health: Mediating and moderating factors. *Ann. Rev. Clin. Psychol.* **2013**, *9*, 723–749. [CrossRef]
39. Deshmukh-Taskar, P.; Nicklas, T.A.; Yang, S.J.; Berenson, G.S. Does food group consumption vary by differences in socioeconomic, demographic, and lifestyle factors in young adults? The Bogalusa Heart Study. *J. Am. Diet. Assoc.* **2007**, *107*, 223–234. [CrossRef]
40. Ares, G.; Machín, L.; Girona, A.; Curutchet, M.R.; Giménez, A. Comparison of motives underlying food choice and barriers to healthy eating among low medium income consumers in Uruguay. *Cad. Saude Publ.* **2017**, *33*, e00213315. [CrossRef] [PubMed]
41. Thompson, F.E.; Subar, A.F. Dietary Assessment Methodology. In *Nutrition in the Prevention and Treatment of Disease*, 3rd ed.; Elsevier: Amsterdam, The Netherlands, 2013; pp. 5–46.
42. Van Assema, P.; Brug, J.; Ronda, G.; Steenhuis, I.; Oenema, A. A short dutch questionnaire to measure fruit and vegetable intake: Relative validity among adults and adolescents. *Nutr. Health* **2002**, *16*, 85–106. [CrossRef] [PubMed]
43. Hair, J.F., Jr.; Anderson, R.E.; Tatham, R.L.; Black, W.C. *Multivariate Data Analysis*, 4th ed.; Prentice Hall: London, UK, 1995.
44. Aertsens, J.; Mondelaers, K.; Verbeke, W.; Buysse, J.; Van Huylenbroeck, G. The influence of subjective and objective knowledge on attitude, motivations and consumption of organic food. *Br. Food J.* **2011**, *113*, 1353–1378. [CrossRef]
45. Markovina, J.; Stewart-Knox, B.J.; Rankin, A.; Gibney, M.; de Almeida, M.D.V.; Fischer, A.; Kuznesof, S.A.; Poínhos, R.; Panzone, L.; Frewer, L.J.; et al. Food4Me study: Validity and reliability of Food Choice Questionnaire in 9 European countries. *Food Qual. Prefer.* **2015**, *45*, 26–32. [CrossRef]
46. Cunha, L.M.; Cabral, D.; Moura, A.P.; de Almeida, M.D.V. Application of the Food Choice Questionnaire across cultures: Systematic review of cross-cultural and single country studies. *Food Qual. Prefer.* **2018**, *64*, 21–36. [CrossRef]
47. Onuorah, C.; Ayo, J. Food taboos and their nutritional implications on developing nations like Nigeria—A review. *Nutr. Food Sci.* **2003**, *33*, 235–240. [CrossRef]
48. Meerman. *Food Environments: Linking Food and Agriculture Policies to Diet Quality. Consultative Workshop on Leveraging FAO's Policy Mapping and Analysis Work to Promote Nutrition*; Food and Agriculture Organization of the United Nations: Rome, Italy, 2015.
49. van Berkum, S.; Dengerink, J.; Ruben, R. *The Food Systems Approach: Sustainable Solutions for a Sufficient Supply of Healthy Food*; Wageningen Economic Research: The Hague, The Netherlands, 2018.
50. Waterlander, W.E.; Mhurchu, C.N.; Eyles, H.; Vandevijvere, S.; Cleghorn, C.; Scarborough, P.; Swinburn, B.; Seidell, J. Food futures: Developing effective food systems interventions to improve public health nutrition. *Agric. Syst.* **2018**, *160*, 124–131. [CrossRef]
51. Colchero, M.A.; Popkin, B.M.; Rivera, J.A.; Ng, S.W. Beverage purchases from stores in Mexico under the excise tax on sugar sweetened beverages: Observational study. *BMJ* **2016**, *352*, h6704. [CrossRef]
52. Hsiao, B.S.; Sibeko, L.; Troy, L.M. A Systematic Review of Mobile Produce Markets: Facilitators and Barriers to Use, and Associations with Reported Fruit and Vegetable Intake. *J. Acad. Nutr. Diet.* **2018**, in press.

sustainability

MDPI

Article

Capturing Social Innovations in Agricultural Transformation from the Field: Outcomes of a Write-Shop

Bram Peters [1], Marion Herens [2] and Jan Brouwers [2,*]

[1] ICCO Cooperation Myanmar; PO Box 8190, 3503-RD Utrecht, The Netherlands; b.peters@icco.nl
[2] Wageningen Centre for Development Innovation (WCDI), Wageningen University & Research (WUR); PO Box 88, 6700-AB Wageningen, The Netherlands; marion.herens@wur.nl
* Correspondence: jan.brouwers@wur.nl; Tel.: +31-613160317

Received: 26 September 2018; Accepted: 9 December 2018; Published: 12 December 2018

Abstract: The aim of this project was to explore the theme of social innovation for nutrition-sensitive and sustainable agriculture, resulting in examples of improved production and consumption of nutritious food. Social innovation refers to the generation and implementation of new ideas about how people organize interpersonal activities, or social interactions, to meet one or more common goals and in the process change basic routines, resources, and decision-making processes. In the country context of Myanmar, this research aimed to capture a variety of social innovation cases related to processes of agricultural transformation. Through the method of a collaborative case study write-shop, Myanmar-specific social innovations were identified, illustrating various forms of social innovation across the cases with citizen engagement processes. The write-shop method, in combination with the embedded expertise of development practitioners, proved to be a promising approach to identify niche innovations, distil insights, reframe actions, and promote critical thinking among different actors.

Keywords: social innovation; agricultural commercialization; nutrition sensitive transformation; Myanmar

1. Introduction

Food systems are undergoing rapid changes in response to economic and market developments, environmental impacts, and dietary changes [1]. Key drivers for these changes often relate to population growth, climate change, urbanisation rates, and globalising economies. Together they create complex changes in systems, institutions, and communities. In low- and middle-income countries, these changes have a profound effect on poverty, livelihoods, and food and nutrition security of poor households and smallholder farmers. In many of these countries both urban and rural households interact with various food system typologies: notably the "modern" agro-industrial system, which is dominated by a few global players with vertical value chains; the "traditional" food system, which is characterised by small-scale production with short supply chains; and the "intermediate" food systems which combine elements of the other two types. It is now seen that in Asia, for example, most consumers interact with intermediate food systems [2].

It is expected that in the coming decades food systems will change even more, influenced by sustainability concerns, changing consumption demands, and social challenges. In parallel, it is becoming clearer that achieving global development goals and tackling wicked problems requires attention to a number of interrelated themes [3]:

- The search for adequate forms of governance fitting with contemporary dynamics such as globalisation and democratisation;

- Growing understanding about the need to deal with complex adaptive systems, incorporating elements of non-linearity, resilience, and constant change;
- Taking into account how the human mind works in response to information streams and in decision-making processes, building on social learning processes [4];
- Exploring innovation as a key way to solve problems, generate new value and transform systems;
- Working in partnerships, building on the fact that these issues cannot be confronted by one or two actors without the collaboration of other organisations or institutions from other parts of societies or sectors.

In many scenarios and partnerships for the development of food systems, commercialisation strategies for smallholder farming are seen as one of the key responses to promote and provide for high-value agricultural products, generate incomes for farming households, and to improve linkages between rural, urban, and global markets [5]. While it is important to explore how smallholders can better participate in existing food system arrangements, it can be argued that deeper or transitional approaches are needed to tackle the themes mentioned above. It is perhaps not enough to only include, but to also empower citizens to change institutions and mental models for better, and sometimes unexpected, outcomes. Thus, in other scenarios, innovation of social, economic, and environmental relationships within food systems is deemed crucial to tackle these challenges and provide sustainable and inclusive transformation. The emerging field of social innovation, drawing from innovation, resilience, social entrepreneurship, and organisational change thinking, seeks to understand how individuals, organisations, and networks can generate new solutions for multiple societal goals [6] or build resilience through transitions [7]. Social innovation departs from the starting point that societal and systemic innovation requires techware (technical elements), software (social/people aspect), and orgware (the institutional organisation and setup). Lasting innovation also or at least partially includes social innovation, combined with technical and institutional innovation. Many innovations did not make it and failed as they underestimated the social component.

Since the 2011 reforms, widespread changes are taking place in Myanmar. In addition to political liberalization, the country has gone through a process of post-socialist economic transition. In economic-institutional terms, Myanmar has been undergoing transformation from a centrally-planned, state-commanded socialist economy, to an open, market-based capitalist economy. The rapid transformation and fragile shift from the political dominance of the military has brought consequences of an unprecedented magnitude. The international community, international organizations, and foreign investors have reengaged in projects with Myanmar, and a new social dynamism is being established, including deeper engagement with globalization. The political reforms implemented over the last years have triggered changes in all parts of the country. It is not clear, however, whether these changes will continue, and how these changes will impact on diverse social groups and across the national space [8].

The research presented here was conducted throughout 2017 as part of Wageningen University and Research programs on Global Food and Nutrition Security and Social Innovation for Value Creation. The goal was to explore the theme of social innovation for the identification of nutrition-sensitive and sustainable agricultural development pathways. Applied to the context of Myanmar, this paper presents the projects' explorative approach and methodology to identify social innovation cases in agricultural development and food and nutrition security programs and initiatives. These cases represent innovative niches that test and refine new arrangements or techniques within the agricultural sector. Innovation scholars in socio-technical transitions have discussed the importance of creating innovative niches, which are understood as safe spaces within an organisation or network. They can also serve as the root of a new organisation [7].

By using the lens of social innovation (SI) and by using write-shop methodology the goal was two-fold: (1) bring together four emergent Myanmar cases related to social innovation, agriculture, and food and nutrition security with a focus on the consumption of nutritious food (Cases 1.0) and turn these into concrete, shareable products (Cases 2.0); and (2) learn from each other's approaches and experiences and utilise the complementary capacities to generate useful principles and insights in relation to SI. In this way, the Myanmar cases contributed to the following research questions, which were identified at the start of the research program:

- What are critical drivers of food choices for farmer households?
- How can farm households manage risks related to climate and market changes?
- What farm household strategies support nutrition sensitive agricultural pathways in commercializing agricultural contexts?

Theoretical Orientation

What Is Social Innovation?

Different types of definitions of SI have generated two main schools of thinking in SI [9]. The first school focuses on new social processes. This relates to exploring changes in social relations, and emphasising changing power balances towards economic equity in society. A definition by Westley and Antadze [10] suggests that "social innovation is a complex process of introducing new products, processes or programs that profoundly change the basic routines, resources and authority flows, or beliefs of the social system in which the innovation occurs. Such successful social innovations have durability and broad impact". Accordingly, authors like Mumford state that social innovation entails complex processes introducing new products, processes or programs that profoundly change the basic routines, resources, and decision-making processes, or beliefs of the system in which the innovation occurs [11]. The second school emphasises understanding of new social outputs and outcomes, and is primarily oriented toward dealing with market failures in the provision of public social goods. In line with this second school the Organisation for Economic Co-operation and Development (OECD) [12] states that: "Social innovation is distinct from economic innovation because it is not about introducing new types of production or exploiting new markets in itself but is about satisfying new needs not provided by the market (even if markets intervene later) or creating new, more satisfactory ways of insertion in terms of giving people a place and a role in production". It has to be noted that not all social innovation is positive as dominant actors may influence the innovation process.

How to Make SI More Concrete?

The difficulty with approaching social innovation is that it is not clearly defined as a concept, and is often both invoked from a strongly actor-oriented, agency perspective on the one hand, but also from a structuralist systems perspective on the other. This is because innovations can be seen to come from individuals, but also from combinations and causal chains of results from external contexts [13]. Scholars attempting to distinguish between agency and structural innovation thinking have identified various levels: innovation of goods and services, institutional innovation geared to reorganising social and economic structures, and system change or radical innovation [14].

It is important to note that these dimensions do not have to suggest yet whether or not these innovations are successful. It rather focuses on the vision of change aimed for. As noted above, SI is not necessarily a good thing, even though the two main definitions seem to suggest this. It can be the case that SI leads to consequences that are beneficial for some while leading to disadvantages and negative effects for others. This can take the form of secret societies and shadow states; unintended consequences that eventually do more harm than good; and that it can fail in its implementation or used for different goals than was intended by other actors in society. It is important to realise this because in SI discussions it is sometimes assumed that win-win and beneficial outcomes are the key elements of SI [9]. Amongst the drivers for SI the need to address so-called wicked problems affecting global societies is central: climate change, growing inequality, demographic transitions, migration, and terrorism among the most important ones. In parallel, the changing of the way society is organised socio-politically may compound these problems: growing nationalism, public sector austerity, financial market complexity, private sector market failures, etc. This has affected all of the traditional spheres of society (public, private, and civil society) [9].

In dealing with the question of who is to deliver public welfare, SI has increasingly taken shape in the form of new partnership models in which public, private, and civil society actors collaborate or become hybrid actors. This is seen to have the potential to increase effectiveness of services and improve performance of control and choice, leads to new stakeholder role divisions and responsibilities, and new ways of co-creation and citizen engagement. New hybrid actors, as sites of social innovation, start to exist at the interface of private/public and civil society sectors. These can take up the ideal-type form of public–private entities, shadow state, and social enterprise, while there is a spectrum of different types of organisations existing between the different sector types [14].

The shifting or rather the blurring of boundaries between actors and organisations from different parts of society is often happening in response to challenges, risks, opportunities, and new mental models that encourage organisations to step out of their traditional role. Businesses seek to find new ways for creating "shared value" and for corporate social responsibility. Civil society actors seek to become more efficient, and realise more sustainable and durable outcomes of their projects by involving more business-like engagement models. Public institutions respond to new public governance trends, incorporating the modernisation and digitalisation of the public sector [9].

Why is SI Relevant for Agricultural Development Pathways for Food and Nutrition Security?

In the field of agriculture and food and nutrition security different forms and cases of SI can be highlighted:

- Processes of resilience and adaptation in production of food;
- Inclusive participation and new roles for stakeholders;
- Community-led organisations and bottom-up initiatives;
- Different interpretations and usages of technologies;
- New kinds of agro-food partnerships;
- Citizen science initiatives.

Various social innovations in this sphere have been captured in the past in Asia. In India, the System of Rice Intensification (SRI) as described by Prasad [15] is one such example. In this case, a combination of factors such as the historical influence of Gandhi's independence teachings to search for alternative narratives, availability of community spaces for exchanging indigenous knowledge, a deep crisis in farmer agency (high productivity rates but low profits leading to a high percentage of farmer suicides), and a heavy emphasis on green revolution technologies led to a counter-response that did not involve genetic modification. The idea was that SRI involves managing rice plants, soil, water, and nutrients with reduced use of material inputs while creating productive and resilient varieties in a collaborative manner. These approaches were tested in collaborations between researchers, civil society organisations, and farmers. While initially meeting with heavy resistance from agricultural research and extension institutes, eventually SRI became an accepted body of knowledge on crop intensification and helped to foster diversity in thinking and renewed valuation of local experimentation and community engagement [15].

A second example piloted in South East Asia refers to "Farmer Field Schools". Piloted by the United Nations Food and Agriculture Organisation since the early 90s. Farmer Field Schools were developed as an approach countering top-down Green Revolution extension methods in places where complex or contradictory problems challenged farmers. The premise of putting farmer peer-to-peer learning and group experimentation ahead of technical knowledge constituted a new approach to farmer capacity development. This approach allowed farmers to investigate, test, and decide for themselves what production methods worked for their environment [16].

In another example, food system researchers have signalled movements that are increasingly calling for "reversed food chain thinking": reshaping relations to follow consumer demand rather than production push or market pull. In these perspectives, the end-consumer is the final judge of food systems. Such an innovation would reorient food chains to food acceptability, safety, health, and use of (nutritious) food. For researchers and value chain actors, such dynamics require a more holistic approach to supply chain problems.

In this research, the theme of "social innovation" for nutrition-sensitive and sustainable agriculture for improved production and consumption of nutritious food in Myanmar is of particular interest. The recent process of opening up to the global economic system, ongoing processes of democratisation in Myanmar, combined with a growing commercialisation in agriculture, represents an interesting setting to explore emergent forms of social innovation [17].

2. Materials and Methods

An SI write-shop was organized in Myanmar. Write-shop methodology [18,19] was applied on four cases that explore and illustrate social innovations in Myanmar. Write-shop methodology focuses on documentation of key findings and lessons learned from practitioners and experts. The challenges posed by the limited uptake of exemplary practices, and the reality that useful knowledge often remained in the mind of field workers or in unpublished documents, prompted the discovery and testing of write-shop approaches [19]. The objective of write-shops is to help make available "hidden" field knowledge and make voices from the field become part of global dialogues on development. With the help of facilitators and editors field knowledge was put on paper (or on another communication medium for that matter). The write-shop method is particularly useful for really sitting down with colleagues and peers, take stock of practice, draw lessons, and work practically on a product that can be used after the workshop. A key driver for organisations to engage in write-shops is the need to document insights and produce shareable and easily consumable materials on certain issues deemed important. Often field practitioners face a situation where information is scattered or not easily understandable even though it could be very relevant to them. Another situation might be that a group of field practitioners has discovered solutions to pressing problems, but do not have the time or skills to capture them fully. In these situations, the departure points are often that information is largely in people's heads; no single person is the expert or has the overview; information is drawn from a broad repository of data; and the information needs to be validated with others [19]. Capturing emergent niche experiments that have the potential to result into sustainable innovation has a focus on reflection and articulating emergent understanding as a learning process. Likewise, the SI niche experiments described in this article all seek to develop sustainable solutions in practice, while integrating social and economic issues [20].

Write-shops generally take the following steps [19]:

- First draft presentation;
- Participants criticise the draft, offer comments, and suggest illustrations;
- Draft re-written and edited;
- Drafts are again reviewed and adapted;
- Final products are developed.

The process of repeated presentations, critiquing, and revising of drafts allows for papers or other products to be reviewed and sharpened substantially, development of new topics, and for topics to be combined, dropped or split into parts. The Myanmar write-shop took place over a two-day workshop in October 2017 (see Figure 1). Key findings and lessons on SI, presented by practitioners, were documented in a workshop report [21].

Four Yangon-based organisations engaged in agricultural innovation were selected to participate based on purposive sampling. The write-shop process started with a participant instruction, prior to the event, to prepare a case for the workshop (bringing pictures, documents, video images, etc.) and to reflect on the kind of desired end-product. In the two-day setup, participants first orally presented their cases on social innovation, followed by a session of questioning, deepening, and critical review.

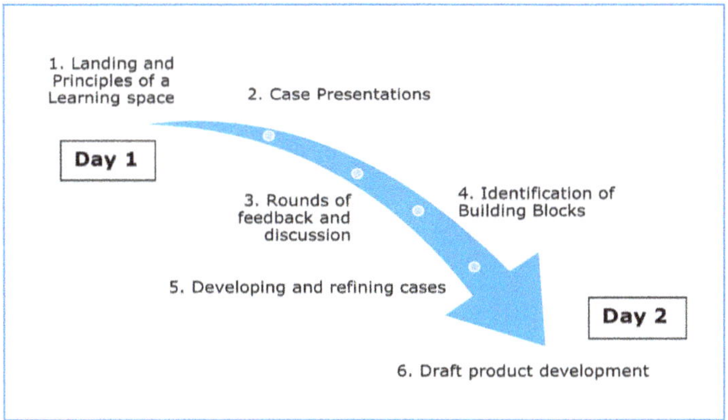

Figure 1. Process followed during the two-day social innovation workshop.

Based on this, a first draft was composed for all four cases, followed by a second round of review and feedback. Draft products were revised, and new topics were developed or combined, or topics dropped. Inputs from all participants were incorporated, taking advantage of the diverse experience and expertise of people present. The cases together formed the basis of the building blocks (or themes) capturing the current processes of SI in Myanmar.

The example of Farmer Field Schools was shared as a social innovation example that started in the 90s and is now institutionalised in various SE Asia countries [16]. This methodology emerged in response to the Green Revolution and the idea that farmers should also have the chance to share and give their opinions and best practices. By letting farmers test and choose key practices and letting them compare between different farming approaches a new way of supporting technology and knowledge uptake was facilitated. After the introduction, participants together translated the concept of SI into Myanmar language and back to English, which provided a more detailed and contextualised understanding of the SI concept.

During an initial scoping visit to Myanmar in January 2017, the team members of Wageningen University & Research had the opportunity to meet with a number of organisations working from diverse angles and expertise on agricultural transformation and food and nutrition security. Some of these organisations showed potential in terms of projects and initiatives that touched upon the abovementioned key concepts of social innovation. Four organisations active in Myanmar brought forward a social innovation case: Fresh Studio, Greenway, Myanmar Heart Development Organisation (MHDO), and Wageningen Centre for Development Innovation (WCDI). The four Myanmar SI cases are presented below:

1. **Fresh Studio** implements the Sustainable and Affordable Poultry for All (SAPA) program in Myanmar. The SAPA program aims at improving the food security and rural incomes of smallholder poultry and maize farmers in Myanmar through a public–private partnership with Dutch, Belgium, and Myanmar parties. One of the key problems SAPA is addressing is the low agriculture productivity in Myanmar in general, and in maize and poultry production specifically. Low agriculture productivity results in low rural incomes and relatively expensive food. With 25 to 50% of rural inhabitants being landless, and often without sufficient income to obtain food, it is crucial that a thriving agri-business sector is developed to generate jobs and lower the cost price of food. The project goals are to improve food security and rural incomes of smallholders in Myanmar, through the introduction of more productive and sustainable poultry and maize farming practices. This will result in lower cost prices and productivity gains, making poultry more affordable, and as the major source of animal protein in Myanmar, contribute to food security.

2. **Greenovator** is a social enterprise launched on 1 May 2011 in Yangon, Myanmar. It was founded by three core members who graduated from the Yezin Agricultural University. The vision of Greenovator is to share online agricultural techniques and information with farmers and to help them improve their agricultural practices and income levels. A key part of the work is the Green Way mobile application. This app serves the information needs of the farming communities by giving access to practical information. Key features include farming practice information; weather forecasts; daily news; Q&A; and daily crop market prices. The Greenovator team actively tries to improve their engagement with farmers and their daily realities, including sharing of practical expertise from farmers themselves.
3. The **Myanmar Heart Development Organisation** (MHDO) is a Myanmar civil society organisation founded in 2006 to create and provide opportunities for improved livelihoods for the needy in Myanmar. The organisational activities include food for education, food for work, food for training and non-food items, and cash for work. The organisation also implements agriculture-related interventions such as integrated farming, livestock rearing, and organic farming techniques. Key areas where the organisation works include Northern Shan State, Magway Region, and Rakhine State. Notably the work conducted in Northern Rakhine State, which was highly unstable and conflict-stricken at the time of the workshop, meant that the organisation had to be very sensitive and adaptive in their balancing of humanitarian and development work. The MHDO case focused on how to best combine activities that contribute to food and nutrition security. One of the methods used was the "Five Colours" approach to teach villagers about nutritional values of fruits and vegetables: each colour represents a different type of nutritious food. Another part of the work is on agricultural development through support in making organic fertilizer.
4. **Wageningen Centre for Development Innovation** researchers generated insights into SI examples in Myanmar during a case study [17,21]. For this write-shop, one village was selected to highlight a number of activities and strategies that could signal the development of socially innovative strategies. In one community in the Pakokku research area (Magway State), interesting activities and dynamics were taking place. In this community forms of strong social cohesion were identified, under the guidance of community leaders, which translated into various economic and social opportunities and goals. Firstly, the community farmers were organising themselves gradually to bypass the role of wholesalers and brokers by collecting their produce together and hiring a truck to bring it to the market themselves. Secondly, it was also seen that, through the support of a non-governmental organisation (NGO), community members had combined a traditional oilseed mortar and pestle with a modern fuel-driven engine to make groundnut oil themselves. This enabled the community to make good-quality oil (free of contaminants they perceived other oil products from the market to have) and at the same time provide a service accessible to the whole community. The third example identified other activities such as collective labour to rebuild dams and water containers, and a strong willingness to participate in, and share the knowledge from trainings given by NGOs, universities, and businesses.

In the write-shop each organisation developed their own needed product and in the process contributed to general learning and insights on SI. In this way, the four participating organisations worked on their case and produced their respective product, while also contributing with insights on SI processes in Myanmar.

3. Insights and Results of the Write-Shop

3.1. Findings and Results Per Case

Fresh Studio—During the write-shop Fresh Studio developed a case study document that highlighted elements of social innovation in their work and practices. This was based on their own poster presentation (Case 1.0) and the SAPA program document, but also their experiences from the past year and a half implementing the program. Based on the iterative feedback process provided in the write-shop, the elements of SI that came forward from the case were the approach of short-cutting the value chains of maize and poultry by more strongly connecting consumers of safe chicken to farmers producing broiler chicken and farmers providing maize for these chickens. Providing a "Myanmar consumer" and a "quality" perspective to food is quite new for Myanmar. It was also noted in the write-shop that whereas there is a technical component to understanding and working with this, there is also a cultural element that is important to pay attention to. Fresh Studio can build on the idea to strengthen the partnership further as it works on this social component. Another key element that was interesting to develop from an SI point of view was the fact that many different types of stakeholders are working together in this Public Private Partnership (PPP). This is a new form of collaboration in Myanmar and has the potential to create opportunities and synergies not considered before. The write-shop resulted in a draft text for a brochure highlighting the SI elements of the SAPA program.

Greenovator—Greenovator wanted to make use of the write-shop to develop a storyboard for a documentary they were planning to make to promote the use of the app. Myanmar farmers often do not have access to agricultural extension and information services—a key systemic problem in the Myanmar food system. The basis of the storyboard idea was to introduce a few farmers who stated that they had really benefited from using the Greenway application. The challenge for the Greenovator group was to identify the elements that make the mobile application a social innovation and to visualize this in a documentary. Going over the exercise of developing a storyboard for a short video or documentary was valuable for the Greenovator team. They experienced that developing a documentary storyboard was not easy, and that the difference between a documentary and a promotional video is not only about the length of the video. The team was challenged to exactly pinpoint wat makes the Greenway application different from other agricultural extension training interventions. These elements, showing the SI potential, had to do with the communication flows between farmers, experts and value chain actors, and the potential for exchanging different forms of knowledge. The write-shop resulted in a storyboard providing a detailed visualized outline of a documentary. If Greenovator is able to make the bridge between expert knowledge and farmer/community practice and traditions, changing the roles of these groups in the process, it can create interesting added value in the Myanmar food system.

Myanmar Heart Development Organisation —During the write-shop MHDO wanted to work on a picture book that illustrated their approach to creating more awareness about food security and nutrition in communities in Northern Rakhine State. Using the inspiration from the Five Colours approach to highlight the nutritious value of products like vegetables and fruits, they drew characters and developed a storyline that tried to tap into the knowledge they already had from the region, the adaptive capacities of communities, and insights on nutrition. In the picture book, the MHDO developed the story of how a development worker arrived in a community in Northern Rakhine and met a community leader. They started talking about good food, healthy food, and nutritious food. The development worker had ideas about what that meant, and the community leader as well. They decided to work on food and nutrition security together, inspired by the Five Colours approach, but also building on the communities' resilience and local agro-ecological circumstances.

Wageningen Centre for Development Innovation—During the write-shop WCDI intended to develop a technical brief highlighting different forms of social innovation identified at the village level. The working title of the case (Case 1.0) was "Kan Zauk, the Prize-Winning Village". The "Prize-Winning Village" concept showed that combining life-course research methodologies with ideas of change in the community could lead to interesting perspectives on social innovation. In the process of developing a Case 2.0, it became clear that though there was quite some interesting information already there, more data needed to be collected to make it into a solid case study. The key message that this story brought was about inspiration and awareness of basic elements and processes of social innovations happening in Myanmar villages. These do not necessarily occur in only this village, but probably in many communities across Myanmar. The activities mentioned, such as the community oil pressing mill, the collective truck or the new partnerships with research and businesses may not seem very inventive from a general development perspective, but in Myanmar these are new opportunities and ideas arising through bottom-up initiatives. For policy-makers this is essential to know about and support.

3.2. Overall Synthesis of Social Innovations Contributing to Nutrition Sensitive Agricultural Transformation

The four cases from Myanmar illustrate how the food system is undergoing a rapid transition, in which existing supply chains are adapting to economic, environmental, market, and dietary change. A synthesis of the findings is presented in Table 1. The table presents a matrix with the different cases and how they illustrate social innovation processes linking agriculture with food and nutrition security (as introduced in Theoretical Introduction):

- Processes of enhancing resilience and adaptation in production of food;
- Inclusive participation and new roles for stakeholders;
- Community-led organisations and bottom-up initiatives;
- Different interpretations and usages of technologies;
- New kinds of agri-food partnerships; and
- Citizen science initiatives.

From the engagement in the process of the write-shop it became clear that quite a few dimensions and key elements of SI are apparent in the work of these organisations. In various ways that is already a good contribution being made to development, but the main added value that is in some ways surprising relates to the fact that (1) these organisations are coming from different angles and interests, yet (unconsciously perhaps) are applying the seeds of SI; and (2) that using a write-shop approach has the potential to bring out these somewhat intangible contributions. Considering this, it became apparent that the two-day setup was perhaps not enough to fully bring out the potential in SI thinking. To capture intangible contributions at least a three-day event seems necessary.

The cases all show examples of the application of new tools and practices that allowed farmers, development practitioners, researchers, and other actors present in the event to understand how small-scale producers respond to a growing process of commercialisation. They also show how individual farmers as well as farmer groups shift from a highly subsistence-oriented production towards more specialised production targeting markets both for their input procurement and output supply. Being flexible and having a diversified livelihood status showed to be a successful strategy applied by farmers to deal with their challenging environment. The cases show how new forms of inclusive community-led organisations are taking root, often in connection with the process of democratisation. However, these social innovation initiatives face the risk of remaining isolated and could miss the opportunity to successfully scale out in Myanmar. These types of bottom-up initiatives can inspire policy-makers, profit and non-profit organizations, and other civil society movements in finding better solutions with respect to the current challenges of Myanmar regarding agricultural development and food and nutrition security.

Table 1. Overview of social innovation (SI) elements within cases.

Organisation	SI Initiative	Elements of Social Innovation				
		Processes of resilience and adaptation in production	*Participation/new partnerships/roles*	*Community-led/bottom-up*	*Different interpretations/use technologies*	*Citizen science*
Fresh Studio Sustainable and Affordable Poultry for All (SAPA) program	- Aim: improving food & nutrition security and incomes of smallholder poultry and maize farmers - Builds on linking two value chains (maize and poultry) - Seeks to connect consumers to farmers producing broiler chicken and the farmers producing the chicken feed (maize)	√	√		√	
Greenovator Greenway App	- Initiated by alumni from agricultural university - Aim: share alternative agricultural techniques with farmers to help them improve their agricultural outputs and income levels - Seeks to actively engage with the lived realities of farmers and find ways to draw from their practical expertise	√	√		√	√
Myanmar Heart Development Foundation Community action for food and nutrition security	- Aim: creating awareness about food security and nutrition in communities in Northern Rakhine State - Example used was the "Five Colours" approach - Seeks to actively build on joint sense making and give meaning to food and nutrition security concerns with community leaders and to translate those into action.	√	√	√		√
Wageningen Centre for Dev. Innovation Life course perspective study	- Prize winning village concept - Seeks to identify concrete examples of SI - Seeks to make explicit that using life-course research methodologies are relevant for identifying significant transformations in households and community	√	√	√		√

3.3. Added Value of the Research and Lessons Learnt

The added value of the present research lies in providing documented material on four social innovation cases as well as a methodology for how this can be done elsewhere. Complexity science points to the importance that scientists understand more about the dynamics of innovation, including the interaction between techware (technical elements), software (social/people aspect), and orgware (the institutional organisation and setup). Responsible Research and Innovation (RRI) points to the importance of trans-disciplinary research and stronger understanding of how civil society can be involved in societal change processes. The present research shows how these types of social initiatives can be identified, understood, documented, and supported, also by researchers. Societal added value for Myanmar and similar emergent economies is provided by showing how innovation processes in the context as described can be more successful by understanding and building on the social component of innovation.

The following lessons can be learnt from the social innovation write-shop in Myanmar:

- In order to be effective, interventions and policies aiming at promoting sustainable food systems have to include smallholder-farmer households' interests on agricultural production and food and nutrition security;
- Local understandings of diets and perspectives on food provide insights on possible entry points for nutritional sensitive agriculture;
- The cases show the importance for policies and interventions to be informed by participatory and holistic baseline assessment where the change perspective and visions of all the relevant stakeholders are taken into account;
- The cases show emergent new social-technical arrangements within the Myanmar food system. Whether they will achieve true change and institutionalise new socio-technical arrangements at scale will depend on further introduction and linkages with actors in the existing regime.

4. Discussion

This research showed how a structured and participative exchange and reflection process as followed in a write-shop can allow to articulate and document how farmers and their partners respond to commercialisation processes in Myanmar. The cases illustrate elements from the emergent intermediate food system in Myanmar, based on interactions between key actors like farmers and small businesses, government, and international businesses. The SAPA program showed a case where the poultry and maize value chains become more connected through PPP linkages, a new type of partnership for Myanmar. Greenovator introduced a new app on agriculture-related information, diminishing dependency of farmers on the rather weak present extension system. In rural parts of Myanmar MHDO is increasing awareness on nutritious food and builds on communities' resilience and the local agro-ecological context for stronger food and nutrition security. The WCDI case illustrated the social innovation dynamics at the village level and how policy-makers can support bottom-up initiatives and accelerate local development. The cases helped to bring to the surface valuable learning for participants and researchers, helping participants to articulate their own practice as well as to understand other SI experiences in a similar context. The study applied write-shop methodology and underscored its importance and potential to help make transformative processes concrete and documented. Write-shop methodology connects with a wide range of techniques and literature on strengthening reflection as part of action research and reflexive monitoring. Caution has to be made, however, in allowing sufficient time in preparing and conducting the event. These types of reflexive and documenting events merit more follow-up support, which was not part of the present research. Before further responsible scaling can be considered and designed, case owners will need to engage with current actors in the Myanmar food system. This process will probably result in further re-design of potential cases with validation and empirical justification before scaling can take place, whereas other cases might not be going to a next stage of development.

Supplementary Materials: The report of the write-shop is available at http://edepot.wur.nl/432091.

Author Contributions: The first two authors organized and conducted the write-shop in Myanmar. The third author contributed with methodological design and integration with social innovation research.

Funding: This research was funded by The Netherlands Ministry of Agriculture, Nature and Food Quality, Knowledge Base Research Funds 22 and 28.

Acknowledgments: We acknowledge the support of the Dutch Embassy and our colleague Jan Verhagen in allowing us to be introduced with potentially interested contacts in Myanmar. We are also thankful for four anonymous reviewers. Their comments and suggestions allowed to make considerable improvements in the first and second drafts. We are also grateful for the comments and suggestions provided by the Academic Editor and the editing MDPI team. Processing their comments and suggestions has considerably strengthened the article.

Conflicts of Interest: The authors declare no conflict of interest. The funders had no role in the design of the study; in the collection, analyses, or interpretation of data; in the writing of the manuscript, and in the decision to publish the results. All draft material of cases are available on request with the authors and have been shared with the participants of the workshop.

References

1. HLPE. Food Losses and Waste in the Context of Sustainable Food Systems. Available online: www.fao.org/3/a-i3901e.pd (accessed on 12 September 2018).
2. Fresco, L.O.; Ruben, R.; Herens, M. Challenges and perspectives for supporting sustainable and inclusive food systems. *GREAT Insights Mag.* **2017**. Available online: http://ecdpm.org/great-insights/sustainable-food-systems/challenges-perspectives-sustainable-inclusive-food-systems/ (accessed on 12 september 2018).
3. Brouwer, H.; Woodhill, J.; Hemmati, M.; Verhoosel, K.; van Vugt, S. *The MSP Guide: How to Design and Facilitate Multi-Stakeholder Partnerships*; Brouwer, H., Woodhill, J., Eds.; Wageningen University—Centre for Development Innovation: Wageningen, The Netherlands, 2015.
4. World Bank. *World Development Report 2015: Mind, Society, and Behavior*; World Bank: Washington, DC, USA, 2015.
5. Westhoek, H.; van Berkum, S.; Özay, L.; Hajer, M. *Food Systems and Natural Resources*; United Nations Environment Programme: Paris, France, 2016.
6. Olsson, P.; Moore, M.-L.; Westley, F.R.; McCarthy, D.D.P. The concept of the Anthropocene as a game-changer. *Ecol. Soc.* **2017**. [CrossRef]
7. Westley, F.; McGowan, K.; Tjörnbo, O. *The Evolution of Social Innovation. Building Resilience through Transitions*; Elgar Publications: Cheltenham, UK, 2017.
8. ADB. *Asian Development Bank 2014: Myanmar Unlocking the Potential. Country Diagnostic Study*; ADB Economics and Research Department: Mandaluyong, Philippines, 2014.
9. Nicholls, A.; Simon, J.; Gabriel, M. *New Frontiers in Social Innovation Research*; Nicholls, A., Simon, J., Gabriel, M., Eds.; Palgrave Macmillan: London, UK, 2015.
10. Westley, F.; Antadze, N. Making a difference: Strategies for scaling social innovation for greater impact. *Innov. J.* **2010**, *15*, 1–19.
11. Mumford, M.D. Social Innovation: Ten Cases from Benjamin Franklin. *Creativity Res. J.* **2002**, *14*, 253–266. [CrossRef]
12. OECD. *The OECD Innovation Strategy. Getting a Head Start on Tomorrow*; OECD: Paris, France, 2010.
13. Howaldt, H.; Kopp, R.; Schwarz, M. *On the Theory of Social Innovations: Tarde's Neglected Contribution to the Development of a Sociological Innovation Theory*; Beltz Juventa: Weinheim, Germany, 2015.
14. Nicholls, A.; Murdock, A. *Social Innovation: Blurring Boundaries to Reconfigure Markets*; Nicholls, A., Murdock, A., Eds.; Palgrave Macmillan: London, UK, 2012.
15. Prasad, S.C. Innovating at the margins: The System of Rice Intensification in India and transformative social innovation. *Ecol. Soc.* **2016**. [CrossRef]
16. FAO. *Building Resilient Agricultural Systems through Farmer Field Schools*; FAO: Rome, Italy, 2015.
17. Gabrielli, M.; Herens, M.; Peters, B. *Farmers' Tales: Adaptive Strategies for Agricultural Commercialisation and Food and Nutrition Security in Myanmar*; Wageningen Centre for Development Innovation: Wageningen, The Netherlands, 2017.

18. Gonsalves, J.; Armonia, R. (Eds.) *Writeshops: A Tool for Packaging and Sharing Fieldbased Experiences (A Guide to Organizing Workshops)*; International Institute of Rural Reconstruction, International Potato Center—Users' Perspective with Agricultural Research and Development: Manila, Philippines, 2010.
19. Mundy, P.; Mathias, E.; Bekalo, I. Out of heads and onto paper. *LEISA Mag.* **2006**, *22*, 26–28.
20. Regeer, B.J.; de Wildt-Liesveld, R.; van Mierlo, B.; Bunders, J.F.G. Reconciling accountability and learning in the evaluation of niche experiments. *Evaluation* **2016**, *22*, 6–28. [CrossRef]
21. Herens, M.; Peters, B. Report: Social Innovation Case Study Write-Shop. Unpublished Report. 2017. Available online: http://edepot.wur.nl/432091 (accessed on 10 September 2018).

sustainability

MDPI

Article

Farmers' Adaptive Strategies in Balancing Commercial Farming and Consumption of Nutritious Foods: Case Study of Myanmar

Marion Herens [1,*], Monica Gabrielli [1], Bram Peters [2], Jan Brouwers [1] and Diane Bosch [1]

1 Wageningen Centre for Development Innovation, Droevendaalsesteeg 1, PO Box 88, 6700 AB Wageningen, The Netherlands; monicagabrielli1@gmail.com (M.G.); jan.brouwers@wur.nl (J.B.); diane.bosch@wur.nl (D.B.)

2 Myanmar ICCO Cooperation, Royal White Elephant Hotel, 8th floor, 11 Kan Street, 10 Quarter, Hlaing Township, Yangon 11051, Myanmar; b.peters@icco.nl

* Correspondence: marion.herens@wur.nl; Tel.: +31-6-2336-9354

Received: 27 September 2018; Accepted: 5 December 2018; Published: 11 December 2018

check for updates

Abstract: Food systems undergo rapid changes in response to economic and market forces, and environmental and dietary changes. This study aimed to disentangle adaptation strategies in farm households balancing interests in the commercial aspects of farming and the consumption of nutritious foods. The area of interest was Central Myanmar, Pakokku region. A literature-based framework was used to identify entry points for adaptation strategies at the farm household level. A purposive sampling strategy was used to select smallholders (<5 acres), engaged in market-oriented agriculture ($\geq$10 years). In 14 households, in-depth interviews were conducted, using a life course perspective depicting the household history in relation to agricultural developments and household food and nutrition security. The narratives of smallholders confirmed that household food and nutrition security was grounded in mixed livelihood strategies, including migration. Diet quality depended largely on income. Supportive strategies were a frugal lifestyle, responsible use of resources, participation in community activities, and different forms of social innovation. The study shows how the understanding of local diets provides insights in entry points for nutrition-sensitive agriculture, and suggests a need for alternative adaptation strategies, replacing those promoting specialization and intensification, for more holistic solutions that reinforce the flexibility and resilience of farmers.

Keywords: Agricultural commercialization; food and nutrition security; salutogenesis; life course perspective; food systems; multi-level; positive deviance; Myanmar

1. Introduction: Food System Transitions and the Case of Myanmar

With the Sustainable Development Goal (SDG) of 'Zero Hunger' [1], much importance has been given to the role of nutrition in reaching the end of hunger for all by 2030. In the past, increasing food production has been the emphasis of agricultural strategies all over the world. However, worldwide, food systems are undergoing a rapid transition, in which existing supply chains are adapting to economic, environmental, market, and dietary change. There is an on-going shift from local food systems characterized by small-scale production by a large group of small holders to a growth of commercial agriculture by fewer, larger farmers and longer and more complex global supply chains [2]. Individual farmers are increasingly involved in processes of commercialization with substantial improvements in agricultural outputs [3], and play a crucial role in a food system as both producers and consumers [4]. However, commercialization of agriculture can have several adverse effects, especially in terms of equity and environmental consequences. With the increase of mechanization, a consistent part of the rural labor force needs to be relocated in the industrial and service sector, with consequent loss of human and social capital, as well as environmental consequences due to the increased use of agricultural chemicals [5,6]. Where property rights are unclear, phenomena, such as land grabbing, can take place. Also, commercialization may lead to a decline in crop diversity for households [7]. In some cases, farmers that invested in cash crops were worse off in terms of nutritional status than subsistence farmers [8]. The persistence of malnutrition (undernutrition, micronutrient deficiencies) in low- and middle-income countries, alongside a worldwide growth in the prevalence of obesity, urges us to further investigate how to simultaneously stimulate individuals' healthy food production as well as consumption [9].

To explore the process and impact of major, and often irreversible, food system transformations, the case of Myanmar is an appropriate context. After 50 years of military rule, a civilian government was installed in 2011, and the first elections were held in November 2015 [10]. Therefore, the country was opened up to the world, allocating large concessions to foreign agribusiness companies [11]. The government expressed its intention to become a full member of the ASEAN (Association of South East Asia Nations) community and more relaxed regulation favored foreign investment. Nevertheless, the key strategies for the government to achieve national food security remained in rice production and local and international agribusiness prioritization [11]. The country faces the contradictory situation of being a net food exporter on the one hand, but experiencing high poverty and malnutrition rates on the other [12]. A major constraint in this regard is access to land: Nearly half of the rural households are officially reported as landless (no ownership). Confiscation of land and conflicts in some areas are two major reasons for landlessness [13]. Until recently, farmers' unions and networks were banned in the country [14]. Even though the interest in nutrition security is on the rise at the policy level, there is still a limited interconnection with the commercialization of agriculture [11].

This study aimed to contribute to a deeper and contextualized understanding of farm household sense-making processes—how people understand and give meaning to life events—in relation to the current rapid food system transition in Myanmar. The study sought to document the views of local smallholders by in-depth analysis of agricultural life stories to identify resilient and emergent strategies, incentives, and innovative practices leading to sustainable agricultural commercialization while achieving household food and nutrition security. The main research question was how do smallholder farmers develop and implement adaptive strategies in response to food system transformations leading to agricultural commercialization, in view of their agricultural livelihoods and diets during their life-course?

1.1. Theoretical Outline

This study used various theoretical entry points. Firstly, the study used a conceptual framework, developed from the literature, to identify and analyze development pathways from agricultural commercialization to nutrition at the household level [15]. The literature showed several pathways through which agriculture-oriented interventions may lead to positive food security and nutrition outcomes: Subsistence-oriented production (source of food); production for sale (source of income); and agricultural policies, affecting supply and demand factors defining the price of marketed food and non-food crops [16,17]. Key elements to define the framework were drawn from existing frameworks to assess food and nutrition security (FNS). For the nutrition components, the United Nations Children's Fund (UNICEF) framework on maternal and child undernutrition [18] and the framework for Actions to Achieve Optimum Fetal and Child Nutrition and Development [19] were used. For FNS, the Food and Agricultural Organization (FAO)'s Food Insecurity and Vulnerability Information and Mapping System [20,21] and the framework for pathways described by Hertforth and Harris [22] were used. For the commercialization components, the frameworks of Von Braun [23] and Kanter et al. [24], describing the linkages between agriculture, food systems, nutrition, and health, were used. The conceptual framework, presented in Figure 1, embraces a multi-level approach taking into account several factors and dynamics that affect farm household livelihood outcomes: Individual level (gender and power dynamics); household level (food production, income generation, food purchase choices, off-farm labor, care practices, access to health care); community level (employment opportunity, collaboration, microfinance, care and social (infra)structure); and regional and macro level (price and trade policy). The ultimate focus of the framework is on the rural agricultural household interactions. In this space, farmers negotiate their assets with the external environment through their decision- and sense-making behaviours. These dynamics generate pathways, which cut across different levels and can take various shapes and forms, potentially leading to changes for famers' livelihoods. It also seeks to include a life course perspective [20], emphasizing the non-linearity of many relations between inputs and outcomes.

Secondly, to disentangle sense-making and decision-making processes happening in response to commercialization, this study was based on three additional theoretical orientations. The salutogenic theory was used, developed by Antonovsky [25,26] for health promotion, which posits that life experiences help shape one's sense of coherence, whereby life is understood as more or less comprehensible, meaningful, and manageable. A strong sense of coherence helps one to mobilize resources to cope with stressors and manage tension successfully. In its more general meaning, salutogenesis refers to a scholarly orientation focusing attention on the study of the origins of health, contra the origins of disease. Salutogenesis is in harmony with developments across the social sciences that seek better understanding of positive aspects of human experience [27]. This theoretical orientation was adopted to guide the analysis of farm household individuals' strategies and coping mechanisms promoting nutritious food consumption and production throughout the life-course [28], using the concept of general resistance resources (resources that can aid resistance to stressors) [29].

The positive deviance theory was used to understand in which way successful farmers can guarantee sustainable livelihoods through commercialization strategies in an environment where others fail [30]. Social innovation theory was used to explore collective dynamics and the interactions between different actors, policies, and interventions [31]. The emerging field of social innovation, drawing on innovation, resilience, entrepreneurship, and organizational change thinking, seeks to understand how individuals, organizations, and networks can generate new solutions for multiple societal goals [32].

1.2. Definitions of Key Concepts Used in the Study

In this study, food system is defined as "a system that embraces all the elements (environment, people, inputs, processes, infrastructure, institutions, markets and trade) and activities that relate to the production, processing, distribution and marketing, preparation and consumption of food and the outputs of these activities, including socio-economic and environmental outcomes" (p. 12) [33].

Commercialization is defined as the agricultural transformation process in which individual farmers shift from a highly subsistence-oriented production towards more specialized production targeting markets both for their input procurement and output supply [34]. Specialization and commercialization could represent a more efficient strategy than subsistence for small farmers [35].

Food and nutrition security (FNS) is defined as "food and nutrition security exists when all people at all times have physical, social and economic access to food, which is safe and consumed in sufficient quantity and quality to meet their dietary needs and food preferences, and is supported by an environment of adequate sanitation, health services and care, allowing for a healthy and active life" [36].

General resistant resources are defined as those resources that can aid resistance to stressors. These can be of a physical nature (e.g., a strong physique, strong immune system, genetic strengths), art factual nature (e.g., money, food, power), cognitive nature (e.g., intelligence, education, adaptive strategies for coping), emotional nature (e.g., emotional intelligence), social nature (e.g., support from friends and/or family), or macrosocial nature (e.g., culture and shared belief systems). General resistance resources can be identified at different levels: Individual-level resources (internal, such as intelligence, religion, and philosophy, genetic, and constitutional); family-level resources (material and emotional support), and community- and society-level (material, knowledge, cultural stability, social support) [37].

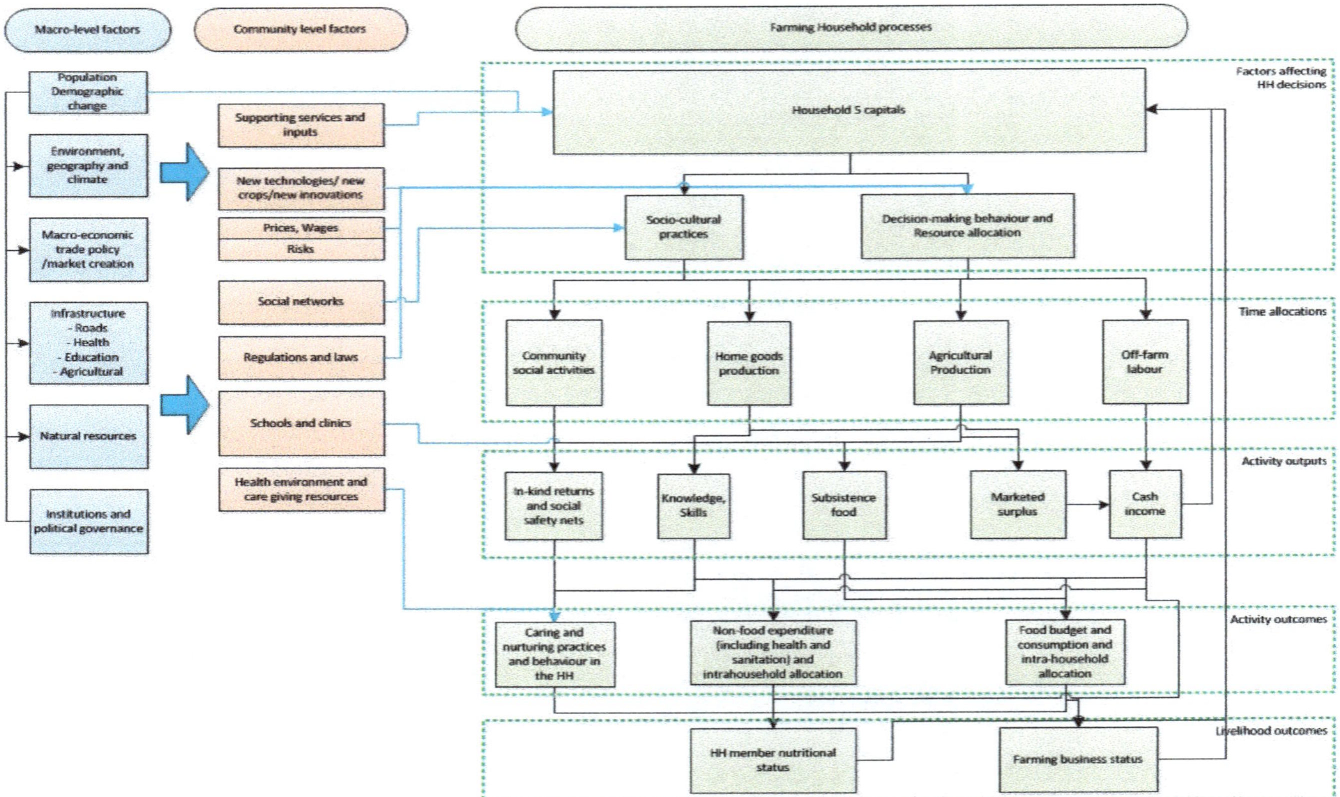

Figure 1. Conceptual framework of development pathways from agricultural commercialization to nutrition [15].

2. Materials and Methods

The Dry Zone of central Myanmar, as shown in Figure 2, was selected for our case study design [38]. This area is generally characterized as one of the most food insecure areas of the country [39]. In 2011, the food security assessment by the World Food Program (WFP) classified 17% of households as severely, and 24% as moderately food insecure [40]. Under-five malnutrition rates showed 7% of wasting and 29% of stunting [41]. Food availability strongly depends on monsoon rains (from May to October). A majority of households rely on rain-fed cultivation on flatland. Farm households generally own their agricultural land, but around 50% of the households in the area are estimated to be landless [42]. Most households grow three or more different types of crops, most commonly pulses, sesame, maize, and groundnuts, alongside animal sourced foods [13]. Over 90% of the households rely on markets for rice [39].

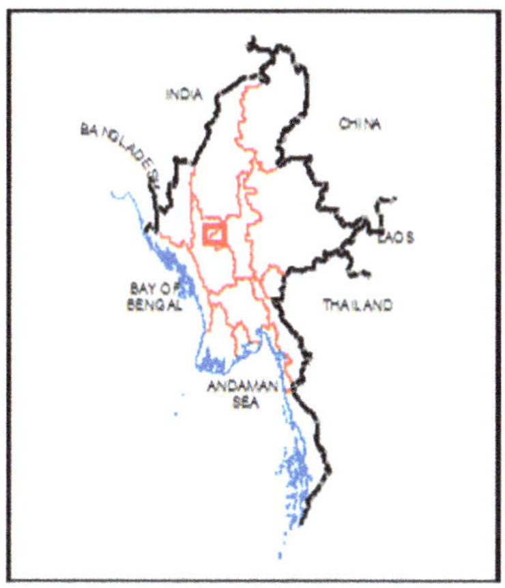

Figure 2. Pakokku township in the Dry Zone.

Income is derived mainly from casual wage labor, farming, small trade, and sales (of livestock). Farmers in the area have limited access to finance and inputs, especially for cultivation that is different from paddies, which are promoted by the government [42]. The region is also characterized by a high presence of female-headed households, due to migration of male family members, mainly to urban centers in Myanmar [39]. Main health issues are poor hygiene practices, poor access to latrines, and use of unprotected water sources, poor drinking water treatment practices, and inappropriate care for sick children. Girls tend to have less access to education than boys [39]. The study area was Pakokku Township in Magway division, in which five villages were selected: Kan Zauk, Sar Kyin, Aung Tha, Oo Yinn, and Yar Lar Lay. Most of the villages count 100 to 200 households [43]. Villages are organized around a group of leaders, who are supposed to actively help organize community activities (i.e., ceremonies) or development actions.

2.1. *Sampling and Household Selection*

A non-probability-based, purposive sampling strategy was used to select positive deviant farm households, i.e., those households reaching optimum results in an environment where the majority fails. To identify positive deviants, we based our inclusion criteria of farmers engaged successfully in commercial agriculture on local consultation with key stakeholders involved in agricultural development and food security in Myanmar, acknowledging that 'positive', 'healthy', and 'successful' are socially constructed concepts. We did so believing these parameters were sufficient for the aim of learning from the positive [44]. The criteria thus defined to select the households for this study were:

- Started farming as landless or smallholder (less than two ha of productive land). This threshold was based on the fact that land is the most commonly used dimension for measuring farm size, although other criteria can also be used. Small is a relative concept, depending on agro-ecological as well as socio-economic considerations, but a 1 or 2 ha threshold is frequently used to designate farms as small [45];
- engage or engaged in the past for at least 10 years in any form of market-oriented farming in the study area; and
- relate directly to current concerns of 'scaling up' of technology, methods, social innovation, and good and best practices.

Household selection was done on site, with support from a local non-governmental organization (NGO). Fifteen households were thus identified, out of which 14 households entered the study. The one drop out was a successful farmer who did not start as a smallholder. The inclusion of female-headed households was emphasized. In each selected household, the household head was asked to participate in the research. In some cases, more than one household member participated in the interview, resulting in 20 individuals interviewed (8 women, 12 men). All respondents had settled in the village where they were born.

Average self-reported farm size early in life was 1.6 acres (range: 0–8 acres), and grew to 7.5 acres (range 4–13 acres) at the time of the study. Average household size was 6.9 members (range: 5–10 members). Average age of the respondents was 51.8 years (range: 31–66 years). Only three households lived off agriculture, all others applied mixed livelihood strategies. Households produced an average of 5.2 commercial crops (range: 3–8), including food and non-food (cotton) crops. All household heads interviewed were literate, eight of them through monastery schools, and in nine out of 14 cases, one of the household members had attended university. The majority of children, however, supported the parents in agriculture activities or in off-farm jobs. In 10 out of the 14 households, migration was common among the youth, some leaving for cities in Myanmar (i.e., Mandalay and Yangon) and others to neighboring countries (i.e., Thailand and China). Table 1 summarizes the household characteristics involved in the study.

2.2. Data Collection

Data were collected during September–October 2017. Data were collected by means of qualitative in-depth interviews using narrative inquiry as the method, including a timeline technique [46]. Narrative inquiry and other forms of qualitative non-structured inquiry have been used to explore food stories in several studies. It has been applied to understand food choice factors [47], to explore food related meanings [48], eating disorders [49,50], relationships with food [51], and healthy eating [52]. Narrative inquiry has also been used to understand whole food systems, including production aspects [53]. The timeline technique was used as it is designed to respect contextual and historical influences, generating data based on stakeholders' individual and collective perceptions, thus reflecting developments over time [54]. In addition, the timeline technique visualizes respondents' perceptions of what matters most, and serves as a graphic tool to guide and summarize the interview while doing it, thus supporting both the researcher and respondent to gain insight and promotes learning on the spot [55,56]. The combination of methods was chosen to:

1. Capture thoughts and emotions of individuals in more depth compared to the traditional interview [57];
2. Capture the meanings attributed by respondents to their lives through the selection of memories [58];
3. Favor self-reflection through the process of expressing their personal life-story [59]; and
4. Favor a reflection on changes in societal and cultural norms from which is it possible to extract time and geographical bound socio-cultural practices [60,61].

Table 1. Respondent characteristics.

Village	# Resp	Sex	Age yrs	Education	# Household Members	Household Livelihoods	# Migrants/Household	Type Commercial Crop	% Product. Sold *	Acres Owned Start Farm **	Acres Owned (2017) **
Kan Zauk	2	F M	60 51	Primary Literate NS ***	8	Mixed ****	3	Groundnut, mung beans, toddy-palm	50%	0	5
	1	F	44	Primary	7	Agriculture	0	Ground nut, mung bean, pigeon peas, sesame, toddy-palm	90%	0	7
	1	M	56	Primary	6	Mixed	1	Ground nut, mung bean, pigeon peas, sesame, lablab bean, potato, tomato, onion	50%	5	5
Sar Kyin	1	M	41	Primary	7	Mixed	1	Pigeon peas, cotton, sesame	50%	0	5.5
	1	M	44	Literate NS	9	Mixed	1	Mung bean, pigeon peas, cotton, maize, sesame	80%	2	7
	2	F M	66 66	Illiterate Literate NS	8	Mixed	3	Ground nut, pigeon peas, cotton, maize, sesame	80%	0	13
	1	M	43	Primary	6	Mixed	0	Pigeon peas, cotton, lablab bean, maize, sesame	80%	0	10
Oo Yinn	2	F M	58 55	Literate NS Primary	5	Mixed	1	Maize, lablab bean, tomato, chili	90%	0	4
	2	F F	60 58	Primary Primary	6	Agriculture	0	Chick pea, maize, bean, chili, eggplant	80%	5	5
	1	M	31	Primary	10	Mixed	2	Maize, chick pea, tomato, eggplant, chilly	60%	8	9
Yae Lar Lay	3	F F M	57 50 58	Literate NS Intermediary Literate NS	5	Mixed	0	Groundnut, sesame, pigeon pea, mung bean, cotton	70%	0	10
Aung Tha	1	M	56	Literate NS	7	Mixed	2	Groundnut, pigeon pea, maize, sesame	70%	3	7
	1	M	41	Secondary	7	Agriculture	0	Chick pea, mung bean, green gram, maize, cotton, sesame, groundnut, pigeon pea	95%	0	9
	1	M	55	Literate NS	6	Mixed	2	Sesame, mung bean, pigeon pea, cotton, chick pea, groundnut and maize	70%	0	9

* Self-reported percentage of agricultural production sold to the market; ** self-reported farm size; *** NS = No formal schooling or education; **** HH Mixed livelihoods refers to on- and off-farm/non-farm income generation.

The interview procedure included the following steps [52]:

1. Introduction and collecting data for respondent and household characteristics.
2. Drawing up the timeline: A timeline was drafted on a flipchart, whereby respondents freely included important moments, transitions, turning points, etc. in relation to agricultural practices and their diets.
3. In-depth interview: Respondents described their personal experience in relation with agriculture and their diets in line with the events graphically plotted on the timeline. Particular attention was given to important stages over the life course and how respondents dealt with challenges and stressors.
4. Reflection on healthy food: Respondents were asked to select an item, which they associated with healthy food, and to explain their choice.

The interviews were conducted in Burmese, were recorded, and then transcribed verbatim to English by the translator.

2.3. Data Analysis

Thematic analysis was applied, using QDA Miner Lite software. The concept of general resistance resources, as defined in the salutogenic theory, were used to develop top down coding to identify adaptive strategies. Both top-down and bottom-up coding were applied. For top-down coding, transcripts of the interviews were coded according to the theoretical framework, departing from the characteristics as described for the general resistant resources. For bottom-up coding, a sample of interviews was taken and analyzed. From these transcripts, salient points were underlined and more elaborated sentences were added as comments on the margin. These sentences expressed a slightly higher level of interpretation and were added to the final list of codes. Thus, a combined list of top-down and bottom up coding was compiled. Two researchers did coding and discussed inconsistencies until consensus on the interpretation was reached. Finally, findings were systematically described upon discussion of the clustering of emerging themes in the research team. Quotes supported the results to transmit unique concepts and meanings.

3. Results: Farm Household Adaptive Strategies over the Life Course

All respondents were involved in agriculture since childhood. They generally had to leave school to help their family in the fields or with other income-generation activities, usually upon finishing primary school. Childhood was generally described as a period of poverty during which food was occasionally scarce and it was difficult to purchase other goods (i.e., clothes). After marriage, the main life events described were related to childbirth and child development. Respondents indicated that, compared to their parents, the current generation of farmers moved from semi-subsistence farming to more market oriented farming. As someone said:

> "*When I was a child, my parents were selling half of the total harvest and we were eating the rest. Now, I store my crops in my house and I wait until market prices are going up. We can also send our crops to warehouses. They will keep it for you and you can sell to them at any time. In the past, we did not have this option. Today, I grow crops with a market-oriented view.*" [Male respondent Sar Kyin]

3.1. Agriculture-Related Events and Adaptive Strategies

During youth, respondents worked for others in order to save money and be able to buy land and start a commercial farm for themselves. A common source of additional income in the area was climbing toddy-palm tree (*Borassus flabellifer* L.) to collect the juice (toddy), practiced by young adults, both men and women. The toddy ferments naturally and is locally popular as a beverage.) Young women were also involved in raising animals, the selling of sweets and vegetables, and cotton fabric production. Young men worked as pond diggers, shepherds, gold diggers, farm laborers, woodcutters, brokers, teachers, and cooperative workers. Some temporarily migrated to neighboring villages.

A major life event was leaving the parental house, with having access to land as an important precondition. Some respondents started their own farm by borrowing land while the majority directly purchased the land. The main push factor to leave the parental house was family growth rather than getting married and the consequent pressure on the parental household resources. Five respondents indicated that they inherited a piece of land from their parents.

Respondents highlighted the flexibility in gender role division whereby women could buy and inherit land, manage the farm, and conduct economic activities as a response to the absence of a productive male member of the family. Female respondents indicated that they were not only involved as a labor force in the fields, but they were actively participating or leading the decision-making processes at the farm (household).

The main events and stressors for agriculture in the Dry Zone, where most farming is rain-fed, relate to harvest loss due to climate or water-related problems, pests and pests or infestations, as is illustrated in Figure 3. These events easily endangered families to become indebted. One respondent recalled:

> *"Around 1999–2000, due to intensive rainfalls, < . . . > there was famine, particularly scarcity of rice and also our crop (mung bean) in the field was damaged by fungus and we could not sell it. Therefore, I borrowed money with high interest rate. I faced debt-burden and it was a very difficult time for me. I tried to raise pigs to have an income. In that period, my husband was bitten by a snake and got sick. This created more difficulties in our family."* [Female respondent Kan Zauk]]

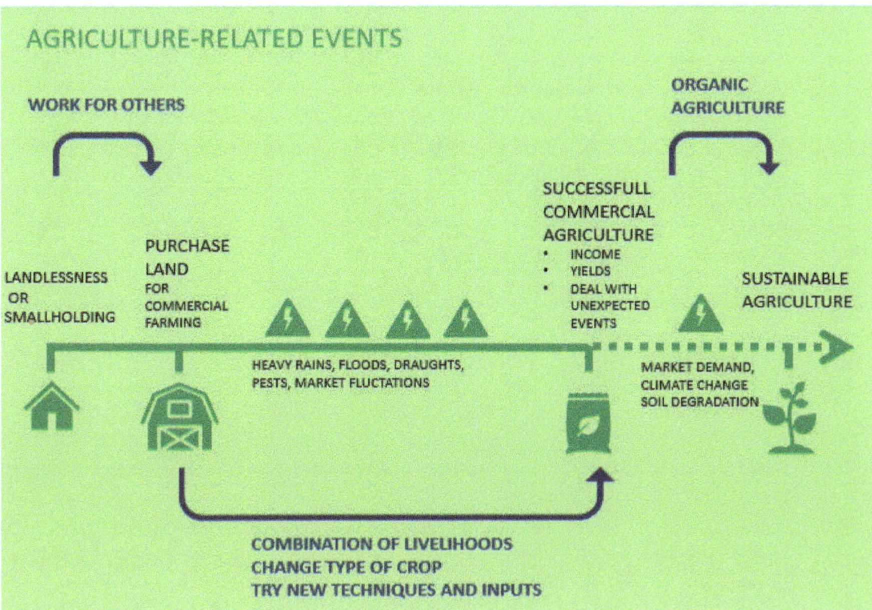

Figure 3. Timeline agriculture-related events; blue line and text indicate adaptive responses.

In response to such stressors, the majority of the respondents engaged in immediate, short term adaptive strategies, such as temporary income generation activities (raising pigs, making handcrafts, off-farm work, etc.); contracting debts, using jewelry as collateral, and selling livestock and carts; or collective action, for instance, to prevent the village from flooding.

More long-term adaptive strategies related predominantly to income stability to provide their family with good living conditions by improving yields and profits, and working with improved inputs and cultivation techniques. Most respondents felt that a stable income could not be achieved by agriculture alone.

Adaptive strategies aiming for improved yields and profits, and being more responsive to price fluctuations were geared through experimentation with new inputs and techniques. Another example was a group of farmers, who organized themselves to rent a truck for selling their products to the market, thus avoiding the intermediation of middlemen, and obtaining better prices and access to new information.

Respondents showed awareness regarding more sustainable ways of food production and consumption, including the possibility to produce organic products. The main problems highlighted related to soil degradation and environment-related health problems. Respondents envisaged continuing to farm, but there were concerns regarding the sustainability of the current agricultural practices. In response, some started to experiment with organic farming. In one village, farmers collaborated to produce organic groundnuts and constructed a mill for common use to produce organic groundnut oil. Others, however, were more skeptical about the economic opportunities, and considered organic farming as going back to methods that are more traditional. Most respondents indicated that they practiced organic food production for household consumption while they were using chemical fertilizers for the crops designated for the market.

Resistance resources used in adaptive strategies for agriculture.

Key resistance resources used in adaptive strategies for agriculture were identified.

- At the individual level, the Buddhist practices and beliefs were mentioned as an important resistance resource to guarantee fortune and good health in the present and in the afterlife. Practicing religion offered support to cope with events beyond people's control. Other individual level resources mentioned were related to (personal) values, knowledge, internal strength, and being healthy. Some respondents explained that they could rely on their own or others' knowledge in the household, and their ability to properly apply it in certain situations.
- At the family-level, family ties were mentioned as the main network of resources, providing material support and a sense of belonging. Through family ties, respondents gained access to different kinds of capital:
    - Natural, i.e., inheritance of family land;
    - Physical, i.e., family assets, such as cattle, carts, bicycles, motorcycles, or agricultural tools;
    - Financial, i.e., credit; relatives were mentioned as a source of financial capital in the form of credit, but also children's remittances and physical help in agriculture were crucial for the sustainment of the household;
    - Human, i.e., educational level of family members. Respondents made efforts to support their children to complete their studies and find jobs outside the agricultural sector, as input from the farm, but also hoping to spare them from the difficulties and struggles experienced by the parents; and
    - Social (i.e., family unity, perseverance); respondents indicated that parental support was particularly relevant in the past, when they were young and lived under the parental roof with the aim to save money for their own future investments.
- At the community level, community or social support from the village or the monasteries were mentioned. Respondents highlighted a sense of unity and solidarity in their communities, which helped farmers to develop their business, for example, farmers organizing to rent a truck to sell directly to the market in order to avoid brokers' intermediation, or villagers mobilizing themselves to secure the riverbank during a flood. Other forms of resistance resources were mutual support among families during hard times, joint production of handcraft, or support of each other by borrowing money at low interest rates. An important community level resource was access to and sharing of agricultural knowledge and information through different sources (traditional

knowledge, observation of other farmers, private companies, trainings by non-governmental organizations, radio, TV, social media, books, university, smart phone). The majority of the respondents indicated that the main source of information on market prices and agricultural inputs and new techniques and inputs stems from fellow farmers.

- At the societal level, respondents mentioned the increased presence of NGOs, private companies, a university, and, to some extent, governmental extension services over the years. These actors provided farmers with trainings (on agriculture, food, or vocational training), access to agricultural inputs (fertilizers, pesticides, seeds, tools, etc.), and market information. The presence of microfinance organizations allowed farmers to access lower interest rates compared to informal moneylenders. Some respondents felt that the support of NGOs was fundamental. Others, however, expressed concerns related to dependency and the real effectiveness of the help received. At the institutional and governance level, the transition towards more democracy created space to form organizations. While, until recently, law in Myanmar forbade meetings of more than five people, during the interviews, it became evident that associations are more common now at the village level. Some respondents expressed the wish to be able to organize more structured farmers' organizations, enabling farmers to improve access to better information and prices and advocate for farmers' rights. Other societal resistance resources were related to improved infrastructure (roads, smartphones, and internet) and inputs (better seeds, fertilizers). Increased mobility of goods, people, and information created better opportunities for the farm households to commercialize their products to the market. Table 2 summarizes an overview of the adaptive strategies in agriculture described by the respondents.

Table 2. Respondents' main adaptive strategies in agriculture.

Stressor	Adaptive Strategies Applied	Goal	Key Resistance Resources
Endangered family living conditions	Agriculture with diversified production (incl. cash crops)	Income stability	**Individual:** Physical health, strength, (tacit) knowledge, perseverance, faith, austere lifestyle **Family:** Parental support, land, financial capital and credit, remittances, labor, education and agricultural knowledge of family members, social support **Community:** Information sharing of agricultural knowledge, sense of unity and solidarity, business collaboration, (social) protection, collaboration for disaster mitigation **Society** agriculture, food, or vocational training by various actors (NGOs, private companies, knowledge institutes), government extension services), microfinance, improved infrastructure, increased opportunities for famers cooperating and organization
	Agriculture and husbandry		
	Agriculture, husbandry, and migration		
	Agriculture, husbandry, and self-employment		
	Agriculture and migration of a family member		
	Agriculture and self-employment		
	Change type of crops in response to climatic conditions		
Harvest loss due to climate or pests	Increase use of chemical pesticides and fertilizers (organic and chemical)	Improve yields and profit	
	Crop rotation		
	Change type of crops in response to market fluctuations	Better inputs and cultivation techniques	
Price fluctuations	Storage of products until prices are favourable to sell	Responsiveness to price fluctuations	
	Improved information on market prices		
	Join forces to facilitate/improve market access		
Viability of farming and food system	Participate in training	Sustainable farming	
	Continue with or re-introduction of rational farming practices/Traditional tillage		
	Organic farming		

3.2. *Farm Households' Diet-Related Events and Adaptive Strategies*

For diet-related life events, respondents described a transition towards better access, availability, and stability of diverse diets. In the past, families used to rely on home production for food. Markets were difficult to access due to road conditions, and could only be reached by car or on foot. Food availability in the past was considered to be even more dependent on seasonality and natural events than today. Rice was rarely cultivated in the area. Three farmers mentioned that the demonetization, leading to political events in 1988, negatively affected the availability of rice and forced them to mix rice to other grains. Most of the respondents (13 out of 14) were used to consuming rice alongside or substituted with other staple foods (i.e., maize, millet), but this was generally associated with poverty and disliked by respondents.

Collecting plants and food from the forest, such as bamboo shoots and mushrooms, was common during childhood. The local food environment offered a wide diversity of non-processed food that was considered stable, available, and accessible by respondents. Traditional crops included beans, rosella, gourd, eggplant, potato, water spinach, groundnut (oil), sesame, tomato, chilly, pumpkin, bitter melon, watermelon, and other plants growing naturally. As was highlighted:

> "*At my parent's time, this area was very poor and vulnerable and we used to eat food in a traditional way. There were available only local vegetables and fruits like beans, rosella and bean leaves that we grow ourselves.*" [Male respondent Kan Zauk]

Respondents indicated that, compared to their childhood, the current diet had improved. They ate fish, meat, eggs, milk, and fruits more often because of increased incomes, improved access to markets, and improved political stability, notably moderating the price of rice (Figure 4). At the moment of the interview, most respondents owned a motorcycle and roads were in good condition even though they were not paved. Meat and (dried) fish were particular food items associated with income increase, but most households also relied on the market for rice.

Respondents were also asked what type of crops were produced for home consumption, and what foods were bought from the market. Main foods grown for home-consumption were maize, sesame, groundnut, pulses and beans, chili, onion, potato, and various vegetables and fruits (bananas, mango, and watermelon). Main foods bought from the market were rice, noodles, chickpeas, yard long beans, spices (garlic, ginger), potato, sweet potato, various vegetables and fruits, oil, eggs, meat, chicken, fish, cookies, tea, and soya chunks. Some farmers were more dependent on the market for access to food. Others made an intentional choice for depending on the market for economic or health-related reasons.

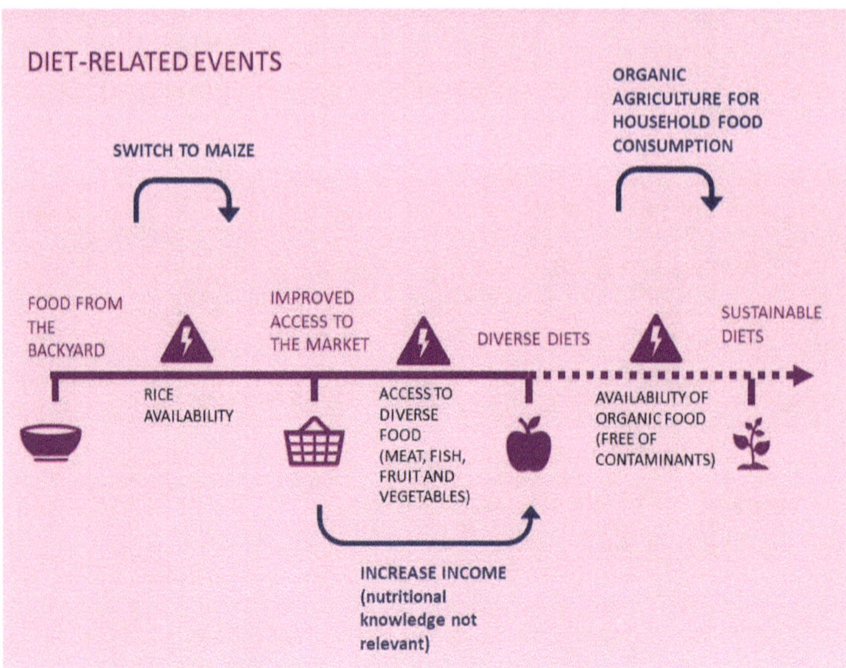

Figure 4. Timeline diet-related events; blue line and text indicate adaptive responses.

When asked about their eating habits, respondents indicated that, traditionally, respondents consumed three meals a day. Meals consisted of a main dish (usually rice or rice noodles) and two or more different curries—made of vegetables, pulses, grains, fish, mushrooms, bamboo shoots, meat—served in small portions and placed in the middle of the table. Women were in charge of food selection and preparation.

Food choice was not primarily guided by scholarly knowledge on nutritious foods or nutrition, oftentimes introduced through training by NGOs. The information of 'declarative'—factual and evidence-based—nutritional knowledge seemed more related to food choice for children, rather than to respondents' choices relating to the production of food and daily choices of food. Fish, meat, eggs, milk, potatoes, and seasonal vegetables were considered good food for children. As someone said:

> *"In the past, we never considered eating healthy food and we were just eating for work and living. We had to eat what we had. Lately we got some money, we buy what we want to eat but without thinking about nutritious food consumption."* [Female respondent Kan Zauk]

Food and nutrition security considerations at the household level were mainly related to the affordability of certain kinds of food (like meat and fish) and access to organic food, understood as access to food free from contaminants. A main concern expressed by respondents connecting health to food related to their awareness around the contamination of food by pesticides, fertilizers, and other products endangering their health. Respondents reported to have learned about it from books, trainings, and from others. For this reason, some of the respondents preferred to grow organic food for their household consumption or buy organic food from other farmers.

Resistance resources used in adaptive strategies for diet-related events.

Key resistance resources used in adaptive strategies for diet-related events were identified, and related to a large extent to what was also found for the adaptive strategies in agriculture. Most prominently:

- At the individual level, practicing religion offered support to cope with events as well as (personal) values, tacit knowledge on food, internal strength, and being healthy.
- At the family-level, the family income as a means to get access to market goods, came out most prominently, but also other kinds of household capital:
    - Physical, i.e., cattle, motorcycles;
    - Financial, i.e., credit; relatives were mentioned as a source of financial capital, children's remittances were crucial for the sustainment of the household;
    - Human, i.e., educational level of family members; and
    - Social (i.e., family unity); respondents indicated that family support was particularly relevant in relation to child and family care. In addition, wives were taking over tasks of the husband in the case of absence or illness.
- At the community level, respondents, as was mentioned for agricultural events, highlighted access to and sharing of knowledge and information on nutrition, health, food preparation, and care, mainly originating from trainings by NGOs, radio, TV, and social media. In addition, knowledge sharing on organic food farming for home consumption was highlighted.
- At the societal level, respondents, as was mentioned for agricultural events, highlighted in particular the role of NGOs as a source of information. In addition, the improved roads, increasing mobility, and access to smartphones and the internet, were re-emphasized. Respondents expressed the wish to be able to organize more structured farmers' organizations. Table 3 summarizes an overview of respondents' adaptive strategies in diet-related events.

Table 3. Respondents' adaptive strategies to diet-related events.

<table>
<tr><th>Stressor</th><th>Adaptive Strategies Applied</th><th>Goal</th><th>Key Resistance Resources</th></tr>
<tr><td rowspan="6">Lack of nutritious food</td><td>Collecting food (from the wild)</td><td rowspan="4">Stability of access to food</td><td rowspan="14">Individual:
Physical health, strength, (tacit) knowledge, faith, austere lifestyle
Family:
Financial capital and credit, remittances, labor, education and tacit knowledge of family members, family care and social support
Community:
Information sharing of knowledge on nutrition, food safety and health, and on organic food production, sense of unity and solidarity
Society
Nutrition and health training by various actors (NGOs, improved infrastructure</td></tr>
<tr><td>Home production of food</td></tr>
<tr><td>Eating less preferred food</td></tr>
<tr><td>Regular meal frequency</td></tr>
<tr><td>Home production of food</td><td rowspan="2">Better dietary diversity</td></tr>
<tr><td>Buying foods from the market, in particular animal sources foods</td></tr>
<tr><td rowspan="5">Lack of market access</td><td>Commercial farming</td><td rowspan="3">Increased income</td></tr>
<tr><td>Remittances</td></tr>
<tr><td>Off-farm labour</td></tr>
<tr><td>Purchase of means of transport</td><td rowspan="2">Better means of transport</td></tr>
<tr><td>Join forces to facilitate/improve market access</td></tr>
<tr><td rowspan="3">Lack of food safety and unhealthy diets</td><td>Participate in training</td><td rowspan="3">Sustainable diets</td></tr>
<tr><td>Continue with or re-introduction of rational farming practices/no contaminants</td></tr>
<tr><td>Organic food production for home consumption</td></tr>
</table>

4. Discussion

The aim of this study was to disentangle adaptive processes in farm households in Myanmar, balancing interests in commercial farming and consumption of nutritious foods, through an in-depth analysis of agricultural life stories (narratives) in order to identify resilient and emergent strategies, incentives, and forms of social innovation.

Relating to the research question 'How do smallholder farmers develop and implement adaptive strategies in response to food system transformations leading to agricultural commercialization, in view of their agricultural livelihoods and diets during their life-course?', our findings indicate that the selected farm households all started as landless or as smallholders, and became successful over time. Transitions had a role in shaping respondents' orientations towards agriculture and food. Most respondents had in common a smooth transition from the parental to the conjugal house, to which the majority attributed an increased sense of wellbeing relating to parental support, as most of them lived under the parental roof, working for others and saving money until they cumulated enough capital to purchase land. These findings are consistent with results reported by Croll [62], who, based on ethnographic studies across East, Southeast, and South Asia, suggested that generations have taken new steps to invest in the intergenerational contract, which has been renegotiated and reinterpreted by both generations in support of a robust and reciprocated cycle of care.

The farm households identified were involved in an agricultural transformation process, shifting from a subsistence-oriented production toward a more market-oriented agriculture. However, during this process of commercialization, farmers did not specialize to become more efficient, as suggested by Jaleta et al. [34], but diversified their agricultural production in order to become more resilient to various types pf stressors. Contrary to the observations of Rerkasem et al. [7], the process of commercialization did not seem to result in a decline in crop diversity per household, as an important adaptive strategy was to change crops in response to climate stressors or market fluctuations. Being flexible and diversifying livelihood strategies emerged as successful strategies to deal with recurrent challenges of different natures. This is consistent with findings reported by Ellis [63]. He found that many of the attributes to diversification as an individual or household level survival strategy might be associated with success at achieving livelihood security under improving economic conditions as well as with livelihood distress in deteriorating conditions rather than just being a strategy of desperation, or a transient phenomenon. Ellis concluded that acquiring the capability to diversify income sources signifies an improvement in the livelihood security and income-increasing capabilities of the rural household, and therefore advocated that policies that reduce constraints to diversification and widen its possibilities are, in general, desirable. Diversification within agriculture to take advantage of new markets is also a desirable policy emphasis.

Farmers' sense-making and decision-making processes considering agriculture engagement were found to differ from those around family food and nutrition security. This is suggested by the finding that a common strategy was to produce (or buy) organic food for household consumption and use chemical inputs for agri-business. Some farmers even expressed a preference for organic, more sustainable ways of agricultural production, but for economic reasons, the majority had to rely on non-organic inputs and practices. The importance of food pricing is an influential factor when it comes to the consumption and production of organic food. Meeker and Haddad had similar observations [17].

Increases in income due to agricultural commercialization and diversified livelihood strategies contributed largely to increased dietary diversity, which is consistent with findings reported by Meeker and Haddad [17]. Income from off-farm sources played an important role for the household wellbeing and especially for accessing food, since all the respondents were dependent from the market for rice. The scarcity of processed food in the local markets induced an increase of fruit, vegetables, meat, and fish consumption. This trend may change in the future, when an influx of processed foods can be expected in response to the on-going transitions in Myanmar.

Consumption patterns did not seem to differ very much from the traditional diets consumed by previous generations, despite the fact that respondents also highlighted an increased access to and availability of (more diverse) foods over their life. This aligns with findings reported by Devine et al. [61] and Rosen [52], who indicated that the positive connotation associated with eating traditional foods might be a result of positive family interactions around food and eating. In addition, as Swan suggested, declarative knowledge (knowing the facts about nutritious food) is less influential for people's diets than procedural knowledge (how to acquire certain skills in relation to food) [64].

Our findings also indicated that Myanmar farmers were able to regain stability and structure after stressful life events, and apply craftiness and fortitude during challenging moments. In the literature, these skills were found to be connected with heathy eating habits [55]. Higher perceived neighborhood collective efficacy is another predictor for healthy eating habits [64].

Our case study showed that women made decisions in relation to both production and consumption of food. Women were in charge of food utilization at the household level. In some cases, they were in charge of the farm and they could buy land. In line with this, women defined themselves as skillful in the art of selecting, purchasing, and preparing good meals for their family. This supports the role of women in ensuring food and nutrition security at the household level [17].

Our findings indicated some evidence for new forms of inclusive community-led organizations taking root, often in connection with the process of democratization in Myanmar. Until recently, in Myanmar, people were not allowed to meet in public at the community level. Limitations for bottom-up forms of organization were still present at the moment of the interview, but some forms of organizations existed and some respondents were actively participating or even leading community groups. Most of the strategies identified in this study, mostly introduced and guided by actors, such as schools, NGOs, or the government, can best be defined as socio-political or socio-organizational innovations derived from the recent possibility of citizens to exercise their rights of free association [65]. Village level organizations promoted grassroots solutions for pressing societal issues (the common mill) and some individuals were willing to organize more systemic innovation involving organizational and institutional frameworks [66].

Overall, this study shows how individuals developed a wide set of adaptive strategies in response to a wide set of stressors (family-, agriculture-, diet-related). Our study confirms that agricultural commercialization and food and nutrition security are interrelated through a set of pathways, which are embedded in local sense-making and decision-making patterns. Turning points that had a positive influence on respondents' ways of producing and consuming food were: (i) Inheritance or purchasing of land; (ii) introduction of better agricultural input; (iii) improved access to the market, and (iv) participation in agricultural and nutritional trainings.

Methodological Considerations

The study was built on a conceptual framework for the analysis of pathways for linking agricultural commercialization to nutritious food consumption [15] to support the analysis at multiple levels. Specifically, the framework helped to disentangle the agricultural household interactions with the external environment. Furthermore, the study combined three theoretical orientations, which proved useful to generate a rich and contextualized description of farm household sense-making processes in relation to the rapid agricultural transitions currently occurring in Myanmar, and the implications for household food and nutrition security. The theoretical lenses of salutogenesis and life-course perspective helped to identify how experiences shaped respondents' connection with agriculture and food. Use of narrative techniques, in particular the timeline technique, generated actor-driven data and generated fruitful discussions on identifying what actually happened over time.

The theoretical lens of positive deviance helped to develop an approach to identify farm households who practiced affordable, acceptable, and sustainable strategies, which might have potential to be adopted and shared within the Myanmar context [67]. Selection criteria for positive deviants, though, are highly dependent on the context of the research [45]. In previous studies involving positive deviants, farmer's income (high), land ownership, and absence of debts were used as criteria for inclusion [68]. For this study, particular importance has been given to farm size [45].

Several limitations to our exploratory study need to be highlighted. Firstly, the case study approach does not support easy generalization of our findings beyond the area of focus. Our sample consisted of respondents selected by local actors. Participation was on a voluntary basis. This may have created a bias in favor of respondents most willing to talk about their lives. Therefore, our findings relating to successful adaptive strategies in response to life-, agriculture-, and diet –related events, cannot simply be extrapolated to other actors or regions. In addition, we have not included non-farm rural households, which may equally suffer from stressors in life, agriculture, and diets, which offers an interesting avenue for future research.

Secondly, the timeline technique builds on techniques for organizational and intercultural learning [69,70], requiring good facilitating skills to manage the conversation and watch over the process of sense making of the actor-driven retrospective recollection of events and the determination of their significance. The fact that we had to work with a local enumerator and with translations of transcripts may have affected the process of capturing all fine details in conversations, thus setting some practical boundaries to our information needs and data collected. In addition, different beliefs about agriculture and food concepts across cultures may have influenced our understanding and interpretation of the information [71].

Thirdly, the positive deviance approach could insufficiently be substantiated with quantitative data for selection of successful farm households, due to a lack of reliable data in Myanmar. Ideally, a case-based qualitative exploratory design, like the one we used, should add or include further quantitative assessments to underpin the contextualized observations [44].

5. Conclusions

This study shows how an understanding of local diets provides insights on possible entry points for nutritionally sensitive agriculture. The diversification of livelihoods and social and emotional components, identified in this study, played a major role in guaranteeing successful outcomes. This suggests a need for alternative strategies moving away from specialization and intensification strategies usually promoted by agri-businesses. This also suggests a need for alternative strategies of (international) NGOs, whose interventionist and project-based approaches usually offer standardized solutions and restricts farmers in mono-directional livelihoods.

This study also shows how important holistic solutions and resilience strategies are for success. Therefore, reinforcing the flexibility and resilience of successful farmers should be a key element to integrate into project strategies. Having diversified livelihood strategies allows farmers to experiment and innovate while holding a strong fallback position represented by other sources of income.

In addition, this study showed how emotional and economic support during youth could represent a solid base for the future. Overall, this study seeks to underline the importance for policies and interventions to be informed by participatory and holistic baseline assessment whereby the theory of change of all the relevant stakeholders is taken in account. As emerged from this study, declarative knowledge transmitted through formative training did not seem to have significant implications for people's food choices. The inclusion of procedural knowledge in food and nutrition security programs and the impact of procedural knowledge transmission rather than declarative could represent an interesting field of research.

6. Ethical Considerations

All participants entered into the research with voluntary consent. They were provided with information about the purpose and contents of the study. Guarantees of confidentiality and anonymity were given prior to each interview. Moreover, participants were able to withdraw from the study at any time for any reason. The authors declare that the study was conducted in accordance with general ethical guidelines for behavioral and social research in the Netherlands. These guidelines stipulate that behavioral research falls outside the scope of the Act on review of medical research involving human subjects (WMO) when a study is not of a medical nature, and subjects do not receive a particular treatment or are asked to behave in a particular way [72].

Author Contributions: M.H. provided overall methodological guidance and facilitated the writing process by providing draft text. M.G. conducted fieldwork in Myanmar. B.P. and J.B. contributed especially with the social innovation part. M.H., B.P. and D.B. were the main authors of the methodological framework presented in Annex A1.

Funding: This research was funded by The Netherlands Ministry of Agriculture, Nature and Food Quality, Knowledge Base Research Funds 22 and 28.

Acknowledgments: We acknowledge the support of the Dutch Embassy and our colleague Jan Verhagen in allowing us to be introduced with potentially interested contacts in Myanmar. In particular, we express our gratitude to Greenovator, Myanmar Heart Development and Zaw Win Kyaw in supporting the fieldwork and data collection.

Conflicts of Interest: The authors declare no conflict of interest.

Acronyms

ASEAN	Association of South East Asia Nations
FAO	United Nations Food and Agricultural Organisation
FNS	Food and Nutrition Security
LMICs	Low and Middle Income Countries
NGOs	Non-governmental Organizations
SDG	Sustainable Development Goal
UNICEF	United Nations Children's Fund
WCDI	Wageningen Centre for Development Innovation,
WFP	World Food Program
WUR	Wageningen University & Research

References

1. USAID. *Convergence and Tension in Nutrition-Sensitive Agricultural Market Development Activities [Draft]*; Discussion Paper; USAID: Washington, DC, USA, 2016.
2. Westhoek, H.; Ingram, J.; Van Berkum, S.; Özay, L.; Hajer, M. *Food Systems and Natural Resources*; United Nations Environment Programme-International Resource Panel: Paris, France, 2016.
3. Carletto, G.; Ruel, M.; Winters, P.; Zezza, A. Farm-Level Pathways to Improved Nutritional Status: Introduction to the Special Issue. *J. Dev. Stud.* **2015**, *51*, 945–957. [CrossRef]
4. Hawkes, C.; Ruel, M. Value Chains for Nutrition. In *Reshaping Agriculture for Nutrition and Health*; IFPRI, Ed.; IFPRI Publications: Washington, DC, USA, 2012.

5. Pingali, P.L. Environmental consequences of agricultural commercialization in Asia. *Environ. Dev. Econ.* **2001**, *6*, 483–502. [CrossRef]
6. Pingali, P.L.; Rosegrant, M.W. Agricultural commercialization and diversification: Processes and policies. *Food Policy* **1995**, *20*, 171–185. [CrossRef]
7. Rerkasem, K.; Lawrence, D.; Padoch, C.; Schmidt-Vogt, D.; Ziegler, A.D.; Bruun, T.B. Consequences of swidden transitions for crop and fallow biodiversity in Southeast Asia. *Hum. Ecol.* **2009**, *37*, 347–360. [CrossRef]
8. Anderman, T.L.; Remans, R.; Wood, S.A.; DeRosa, K.; DeFries, R.S. Synergies and tradeoffs between cash crop production and food security: A case study in rural Ghana. *Food Secur.* **2014**, *6*, 541–554. [CrossRef]
9. Miller, D.D.; Welch, R.M. Food system strategies for preventing micronutrient malnutrition. *Food Policy* **2013**, *42*, 115–128. [CrossRef]
10. Fisher, J. *Myanmar's 2015 Landmark Elections Explained*; BBC, Ed.; BBC: London, UK, 2015.
11. Haggblade, S.; Boughton, D.; Cho, K.M.; Denning, G.; Kloeppinger-Todd, R.; Oo, Z.; Sandar, T.M.; Than, T.M.; Wai, N.E.M.A.; Wilson, S.; et al. Strategic choices shaping agricultural performance and food security in Myanmar. *J. Int. Aff.* **2014**, *67*, 55–71.
12. Rammohan, A.; Pritchard, B. The role of landholding as a determinant of food and nutrition insecurity in rural Myanmar. *World Dev.* **2014**, *64*, 597–608. [CrossRef]
13. Wilson, S.; Aye Wai, N.E.M. *Background Paper No. 4: Food and Nutrition Security in Burma*; MDRI-CESD, Ed.; USAID: Washington, DC, USA, 2013.
14. Mercy Corps. *What Holds Vegetables Farmers Back? Conflict, Governance and Markets Assessment. Making Vegetable Markets Work for Smallholders Program*; LIFT Fund: Yangon, Myanmar, 2015.
15. Herens, M.; Peters, B.; Brouwers, J.; Bosch, D.; Van der Maden, E.; Linderhof, V.; Verhagen, J. *Commercially Viable Agriculture and Consumption of Nutritious Foods: A Framework for Identifying Development Pathways*; Wageningen Centre for Development: Wageningen, The Netherlands, 2018; p. 53.
16. Hawkes, C.; Ruel, M. *From Agriculture to Nutrition: Pathways, Synergies and Outcomes*; World Bank Agriculture and Rural Development Department: Washington, DC, USA, 2007.
17. Meeker, J.; Haddad, L. *A State of the Art Review of Agriculture-Nutrition Linkages*; An AgriDiet Position Paper; Department of Food Business and Development, University College Cork: Cork, Ireland, 2013.
18. UNICEF. *Conceptual Framewrok for Nutrition*; UNICEF: New York, NY, USA, 1991.
19. Black, R.E.; Victora, C.G.; Walker, S.P.; Bhutta, Z.A.; Christian, P.; de Onis, M.; Ezzati, M.; Grantham-McGregor, S.; Katz, J.; Martorell, R. Maternal and child undernutrition and overweight in low-income and middle-income countries. *Lancet* **2013**, *382*, 427–451. [CrossRef]
20. FAO/FIVIMS. *FAO/FIVIMS Framework: Linkages between the Overall Development Context, the Food Economy Households, and Individual Measures of Well-Being*; FAO: Rome, Italy, 2008.
21. Webb, P. *Impact Pathways from Agricultural Research to Improved Nutrition and Health: Literature Analysis and Research Priorities*; Background Paper Prepared for the ICN2; FAO: Rome, Italy, 2013.
22. Herforth, A.; Harris, J. Understanding and Applying Primary Pathways and Principles Improving Nutrition through Agriculture. In *Technical Brief Series Strengthening Partnerships, Results, and Innovations in Nutrition Globally (SPRING) Project*; USAID: Arlington, VA, USA, 2014.
23. Von Braun, J. Agricultural commercialization: Impacts on income and nutrition and implications for policy. *Food Policy* **1995**, *20*, 187–202. [CrossRef]
24. Kanter, R.; Walls, H.L.; Tak, M.; Roberts, F.; Waage, J. A conceptual framework for understanding the impacts of agriculture and food system policies on nutrition and health. *Food Secur.* **2015**, *7*, 767–777. [CrossRef]
25. Antonovsky, A. The salutogenic model as a theory to guide health promotion. *Health Promot. Int.* **1996**, *11*, 11–18. [CrossRef]
26. Antonovsky, A. *Health, Stress and Coping*; Jossey-Bass: San Francisco, CA, USA, 1979.
27. Mittelmark, M.B.; Bauer, G.F. The Meanings of Salutogenesis. In *The Handbook of Salutogenesis*; Mittelmark, M.B., Sagy, S., Eriksson, M., Bauer, G.F., Pelikan, J.M., Eds.; Springer International Publishing: Cham, Switzerland, 2017; pp. 7–13.
28. Lindström, B.; Eriksson, M. *The Hitchhiker's Guide to Salutogenesis: Salutogenic Pathways to Health Promotion*; Folkhälsan Research Center: Helsinki, Finland, 2010.
29. Mittelmark, M.B.; Sagy, S.; Eriksson, M.; Bauer, G.F.; Pelikan, J.M. *The Handbook of Salutogenesis*; Springer: Cham, Switzerland, 2017.

30. Pascale, R.T.; Sternin, J.; Sternin, M. *The Power of Positive Deviance: How Unlikely Innovators Solve the World's Toughest Problems*; Harvard Business Press: Boston, MA, USA, 2010; Volume 1.
31. Nicholls, A.; Simon, J.; Gabriel, M.; Whelan, C. *New Frontiers in Social Innovation Research*; Springer: Berlin, Germany, 2016.
32. Olsson, P.; Moore, M.-L.; Westley, F.R.; McCarthy, D.D.P. The Concept of the Anthropocene as a Game-Changer. *Ecol. Soc.* **2017**, *22*, 31. [CrossRef]
33. HLPE. *Food Losses and Waste in the Context of Sustainable Food Systems*; FAO, UN Committee on World Food Security: Rome, Italy, 2014.
34. Jaleta, M.; Gebremedhin, B.; Hoekstra, D. Smallholder Commercialization: Processes, Determinants and Impact. In *Improving Productivity and Market Success of Ethiopian Farmers Improving Market Opportunities*; International Livestock Research Institute (ILRI): Addis Ababa, Ethiopia, 2009.
35. Gebremedhin, B.; Jaleta, M. *Commercialization of Smallholders: Does Market Orientation Translate into Market Participation?* ILRI (aka ILCA and ILRAD): Nairobi, Kenya, 2010.
36. UNSCN. *Meeting of the Minds on Nutrition Impact of Food Systems*; United Nations Standing Committee on Nutrition (UNSCN): Geneva, Switzerland, 2013.
37. Griffiths, C.A.; Ryan, P.; Foster, J.H. Thematic analysis of Antonovsky's sense of coherence theory. *Scand. J. Psychol.* **2011**, *52*, 168–173. [CrossRef] [PubMed]
38. Bowling, A.; Ebrahim, S. *Handbook of Health Research Methods: Investigation, Measurement and Analysis*; McGraw-Hill Education: London, UK, 2005.
39. Ah Poe, C. Food Security Assessment in the Dry Zone Myanmar. 2011. Available online: http://documents.wfp.org/stellent/groups/public/documents/ena/wfp234780.pdf?iframe (accessed on 14 May 2017).
40. Shaw, J.E. Assessment of Stakeholder Interventions for Sustainable Agriculture in Myanmar's Dry Zone. 2016. Available online: http://uncapsa.org/?q=publications/assessment-stakeholder-interventions-sustainable-agriculture-myanmar%E2%80%99s-dry-zone (accessed on 14 May 2017).
41. Ministry of Health and Sports and ICF. Myanmar Demographic and Health Survey 2015-16. In *The DHS Program*; MoHS and ICF: Naypyidaw, Myanmar, 2017.
42. Thorpe, J. *Delivering Prosperity in Myanmar's Dry Zone. Lessons from Mandalay and Magwe on Realizing the Economic Potential of Small-Scale Farmers*; Oxfam Briefing Paper; Oxfam GB: Oxford, UK, 2014.
43. MHDO. *Dry Zone Food for Asset Creation Project WFP*; Pakokku & Ngape Townships, Magwe Region, WFP, Eds.; Mynamar Heart Development Organization (MHDO): Yangon, Myanmar, 2017.
44. Biggs, S. Learning from the positive to reduce rural poverty and increase social justice: Institutional innovations in agricultural and natural resources research and development. *Exp. Agric.* **2008**, *44*, 37–60. [CrossRef]
45. Lowder, S.K.; Skoet, J.; Raney, T. The number, size, and distribution of farms, smallholder farms, and family farms worldwide. *World Dev.* **2016**, *87*, 16–29. [CrossRef]
46. Wielinga, E.; Vrolijk, M. Language and Tools for Networkers. *J. Agric. Educ. Ext.* **2009**, *15*, 205–217. [CrossRef]
47. Edquist, S.; Koch, P.; Contento, I.R. Understanding Adolescent Food Choices through Personal Food Stories. *J. Nutr. Educ. Behav.* **2009**, *41*, S16–S17. [CrossRef]
48. Newcombe, M.A.; McCarthy, M.B.; Cronin, J.M.; McCarthy, S.N. "Eat like a man". A social constructionist analysis of the role of food in men's lives. *Appetite* **2012**, *59*, 391–398. [CrossRef]
49. Papathomas, A.; Lavallee, D. Narrative Constructions of Anorexia and Abuse: An Athlete's Search for Meaning in Trauma. *J. Loss Trauma* **2011**, *17*, 293–318. [CrossRef]
50. McCormack, L.; Lewis, V.; Wells, J.R. Early Life Loss and Trauma: Eating Disorder Onset in a Middle-Aged Male—A Case Study. *Am. J. Mens Health* **2014**, *8*, 121–136. [CrossRef] [PubMed]
51. Busanich, R.; McGannon, K.R.; Schinke, R.J. Expanding understandings of the body, food and exercise relationship in distance runners: A narrative approach. *Psychol. Sport Exerc.* **2012**, *13*, 582–590. [CrossRef]
52. Rosen, L. Salutogenic Lives: Stories of Healthy Eaters. In *Health and Society, Social Sciences Department*; Wageningen University: Wageningen, The Netherlands, 2015.
53. O'Kane, G.; Pamphilon, B. The importance of stories in understanding people's relationship to food: Narrative inquiry methodology has much to offer the public health nutrition researcher and practitioner. *Public Health Nutr.* **2016**, *19*, 585–592. [CrossRef] [PubMed]
54. Wielinga, E.; Herens, M. Monitoring Creative Processes: The Timeline and Learning History. In *21st European Seminar on Extension Education*; Routledge: Antalya, Turkey, 2013.

55. Hughes, I. Action research in healthcare. In *The Sage Handbook of Action Research: Participative Inquiry and Practice*; Reason, P., Bradbury, H., Eds.; Sage: London, UK, 2008; pp. 381–393.
56. Koelen, M.A.; van den Ban, A.W. *Health Education and Health Promotion*; Wageningen Academic Publishers: Wageningen, The Netherlands, 2004; p. 285.
57. Savin-Baden, M.; Niekerk, L.V. Narrative Inquiry: Theory and Practice. *J. Geogr. High. Educ.* **2007**, *31*, 459–472. [CrossRef]
58. Polkinghorne, D.E. Validity Issues in Narrative Research. *Qual. Inq.* **2007**, *13*, 471–486. [CrossRef]
59. Sheridan, J.; Chamberlain, K. The Power of Things. *Qual. Res. Psychol.* **2011**, *8*, 315–332. [CrossRef]
60. Devine, C.M. A life course perspective: Understanding food choices in time, social location, and history. *J. Nutr. Educ. Behav.* **2005**, *37*, 121–128. [CrossRef]
61. Devine, C.M.; Connors, M.; Bisogni, C.A.; Sobal, J. Life-Course Influences on Fruit and Vegetable Trajectories: Qualitative Analysis of Food Choices. *J. Nutr. Educ.* **1998**, *30*, 361–370. [CrossRef]
62. Croll, E.J. The Intergenerational Contract in the Changing Asian Family. *Oxf. Dev. Stud.* **2006**, *34*, 473–491. [CrossRef]
63. Ellis, F. Household strategies and rural livelihood diversification. *J. Dev. Stud.* **1998**, *35*, 1–38. [CrossRef]
64. Swan, E.; Bouwman, L.; Hiddink, G.J.; Aarts, N.; Koelen, M. Profiling healthy eaters. Determining factors that predict healthy eating practices among Dutch adults. *Appetite* **2015**, *89*, 122–130. [CrossRef] [PubMed]
65. Wigboldus, S. *Ten Types of Social Innovation—A Brief Discussion Paper*; Wageningen CDI Working Paper; WCDI, Ed.; Wageningen University and Research: Wageningen, The Netherlands, 2016.
66. Van der Have, R.P.; Rubalcaba, L. Social innovation research: An emerging area of innovation studies? *Res. Policy* **2016**, *45*, 1923–1935. [CrossRef]
67. Marsh, D.R.; Schroeder, D.G.; Dearden, K.A.; Sternin, J.; Sternin, M. The power of positive deviance. *BMJ* **2004**, *329*, 1177–1179. [CrossRef] [PubMed]
68. De Adelhart Toorop, R.; Gosselink, K. *Analysis of Positive Deviants Among Organic Dairy Farmers in The Netherlands*; Wageningen University: Wageningen, The Netherlands, 2013.
69. Brunton, M.A.; Jeffrey, L.M. Using the critical incident technique for triangulation and elaboration of communication management competencies. *J. Vocat. Educ. Train.* **2010**, *62*, 239–255. [CrossRef]
70. McAllister, L.; Whiteford, G.; Hill, B.; Thomas, N.; Fitzgerald, M. Reflection in intercultural learning: Examining the international experience through a critical incident approach. *Reflective Pract.* **2006**, *7*, 367–381. [CrossRef]
71. Bonevski, B.; Randell, M.; Paul, C.; Chapman, K.; Twyman, L.; Bryant, J.; Brozek, I.; Hughes, C. Reaching the hard-to-reach: A systematic review of strategies for improving health and medical research with socially disadvantaged groups. *BMC Med. Res. Methodol.* **2014**, *14*, 42. [CrossRef] [PubMed]
72. CCMO. Manual for the review of medical research involving human subjects. In *Behavioural Research and the WMO*; CCMO [Central Committee on Human Related Research]: The Hague, The Netherlands, 2002; pp. 66–69.

sustainability

MDPI

Article

Intensification and Upgrading Dynamics in Emerging Dairy Clusters in the East African Highlands

Jan van der Lee [1,*], Laurens Klerkx [2], Bockline Omedo Bebe [3], Ashenafi Mengistu [4] and Simon Oosting [5]

1 Wageningen UR Livestock Research, PO Box 338, 6700 AH Wageningen, The Netherlands
2 Knowledge, Technology and Innovation Group, Wageningen University, 6700 EW Wageningen, The Netherlands; laurens.klerkx@wur.nl
3 Department of Animal Sciences, Egerton University; Njoro, 20115 Egerton, Kenya; bbebe@egerton.ac.ke
4 College of Veterinary Medicine and Agriculture, Addis Ababa University, Bishoftu, Ethiopia; mengistuashenafi@gmail.com
5 Department of Animal Science, Wageningen University; 6708 WD Wageningen, The Netherlands; simon.oosting@wur.nl
* Correspondence: jan.vanderlee@wur.nl; Tel.: +31-317-481348

Received: 10 October 2018; Accepted: 14 November 2018; Published: 21 November 2018

check for updates

Abstract: Based on farmer and value chain actor interviews, this comparative study of five emerging dairy clusters elaborates on the upgrading of farming systems, value chains, and context shapes transformations from semi-subsistent to market-oriented dairy farming. The main results show unequal cluster upgrading along two intensification dimensions: dairy feeding system and cash cropping. Intensive dairy is competing with other high-value cash crop options that resource-endowed farmers specialize in, given conducive support service arrangements and context conditions. A large number of drivers and co-dependencies between technical, value chain, and institutional upgrading build up to system jumps. Transformation may take decades when market and context conditions remain sub-optimal. Clusters can be expected to move further along initial intensification pathways, unless actors consciously redirect course. The main theoretical implications for debate about cluster upgrading are that co-dependencies between farming system, market, and context factors determine upgrading outcomes; the implications for the debate about intensification pathways are that they need to consider differences in farmer resource endowments, path dependency, concurrency, and upgrading investments. Sustainability issues for consideration include enabling a larger proportion of resource-poor farmers to participate in markets; enabling private input and service provision models; attention for food safety; and climate smartness.

Keywords: agribusiness cluster; commercialization; sustainable intensification; dairy value chain; farming system; service arrangements; Ethiopia; Kenya

1. Introduction

Upgrading of dairy farming and value chains has been promoted by policy makers and development practitioners as a promising pathway to deal with the sustainability challenges of mixed crop–livestock systems [1,2]. These challenges include alleviation of rural poverty, supply of sufficient and safe food to growing urban populations, alleviation of rural poverty, and making farming climate-smart [3]. Of all livestock farming systems in the world, mixed crop–livestock systems produce the majority of livestock output and constitute the majority of livestock-keeping households, often smallholders [3,4]. Therefore, prospects for these systems to become more market-oriented and sustainably intensify are matters of academic, political, and societal interest [3,5].

Studies on the commercialization of milk production repeatedly show the complexity of the transition from semi-subsistence to market-oriented dairy farming, which is often associated with intensification and specialization [6–10]. For this transition, many farm practices may need innovation in areas such as feeding, housing, and output marketing. These innovations contribute to upgrading, defined as changes in the production process to increase productivity and added value and to improve product quality [1,2]. They require higher input levels, for which farmers need sufficient access to external resources, inputs, and services, both pre- and post-production [3,6]. In practice, upgrading occurs in so-called agribusiness clusters, i.e., geographic concentrations of producers and other actors engaged in the same subsector that facilitate the required linkages to input and output markets [11]. In clusters, the range and types of input–output connections for dairy farms and small and medium enterprises are increased, positively influencing knowledge creation and transfer between actors, enabling them to benefit from economies of scale (e.g., volumes of inputs and outputs) and scope (e.g., use of imported semen and sale of milk to new markets) [12–15].

Many studies have focused on understanding the drivers and bottlenecks affecting upgrading of dairy farming systems and value chains. These drivers include breeds; farm size; access to capital, inputs, and services; demand for dairy products; collective action; infrastructure and policies [7–10,16–18]. Literature yields limited analysis, however, of how these upgrading processes facilitate dairy cluster emergence and transformation to more market-oriented dairy farming, as most studies focus on a particular type of upgrading, on partial processes, or on single cases. Moreover, various authors have indicated that looking at the socioeconomic context aids comprehension of changes in agricultural practices and upgrading of farming systems [19–21]. It is apparent that understanding the complex dynamics of dairy farming systems requires assessment of upgrading in three domains: farming system, market, and context (including biophysical, institutional, and social conditions) [22,23]. However, empirical analysis of these dynamics remains limited. A comprehensive analysis of multiple clusters in comparable transition trajectories is expected to offer insights into the upgrading dynamics, causes of variation, and interactions between the three domains.

The present study, therefore, explores how interactions of the farming system with market and context determine upgrading pathways and outcomes. In particular, it (1) describes the present status of regional clusters; (2) assesses upgrading pathways; and (3) analyzes how interactions affect pathways and outcomes of upgrading. It compares five emerging clusters in the Kenyan and Ethiopian tropical highlands that vary in upgrading status. In all these clusters, dairy farmers face the question of whether or not to transition from 'marketing of small surplus to local markets' to 'commercial supply to wholesale chains' [3].

By looking systemically at these interactions, this paper contributes to the debates about upgrading in clusters, value chains, and farming systems; inclusion of smallholders in markets; system jumps; and pathways to sustainable intensification. The results can be used in devising future scenarios for system development and in co-design of interventions, as outlined by Martin et al. [24]. They inform strategic upgrading options for farmers and other value chain actors by pointing at the future shape of farm operations and the markets to supply to.

2. Methodology

2.1. Analytical Framework—Two Subsystems in Context

The analytical framework for this study considers that farming systems evolve because of the interaction with the market and context within a cluster (Figure 1). We take the dairy farming system within an emerging cluster as the main unit of analysis (A), from which we analyze linkages with and influences from the other two domains—market system [25] (B) and context (C)—taking into account inter-farm variation within clusters. Upgrading, defined above, can occur in all three domains and in this study is respectively distinguished as technical, value chain, and institutional upgrading [1,2]. Upgrading leads to system change (transition) and ultimately to alternative system state (transformation). The three forms of upgrading collectively can lead to commercial dairy farming and to the emergence of dairy clusters [26]. Transformation to a next development stage requires significant upfront investments in new practices, technologies, innovation system, etc. [3,27].

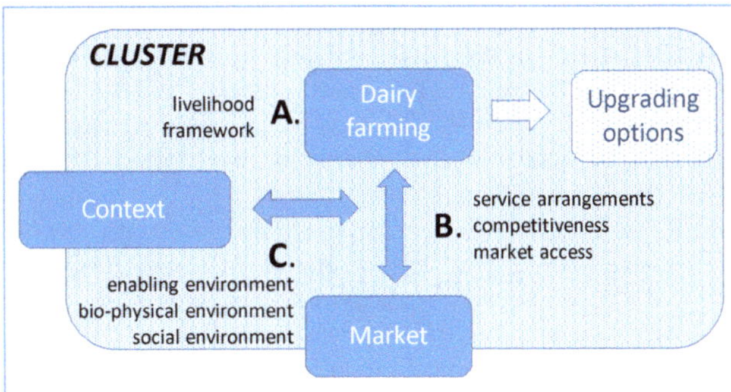

Figure 1. Dairy farming system upgrading options as a result of interaction between farm, market, and context within a dairy cluster.

We view the farming system and market system as two interacting, co-evolving systems within dairy clusters, each of which may experience 'system jumps' between development stages [3]. Various system behaviors can be expected, as described by Schiere et al. [27], depending on the specific farm, market, and context factors that influence farmers' livelihood strategy choices. These may include 'adaptive cycles', where change is episodic and periods of slow accumulation of capital (e.g., nutrients) are punctuated by release of capital and reorganization, for example by a forest fire or an epidemic; and 'lock-in', where systems get used to particular routines [27].

We build on two approaches for farming system analysis: (1) The farmers' perspective of Oosting et al. [3], who in their LIVCAF model describe the transition from 'rural farmers supplying to rural consumers' to 'rural farmers supplying to urban consumers'; and (2) The market quality perspective of Duncan et al. [28], who found that well-developed markets with good procurement and support service arrangements are key to sustainable dairy intensification, and that better market quality is associated with a higher proportion of improved cows that are better fed (sustainability here is used in the blended approach advocated by Mockshell and Kamanda [5]).

In all clusters, the primary driver for upgrading is the decline in livelihood due to diminishing farm size, mainly as a result of population growth [29]. This requires intensification, i.e., the increased use of external inputs and services to increase outputs per unit of input [6], in this case land use. We analyze upgrading dynamics by identifying and exploring changes in farming and marketing practices, as well as the secondary drivers that influence these; these act as accelerators of upgrading if present and as inhibitors if absent.

Analysis of upgrading dynamics thus includes three components:

A. Farming system factors—Technical upgrading of the farming system is explored based on the sustainable livelihoods framework [30]. This considers how farmers combine the different types of livelihood *resources* they own or can get access to into livelihood *activities*, such as food and cash cropping, livestock-keeping, and off-farm activities, using a variety of *practices*, which often reinforce each other [31,32]. Farmers optimize several objectives into a livelihood strategy [33]. We thus assessed dynamics in the current mixed crop–dairy farming systems by looking at changes in the livestock and crops grown and at their functions in the farm, e.g., livestock for meat, milk, manure, draft power, social functions, household food, or sale; crops for food or sale [6].

B. Farm–market interaction—Value chain upgrading changes the way a farm interacts with the market. Following the Windmill approach of Leonardo et al. [34], we explored the influence of the various service arrangements that determine farmers' options for marketing their produce. We looked at farmers' access to markets, associated transaction costs, and fit of service arrangements with particular degrees of market integration [13,17]. The service arrangements offer varying degrees and combinations of the horizontal (between farmers) and vertical (with input and output side chain actors) coordination that are necessary to effectively integrate smallholders into markets [11,35]. Market-integrated dairy requires a large variety of pre-production inputs and professional services, so this typology needs to cover service arrangements on both the input and output side.

C. Context influence on farm–market interaction—Lastly, several context factors significantly influence farm–market interaction and determine the need for *institutional upgrading*, i.e., the improvement of institutional voids that constrain value chain operations [1,23]. We considered three types of factors: (1) factors in the biophysical environment, which include land-use patterns, infrastructure (roads and utilities), climate and weather, animal and crop pests and diseases, risks of natural and human-induced disasters (such as droughts and wars), seasonality of production, and environmental impact of farming, including effects of agro-chemical use [18,36]; (2) factors in the enabling context, i.e., the regulatory framework elements and their enforcement (such as agricultural policies, subsidies, access to finance, property rights, and quality standards) that determine whether the institutional context enables upgrading [16,23,29,37,38]; and (3) factors in the social environment, i.e., social identity and (dairy) farming history [39].

2.2. Data Collection and Analysis

Case study sites were selected from the highlands of Ethiopia and Kenya, home to significant dairy production on a large number of smallholder mixed farms and a smaller number of medium- and large-scale dairy farms. The two countries differ in terms of sociopolitical context. The presence, reliability, and attractiveness of market service arrangements for pre- and post-production inputs and services vary between and within countries, leading to differences in market quality [28].

Sub-regional administrative units of roughly similar size were chosen as starting points for cluster selection: Ethiopian Zones and Kenyan Counties. Based on a scoping exercise and team knowledge, in each country two emerging clusters were selected that have good and comparable agro-ecological potential for dairying (located between 1750 and 3000 m above sea level) but differ in market quality (see Figure 2). Milk production differs widely between clusters. For example, while Nyandarua and Nandi counties are roughly equivalent in terms of arable land, human population, and cattle herd size, the annual milk production in Nyandarua is nearly three times that of Nandi [40,41] (see Supplementary Material S1 for more detail). Due to two distinctly different milk-marketing situations within Nandi County, Nandi was divided into two clusters. To capture the within-cluster variety in market quality, six villages were selected per cluster, with the exception of East Shoa and Nyandarua clusters, where three and nine villages were selected respectively (see Figure 2 for location of study sites). Villages vary in access to rural service centers and end markets, with one-third each having good, medium, and poor access to a service center, located at zero, one, and two hours' walk from a service center respectively.

Interviews with farmer groups and with other value chain actors occurred between September 2016 and May 2017. Dairy farmer group interviews (FGIs) were held in all thirty villages, with group numbers ranging from five to eleven participants, averaging eight. In Arsi, East Shoa and Nyandarua clusters, all farmers who had been interviewed as part of a previous study [42] were invited; in Nandi North and Nandi South, a new sample was invited to participate in FGIs. Farmers were purposively sampled to represent the range of dairy farm sizes in the village. The FGIs used a questionnaire with open questions for discussion and a number of participatory ranking exercises, focusing on both current situation and historic developments. The latter used either importance ranking or the ten seed technique [43], which was modified to use twenty seeds in case answers exceeded five items. Farm classification categories offered by FGIs were harmonized, as categories such as 'small scale' and 'medium scale' are context-specific; some categories were combined. Questions about dairy experience, farm acreage, number of dairy cows and main crops grown were included in the FGIs in Nandi; for other clusters, these data were derived from previous dairy farmer interviews [42].

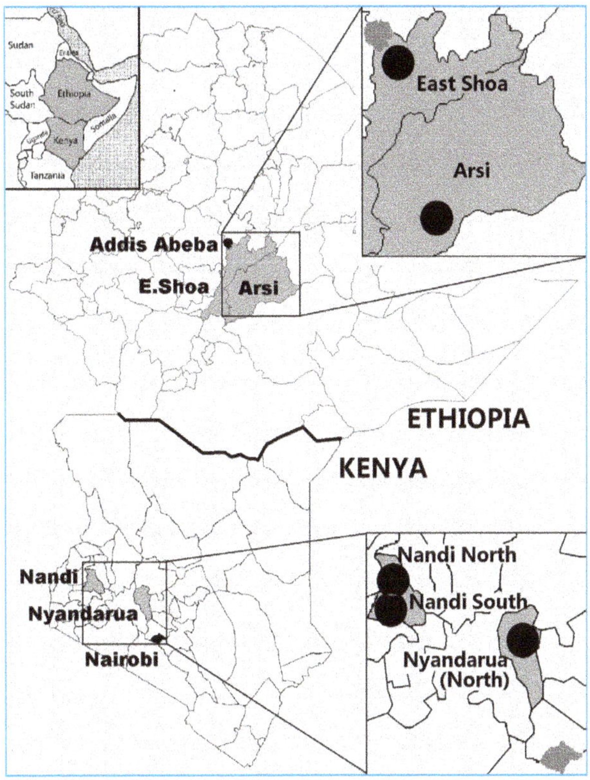

Figure 2. Map of Ethiopia and Kenya with study clusters and study sites.

For value chain actor interviews (VCAIs), dairy actors were selected by using information from earlier farmer interviews [42] and by snowballing. A broad range of value chain actors was covered: private and public suppliers of pharmaceutics, agro-chemicals, semen, feeds, forage, and equipment; private and public providers of artificial insemination (AI), veterinary, extension, and financial services; milk and butter traders, transporters, and dairy processors; cooperative societies and farmer groups; and development agencies and knowledge institutes (see Supplementary Material S1 for portrayal of dairy value chains in Ethiopia and Kenya). VCAIs numbered 118 in total (18 in East Shoa, 20 in Arsi, 43 in Nyandarua, 18 in Nandi North, 10 in Nandi South and 9 with multi-county actors in Kenya).

Secondary factors assessed in the FGIs and VCAIs—which act as drivers of upgrading and transition if present and as bottlenecks if absent—were derived from literature [11,16,18,28,30,44–46]:

- Farming system internal factors: Changes in farmer livelihood strategies, practices, outcomes, and resources (also called capitals or assets) including natural (land acreage and soils, water, climate and weather, herd size and genetics, functions of and interaction between livestock and crops used); economic (capital); physical (farm structures, equipment); human (labor, knowledge and skills); and social resources (networks, groups)

- Market factors: Dairy pre- and post-production service arrangements and service offer; farmer utilization and satisfaction; demand for dairy products (product, price, place); scarcity of inputs, services, and production factors; key marketing institutions, such as competition, role division in service supply, availability of market information, actor relationships, and milk quality assurance
- Context factors: Collective action; dairy history and identity; consumer preferences; conducive infrastructure; access to production factors; regulatory space for private services; policy priority/instruments, public services, and subsidies; social inclusion and environmental impact.

Analysis—FGI and VCAI recordings were transcribed. Along with notes made during FGIs, they were analyzed in Atlas.ti using secondary factors as codes. Differences between clusters were rated by the first author based on data analysis. Results from FGI ranking exercises were translated into percentages and tabulated along with quantitative data; simple statistics were calculated.

3. Results

3.1. Cluster Description

The five clusters selected are briefly described using the schematic positioning of their specialization and upgrading dynamics along two axes (Figure 3): feeding system and cash crop types. These axes denote the variation and recent upgrading in farming systems that, under pressure of land shortage, intensify in different ways along two directions (as observed in clusters studied): a feeding system transition from 'grazing with crop residue use' (low dairy intensity—L_d) to 'zero-grazing with planted forage' (high dairy intensity—H_d) and a cash crop transition from 'grains' (low cropping intensity—L_c) to 'horticulture and/or perennials' (high cropping intensity—H_c).

The clusters are thus characterized as (Table 1):

I. Dairy clusters—H_dH_c Nandi North and Nyandarua gradually specialize to dairy and become increasingly market-oriented; there is significant milk collection by cooperatives and processors; increasingly sophisticated types of service arrangements exist; other cash crops or livestock products are produced as a second activity; Nyandarua enjoys high demand for milk from processors and traders; 98% of the dairy farm herd is either crossbred or purebred exotic; potatoes come second after dairy; Nandi North has more non-dairy farmers and more medium- and large-scale farms; the choice of dairy over horticulture or perennials is still tentative.

II. Grain and fattening cluster—L_dL_c Arsi specializes in barley and wheat as cash crops, enabled by farm sizes that still allow such relatively extensive crops; for a long time, poor roads limited market access for dairy; just before roads improved around 2012, farmers adopted improved grain crop packages promoted by government and agribusiness; as a result, farmers focus on livestock activities, other than dairy, that utilize cash crop residues, but do not require daily marketing, i.e., beef, mutton, and heifer production; dairy development interventions have been occurring since the 1950s.

III. Perennial and horticultural crop cluster—L_dH_c Nandi South saw a diminishing role for dairy, as a move to high-value/ha activities occurred; farmers specialize in tea due to better support services; milk collection is almost only informal; cattle are being replaced by small livestock; semi-subsistence farming with extensive livestock and off-farm labor continues in areas unsuitable for tea and vegetable marketing.

IV. Mixed cluster—L_dH_c East Shoa, some farmers specialize in dairy (Type I), others in horticulture (Type III), while in more remote areas grains prevail (Type II). In the dairy herds of interviewed farmers, only 34% of animals are crossbred or purebred exotic; both subsectors benefit from fresh food demand in the nearby metropolis; competition for land occurs between the two and with export-oriented flower farming and urban development.

In all of the five clusters, intensification pressure is high. Over the past decades, farm sizes have shrunk due to customary intergenerational subdivision of land. In addition, the Ethiopian clusters reported land scarcity due to significant withdrawal of farm land for town and infrastructure development (past two decades) and due to allocation of land to state farms (L_dL_c Arsi cluster, 1980s) and flower farms (L_dH_c East Shoa cluster, 1990s–2000s).

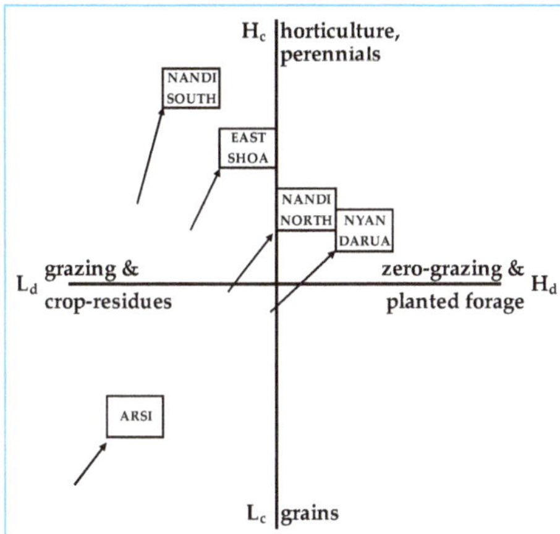

Figure 3. Schematic positioning of specializing clusters along cash crop and feeding system intensity scales.

Table 1. Key characteristics of dairy farming and marketing in five Ethiopian and Kenyan clusters.

Country:	Ethiopia			Kenya	
Characteristics Cluster Type: / Cluster Name:	L_dL_c Arsi	L_dH_c East Shoa	L_dH_c Nandi S	H_dH_c Nandi N	H_dH_c Nyandarua
Average farm size (ha)	3.2	4.0	0.8	1.6	2.9
Proportion improved cattle	55%	34%	n.a.	95%	98%
Feeding system (1)	grazing and residues	grazing and residues	grazing and residues	residues + planted fodder	residues + planted fodder
Main cash crop(s) (2)	grains	various	tea	various	Potatoes
Main marketing channel	traders	processors and coops	traders	coops	coops
Milk demand	low	medium	low	medium	high
Average est. milk sales (US$/yr)	859	2384			1621
Input service offer	low	low–med.	low	med–high	high
Main service providers	public	public	private	private	private

(1) In all clusters, urban farms mostly practice zero-grazing. (2) 'Various' indicates that no crop is dominant.

3.2. Analysis of Upgrading in Three Domains

Figure 4 lists the main secondary factors that were identified in this study as influencing upgrading dynamics in the clusters. Upgrading in all three domains is most advanced in H_dH_c clusters, especially in Nyandarua, as Table 2 shows. While a number of context conditions in L_dH_c Nandi South are good, specialization toward high-value cash crops is at the cost of upgrading in dairy. In L_dH_c East Shoa, competition with cash crops explains upgrading limitations for dairy. In L_dL_c Arsi, market constraints clearly affect dairy prospects. In the latter two clusters, less favorable context factors also dampen upgrading. Observed dynamics related to these factors are described in the next sections, following steps A–C from Figure 1. Factors with less apparent effect on upgrading dynamics were considered, but generally not described. A more detailed description of upgrading dynamics in each cluster is included as Supplementary Material S2.

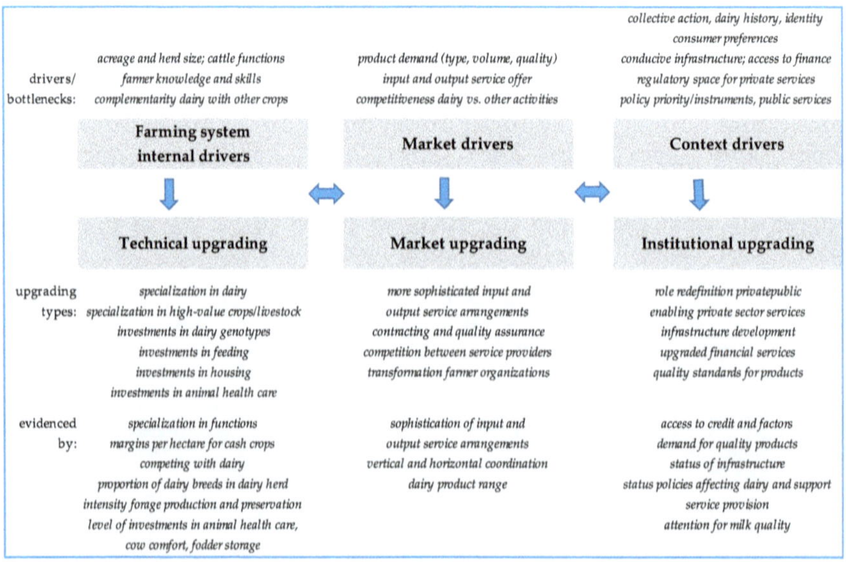

Figure 4. Causal relationships between secondary drivers and upgrading types.

Table 2. Technical, value chain and institutional upgrading in five clusters.

Country:	Ethiopia		Kenya		
Cluster type:	L_dL_c	L_dH_c	L_dH_c	H_dH_c	H_dH_c
Upgrading Type / Cluster Name:	Arsi	East Shoa	Nandi South	Nandi North	Nyandarua
Technical upgrading					
- specialization in 'dairy as business'	+	++	+	++	+++
- investments in dairy genotypes	++	++	+	++	+++
- investments in feeding	+	+	+	++	++
- investments in housing	+	+	+	++	+++
- investments in animal health care	+	++	+	++	+++
- *specialization in high-value crops/livestock i.o. dairy*	++	++	+++	++	+
Value chain upgrading					
- more sophisticated input and output service arrangements		+	+	++	+++
- contracting and quality assurance		++	+	+++	+++
- competition in service provision		++	+	++	+++
- transformation farmer organizations		+	+	++	+++
Institutional upgrading					
- role redefinition private–public			+	++	+++
- enabling private sector services	+	+	+	++	+++
- infrastructure development	+	++	++	++	+++
- upgraded financial services	+	+	++	+++	+++
- quality standards for products		+		++	++

N.B. Number of + denotes degree of upgrading: one + means some upgrading, additional + means more upgrading than in other clusters; no + means no upgrading identified.

The examined clusters are under land-scarce conditions, which means that farm acreage and stocking rate (livestock units per hectare) are key indicators to observe when assessing intensification and upgrading status. A number of additional parameters—suggested by this study as potential indicators for upgrading in the three domains that score resource base, intensity of production, and market—are shown in Figure 4.

3.2.1. Farming System Factors (A)

This section describes technical upgrading dynamics identified in the farming systems domain. The data in Table 3 offers insight into the ongoing changes in farming and the similarities and differences between clusters.

Specialization in dairy: smaller herds and less cattle functions—With farm size dropping to an average of three to four hectares, farmers in the Ethiopian FGIs reported that they specialize and reduce herd sizes, focusing on productivity rather than number of animals by crossbreeding with exotic dairy types: 'Two improved cows compare to ten local cows, but they need intensive care.' Farmers did not consider classification based on cattle number or land acreage to be meaningful; rather, they classified dairy farms based on market orientation and management level (see Table 3). This points to the ongoing transition in cattle functions in the farming system, from multipurpose (with local cattle for draft power, beef, manure, savings, social functions such as dowry, household consumption, and a small surplus for market) to more dairy-oriented, with fewer but specialized dairy cows. In Kenya, where average farm size is already well below three hectares and nearly all dairy cows have exotic blood, farmers specialize further to increase income per hectare. Breed choice is mainly between Friesian (higher producer) and Ayrshire (more disease-resistant and less heavy feeder). Entrepreneurial entrants, who have accumulated resources through employment or business, are investing in medium- to large-scale commercial farms and in advanced technology for feeding, housing, reproduction, etc., but often without commensurate investment in high quality farm labor.

Specialization in high-value crops/livestock/off-farm activities—Due to ongoing pressure on land, farmers reported that they choose livestock types and cash crops with shorter maturation time and higher margin per hectare, to offset rising land costs. Choice of crops/livestock types depends on how available options 'fit' within the farm, market, and context, including personal preferences and identity: especially in the Nandi clusters, farmers consider cattle-keeping an inalienable part of their identity. This brings important experience and skills, but also explains why farmers continue with dairy cattle even where the farm size barely allows for it (see Table 3) and when competitive advantages of other livestock and crops as livelihood options outweigh those of dairy. Until some decades ago, sale of fresh milk and dairy products was subject to taboos (e.g., in L_dL_c Arsi cluster) that are only gradually losing their impact as milk undergoes commodity individuation [47].

While dairy is being upgraded in H_dH_c Nyandarua, H_dH_c Nandi North and L_dH_c East Shoa clusters, it is being replaced by smaller species (such as goats, sheep, chickens, or rabbits) in L_dH_c Nandi South and by heifer production and/or fattening in L_dL_c Arsi and remote parts of Nandi and Nyandarua. Farmers increase roots/tubers/bananas and horticulture (in all clusters but Arsi) and perennials (tea, fruit trees and sugarcane, in Nandi), largely at the expense of grains. Due to more favorable market service arrangements for tea, since the 1980s 30–40% of farmers in L_dH_c Nandi South cluster have planted tea; this crowds out dairy, as tea plantations do not offer edible crop residues nor sufficient space for forage. In the Nandi clusters, mechanized land preparation is being replaced by manual work due to declining farm sizes and shift to perennials. In Ethiopia, draft animals are starting to be replaced by equipment such as broad bed makers and combine harvesters, due to scarcity of feed resources for draft animals. Nevertheless, the presence of draft animals explains why only one in three animals in the dairy herd is a dairy cow, compared to two in three in Kenya.

Table 3. Farm characteristics for the five clusters.

	Clusters (*with sub-Counties/Districts Where Study Sites Are Located*)	No. of Villages (FGI Part.)	Management Level (% of Dairy Farms) (1)				Farm Size (ha) (2)	Herd Size	Milking Cows as % of herd	% Improved Cattle in Herd (2)	Main crops (2),(3) (Dominant Crops in Bold)	% hh w/Off-farm Activ. (2)	Non-Dairy Farms (% of hh.)
			Low	Medium	High								
	Breed type		*Local*	*Local and improved*	*Improved*								
	Market focus		*Within village*	*Mixed*	*Outside village*								
L_dL_c	Arsi (*Limu-Bilbilo and Digalu-Tiyo*)	6 (42)	61		39		3.2	6.5	31	55	**Cereals**, pulses, vegetables	42	13
L_dH_c	East Shoa (*Ada'a*)	3 (21)	67		33		4.0	7.6	31	34	Cereals, pulses, oil seed, veg.	35	40
Average for clusters Ethiopia			**64**		**36**		**3.4**	**6.9**	**31**	**45**		**40**	**27**
			Farm Size (% of Dairy Farms)										
			Very sm.	**Small**	**Medium**	**Large**							
	Farm size (ha)		*<2.5*	*<2.5*	*2.5–8*	*>8*							
	No.of dairy cows		*1–2*	*3–5*	*6–20*	*>20*							
L_dH_c	Nandi South (*Aldai, Nandi South*)	6 (63)	78	16	6	0	0.8	2.2	66	n.a.	**Tea**, maize, RTB, veg., fruits	n.a.	19
H_dH_c	Nandi North (*Chesumei, Nandi North*)	6 (47)	76		20	4	1.6	3.3	64	n.a.	Maize, sugarc., tea, veg., fruits	n.a.	25
H_dH_c	Nyandarua (*Ol Kalou, Kipipiri*)	9 (63)	92		8	0	2.9	4.3	65	98	**Potatoes**, cereals, pulses, veg.	40	13
Average for clusters Kenya			**87**		**11**	**1**	**1.8**	**3.3**	**65**				**19**

(1) In the Ethiopian clusters, FGIs distinguished between 'low','medium' and 'high' management levels; numbers for the first two categories have been combined, as not all villages identified an intermediate management level. (2) From dairy farmer interviews for East Shoa (n = 37), Arsi (n = 85), Nyandarua (n = 91) [42]; from focus group discussions for Nandi North and South; different methodologies may affect results; farm size in Nyandarua was 2.88 ha according to FGI, 3.59 ha according to dairy farmer interviews [42]. (3) RTB = roots/tubers/bananas; here it includes Irish potatoes, sweet potatoes, arrowroot, cassava and bananas.

Table 4. Support service arrangements and input and output service offer in five clusters.

	Output Service Arrangements with Integrated Input Services (If Any)							Input Service Arrangements (5)				
Service Delivery Model:	**Local Market**	**Trader/Restaur. (1),(2)**	**Self-help group (3)**	**Cooperative (4)**	**Processor**	**Cooperative Company (4)**	**Processor w/Services**	**Independ. Suppliers**		**Dept. of Livestock**		**Dev. Agencies**
Arrangement type:	Spot market	Relational contract	Formal contract	Multilateral contract	Formal contract	Equity participation	Vertical integration	Formal contract		Public services		Various
Prevalence in:												
Ethiopia												
Arsi	++	++		+				+/-		++		+
East Shoa	+	+		+	++	+/-		+		++		+
Kenya												
Nandi South	+/-	++		+/-				+		+/-		+/-
Nandi North	+/-	+	+	++	+	++	+/-	++		+/-		+/-
Nyandarua	+/-	+	+	++	+	++	+	+++		+		+/-
Output services offered to farmers												
Collection from farm		+	+	+/-	+/-	+/-	+/-					
Chilling			+/-	+/-	+/-	+/-	+/-					
Transportation MCC to plant				+/-	+	+/-	+					
Quality testing at collection				+/-	+	+	+					
Input services offered to farmers								**Eth**	**Ken**	**Eth**	**Ken**	
Farm advice						+/-			+/-	+	+/-	+/-
AI						+/-		+/-	+	+	+/-	
Veterinary services				+/-		+/-		+/-	+	+	+	
Feed, forage, and drugs				+/-		+		+	+	+/-		
Linking to input suppliers				+/-		+/-	+/-				+	+/-
Facilitate access to finance				+/-		+	+/-					+/-
Input advancing (on credit)				+/-		+	+/-					

+ = commonly provided; number of + denotes relative dominance of service arrangement; +/- = provided by some actors only or to some farmers only. (1) Direct supply to restaurants primarily by peri-urban farmer. (2) In Ethiopia: butter traders; in Kenya: private milk collectors buying from farmers and selling to retailers, restaurants, consumers, and MCCs. (3) In Ethiopia we encountered no farmer groups (less formal than cooperatives) supplying milk. (4) Some cooperatives also process, which adds additional output services. (5) In both countries, private companies, public agencies and NGOs/development projects play a role in provision of (subsidized) inputs and services. As these affect other service arrangements, we include them as separate categories.

Farmers reported an increase of private business activities and casual labor in agriculture, construction, and transportation services. Around 40% of farmers indicated that they are engaged in off-farm activities, primarily in formal employment, private business, and trade. Households with jobs in the public or civil society sector are generally involved in private business as well, in which they invest their salaries.

Changes in dairy practices—The specialization mentioned above plays out in a number of 'technology upgrades' in terms of farming practices. Only some farmers make these changes, and there are large differences between clusters. The highest proportions of farmers who make changes are in H_dH_c Nyandarua and Nandi North clusters and in dairy farms in or close to towns in all clusters:

- Investments in dairy genotypes using AI or improved bulls. This breed-replacement process is ongoing in Ethiopia and mostly completed in Kenya; except for in some remote, barely specialized villages, farmers in Kenya overwhelmingly keep purebred or crossbred Ayrshire, Friesian, Jersey, and Guernsey
- Investments in feeding practices follow a standard pattern over time: (1) grazing and crop residues are supplemented with industrial by-products and mixed rations; (2) grazing land is paddocked; (3) investments are made in production and preservation of planted forages such as oats, maize, and Napier and Rhodes grass to counter forage shortages
- Investments in animal housing in Ethiopia include new barns to house improved breeds; in Kenya, zero-grazing units and feed storage are used when intensifying further
- Investments in animal health care increase; due to the failure of communal cattle dips to control tick-borne diseases, in Kenya many farmers have moved to individual spraying and some vaccination for East Coast Fever; treatment by veterinary workers is increasing, as is self-administration of drugs purchased from agro-veterinary shops, especially de-wormers; in Ethiopia, farmers use government veterinary personnel, who often provide better private service on the side.

3.2.2. Farm–Market Interaction (B)

The data in Tables 4 and 5 reflect upgrading dynamics stemming from the interaction between farming system and market, which become particularly clear when comparing clusters. As input service arrangements are important in more intensive dairy and become increasingly integrated with output service arrangements, Table 4 includes both input and output service arrangements identified. This description follows the value chain upgrading categories of Table 2.

More sophisticated input and output service arrangements, tailored to farmer types—Dominant service arrangements range from local markets and traders in the limited market conditions of L_dL_c Arsi and Nandi South clusters to cooperative companies and processors, with increasingly integrated services in H_dH_c Nyandarua. In L_dH_c East Shoa cluster, processors and cooperatives are replacing the first two output service arrangements, as yet without significant upgrading in input service arrangements. In H_dH_c Nandi North and Nyandarua clusters, service arrangements of cooperative companies (i.e., upgraded cooperative societies) are being upgraded to integrated input and output service packages. Processors here, who source from farmer organizations and larger farms, are experimenting with integrated input and output service arrangements as well, more so in H_dH_c Nyandarua where competition for milk and service provision is fiercer.

Service arrangement use by farmers depends on their market integration and milk sales volumes. Table 5 shows how different service delivery models cater to different farmer categories. Interviews revealed a strong relation between farmers' choice of service arrangements and farm household resource level, which in turn is related to off-farm activities. For resource-poor farmers, payment conditions are most important. They mainly sell to traders, as they need today's milk money for today's food, and they often lack the cash to acquire external inputs and services. Smallholders with more resources tend to sell to cooperatives and processors (sometimes through self-help groups), to benefit from larger two-weekly or monthly payments that can be used for inputs and investments.

However, they usually sell at least some milk to traders to benefit from higher prices and to satisfy immediate cash needs. In Kenya, the resource-endowed smallholders selling to cooperatives can benefit from input and service advancing through widespread 'check-off' systems, in which costs for inputs and services advanced are deducted from the next milk payment. Medium-scale farms in both countries seem to use any of the output service arrangements and mainly consider price, buyer dependability and transaction costs.

Table 5. Factors affecting choice of service arrangement by farmer category.

Service Arrangement:	Local Market	Trader + Restaur.	Self-help group	Cooperative	Processor	Cooperative company	Processor w/ Serv.
Dominant farm size	*Smallholders (peri-urban)*	*Resource-poor smallholders*	*Resource-endowed smallholders*	*Resource-endowed smallholders*	*Larger farmers, organized smallholders*	*Resource-endowed smallholders*	*Larger farmers, organized smallholders*
Factors affecting choice							
Payment period (days) (1)	direct	negotiable	<45	<45	<45	<45	<45
Farmgate price (US/kg) (2)							
- Ethiopia: milk	0.35–0.90	0.55–0.75	-	0.35–0.65	0.35–0.70	-	-
- Ethiopia: butter	3–12	7–13	-	-	-	-	-
- Kenya: milk	0.30–0.45	0.30–0.50	0.28–0.37	0.26–0.34	0.26–0.37	0.26–0.34	0.26–0.37
Milk buyer advances	-	cash	-	(inputs)	-	inputs	inputs
Proximity to services	<1 h	farmgate		————	*depending on location*	————	

(1) With the exception of one processor in Kenya, whose terms are 90 days. (2) Using 2016 prices and exchange rates of ETB 20:USD 1 and KES 100:USD 1; incl. dairy farmer interv. data [42].

Interviews in both of the countries further indicated that increases in productivity and marketed milk volumes are necessary to be able to pay for the extra inputs and services. Farmers in Ethiopia mentioned a break-even point of 9 L/cow/day.

Chain contracting arrangements and quality assurance—Low levels of trust in the chain form a strong inhibitor to upgrading, especially in Kenya. This is evidenced by significant 'side-selling' of milk: farmers and farmer organizations hedge marketing risks by selling to multiple clients. Processors do the same by contracting fixed volumes with suppliers. The result is a supply network rather than a supply chain, with associated high production and transaction costs. Marketing is volume- rather than quality-driven. Marketing relationships are complicated by the stark seasonality of production, with a slump in production during the dry season, and by the seasonality of consumption due to Orthodox Christian fasting seasons in Ethiopia.

Competition in service provision—In Ethiopian clusters, government agencies are the primary input and service providers. Although the main product in L_dH_c Nandi South, Kenya, is fresh milk rather than butter, the output service arrangements are unsophisticated, as in L_dL_c Arsi. Stronger competition leads to more sophisticated arrangements with higher degrees of horizontal and vertical coordination, as observed in H_dH_c Nyandarua cluster. Here, improved service levels were reported in milk contracting, milk collection, value chain financing, feed supply, drug supply, and AI services, but less so in curative health care and hay supply. Use of own bulls rather than AI services is diminishing, but still common in all clusters, pointing to issues with the quality of AI services (proportion of farmers using bulls is lowest in H_dH_c Nyandarua, at around 40%).

Transformation of farmer organizations—The poor track record of cooperatives in both countries in terms of governance, efficiency, and sustainability makes many farmers wary of investing heavily in them; many regard cooperatives primarily as channels for public and NGO subsidies. The more entrepreneurial smallholders in Kenya circumvent these issues by forming less formal 'self-help' groups that aggregate milk and supply directly to processors. Cooperative companies, generally initiated with support from development agencies such as Heifer and partners, add a variety of services to these inputs, including access to credit lines (see Table 4). In Ethiopia, such systems are much less developed.

3.2.3. Context Influence on Farm–Market Interaction (C)

This section describes identified upgrading dynamics stemming from interaction with the context. Institutional upgrading (or the absence of it) may have a synergistic, antagonistic, or inconsequential influence on technical and value chain upgrading. The main context factors identified in interviews are presented in Table 6 and are described here following the institutional upgrading categories of Table 2. A more elaborate description of policy dynamics is included in Supplementary Material S3.

Impact of role division between private and public actors on service arrangements—Both countries have a turbulent history of public influence on agricultural service provision, contributing to large changes in Kenya and stagnation in Ethiopia. In Ethiopia, public actors play an overriding role in access to inputs, services, and land. In Kenya, 25 years of significant policy changes have affected dairy in diverse ways: very significant cuts in public services in the early 1990s resulted in a collapse of the dairy sector, evidenced by the bankruptcy of many cooperatives and the state processor KCC (1999); market liberalization policy only gradually resulted in private service delivery [48]; and the enabling environment now varies from county to county [49].

In both countries, many interviewees complained about the inconsistency and inadequacy of public services for dairy. Minimization of dairy extension services in Kenya in the 1990s resulted in declining farmer skills and ultimately in declining yields. Public agencies have a (virtual) monopoly on vaccination for notifiable diseases in Kenya and on vaccination, AI, veterinary, and extension services in Ethiopia. The regulatory gaps for private AI, animal health services, and quality assurance of feed and the low policy priority for dairy compared to crops and meat received strong negative feedback. Relatively large positive impact was attributed to development projects.

In both countries, governments use subsidies to promote uptake of more market-oriented practices and to make services more accessible to farmers in remote locations and/or with fewer resources. In Kenya, interviewees mentioned many downsides to subsidized services. In Ethiopia, public monopolies on most inputs and services lead to an insensitivity toward demand, favoritism and lack of a level playing field for private providers. In both countries, subsidies seem to have created dependency on chemical fertilizers, leading to soil fertility issues.

Space for private sector service provision—The above indicates a number of bottlenecks for private service provision, even in Kenya where market liberalization is standing policy. In Ethiopia, regulatory space for private service providers primarily results in private agro-input shops (feed, drugs) and milk/butter trade; in Kenya, it results in agro-input shops and milk trade, as well as AI, veterinary, and advisory services. In both countries various business licenses are required, but monitoring of licenses is lax in Kenya.

Infrastructure development—Infrastructure, in terms of roads and utilities, was improving in all clusters. Market access for remote villages was more restricted by poor roads in Ethiopia than it was for remote villages in Kenya, as was least restricted in H_dH_c Nyandarua, where authorities have invested more in roads. While road upgrading in L_dL_c Arsi did improve access to markets, in L_dH_c East Shoa cluster it was mostly seen as taking away land from farming.

Financial services, factor access and information supply—In Ethiopia, poor access to finance is a significant bottleneck for upgrading of dairy farms and support services; farmers primarily rely on community savings and community credit institutions such as '*ekub*'. This is less of an issue in Kenya, where people who are connected to more formal value chains benefit from chain financing mechanisms, cooperative savings and credit institutions, and easier access to bank loans. Capping of interest rates at 14% per year for agricultural loans was applauded by Kenyan farmers. Access to labor is impeded by the image of dairy as involving much heavy and dirty labor. Access to information is increased by the presence of private advisory service providers next to public ones, and local language radio and TV programs about agriculture are highly appreciated by farmers.

Quality standards for products—In Kenya, demand for dairy products is strong and growing (annual consumption exceeds 110 L/capita [50]). Consumer preference for raw milk gives the informal market a strong advantage. Its market share remains over 70%, despite many decades of formal chain development efforts and presence of product standards [50,51]. In Ethiopia, annual consumption is much lower, at around 20 L/capita, and the informal market trades over 98% of the volume [50]; here, cooperatives and processors find it difficult to deal with seasonality of consumer demand resulting from long fasting seasons (on top of seasonality of production), although interviewees may have been using this as a metaphor for the difficult business climate.

Table 6. Conduciveness of context factors in five study clusters.

	Country:	Ethiopia		Kenya		
	Cluster Type:	L_dL_c	L_dH_c	L_dH_c	H_dH_c	H_dH_c
Context factors	**Cluster Name:**	**Arsi**	**East Shoa**	**Nandi S**	**Nandi N**	**Nyandarua**
Biophysical						
Climate/weather		+++	++	++	++	+++
Absence of disease threat		+	+			++
Infrastructure		+	+	+	++	+++
Enabling environment						
Policies promoting dairy				+	++	+++
Policy space for private service prov.		+	+	++++	++++	++++
Public disease prevention services		++	++	+	+	+
Research–extension–farmer linkages		++	++	+	+	+
Enforcement of service quality				+	++	+
Enforcement of milk quality				+	+	+
Access to finance				+	++	++
Chain upgrading facilitators		++	++	++	++	++
Social environment						
Dairy history and culture		+++	+++	+++	+++	+++
Dairy seen as business		+	+	++	+++	++++
Milk consumption		+	+	+++	+++	+++
Land availability		+	+			+
Labour availability		+++	++	++	++	++

N.B. The number of +'s indicate how conducive the situation is in comparison with other clusters.

4. Discussion

4.1. Present Upgrading Status of Farming and Clusters

This comparative assessment between clusters clearly draws out important differences in upgrading of farming systems that emanate from farm–market–context interactions. It reveals that all five clusters show clear evidence of technical, value chain and/or institutional upgrading of 'typical' semi-subsistence mixed crop–livestock systems to more market-oriented systems. The need for higher returns per hectare requires specialization and commercialization, in order to maintain or increase farm yields and household incomes. Technical, value chain and institutional upgrading are most pronounced in the H_dH_c clusters and least in the L_dL_c cluster, where the market system showed little to no upgrading (see Table 2). Degrees of upgrading are clearly related to secondary drivers that act as accelerators and inhibitors.

The current status of each cluster is the result of diverging pathways along dairy feeding system and cash crop intensification dimensions. These lead to increased market orientation of farmer livelihood strategies, marketed volume, and use of pre- and post-production inputs and services (Figure 3), but for different commodities and to different degrees. More intensive dairy can thus be considered to be one of the high-value 'cash crop' options that farmers can specialize in when market and context conditions are right; so are other intensive livestock activities, such as commercial poultry. This makes the Windmill approach, postulated for crop commodities by Leonardo et al. [34], to be applicable to livestock commodities as well. However, ample attention is needed for input service arrangements, which need to be especially elaborate for livestock ventures.

4.2. Cluster Upgrading Pathways Toward the Future

Cluster upgrading directions diverge as a result of different specialization choices. The different clusters react differently to the primary driver of land-use intensification, which requires higher productivity and higher returns per hectare. Choice of either intensive dairy or horticultural and perennial cash crops will be at the expense of the other option (L_dH_c vs. H_dH_c). Most clusters can be expected to move further along the intensification pathway type started, unless actors consciously redirect course:

- H_dH_c—Dairy clusters. Dairy is competitive against other commodities; service arrangements become increasingly sophisticated and competitive; private and/or cooperative actors play a strong role. Continued development of H_dH_c clusters toward dairy seems likely, provided upgrading in farming, market, and context progresses. Further specialization may lead to singular focus on dairy (H_dL_c). This expected further upgrading of the H_dH_c dairy clusters contradicts modeling outcomes of Herrero et al. [4], who only foresaw such upgrading for peri-urban dairy in Kenya, and may warrant review of their modeling assumptions.
- L_dL_c—Grain and fattening cluster. Strong public policy directions and public–private collaboration made grains in L_dL_c Arsi cluster more competitive than dairy. Future development of L_dL_c clusters toward dairy depends on serious value chain and institutional upgrading, if dairy is to effectively compete with cash crops. For the time being, available farmer expertise and presence of improved dairy breeds in L_dL_c Arsi keep the door open for upgrading of dairy, but heifer production and commercial forage production for supply to other dairy clusters seem to be more attractive alternatives. These alternative opportunities are enhanced by (1) the competition for fodder between dairy and draft animals in Ethiopia; and (2) the low capacity of intensifying tropical dairy systems to produce sufficient replacement stock and fodder [52], which results in high prices for dairy heifers and fodder.
- L_dH_c—Perennial and horticultural crop cluster. Severe land scarcity affects these clusters, with specialization toward perennials, horticulture, and intensive livestock. Due to strong path dependency, further upgrading and specialization of L_dH_c clusters around perennials and horticulture are most likely, along with intensive non-dairy livestock-keeping in areas not suitable for perennials and horticulture. It will be interesting to watch whether farmers with a strong 'cattle identity' will give up dairy.

Prospects for the remaining L_dH_c East Shoa cluster are still uncertain. It could either move toward intensive dairy, toward horticulture or toward other high-value commodity options. Upgrading prospects for dairy depend on how relative competitiveness of each venture is affected by dynamics in its respective markets (e.g., conduciveness of service arrangements for each option) and context (e.g., spatial planning and enabling policies).

An interesting next step would be to quantify the degree of specialization and intensification of (dairy) farming in clusters, building on recent work in Europe and West Africa [53,54].

4.3. Upgrading Options at Farm Level

To explore upgrading options for dairy farmers in different clusters, we draw attention to *path dependency*, farmer *feasibility space* and *aspirations*. Path dependency [27] as system behavior applies at cluster, value chain, and farm level: past investments in an established commodity favor its current competitiveness. A 'new' commodity still needs to build up its capitals and is competing against stakes in the established commodity. This path dependency becomes stronger the more intensive the competitive crop or livestock activity. When dairy is being compared against tea and against barley as an investment choice, investments in technical and value chain upgrading for dairy need to be higher to beat tea than to beat barley, as tea has a higher expected return per hectare. Waithaka et al. [55] suggested that the intensification of farms in Nandi South could increase milk production on purchased feed, but the present study shows that in this L_dH_c cluster, the suite of service arrangements required for entrepreneurial dairy are lacking, whereas they are present for tea. While path dependency is expected to be stronger for H_dH_c and L_dH_c than for L_dL_c type clusters, it can influence upgrading pathways in any cluster. For example, the ongoing reliance on draft oxen rather than on machines in Ethiopia appears to be a significant barrier for transition to market-oriented dairy, as a large proportion of the fodder biomass is fed to oxen and (local) oxen dams, limiting fodder availability for dairy cows.

Differences in *farmer livelihood strategies* help explain the presence of multiple types of service arrangements coexisting within the same cluster (Table 4). These cater to different farmer groups: the supply conditions of the formal arrangements are suiting resource-endowed farmers with more intensive dairy farming but are unfavorable to resource-poor farmers (Table 5). To them the informal arrangements offer a flexible and convenient market outlet with a competitive milk price, at an input level they can afford (Table 5). For policy makers and development actors who aim to connect more smallholders to (formal) markets, an important consideration should be that farmer livelihood strategies are the result of *feasibility space* and *aspirations*, which do not necessarily go hand in hand.

Farmers' *feasibility space* expands along with their resource base, access to production factors, presence of service arrangements, and conducive context factors [27]. Resource-endowed farmers can intensify crop or livestock activities; utilize upgraded service arrangements; and access land, labor, credit, and information. In contrast, due to limited feasibility space, resource-poor smallholders are likely to choose *autonomy* and *risk aversion*, reducing external input and service use and using informal service arrangements.

Farmer *aspirations* determine the livelihood strategy choices made within this feasibility space. The less sophisticated informal service arrangements better fit with the livelihood strategies of resource-poor smallholders, for whom dairy likely serves food security, savings, and consumption assets objectives rather than income generation [21]. A growing feasibility space will not necessarily be used to produce more milk (or other produce) for the market, let alone to make the significant changes to farming practices that are required for intensive dairy farming [6]. The effect of farmer aspirations is also apparent in the presence of 'positive deviants', those who actually utilize their feasibility space for dairy development. They are recognized by peers as 'serious farmers' (Kenya) with 'good management' (Ethiopia). These households achieve higher productivity and income levels with intensive dairy farming, utilizing more inputs and services, and marketing through formal channels. They adopt suitable upgrading options, such as investments in zero-grazing units, planted forage, feed rationing, mechanization of milking and forage production, and stronger contracting with milk buyers, which may also involve quality control of milk, inputs, and services [21].

4.4. Sustainability of Intensification Pathways

We now address the question of whether the identified transition pathways do actually contribute to the sustainability challenges mentioned in the introduction.

Alleviation of rural poverty—Social inclusion of smallholders in agricultural markets is a policy priority in both countries. It is enacted through infrastructure development, support to cooperatives, and public facilitation of pre-production inputs and services. To contribute to poverty alleviation, these services need to reach the rural poor, i.e., smallholder dairy farmers, and need to support upgrading of dairy farming. While not intending to evaluate public dairy interventions, this case study yielded the following insights:

(i) Market access for resource-poor farmers can be positively impacted by policy support instruments and development interventions; these have their own dynamics, which often appear to be at odds with the space for private service provision. Long-term impact assessment is critical, as their effects are often slow and not very noticeable [56].

(ii) Cooperatives offer no panacea for upgrading. In less sophisticated markets, cooperatives with a basic service offer can stimulate market orientation. In intermediate market conditions, they serve as collection and aggregation centers that are highly valued. In more sophisticated markets, however, in order to stay competitive they have to move beyond being what Royer, et al. [57] call a 'claim group' and develop into more efficient service providers.

(iii) As membership of cooperatives consists of resource-endowed smallholders with a relatively large feasibility space, supporting them through the cooperatives has a large potential to grow agricultural output [58] but excludes resource-poor smallholders.
(iv) The quality of public services generally is insufficient for dairy farming upgrading, which requires dependable pre-production inputs and services [6]. While in Ethiopia authorities unintentionally hamper dairy farming upgrading by monopolizing key support services, authorities in Kenya at times hinder private service delivery development by subsidizing inputs and services to farmers who have sufficient purchasing power.

The question thus remains: How can authorities effectively support market inclusion of the resource-poor, offering them options to *step up* or *step out*, rather than *hang in* [21]? This study illustrates the urgency of this issue by the observation that in areas such as L_dH_c Nandi South, the size of many farms is close to or already below the 0.4 ha that farmers consider the threshold for a viable livelihood, according to Waithaka et al. [55].

Supply of sufficient and safe food—In terms of quantity and product range, the Kenyan dairy sector is meeting demand [51]. Focus on quantity rather than quality leads, however, to increased concern about safety of milk and dairy products. These need to be addressed through upgrading of quality assurance practices in all three domains. In Ethiopia, the sector cannot meet demand in terms of either quantity or quality, as is evidenced by high prices and growing imports [17,50].

Upgrading should lead to higher marketed milk volumes, higher farmer incomes, and marketing of safe food. This confirms findings of Duncan et al. [28] and Murage et al. [59]. 'Jumps' in production are achieved by specialization, which requires investments of different kinds, including management focus. Specific upgrading options are relevant within specific cluster conditions. For example, the hub concept described by Kilelu et al. [11] may work best under smallholder conditions with competitive demand for services and competition for milk; moreover, context conditions for hub success include policy priority for smallholder dairy development, ample space for private service provision, and presence of a third-party innovation intermediary [56].

Making farming climate-smart—Regarding environmental impact, interviewees in both countries showed concern for the imbalanced use of fertilizer, leading to acidification and leaching of soils, and for the injudicious use of agro-chemicals that can affect human health, water quality, and product quality. The results suggest that farmers do worry about increasingly erratic weather—indicating the need for climate adaptation—but did not connect climate change with their own practices. These results show that before dairy sector actors will take action, climate change mitigation does require carefully designed policy regulations that address both farmer and public interests, as was also illustrated by Paul et al. [60].

We conclude that, in both countries, progress is centered around poverty alleviation objectives, which aligns well with current policy interests. Sustainable upgrading pathways require more attention for food safety and climate-smart criteria.

4.5. Upgrading Dynamics as Result of Farm–Market–Context Interactions

This study builds on three approaches for analysis of a farming system and its interaction with the market: the *farmers' perspective* of Oosting et al. [3], the *market quality* perspective of Duncan et al. [28] and the *sales arrangement/Windmill* perspective of Leonardo et al. [34]. Our exploration of the co-evolution of farming systems and service arrangements offers new insights in three areas.

Firstly, this study sheds light on the reasons particular types of farmers participate in particular chains: upgrading of service arrangements within a dairy cluster offers technical upgrading opportunities and enlarges farmers' feasibility space, but each individual farmer needs to master the resources required and aspire to upgrade. As farm resource endowments differ, a gradual and incomplete shift of farmers to upgraded chain and farming practices is apparent. This study shows that not only urban farmers but also rural farmers participate in multiple chains as a risk-reduction strategy where service arrangements are insufficiently dependable. The traders' arrangement connects rural farmers in all clusters to both rural and urban consumers, while in the more dairy-oriented clusters, farmers sell to both traders and processors. This suggests that farmers can be part of both chains for a large part of the transformation trajectory from 'semi-subsistence with small surplus to local markets' to 'commercial supply to wholesale chains'. The transition described by Oosting et al. [3] of 'rural farmers supplying to rural consumers' to 'rural farmers supplying to urban consumers' can apparently last for decades when market and context conditions are sub-optimal.

Secondly, this study sheds light on dynamics of co-evolution between farming system and service arrangements. It adds five insights to the findings by Duncan et al. [28]: (1) technical upgrading of housing and health care practices accompanies upgrading of breeding and feeding; (2) relations with off-farm activities appear to be complex: while income from off-farm business and employment is important to finance dairy investments and to supplement farm income, the proportion of households engaging in off-farm activities in this study did not change with market quality; further research is warranted into the patterns of such investment and its impact on dairy upgrading; (3) it shows the competition between farming activities in the specialization process: in clusters where dairy support services remain less conducive, farmers specialize into cash crops and short maturity livestock production activities at the expense of dairy; (4) it shows the propelling role of competition between service providers in the co-evolution between farming system and service arrangements; (5) it shows the correlation between farming system upgrading and the activity of innovation intermediaries; various authors [11,26] have shown the important roles of innovation intermediaries in upgrading. While this study identified activities and impact of intermediaries in the various service arrangements—dairy cooperatives, processing companies, public–private collaboration and development agencies—further description goes beyond the scope of this paper.

Lastly, this study sheds light on system behaviors such as system jumps and adaptive cycles [3,27]. We postulate that co-dependencies between farm, market, and context are key to understanding the adaptive cycle dynamics of system upgrading, including system jumps, stagnation, and collapse. Section 4.6 further elaborates on these system dynamics.

4.6. Positive and Negative Co-Dependencies in Relation to System Jumps

The marked differences in upgrading status between clusters can be attributed to co-dependencies between technical, value chain, and institutional upgrading processes. Co-dependencies make upgrading in one domain dependent on that in another. An example of strong co-dependency is when farmers can only adopt a new forage crop with commensurate investments in skills, (imported) seed, and equipment, if service providers simultaneously invest in providing the inputs and services necessary to grow the crop and if policy makers ensure adequate advisory services, as well as regulations for importation and control of seed and equipment. We coin the concept of 'concurrency' to describe this mutual dependency in terms of timing of synergistic upgrading in different domains.

Upgrading in all three domains can be expected to occur when 'all lights are green', i.e., drivers in all three domains work as accelerators. *Positive feedback loops* [27] propel upgrading, potentially leading to significant transitions. For example, farmers who consistently supply to formal milk buyers can use their supply records to more easily get credit from financial institutions. This enables investments in higher production capacity, which further improves access to services. This bankability cycle may be initiated by infusion of capital from other income sources, such as employment and/or business, and is more apparent in Kenya than in Ethiopia, where banks rarely provide (scarce) credit to dairy farms due to dairy's low rate of capital turnover.

Concurrency and positive feedback loops will not occur, however, when one or more drivers 'throw a spanner in the works', consecutively inhibiting upgrading in the three domains. In such cases, co-dependencies cause *negative feedback loops* [27] that lead to stagnation and may be hard to break. For example, the uncertainty about price and payment conditions pushes farmers to lower external input levels, leading to lower production levels and higher seasonality of production. These in turn inhibit processors from offering good payment conditions. Where other livelihood opportunities have a significant competitive advantage, farmers can be expected to turn to those. In their absence, decreasing farm size will lead to stagnation and declining wealth. Where dairy is hard to combine with new livelihood activities, as in the case of tea, dairy may collapse and the farming system will transform to a system without dairy; total disappearance of dairy in the L_dH_c cluster has so far been prevented by the strong 'cattle identity' of the Nandi farmers.

Progressive upgrading may lead to *transformation* of the farming system and/or market system. Farmers in all clusters noticed the final stages of the transformation from grazing land to farmland for crops. The L_dL_c cluster only recently completed this transformation 'from grazing to grain', following public promotion of improved grain variety packages in the 2010s. In the meantime, the most upgraded H_dH_c cluster appears to be facing another transformation that will manifest in upgraded feeding strategies: 'from grazing with crop residues to zero-grazing with planted forage'. However, this is co-dependent on further value chain upgrading that will ensure supply certainty and improved access to and quality of inputs and services.

When a sizable number of upgrades needs to occur concurrently, a *system jump* can be expected when reaching a certain threshold—or tipping point—of pressure to transform between alternative system states [61]. This study illustrates this for two scale levels: (1) semi-subsistence clusters transforming to more commercial intensive systems (dairy or horticulture) mentioned above; and (2) households shifting their milk supply from traders to wholesale chains. At both levels the jump requires concurrent synergistic upgrading and build-up of resources. In H_dH_c clusters in Kenya, a number of positive dynamics occur that may lead to such a system jump, once the current lock-in of farming and market systems can be overcome. That system lock-in is evident in chain fragmentation, high costs of production and transactions, and disregard for quality assurance of milk and inputs. We speculate that the pressure to upgrade gradually builds up and forces a number of concurrent technical, value chain, and institutional upgrades to suddenly take place. Time will tell whether lock-in will be overcome by a system jump through upgrading, or whether it will persist by protection of vested interests, perpetuating the current situation until a crisis causes system collapse.

5. Conclusions

This comparative case study of five emerging dairy clusters in the East African highlands aimed to explore how interaction of the farming system with market and context shape cluster emergence and transformation from semi-subsistent to market-oriented dairy farming. Key findings of this study add to debates about upgrading in clusters, value chains and farming systems; inclusion of smallholders in markets; system jumps; and sustainable intensification pathways. They include:

- Co-dependencies between technical, value chain, and institutional upgrading processes are key to understanding the adaptive cycle dynamics of farming- and market-system upgrading, including system jumps, stagnation, and collapse. We coin the concept of 'concurrency' to describe co-dependency in terms of timing of synergistic upgrading in different domains. When a sizable number of upgrades needs to occur concurrently, a system jump can be expected upon reaching a certain threshold of pressure to transform. The implications for studies of technical upgrading in farming systems are that synergies between internal (farming system) and external (market and context) factors determine upgrading outcomes.
- The upgrading status of dairy clusters results from diverging pathways along two dimensions: feeding system intensification and cash crop intensification. Intensive dairy is competing with other high-value cash crop options—intensive livestock activities, horticulture, and perennials—that farmers specialize in depending on market and context conditions. Clusters can be expected to move further along the intensification pathway started, unless actors consciously influence direction through investments in upgrading conditions. The implications for the debate on cluster upgrading are that (1) transition emerges from synergistic technical, value chain, and institutional upgrading; and (2) evaluation of upgrading options needs to consider notions of path dependency, concurrency, and investments in upgrading conditions.

- Farmers' feasibility space for participation in transition expands along with their resource base, access to production factors, presence of service arrangements, and conduciveness of context factors. Resource endowment levels help explain why particular farmers participate in particular chains. Transition from 'semi-subsistent farmers supplying to local markets' to 'market-oriented farmers supplying to urban markets' may take decades when market and context conditions are sub-optimal. This adds to earlier work on inclusiveness of connecting resource-poor farmers to markets.
- The most upgraded H_dH_c cluster appears to be facing another transformation that will manifest in upgraded feeding strategies and further value chain upgrading, which will ensure supply certainty and improved access to and quality of inputs and services. Studies of such real-life system transformation cases will add to understanding of system jumps.

Further research may focus on quantification of the degree and thresholds of specialization and intensification of (dairy) farming in clusters and on the impact of different service arrangements and vertical coordination mechanisms on local economic development.

In both countries dairy development objectives are centered around poverty alleviation, which aligns well with current policy interests. We recommend that policy makers and cluster development planners carefully design sustainable intensification pathways for competitive commodities. Sustainability issues to be considered include: (1) enabling a larger proportion of resource-poor farmers to participate in markets; (2) at the same time, enabling private input and service provision models that can last; and (3) more attention for food safety and climate smartness of agricultural development.

Limitations to this study—The two x three villages sampling scheme used appears to sufficiently capture variation within clusters to assess upgrading dynamics and transitions. While the small number of one x three study villages in East Shoa cluster may insufficiently capture variation in the zone, the study area can be considered representative for the peri-urban half of the zone. The three x three scheme used in Nyandarua did not yield significantly more insight than the two x three scheme used elsewhere.

The sub-regional administrative units taken as starting points for cluster boundaries allow a researchable unit in which farm, market and context can show sufficient homogeneity and variation. However, clusters do not necessarily coincide with such units. Nandi County in Kenya shows such distinct differences that we can speak of two clusters, each appearing to be part of multi-county clusters with Eldoret and Kisumu as centers. Further research will benefit from clearer delineation of clusters. This will also improve sampling of study sites.

The retrospective interview tools, which explored timelines and past changes in farming practices, did provide considerable insight in developments since the 1980s. Nevertheless, overcoming the bias inherent in a snapshot approach when looking at time-based processes may only be possible through longitudinal or historic research.

While this study analyzed interaction between two systems—farming and market—the farming system was analyzed in more detail. Additional analysis of the market system may add valuable insights, as suggested by Reardon [62], although it risks making the analysis too complex. Using a food systems approach may be useful.

Additional studies may explore the impact of different service arrangements and vertical coordination mechanisms on local economic development. Out-of-cluster service providers such as processors and input suppliers may play a key role in upgrading dynamics, but may also capture a significant part of the benefits of transition.

Supplementary Materials: The following is available online at http://www.mdpi.com/2071-1050/10/11/4324/s1, S1: Basic Data on Study Areas; S2: Cluster-Wise Description of Upgrading Dynamics; S3: Context Conduciveness—The Impact of Dairy Policy.

Author Contributions: Conceptualization and Methodology, J.v.d.L., L.K., S.O., B.O.B., A.M.; Data Collection, Data Curation, Formal Analysis, Writing–Original Draft Preparation, and Visualization, J.v.d.L.; Writing–Review and Editing, L.K., S.O., B.O.B., A.M.; Supervision, L.K., S.O.; Funding Acquisition and Project Administration, J.v.d.L., L.K.

Funding: This research was funded by NWO–WOTRO (project W 08.260.301).

Acknowledgments: The authors would like to thank Feki Misbah, Alemayehu Asrat, Felix Opinya, Linnet Mwangi and Ida Rademaker for help in data collection and transcription; Hans Schiere, Jos Bijman, Jan Helder and Jim Groot for stimulating discussions and input.

Conflicts of Interest: The authors declare no conflict of interest.

References

1. Kilelu, C.; Klerkx, L.; Omore, A.; Baltenweck, I.; Leeuwis, C.; Githinji, J. Value chain upgrading and the inclusion of smallholders in markets: Reflections on contributions of multi-stakeholder processes in dairy development in Tanzania. *Eur. J. Dev. Res.* **2017**, *29*, 1102–1121. [CrossRef]
2. Giuliani, E.; Pietrobelli, C.; Rabellotti, R. Upgrading in global value chains: Lessons from Latin American clusters. *World Dev.* **2005**, *33*, 549–573. [CrossRef]
3. Oosting, S.J.; Udo, H.M.J.; Viets, T.C. Development of livestock production in the tropics: Farm and farmers' perspectives. *Animal* **2014**, *8*, 1238–1248. [CrossRef] [PubMed]
4. Herrero, M.; Thornton, P.K.; Bernués, A.; Baltenweck, I.; Vervoort, J.; van de Steeg, J.; Makokha, S.; van Wijk, M.T.; Karanja, S.; Rufino, M.C. Exploring future changes in smallholder farming systems by linking socio-economic scenarios with regional and household models. *Glob. Environ. Chang.* **2014**, *24*, 165–182. [CrossRef]
5. Mockshell, J.; Kamanda, J. Beyond the agroecological and sustainable agricultural intensification debate: Is blended sustainability the way forward? *Int. J. Agric. Sustain.* **2018**, *16*, 127–149. [CrossRef]
6. Udo, H.M.J.; Aklilu, H.A.; Phong, L.T.; Bosma, R.H.; Budisatria, I.G.S.; Patil, B.R.; Samdup, T.; Bebe, B.O. Impact of intensification of different types of livestock production in smallholder crop-livestock systems. *Livest. Sci.* **2011**, *139*, 22–29. [CrossRef]
7. Burke, W.J.; Myers, R.J.; Jayne, T.S. A triple-hurdle model of production and market participation in Kenya's dairy market. *Am. J. Agric. Econ.* **2015**, *97*, 1227–1246. [CrossRef]
8. Novo, A.M.; Slingerland, M.; Jansen, K.; Kanellopoulos, A.; Giller, K.E. Feasibility and competitiveness of intensive smallholder dairy farming in Brazil in comparison with soya and sugarcane: Case study of the Balde Cheio programme. *Agric. Syst.* **2013**, *121*, 63–72. [CrossRef]
9. Olwande, J.; Smale, M.; Mathenge, M.K.; Place, F.; Mithöfer, D. Agricultural marketing by smallholders in Kenya: A comparison of maize, kale and dairy. *Food Policy* **2015**, *52*, 22–32. [CrossRef]
10. Didanna, H.L.; Wossen, A.M.; Worako, T.K.; Shano, B.K. Factors influencing intensification of dairy production systems in Ethiopia. *Outlook Agric.* **2018**, *47*, 0030727018770463. [CrossRef]
11. Kilelu, C.W.; Klerkx, L.; Leeuwis, C. Supporting smallholder commercialisation by enhancing integrated coordination in agrifood value chains: Experiences with dairy hubs in Kenya. *Exp. Agric.* **2016**, *53*, 269–287. [CrossRef]
12. Dirven, M. Dairy clusters in Latin America in the context of globalization. *Int. Food Agribus. Manag. Rev.* **2001**, *2*, 301–313. [CrossRef]
13. Brasier, K.J.; Goetz, S.; Smith, L.A.; Ames, M.; Green, J.; Kelsey, T.; Rangarajan, A.; Whitmer, W. Small farm clusters and pathways to rural community sustainability. *Community Dev.* **2007**, *38*, 8–22. [CrossRef]
14. Morosini, P. Industrial clusters, knowledge integration and performance. *World Dev.* **2004**, *32*, 305–326. [CrossRef]
15. van Dijk, M.P.; Sverrisson, Á. Enterprise clusters in developing countries: Mechanisms of transition and stagnation. *Entrep. Reg. Dev.* **2003**, *15*, 183–206. [CrossRef]
16. Kebebe, E.G.; Duncan, A.J.; Klerkx, L.; de Boer, I.J.M.; Oosting, S.J. Understanding socio-economic and policy constraints to dairy development in Ethiopia: A coupled functional-structural innovation systems analysis. *Agric. Syst.* **2015**, *141*, 69–78. [CrossRef]
17. Ruben, R.; Dekeba Bekele, A.; Megersa Lenjiso, B. Quality upgrading in Ethiopian dairy value chains: Dovetailing upstream and downstream perspectives. *Rev. Soc. Econ.* **2017**, *75*, 296–317. [CrossRef]
18. Omiti, J.; Otieno, D.; Nyanamba, T.; McCullough, E. Factors influencing the intensity of market participation by smallholder farmers: A case study of rural and peri-urban areas of Kenya. *Afr. J. Agric. Resour. Econ.* **2009**, *3*, 57–82.

19. Duru, M.; Therond, O. Livestock system sustainability and resilience in intensive production zones: Which form of ecological modernization? *Reg. Environ. Chang.* **2015**, *15*, 1651–1665. [CrossRef]
20. Vroegindewey, R.; Hodbod, J. Resilience of agricultural value chains in developing country contexts: A framework and assessment approach. *Sustainability* **2018**, *10*, 916. [CrossRef]
21. Dorward, A.; Anderson, S.; Bernal, Y.N.; Vera, E.S.; Rushton, J.; Pattison, J.; Paz, R. Hanging in, stepping up and stepping out: Livelihood aspirations and strategies of the poor. *Dev. Pract.* **2009**, *19*, 240–247. [CrossRef]
22. Poulton, C.; Dorward, A.; Kydd, J. The future of small farms: New directions for services, institutions, and intermediation. *World Dev.* **2010**, *38*, 1413–1428. [CrossRef]
23. Medina, G.; Almeida, C.; Novaes, E.; Godar, J.; Pokorny, B. Development conditions for family farming: Lessons from Brazil. *World Dev.* **2015**, *74*, 386–396. [CrossRef]
24. Martin, G.; Allain, S.; Bergez, J.-E.; Burger-Leenhardt, D.; Constantin, J.; Duru, M.; Hazard, L.; Lacombe, C.; Magda, D.; Magne, M.-A. How to address the sustainability transition of farming systems? A conceptual framework to organize research. *Sustainability* **2018**, *10*, 2083. [CrossRef]
25. Campbell, R. A framework for inclusive market systems development. In *LEO Brief*; USAID: Washington, DC, USA, 2014.
26. Ramirez, M.; Clarke, I.; Klerkx, L. Analysing intermediary organisations and their influence on upgrading in emerging agricultural clusters. *Environ. Plan. A* **2017**. [CrossRef]
27. Schiere, J.B.; Darnhofer, I.; Duru, M. Dynamics in farming systems: Of changes and choices. In *Farming Systems Research into the 21st Century: The New Dynamic*; Darnhofer, I., Gibbon, D., Dedieu, B.E., Eds.; Springer: Berlin, Germany, 2012; pp. 337–363.
28. Duncan, A.; Teufel, N.; Mekonnen, K.; Singh, V.; Bitew, A.; Gebremedhin, B. Dairy intensification in developing countries: Effects of market quality on farm-level feeding and breeding practices. *Animal* **2013**, *7*, 2054–2062. [CrossRef] [PubMed]
29. Veldwisch, G.J.; Beekman, W.; Bolding, A. Smallholder irrigators, water rights and investments in agriculture: Three cases from rural Mozambique. *Water Altern.* **2013**, *6*, 125–141.
30. Department for International Development (DFID). *Sustainable Livelihoods Guidance Sheets*; Department for International Development: London, UK, 1999.
31. Darnhofer, I.; Gibbon, D.; Dedieu, B.E. *Farming Systems Research into the 21st Century: The New Dynamic*; Springer: Berlin, Germany, 2012; p. 490.
32. Kebebe, E.G. Understanding Factors Affecting Technology Adoption in Smallholder Livestock Production Systems in Ethiopia: The Role of Farm Resources and the Enabling Environment. Ph.D. Thesis, Wageningen University, Wageningen, The Netherlands, 2015.
33. Bosman, H.; Moll, H.; Udo, H. Measuring and interpreting the benefits of goat keeping in tropical farm systems. *Agric. Syst.* **1997**, *53*, 349–372. [CrossRef]
34. Leonardo, W.J.; Bijman, J.; Slingerland, M. The windmill approach: Combining transaction cost economics and farming systems theory to analyse farmer participation in value chains. *Outlook Agric.* **2015**, *44*, 207–214. [CrossRef]
35. Bijman, J.; Muradian, R.; Cechin, A. Agricultural cooperatives and value chain coordination. In *Value Chains, Social Inclusion and Economic Development: Contrasting Theories and Realities*; Helmsing, A., Vellema, S., Eds.; Routledge: Abingdon, UK, 2012; pp. 82–101.
36. Odhong, C.; Wahome, R.; Vaarst, M.; Kiggundu, M.; Nalubwama, S.M.; Halberg, N.; Githigia, S. Dairy cattle management, health and welfare in smallholder farms: An organic farming perspective. *J. Org.* **2015**, *2*, 3–20.
37. Arias, P.; Halam, D.; Krivonos, E.; Morrison, J. *Smallholder Integration in Changing Food Markets*; FAO: Rome, Italy, 2013.
38. Zeleke, T.; Awulachew, S.B. *Policy Analysis of Water for Productive Use among Smallholder Irrigators in Ethiopia*; OXFAM America: Boston, MA, USA; Addis Abeba, Ethiopia, 2014.
39. Poole, N.D.; Chitundu, M.; Msoni, R. Commercialisation: A meta-approach for agricultural development among smallholder farmers in Africa? *Food Policy* **2013**, *41*, 155–165. [CrossRef]
40. Kenya National Bureau of Statistics (KNBS). *Nandi County Statistical Abstract*; Republic of Kenya, Kenya National Bureau of Statistics: Nairobi, Kenya, 2015.
41. Kenya National Bureau of Statistics (KNBS). *Nyandarua County Statistical Abstract*; Republic of Kenya, Kenya National Bureau of Statistics: Nairobi, Kenya, 2015.

42. van der Lee, J.E.A. Influence of proximity to input and output markets on dairy farming in Ethiopia and Kenya. in preparation.
43. Jayakaran, R. *The Ten Seed Technique*; World Version China: Beijing, China, 2002.
44. Francesconi, G.N. *Cooperation for Competition: Linking Ethiopian Farmers to Markets*; Wageningen Academic Publishers: Wageningen, The Netherlands, 2009.
45. Amankwah, K.; Klerkx, L.; Oosting, S.J.; Sakyi-Dawson, O.; van der Zijpp, A.J.; Millar, D. Diagnosing constraints to market participation of small ruminant producers in northern Ghana: An innovation systems analysis. *NJAS Wageningen J. Life Sci.* **2012**, *60–63*, 37–47. [CrossRef]
46. Gebremedhin, B.; Jaleta, M. Commercialization of smallholders: Is market participation enough? In Proceedings of the Joint 3rd African Association of Agricultural Economists (AAAE) and 48th Agricultural Economists Association of South Africa (AEASA) Conference, Cape Town, South Africa, 19–23 September 2010.
47. Pearson, O.; Schmidt, M. Commodity individuation of milk in the Somali region, Ethiopia. *Area* **2017**, *50*, 213–221. [CrossRef]
48. Kijima, Y.; Yamano, T.; Baltenweck, I. Emerging markets in the post-liberalisation period: Evidence from the raw milk market in rural Kenya. *J. Afr. Econ.* **2010**, *19*, 88–110. [CrossRef]
49. Recha, C.W. *Local and Regional Variations in Conditions for Agriculture and Food Security in Kenya*; AgriFoSe2030 Report 7; Lund University, Sweden Egerton University: Njoro, Kenya, 2018.
50. Makoni, N.; Mwai, R.; Redda, T.; van der Zijpp, A.J.; van der Lee, J. *White Gold: Opportunities for Dairy Sector Development Collaboration in East Africa*; Centre for Development Innovation, Wageningen UR: Wageningen, The Netherlands, 2014; p. 205.
51. KDB. Website Kenya Dairy Board. Available online: www.kdb.co.ke (accessed on 15 September 2018).
52. Bebe, B.O. Assessing potential for producing dairy replacements under increasing intensification of smallholder dairy systems in the Kenya highlands. *Livest. Res. Rural Dev.* **2008**, *20*, 24.
53. Roessler, R.; Mpouam, S.E.; Muchemwa, T.; Schlecht, E. Emerging development pathways of urban livestock production in rapidly growing West Africa cities. *Sustainability* **2016**, *8*, 1199. [CrossRef]
54. Gonzalez-Mejia, A.; Styles, D.; Wilson, P.; Gibbons, J. Metrics and methods for characterizing dairy farm intensification using farm survey data. *PLoS ONE* **2018**, *13*, e0195286. [CrossRef] [PubMed]
55. Waithaka, M.M.; Thornton, P.K.; Herrero, M.; Shepherd, K.D. Bio-economic evaluation of farmers' perceptions of viable farms in western Kenya. *Agric. Syst.* **2006**, *90*, 243–271. [CrossRef]
56. Reardon, T. The hidden middle: The quiet revolution in the midstream of agrifood value chains in developing countries. *Oxf. Rev. Econ. Policy* **2015**, *31*, 45–63. [CrossRef]
57. Lie, H.; Rich, K.M.; van der Hoek, R.; Dizyee, K. An empirical evaluation of policy options for inclusive dairy value chain development in Nicaragua: A system dynamics approach. *Agric. Syst.* **2018**, *164*, 193–222. [CrossRef]
58. Royer, A.; Bijman, J.; Bitzer, V. Linking smallholder farmers to high quality food chains: Appraising institutional arrangements. In *Quality and Innovation in Food Chains: Lessons and Insights from Africa*; Wageningen Academic Publishers: Wageningen, The Netherlands, 2016; pp. 359–381.
59. Mellor, J.W. High rural population density Africa—What are the growth requirements and who participates? *Food Policy* **2014**, *48*, 66–75. [CrossRef]
60. Murage, P.; Ngigi, M.; Bebe, B.; Ndambi, O. Technical efficiency of smallholder dairy farms across dairy business models and associated determinants in the Kenya highlands. *J. Rural Stud.* **2018**. submitted.
61. Paul, B.K.; Frelat, R.; Birnholz, C.; Ebong, C.; Gahigi, A.; Groot, J.C.J.; Herrero, M.; Kagabo, D.M.; Notenbaert, A.; Vanlauwe, B.; et al. Agricultural intensification scenarios, household food availability and greenhouse gas emissions in Rwanda: Ex-ante impacts and trade-offs. *Agric. Syst.* **2018**, *163*, 16–26. [CrossRef]
62. Tittonell, P. Farm typologies and resilience: The diversity of livelihood strategies seen as alternative system states. In Proceedings of the 5th World Congress on Conservation Agriculture, Brisbane, Australia, 26–29 September 2011.

sustainability

MDPI

Article

Assessing Sustainable Food and Nutrition Security of the EU Food System—An Integrated Approach

Monika Zurek [1,*], Aniek Hebinck [1,2], Adrian Leip [3], Joost Vervoort [1,4], Marijke Kuiper [5], Maria Garrone [6], Petr Havlík [7], Thomas Heckelei [8], Sara Hornborg [9], John Ingram [1], Anneleen Kuijsten [10], Lindsay Shutes [5], Johanna M. Geleijnse [10], Ida Terluin [5], Pieter van 't Veer [10], Jo Wijnands [5,†], Andrea Zimmermann [8,11] and Thom Achterbosch [5]

1 Environmental change Institute, University of Oxford, Oxford OX1 3QY, UK; aniek.hebinck@su.se (A.H.); j.m.vervoort@uu.nl (J.V.); john.ingram@eci.ox.ac.uk (J.I.)
2 Stockholm Resilience Centre, Stockholm University, 10405 Stockholm, Sweden
3 European Commission, Joint Research Centre, I-21027 Ispra (VA), Italy; adrian.leip@ec.europa.eu
4 Copernicus Institute of Sustainable Development, University of Utrecht, 3584 CB Utrecht, The Netherlands
5 Wageningen Economic Research, 2595 BM The Hague, The Netherlands; marijke.kuiper@wur.nl (M.K.); lindsayjshutes@gmail.com (L.S.); ida.terluin@wur.nl (I.T.); thom.achterbosch@wur.nl (T.A.)
6 LICOS—Centre for Institutions and Economic Performance, KU Leuven University, 3000 Leuven, Belgium; maria.garrone@kuleuven.be
7 Ecosystems Services and Management Program, International Institute for Applied Systems Analysis, 2361 Laxenburg, Austria; havlikpt@iiasa.ac.at
8 Institute for Food and Resource Economics, University of Bonn, 53115 Bonn, Germany; thomas.heckelei@ilr.uni-bonn.de (T.H.); andrea.zimmermann@fao.org (A.Z.)
9 Agrifood and Bioscience, RISE Research Institutes of Sweden, 40229 Gothenburg, Sweden; sara.hornborg@ri.se
10 Division of Human Nutrition and Health, Wageningen University, 6700 AA Wageningen, The Netherlands; anneleen.kuijsten@wur.nl (A.K.); marianne.geleijnse@wur.nl (J.M.G.); pieter.vantveer@wur.nl (P.v.V.)
11 Trade and Markets Division, Food and Agriculture Organization of the United Nations, 00153 Rome, Italy
* Correspondence: monika.zurek@eci.ox.ac.uk; Tel.: +44-(0)1865-285531
† Jo Wijnands passed away.

Received: 29 September 2018; Accepted: 13 November 2018; Published: 19 November 2018

Abstract: Steering the EU food system towards a sustainability transformation requires a vast and actionable knowledge base available to a range of public and private actors. Few have captured this complexity by assessing food systems from a multi-dimensional and multi-level perspective, which would include (1) nutrition and diet, environmental and economic outcomes together with social equity dimensions and (2) system interactions across country, EU and global scales. This paper addresses this gap in food systems research and science communication by providing an integrated analytical approach and new ways to communicate this complexity outside science. Based on a transdisciplinary science approach with continuous stakeholder input, the EU Horizon2020 project 'Metrics, Models and Foresight for European SUStainable Food And Nutrition Security' (SUSFANS) developed a five-step process: Creating a participatory space; designing a conceptual framework of the EU food system; developing food system performance metrics; designing a modelling toolbox and developing a visualization tool. The Sustainable Food and Nutrition-Visualizer, designed to communicate complex policy change-impacts and trade-off questions, enables an informed debate about trade-offs associated with options for change among food system actors as well as in the policy making arena. The discussion highlights points for further research related to indicator development, reach of assessment models, participatory processes and obstacles in science communication.

Keywords: food systems; metrics; interdisciplinarity; sustainable food and nutrition security; food system assessment; participatory approach; SUSFANS

1. Introduction

Food systems are complex entities, described as consisting of many different actors, their activities and interactions, the driving forces shaping these activities and the outcomes produced at the individual and system's level. This complexity makes it difficult to analyze, govern or change food systems. As such, food systems research asks for an integrated approach for analysis and new ways to communicate this complexity outside the research domain [1,2]. Food systems currently do not deliver the expected food and nutrition security outcomes, have many social inequalities embedded in them and are regarded as a major driver of global environmental change [3–5]. The need to change our food systems to be more sustainable (including in health, environmental, social and economic terms) is widely acknowledged in research and policy circles [6–9]. Governance is considered an important lever in changing these systems to becoming more sustainable and to reaching these goals [10–14], but food systems encompass many different actors, each with their own set of driving forces and goals [15–17]. These need to be considered when discussing and negotiating change in the system. To govern effectively, decision makers need both an understanding of food systems elements and the effect policy, financial, or technical innovations (e.g., new technologies, measures to change consumer behavior) might have on specific actors as well as overall food system outcomes. Only then a differentiated debate across food system actors and policy makers and actors on what type of systems' change is needed and desirable will be enabled.

In this paper we present an integrated approach to food systems assessments applied to the European Union. This approach was developed as a basis for the work of the Horizon2020 EU-funded project 'Metrics, Models and Foresight for European SUStainable Food And Nutrition Security' (SUSFANS, www.susfans.eu). SUSFANS' overall objective is to develop the conceptual understanding, tools, foresight work and the evidence base to inform EU-wide and member state food and agriculture policies towards a transformation of the food system with respect to its impacts on public health, nutrition and diets, environmental, economic and social outcomes. Central to the project's approach is the assessment of the status of sustainable food and nutrition security (SFNS) in the EU and innovation options for enhancing SFNS in the EU. SFNS can be defined as the capacity of a food system to deliver food and nutrition security in an environmentally, economically and socially sustainable manner. This concept combines nutrition and health with a social-ecological systems perspective. The project aims to contribute to an integrated analysis of and effective decision making for food system change in the EU and its member states [2].

This paper is the first in a series of papers that describe the project work with respect to assessing food system status and transformation options. Here we lay out the conceptual foundations and steps of the project's approach. Subsequent papers will describe food system metrics work in more detail and how the approach was operationalized for the EU and for the project's four case study countries. These papers will also describe the experiences and challenges with the approach. This paper is structured as follows: After exploring what SFNS entails, we present a set of coherent steps that can enable an integrated debate around food systems change. This includes the creation of a participatory space; the development of a conceptual framework for mapping the EU food system (we recognize that there is no 'EU food system' as such, but rather that it is a set of local, regional, national and global interconnected systems and dynamics. We are using the term as a catch all term for the interlocked systems and their actors that put food on people's plates within the geographical unit that is the EU); the development of a set of metrics to assess SFNS and innovation pathways for food system change; the development of a modelling strategy based on the metrics; and the design of an integration and visualization tool across the metrics particularly geared towards policy makers. The steps are illustrated by the work done in the SUSFANS project (hereafter referred to as 'the project') to show how the conceptual approach could be operationalized.

2. Defining Sustainable Food and Nutrition Security

Debates on how to define and achieve food security have shifted substantially over the last decades. Various terms and terminologies have been used to capture the complexities of describing the outcomes of a food system. These terms have come from different scientific disciplines, such as the agricultural, development, or nutrition communities, and reflect the main discourses prevalent at the time. Born out of the experiences of the famines after the Second World War, the definition of food security focused on the supply of enough food and, thus, efforts to achieve it mainly emphasized increasing agricultural production, specifically around staple foods [18–20]. The work of Amartya Sen added the notion of entitlements to food, emphasizing accessibility, availability, affordability and utilization of food [21,22]. The nutrition community complemented the discussion on food security by emphasizing the need for an adequate nutritional status (in terms of under and over nutrition) in the 1990s [23,24], thus enlarging the concept by bringing in micro-level aspects of nutrition and health and becoming more applicable for nutrition planning. With the realization of the environmental footprint of our food system, the notion of 'sustainable diets' [13,25–31] describes the individual diet and its associated environmental sustainability implications, calling for a better matching of food consumption patterns with their environmental impacts. Since the 2000s, the question of how to change agriculture has evolved into the notion of how to govern and change a complex system of interconnected food actors (including farmers, food processors, retailers, consumers, and more). In short, aiming to put enough calories on people's plates as the key food security goal has morphed into several goals that encompass nutritional adequacy, environmental issues as well as economic, livelihoods and equity considerations. The SUSFANS project acknowledges this variety of goals for today's food systems in the notion of 'sustainable food and nutrition security'.

The 'food system' concept is increasingly used today as a conceptual and analytical tool to describe the processes and actors in the food sector. A food system is made up of food system activities (growing, harvesting, processing, packaging, transporting, marketing, consuming, and disposing of food and food-related items), and food system actors, all influenced by certain 'drivers' (processes determining how these activities are performed). These activities by actors in the food system, result in a number of outcomes, that, in turn, feed back to environmental and socio-economic drivers [15,17,32–34]. A food systems approach is "seen as the most effective strategy to enhance [food and] nutrition security in a more sustainable manner" [24], for two main reasons. First, by focusing on impacts and leverage points in the different domains it allows for an integrated assessment. Second, and building on the first, through this integrated assessment it can provide a framework to structure the debate of a complex issue [15].

Central to the integrated food system approach presented in this paper is 'Sustainable Food and Nutrition Security' (SFNS). This partly captures earlier work around food systems that embodies divergent theoretical and ontological framings [35,36] and allows for the embedding of sustainability dimensions into the evaluation of the food systems. This means that the approach also includes the assessment of environmental, social and economic outcomes of the food system. One of the core objectives of this integrated approach here is to support EU decision makers in fostering the EU food policy goals described in various EU, member stated and food industry policies for the EU food system. These are: (1) Deliver a balanced healthy diet to consumers, (2) reduce the systems' negative environmental impact, (3) build a viable, competitive and socially balanced agri-food sector, and (4) contribute to social equity goals and global food security.

3. An Integrated Approach

Answering questions on how to achieve sustainable food and nutrition security, which is a multi-faceted concept, requires answers that are built on the insights that various disciplines (e.g., nutritional, agricultural, economic, social sciences, etc.) can bring to the table and the development of integrative tools that bring these together in accessible ways. In addition, the options for change need to be grounded in the realities of the actors within the system as they need to be implementation partners for new solutions.

A commonly used tool to communicate the status of a complex system—such as a food system—is the use of indicators [23,24,37]. They are regarded as functional information tools that can indicate the state of a certain policy goal, as they harness systems' complexities by making data insightful to people outside a certain research discipline [24,38]. Ranging from descriptive (pure data) to aggregated (consisting of multiple indicators) indicators, they are used to explain "a given situation or underlying reality which is difficult to quantify directly" [19]. The development of improved metrics and data for the assessment of food systems is essential to better inform policymakers [39]. While there is a general impression that indicators reflect reality and are 'neutral', assumptions embedded in underlying conceptual frameworks remain hidden. Especially when indicators are developed by experts alone the process can be opaque, not allowing for stakeholder-input [38,40]. Without transparency on the construction of indicators, they have the potential to become a tool of control to the already powerful, rather than empower all stakeholders.

The project developed several steps to deliver an integrated approach for assessing SFNS of the EU food system and innovation options for food system transformation. These included:

1. The creation of a participatory space;
2. The development of a conceptual framework mapping out the driving forces, actors, outcomes and goals for the EU food system [41];
3. An approach to devising a set of performance metrics for assessing the food system's status with respect to achieving SFNS and innovation options across four key policy goals formulated by food system actors [42];
4. A modelling strategy for quantifying the sustainability status of FNS in the EU [43];
5. A visualization tool that allows food system actors to assess the outcomes and associated trade-offs of possible innovation options in an integrated manner across the policy goals (the SFNS visualizer) [44].

3.1. Step 1: The Creation of a Participatory Space

To achieve a food system assessment that can be an aid for decision making, it is essential to create a participatory space [45,46], ideally involving the decision makers in the discussions. This includes building on the knowledge, experiences and values of the many actors embedded in the system. Creating a participatory space can be done in various ways but it is important that the stakeholders come together from all over the food system and represent varying 'environments' and world views.

We invited a broad range of stakeholders in the EU food system into a so-called Stakeholder Core Group that acted as an advisory group to the project, reviewing and refining research ideas, the case study work and preliminary results. The selection was done based on an initial mapping of the EU food system while preparing the project's conceptual framework. The group consisted of the three main types: Actors associated with food systems activities, food system policy makers, and food system influencers (also see Figure 1 for details). The actual group consisted of 30 selected experts and decision makers representing primary producers, the food industry, retail, consumer groups, investors, regulators, policymakers and academics. The Stakeholder Core Group was brought together in four annual, interactive workshops in which the team discussed and received input on ideas and on-going work, particularly on the project's conceptual framework, metrics selection, case studies on innovation options and foresight work (Details on the project's stakeholder engagement work can be found on the project website, work package 6. www.susfans.eu). The input was then brought back into the research design and shaped various research outputs (e.g., the stakeholders suggested to establish equity considerations as one of the food system policy goals. This suggestion was then brought into the conceptual framework design, metrics work and visualizer). As such, the involvement of the Stakeholder Core Group was an intricate part of the project.

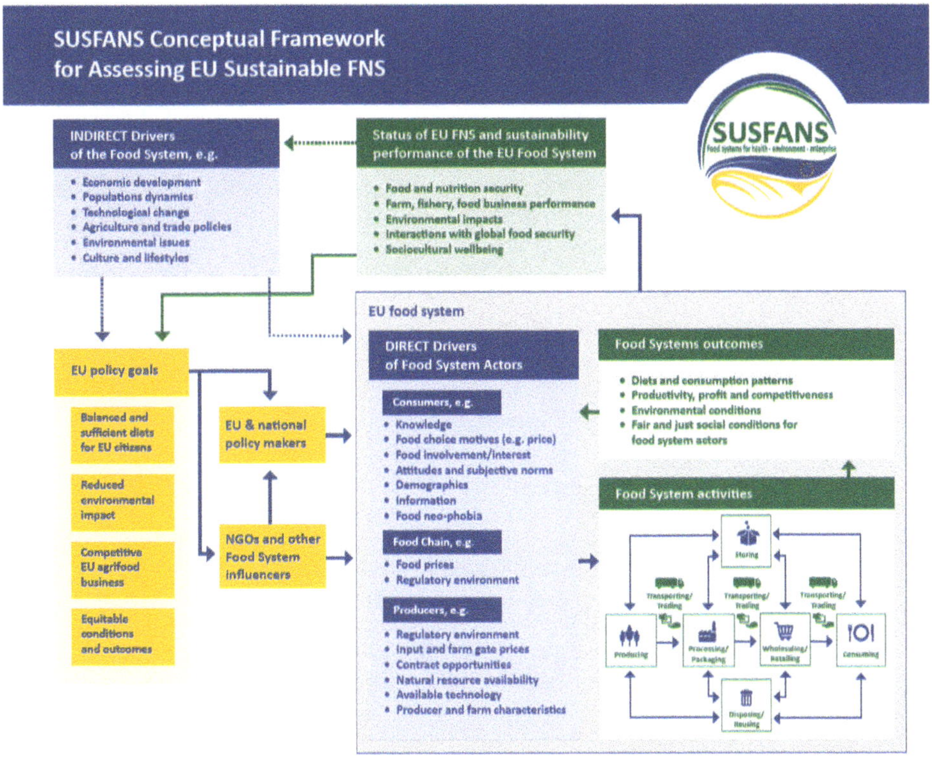

Figure 1. A visual representation of the conceptual framework that was developed based on an iterative consultation process with the Stakeholder Core Group.

3.2. Step 2: A Conceptual Framework of the Food System

As the food system is a complex system and consists of various elements, a prerequisite for a proper analysis and the assessment of its status is a conceptual framework that ensures that all the important elements of the system are included. The framework provides a logical structure for analysis together with a visual and written representation of the system, gives appropriate weight to its components and shows important assumptions about the interactions within the system as well as gaps in understanding. The most important purpose of the conceptual framework is to visualize the joint understanding of the food system across academics as well as stakeholders to provide a common ground for the interdisciplinary work of the project team as well as for the stakeholder work. The framework helps to identify possible entry points for improving system's performance and outcomes.

The conceptual framework (see Figure 1) was created with the input of all the scientific partners and discussed in detail with the Stakeholder Core Group (Details on the project's stakeholder engagement work can be found on the project website, work package 6. www.susfans.eu). For that, the project team provided the Core Group with a basic set up which was then populated, discussed and revised jointly in interactive meetings. The stakeholder core group was later also consulted about the chosen metrics (Step 3) whose selection depended on the conceptual framework. The basic components of the conceptual framework (Figure 1) describing the EU food system are (a detailed description of the conceptual framework can be found in project deliverable D1.1 on the website: www.susfans.eu):

- The diverse sets of actors that are connected to the EU food system;
- The direct and indirect factors driving the behavior of food system actors and therefore influencing change within the food system (drivers of change);
- The outcomes that are related to the EU food system and its activities;
- The goals at the EU level that are shaping the drivers and the EU wide and national policies affecting the food system;
- The interactions and feedback loops that take place among the aforementioned food system components.

Ultimately, the framework allows for the identification of a wide set of policy and technical recommendations which can be assessed not just against a number of specific objectives or for specific actors, but which can also be examined against their effects when rippling through the whole system and their resulting potential unintended consequences/trade-off effects. Furthermore, the framework is intended to highlight the dynamic aspects of the system by laying out the system's drivers as well as the system's interactions and various feedback mechanisms.

3.3. Step 3: Sustainability Metrics for Assessing the Food System and Innovation Options

A meaningful set of metrics to assess the performance of a food system as well as how the system might transform with the introduction of possible innovations must give information with respect to one or more system elements, ideally as presented in the conceptual framework. The project decided in consultation with its Stakeholder Core Group to focus on the four policy goals as laid out in the conceptual framework. The group wanted to understand how close the current system is to achieving SFNS, defined in the four goals, and what possible innovations would do to help achieve the goals. To that end, the project developed a Hierarchical Approach with four different layers (Figure 2). The hierarchical approach intends to combine the idea that decision makers prefer a small, yet comprehensive set of metrics that can communicate findings of an assessment, together with the need to base these metrics on data from many sources that are easily accessible in a transparent way. The hierarchy of the approach is depicted in Figure 2, using the following terminology:

- Individual variable: A measure that can be quantified and/or counted in commonly used standards (e.g., hectares, kg).
- Derived variable: Combines a number of *individual variables* to devise a new measure (e.g., nitrogen input vs. output, ratio of energy intake vs. expenditure,). Sometimes additional information is required to derive the variable (e.g., conversion of individual Greenhouse Gas Emissions to total CO_2 equivalents).
- Aggregate indicator: Combines one or several derived variables and assesses them against a particular goal (e.g., marine biological diversity, reduction of nitrogen surplus, food access).
- Performance metric: Combines a number of aggregated indicators and evaluates them against performance of EU goals (e.g., climate stabilization, balanced diet for EU citizens).

The sets of performance metrics describing each policy goal are described in Table 1. A draft set of individual and derived variables, substantiating the aggregate indicators and performance metrics, as well as objectives for aggregate indicators is given in Reference [42].

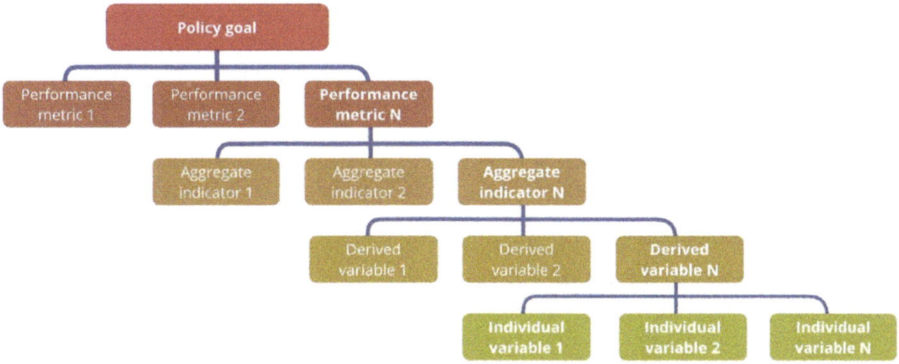

Figure 2. Visualization of the layers that make up the hierarchical approach to devising performance metrics in the Horizon2020 EU-funded project 'Metrics, Models and Foresight for European SUStainable Food And Nutrition Security' (SUSFANS) project. 'N' indicates that the number of variables, indicators and metrics can vary per policy goal.

Table 1. The policy goals used by the project to describe sustainable food and nutrition security (SFNS) and their underlying performance metrics. For a full set of metrics, showing the full hierarchal approach, see Reference [42].

Policy Goals	Performance Metrics
1. Balanced and sufficient diets for EU citizens	Food based intake summary Nutrient based intake summary Energy balance
2. Reduced environmental impacts of the EU food system	Climate stabilization Clean air and water Biodiversity conservation Preservation of natural resources
3. Competitiveness of EU agri-food businesses	Relation between production and trade Orientation and specialization of trade Economic performance and productivity
4. Equitable outcomes and conditions of the EU food system	Between producers and chain actors Among consumers with regards to system conditions Among consumers with regards to system outcomes Footprint of food

3.4. Step 4: Modelling Strategy for Quantifying Performance Metrics for the Food System

Quantification and modelling of the metrics allows for a ground-truthing in available data and establishing relationships across different metrics. However, current models have either a specific focus on one section of a food system or do not cover all the different parts, such as seafood production [47]. Especially the latter is a major gap in food system assessments, as most modelling efforts up to date focus on crop and livestock production, often excluding seafood. Using the integrated set of metrics to assess SFNS in food systems in a forward-looking manner requires an integrated modelling strategy that is currently missing. Thus, existing models with different characteristics and that can produce different sets of outputs related to SFNS need to be connected in order to achieve a full picture of the integrated set of metrics developed in consultation with the stakeholders.

In the project, this was done by building on already existing state-of-the art models such as MAGNET [48], DIET [49], CAPRI [50,51] and GLOBIOM [52,53] and the development of a new tool to define model diets that perform well on sustainability and health objectives (SHARP) [54]. However, each of these have been developed to serve a specific sectoral policy—e.g., CAPRI has been developed for the EU agricultural policy, and DIET was designed for the purpose of assessing the social welfare impact of dietary change in the population at country level (see Figure 3). Taken separately, they are not designed for integrated assessments such as we argue for in this paper. Using the models in combination allows a consistent assessment across the different domains covered by each individual model. In addition, the combined models better capture diets, include seafood production and aim to portray synergies between sustainability and nutrition outcomes. For more detailed information about the developed integrated modeling strategy see References [55,56]. Regardless, three important deficiencies for food system assessment remain, namely actors behaving rationally by responding to 'perfect information' (see Section 4) are key for the models, food chain actors are not well represented in all models, and not all metrics can so far be quantified.

Conscious of the deficiencies that are embedded in the different modelling tools and data availability (more on this in Section 4), a modelling toolbox has been developed. This toolbox serves mainly for forward looking exercises. It combines the economy-wide but aggregate assessments of MAGNET with the European detail in CAPRI, and environmental assessments of GLOBIOM for consistent projections of developments in the global food system. Results in terms of prices, income and sustainability indicators are mapped to DIET for assessing consumer utility and to evaluate social welfare impact, and to SHARP to assess at detailed product level the nutrition, health and sustainability of diets. The toolbox also allows a reverse set-up, where an optimal diet derived by SHARP is imposed in the other models to assess the food system changes it implies at aggregate level.

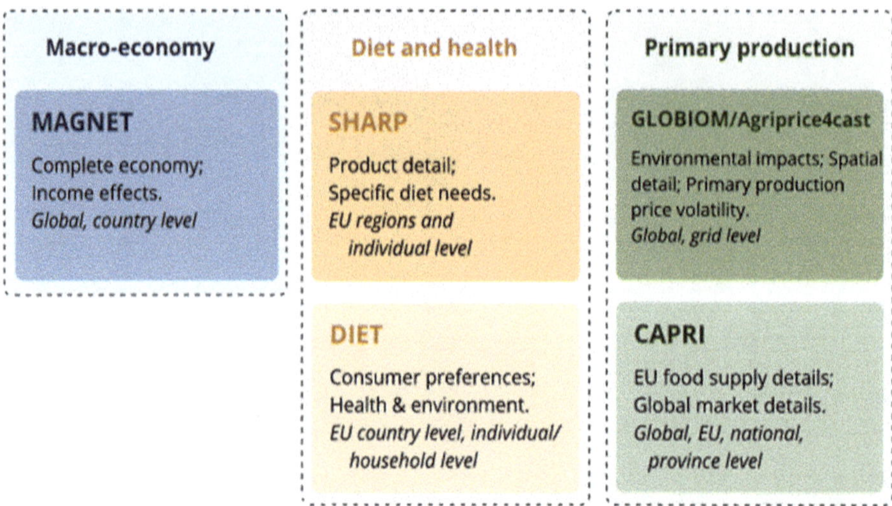

Figure 3. Existing models with different characteristics used to achieve a full picture of the integrated set of food system assessment metrics. The figure indicates differences between the models in terms of sectors (economy, diet, or production), differences in focus within these sectors and the scale level of their assessment.

3.5. Step 5: Integration Across Metrics: The SUSFANS SFNS-Visualizer

The last step is to visualize change in the status of SFNS of the food system. Making trade-offs visible and transparent is vital for decision making. A systemic perspective allows for evaluation of how particular actions or decisions could affect people or dynamics in different ways, and what trade-offs are key to making decisions on moving forward into more sustainable trajectories [57]. One of the key objectives of the SFNS visualizer is to communicate the systems complexities embedded in the SUSFANS sustainability assessment framework to a non-technical audience. The tool informs a discussion about the current sustainability performance of the EU food system, and pathways for transforming the EU food system by means of public and private policy reform and innovation. In doing so, it can highlight potential synergies between several goals in the EU food systems, while avoiding trade-offs. As such, the decision maker can see the potential impact for the key goal they are interested in, as well as for the other goals associated with sustainable food system change. An interactive spider diagram called the 'SFNS visualizer' was developed (see Figure 4), aiming to:

- Show the status of the EU food system with respect to policy goals today;
- Allow the user to look across all policy goals at the same time;
- Allow the user to assess changes to the food system's performance when introducing innovations;
- Visualize synergies and trade-offs across policy goals for the selected innovations to enable an informed discussion about which innovations to pursue.

Figure 4 depicts a mock-up of the visualizer showing a hypothetical assessment of the current status of the EU food system. The visualizer is constructed through the aggregation of variables and indicators into performance metrics as described in the hierarchical approach in Step 3, see Figure 5. The closer the red wedge in the middle gets to the performance metrics (i.e., the bigger the wedge gets) the closer the food system gets towards achieving the formulated goals and their targets. The hypothetical assessment shows, for example, that the project hypothesized that the EU is doing relatively well with respect to a healthy diet and the food sector is quite competitive, while the food system nevertheless produces high environmental impacts and could improve with respect to equity considerations (the on-going modelling work is testing these hypotheses). If innovations are introduced to the system (e.g., changes in fishery practices, more stringent nutrient management in livestock systems or social innovations that help reduce food waste) these will result in changes in one or more food system outcomes, i.e., red wedges. This will allow the user to see the effect of a particular innovation option across all policy goals, thus visualizing synergies and trade-offs across policy goals.

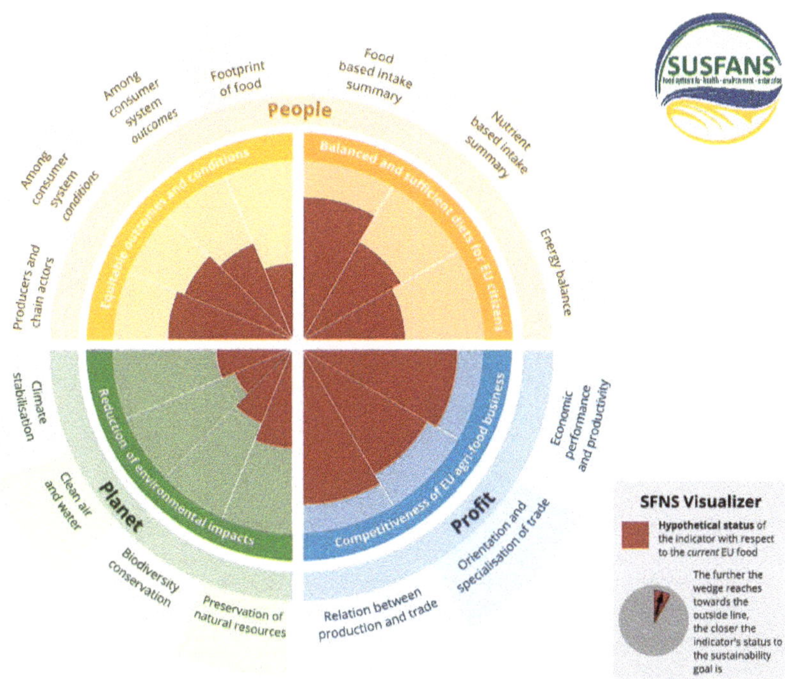

Figure 4. The SUSFANS SFNS-impact visualizer showing a hypothetical assessment of the current EU food system's SFNS status.

How the assessment shown in the visualizer is constructed through the aggregation of variables and indicators, as described in Step 3, can be seen in Figure 5. Different stakeholders, countries, or individuals will differ in their priorities with respect to the food system outcomes. For example, one could prioritize social equity outcomes over environmental ones. While this is a static figure, the next version of the SFNS visualizer will be able to show this change at a click of a button in an interactive web platform. This will allow for the visualization of different world views, but also to 'play' with the weights given to different performance metrics and see the effects this might have on achieving certain objectives.

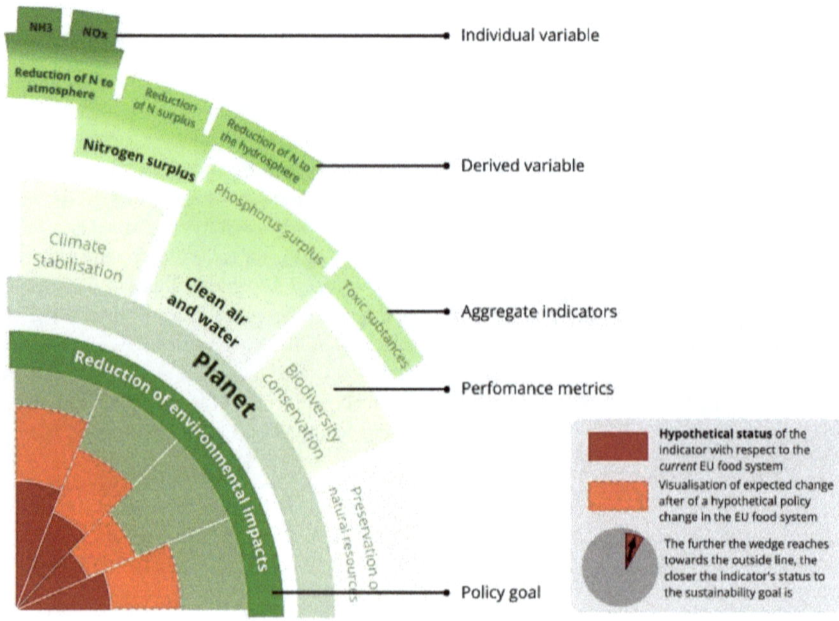

Figure 5. The suggested design for the SFSNS-visualizer when in use. Unfolding the hierarchical approach that underlies a performance metric, in an attempt to make the complexity and assumptions that are embedded in the integrated approach visual to policy-makers or other non-metric experts.

4. Discussion

4.1. Food Systems as 'Transformative Space Making'

The food systems lens is particularly useful in "making of a transformative space" [58]. As argued elsewhere [45,46,58], transdisciplinary research is increasingly becoming a bridge between the worlds of science and practice, while being highly problem-oriented. The use of an inherently transdisciplinary perspective such as food systems, allows for the convening of people that would otherwise be less likely to enter into discussions about the future of food. The use of the food systems lens, breaking something as complex and dynamic as the food system down into activities, actors and drivers, allows for improved collaboration between stakeholders. This contributes to a mediation of the many different, and sometimes contesting, discourses actors hold about the trajectory of food systems [59]. Consequently, a food systems perspective leads to more effective discussions and an ability to come to a shared understanding.

As this approach is centered on participatory processes, where stakeholders and academics share their ideas about activities and drivers within the system, knowledge about and experience with the diverse food systems components is likely to be varied. Moving through a participatory process to map food systems also increases the awareness of complexity and dynamics that are associated with the food system [60]. These characteristics lead us to the argument that the food systems lens is a transformative space making tool.

4.2. Availability of Indicators and Metrics

Central to the approach is integration of different types of knowledge and ways of conceiving the food system and its challenges. The move towards incorporation of social equity was the result of an iterative process to co-construct the conceptual framework with the stakeholders. Consultation with food sovereignty experts led to the realization that the availability of social equity or food sovereignty metrics and indicators are less straight forward. While we have started a modest attempt at formulating such metrics, it signals a deeper issue in food systems research. A disconnect between research communities is splitting crucial themes of food systems research between a metrics and modelling-centered side of research and more qualitative equity-centered side of research. Consequently, this has resulted in a disconnect between the two streams, partly explaining the lack of metrics-based food systems assessments that include social equity components [61]. There is an urgent need for research to focus on the conceptualization of social equity assessment mechanisms, as of yet there are only few examples that are moving in this direction [62,63].

There are some obstacles to the formulation of such metrics, such as cross-scale interactions, contextual and situational differences. The global and cross-scale interactions that make up food systems complicate social equity assessment. For example, it is difficult to establish whether the effect of a policy change within EU boundaries is solely responsible for an improvement or deterioration of social equity among food chain actors in a non-EU country. Nevertheless, interaction between the metrics and modelling oriented, and the qualitative equity-centered, research communities is vital for integrated assessment and sustainable development of food systems. Secondly, availability of data remains a common challenge. Data for the EU, especially considering environmental, economic and health data, is generally well documented and readily available via open access statistics. Although, data for seafood is often more difficult to access and is generally treated separately from broader food system issues [47]. For effective assessment of food systems, approaches will need to be available on a global scale [64]. However, the availability of data is political, as it partly relies on strong governance mechanisms to enforce stable monitoring systems and requires overall stable circumstances within a country. Especially when the latter is lacking, as for example could be the case after natural disasters, monitoring and data collection become less of a priority.

4.3. Learning from the SUSFANS Approach

Central to the integrated approach is participation of stakeholders, as the entire approach builds on the development of a common framing around food systems. A crucial and difficult first step is the ability to attract a broad group of stakeholders that can represent the different communities and framings (industry, Non-governmental Organizations, consumers, farmers etc.) together with group of academics covering various disciplines. Anyone aspiring to develop a similar integrated approach is required to perform several tasks to create the needed participatory space. This includes the selection of participants that are able to reflect or represent stakeholders for the participatory work; engagement with different mechanisms to get these partners on board and; offering of space for consultation and evaluation that also benefits these partners in some way (be it in gaining a network, or achieving the actual objective of the approach). In other words, the development of a successful integrated approach hinges on the ability to create such a participatory space.

With respect to building an integrated modelling framework the project encountered three different issues with the current selection of models. First, prices (and incomes) driving rational behavior of actors having perfect information (i.e., all and correct information) play a key role in the partial (DIET, CAPRI, GLOBIOM) and general (MAGNET) equilibrium model responses [65]. Thus, prices and income weigh heavily in the quantification of changes in the metrics. Non-price considerations (like taste, social norms, convenience, etc.) are partly captured by estimated responses by producers and consumers to changes in prices. Using the elasticities in models, these are then summarized to quantify reactions to potential price changes. In other words, there is a high focus on prices because data on the responses of consumers and producers to non-price incentives are lacking. As such, it is difficult to separate the impact of non-price incentives on consumption patterns and production decision. Complementary analyses on how non-price incentives affect aggregate behavior are needed to be able to assess their implications for the food system. Second, most models cover both the primary producer as the consumer-end of the food system well, while the food chain actors are not very well represented either because the partial equilibrium models (CAPRI, GLOBIOM) define demand in terms of primary products [50,52], focus on the consumer (DIET, SHARP) or have a very aggregated representation of the supply chain (MAGNET) [48,54]. This restricted depiction of the value chain can affect the accuracy of the quantification of a number of performance metrics. This can for example lead to overlooking a number of processes, and therewith miss the associated emissions in the environmental impact calculations. Current data availability and state of research are reflected in these models. It is challenging to address the large heterogeneity that exists across supply chains in applied simulation models. Third, there are a small number of variables—and with that performance metrics—where current models are not able to quantify all variables. This is particularly associated to the goal of 'equitable outcomes and conditions of the EU food system', as this is still novel terrain when it comes to metric development [61,62,66].

Integral to trans-disciplinary science is communication of its results to decision making and policy-makers. This becomes especially important when scientific advice regarding complex dynamic systems, such as a food system, is plural and conditional [67]. In the development of the SFNS-visualizer we have attempted to present the complexity of the food system in a manner that is palatable to a broad audience of decision makers. We were aware of these issues engrained in traditional science-policy communication that are directed at communicating something that is complex in a brief and 'simple' manner. As such, we attempted to design the SFNS-visualizer such that it 'opens up' discussions around food systems, as it will be able to show various states of the system, rather than a single state [68]. It aims to encourage a more informed dialogue rather than one-way, singular science-policy communication.

5. Conclusions

The paper has presented an integrated approach to assessing food systems and possible innovations based on a conceptual framework of the EU food system, divided into five steps. There are a number of novelties to this approach: Firstly, this approach is based on a set of integrated performance metrics that allow for assessment of such innovations from various angles, thereby revealing synergies and trade-offs. Secondly, it puts forward a transdisciplinary, multidimensional model that emphasizes usability for a wide range of food system actors. Lastly, it is the first approach that explicitly includes social equity as part of food system assessments: All four components (nutrition and diet, environmental and economic outcomes, and social equity) are essential for an integrated assessment of food systems and development of more sustainable trajectories.

Presented in the paper is the 'SFNS-visualizer' which can show the status of the EU food system with respect to reaching four key policy goals for food system outcomes, as formulated by stakeholders and actors in the system. Brought together in this visualizer are the integrated metrics that are developed based on the designed conceptual framework. These metrics are backed by a modelling strategy that combines several models that span across the food system components. The latter gives way to a 'dynamic' visualizer. The planned final visualizer will be an online tool that allows users to make changes and see the impact of these changes. This contributes to an informed dialogue around desired changes and the possible forms of action to be taken to achieve these.

Lastly, we have highlighted some limitations regarding the food systems assessment that require further research. The first is the development of social equity metrics, due to lack of prior developed metrics around the topic and an overall difficulty to obtain data on possible equity indicators. We urgently suggest exploration of social equity metrics development and more broadly, given the lack of data, metrics that can be gathered on a large scale. Secondly, we emphasize the difficulty, but necessity, of communicating the plurality and conditionality of complex, dynamic systems research to an audience of policy-makers. With this paper we have aimed to set a standard for food systems research that provides a starting point for further refinement and development.

Author Contributions: Funding acquisition, T.A.; Investigation, M.Z., A.H., A.L., J.V., M.K., M.G., P.H., T.H., S.H., A.K., L.S., J.M.G., I.T., P.v.V., J.W. and A.Z.; Methodology, M.Z., A.L., M.K. and John Ingram; Project administration, T.A.; Supervision, M.Z.; Visualization, A.H.; Writing–original draft, M.Z., A.H. and M.K.; Writing–review & editing, M.Z., A.H., A.L., J.V., T.A., M.G., P.H., T.H., S.H., J.I., A.K., L.S., I.T., J.W. and A.Z.

Funding: The research leading to the ideas expressed in this paper has received funding from the European Union's Horizon 2020 framework programme under grant no. 633692 (SUSFANS).

Acknowledgments: We are grateful to all members and advisers of the SUSFANS project, to the participants of four successive workshops of the SUSFANS Stakeholder Core Group between October 2015 and June 2018, and the FOOD 2030 congress (Hohenheim, October 2018) for useful comments and suggestions. The authors would like to thank the anonymous reviewers and Luigi Piemontese for insightful feedback to the paper.

Conflicts of Interest: The authors declare no conflict of interest.

References

1. Chaudhary, A.; Gustafson, D.; Mathys, A. Multi-indicator sustainability assessment of global food systems. *Nat. Commun.* **2018**, *9*, 848. [CrossRef] [PubMed]
2. Rutten, M.; Achterbosch, T.J.; de Boer, I.J.M.; Cuaresma, J.C.; Geleijnse, J.M.; Havlík, P.; Heckelei, T.; Ingram, J.; Leip, A.; Marette, S.; et al. Metrics, models and foresight for European sustainable food and nutrition security: The vision of the SUSFANS project. *Agric. Syst.* **2018**, *163*, 45–57. [CrossRef]
3. Foley, J.A.; Ramankutty, N.; Brauman, K.A.; Cassidy, E.S.; Gerber, J.S.; Johnston, M.; Mueller, N.D.; O'Connell, C.; Ray, D.K.; West, P.C.; et al. Solutions for a cultivated planet. *Nature* **2011**, *478*, 337–342. [CrossRef] [PubMed]
4. Garnett, T. Where are the best opportunities for reducing greenhouse gas emissions in the food system (including the food chain)? *Food Policy* **2011**, *36*, S23–S32. [CrossRef]
5. McKeon, N. Global food governance in an era of crisis: Lessons from the United Nations Committee on World Food Security. *Can. Food Stud.* **2015**, *2*, 328. [CrossRef]
6. Gordon, L.J.; Bignet, V.; Crona, B.; Henriksson, P.J.G.; van Holt, T.; Jonell, M.; Lindahl, T.; Troell, M.; Barthel, S.; Deutsch, L.; et al. Rewiring food systems to enhance human health and biosphere stewardship. *Environ. Res. Lett.* **2017**, *12*, 100201. [CrossRef]
7. Brunori, G.; Bartolini, F.; Avermaete, T.; Brzezina, N.; Mathijs, E.; Marsden, T.; Faus, A.M.; Sonnino, R.; Hebinck, P.; Oostindie, H.; et al. *Creating Resilient Food Systems for Enhancing Food and Nutrition Security*; TRANSMANGO: EU KBBE.2013.2.5-01 Grant Agreement No. 613532; TRANSMANGO: Leuven, Belgium, 2017.
8. IPES Food. *The New Science of Sustainable Food Systems: Overcoming Barriers to Food Systems Reform*; IPES Food: Brussels, Belgium, 2015. Available online: http://www.ipes-food.org/reports (accessed on 15 November 2018).
9. Milan Expo. *Milan Urban Food Policy Pact*; Milan Expo: Milan, Italy, 2015. Available online: http://www.milanurbanfoodpolicypact.org/ (accessed on 15 November 2018).
10. Duncan, J. "Greening" global food governance. *Can. Food Stud.* **2015**, *2*, 335. [CrossRef]
11. Haysom, G. Food and the city: Urban scale food system governance. *Urban Forum* **2015**, *26*, 263–281. [CrossRef]
12. Lang, T.; Barling, D. Food security and food sustainability: Reformulating the debate. *Geogr. J.* **2012**, *178*, 313–326. [CrossRef]
13. Lang, T.; Barling, D.; Caraher, M. *Food Policy: Integrating Health, Environment and Society*; Oxford University Press: Oxford, UK, 2009; ISBN 0191015717.
14. Candel, J.J.L.; Pereira, L. Towards integrated food policy: Main challenges and steps ahead. *Environ. Sci. Policy* **2017**, *73*, 89–92. [CrossRef]
15. Ingram, J. A food systems approach to researching food security and its interactions with global environmental change. *Food Secur.* **2011**, *3*, 417–431. [CrossRef]

16. Ericksen, P.J. Conceptualizing food systems for global environmental change research. *Glob. Environ. Chang.* **2008**, *18*, 234–245. [CrossRef]
17. Ericksen, P.J. What is the vulnerability of a food system to global environmental change? *Ecol. Soc.* **2008**, *13*, 14. [CrossRef]
18. CFS. *Coming to Terms with Terminology—Food and Nutrition Security*; Thirty-Ninth Session, Item V.a; Commitee on World Food Security: Rome, Italy, 2012. Available online: http://www.fao.org/docrep/meeting/026/MD776E.pdf (accessed on 15 November 2018).
19. Pangaribowo, E.H.; Gerber, N.; Torero, M. Food and Nutrition Security Indicators: A Review. Available online: https://www.wecr.wur.nl/WECRGeneral/FoodSecurePublications/05_Pangaribowo%20Gerber%20Torero_FNS%20Indicators.pdf (accessed on 15 November 2018).
20. Pinstrup-Andersen, P. Food security: Definition and measurement. *Food Secur.* **2009**, *1*, 5–7. [CrossRef]
21. Sen, A. *Hunger and Entitlements*; World Institute for Development Economics Research: Helsinki, Finland, 1987; ISBN 9519991506.
22. Sen, A. *Poverty and Famines: An Essay on Entitlement and Deprivation*; Clarendon Press: New York, NY, USA; Oxford, UK, 1981; ISBN 0198284632.
23. Acharya, T.; Fanzo, J.; Gustafson, D.; Ingram, J.S.I.; Schneeman, B. *Assessing Sustainable Nutrition Security: The Role of Food Systems*; The International Life Sciences Institute, Research Foundation, Center for Integrated Modeling of Sustainable Agriculture and Nutrition: Washington, DC, USA, 2014.
24. Gustafson, D.; Gutman, A.; Leet, W.; Drewnowski, A.; Fanzo, J.; Ingram, J. Seven food system metrics of sustainable nutrition security. *Sustainability* **2016**, *8*, 196. [CrossRef]
25. Allen, T.; Prosperi, P. *Metrics of Sustainable Diets and Food Systems*; Workshop Report; Bioversity International & CIHEAM-IAMM: Montpellier, France, 2014.
26. Fischer, C.G.; Garnett, T. *Plates, Pyramids and Planet*; FAO; The Food Climate Research Network at The University of Oxford: Oxford, UK, 2016; ISBN 9789251092224.
27. Lukas, M.; Rohn, H.; Lettenmeier, M.; Liedtke, C.; Wiesen, K. The nutritional footprint—Integrated methodology using environmental and health indicators to indicate potential for absolute reduction of natural resource use in the field of food and nutrition. *J. Clean. Prod.* **2016**, *132*, 161–170. [CrossRef]
28. Ranganathan, J.; Vennard, D.; Waite, R.; Dumas, P.; Lipinski, B.; Searchinger, T. *Shifting Diets for a Sustainable Food Future*; Installment 11 of "Creating a Sustainable Food Future"; World Resources Institute: Washington, DC, USA, 2016.
29. Lang, T.; Heasman, M. *Food Wars: The Global Battle for Mouths, Minds and Markets*; Earthscan: London, UK, 2004.
30. Jones, A.D.; Hoey, L.; Blesh, J.; Miller, L.; Green, A.; Shapiro, L.F. A Systematic Review of the Measurement of Sustainable Diets. *Adv. Nutr. Int. Rev. J.* **2016**, *7*, 641–664. [CrossRef] [PubMed]
31. Gussow, J.D. Mediterranean diets: Are they environmentally responsible? *Am. J. Clin. Nutr.* **1995**, *61*, S1383–S1389. [CrossRef] [PubMed]
32. Tendall, D.M.; Joerin, J.; Kopainsky, B.; Edwards, P.; Shreck, A.; Le, Q.B.; Kruetli, P.; Grant, M.; Six, J. Food system resilience: Defining the concept. *Glob. Food Secur.* **2015**, *6*, 17–23. [CrossRef]
33. FAO. *Climate Change and Food Security: A Framework Document*; Food and Agriculture Organization of the United Nations: Rome, Italy, 2008.
34. Ingram, J.; Ericksen, P.; Liverman, D. *Food Security and Global Environmental Change*; Earthscan: London, UK; Washington, DC, USA, 2010.
35. Foran, T.; Butler, J.R.A.; Williams, L.J.; Wanjura, W.J.; Hall, A.; Carter, L.; Carberry, P.S. Taking complexity in food systems seriously: An interdisciplinary analysis. *World Dev.* **2014**, *61*, 85–101. [CrossRef]
36. Brinkley, C. Avenues into food planning: A review of scholarly food system research. *Int. Plan. Stud.* **2013**, *18*, 243–266. [CrossRef] [PubMed]
37. Prosperi, P.; Allen, T.; Cogill, B.; Padilla, M.; Peri, I. Towards metrics of sustainable food systems: A review of the resilience and vulnerability literature. *Environ. Syst. Decis.* **2016**, *36*, 3–19. [CrossRef]
38. Lehtonen, M.; Sébastien, L.; Bauler, T. The multiple roles of sustainability indicators in informational governance: Between intended use and unanticipated influence. *Curr. Opin. Environ. Sustain.* **2016**, *18*, 1–9. [CrossRef]
39. Global Panel. *Improved Metrics and Data Are Needed for Effective Food System Policies in the Post-2015 Era*; Technical Brief; Global Panel on Agriculture and Food Systems for Nutrition: London, UK, 2015.

40. Lehtonen, M. Indicators: Tools for informing, monitoring or controlling? In *The Tools of Policy Formulation: Actors, Capacities, Venues and Effects*; Jordan, A.J., Turnpenny, J.R., Eds.; Edward Elgar Publishing Ltd.: Cheltenham, UK, 2015; pp. 76–99. ISBN 9781783477036.
41. Zurek, M.; Ingram, J.; Zimmermann, A.; Garrone, M.; Rutten, M.; Tetens, I.; Leip, A.; Veer, P.; Verain, M.; Bouwman, E.; et al. *D1.1 A Framework for Assessing and Devising Policy for Sustainable Food and Nutrition Security in EU: The SUSFANS Conceptual Framework*; SUSFANS, GA No. 633692; SUSFANS: The Hague, The Netherlands, 2016.
42. Zurek, M.; Leip, A.; Kuijsten, A.; Wijnands, J.; Terluin, I.; Shutes, L.; Hebinck, A.; Zimmermann, A.; Götz, C.; Hornborg, S.; et al. *D1.3 Sustainability Metrics For The EU Food System: A Review Across Economic, Environmental And Social Considerations*; SUSFANS, GA No. 633692; SUSFANS: The Hague, The Netherlands, 2017.
43. Kuiper, M.; Zurek, M.; Havlik, P.; Deppermann, A.; Valin, H.; Kuijsten, A.; Geleijnse, M.; Veer, P.; Heckelei, T.; Götz, C.; et al. *A Modelling Strategy for Quantifying the Sustainability of Food and Nutrition Security in the EU: Deliverable 1.4*; SUSFANS, EU Grant Agreement 633692; SUSFANS: The Hague, The Netherlands, 2017.
44. Zurek, M.; Hebinck, A.; Leip, A.; Achterbosch, T.; Garrone, M.; Havlik, P.; Heckelei, T.; Ingram, J.; Soler, L.G.; van't Veer, P.; et al. *D1.5 An Inegrated Set of Metrics for Assessing the Overall Sustainability of Food and Nutrition Security in the EU*; SUSFANS, GA No. 633692; SUSFANS: The Hague, The Netherlands, 2017.
45. Pereira, L.; Karpouzoglou, T.; Doshi, S.; Frantzeskaki, N. Organising a safe space for navigating social-ecological transformations to sustainability. *Int. J. Environ. Res. Public Health* **2015**, *12*, 6027–6044. [CrossRef] [PubMed]
46. Hebinck, A.; Vervoort, J.M.; Hebinck, P.; Rutting, L.; Galli, F. Imagining transformative futures: Participatory foresight for food systems change. *Ecol. Soc.* **2018**, *23*, 16. [CrossRef]
47. Farmery, A.K.; Gardner, C.; Jennings, S.; Green, B.S.; Watson, R.A. Assessing the inclusion of seafood in the sustainable diet literature. *Fish Fish.* **2017**. [CrossRef]
48. Woltjer, G.; Kuiper, M.; Kavallar, A.; van Meijl, H.; Powell, J.; Rutten, M.; Shutes, L.; Tabeau, A. *The MAGNET Model: Module Description;* LEI Report No. 14-057; LEI—Wageningen UR: Wageningen, The Netherlands, 2014.
49. Irz, X.; Leroy, P.; Réquillart, V.; Soler, L.G. Economic assessment of nutritional recommendations. *J. Health Econ.* **2015**. [CrossRef] [PubMed]
50. Britz, W.; Van Ittersum, M.; Lansink, A.O.; Heckelei, T. Tools for integrated assessment in agriculture. State of the art and challenges. *Bio-Based Appl. Econ.* **2012**, *1*, 125–150. [CrossRef]
51. Leip, A.; Britz, W.; Weiss, F.; De Vries, W. Farm, land, and soil nitrogen budgets for agriculture in Europe calculated with CAPRI. *Environ. Pollut.* **2011**, *159*, 3243–3253. [CrossRef] [PubMed]
52. Havlik, P.; Valin, H.; Herrero, M.; Obersteiner, M.; Schmid, E.; Rufino, M.C.; Mosnier, A.; Thornton, P.K.; Bottcher, H.; Conant, R.T.; et al. Climate change mitigation through livestock system transitions. *Proc. Natl. Acad. Sci. USA* **2014**, *111*, 3709–3714. [CrossRef] [PubMed]
53. Havlík, P.; Schneider, U.A.; Schmid, E.; Böttcher, H.; Fritz, S.; Skalský, R.; Aoki, K.; De Cara, S.; Kindermann, G.; Kraxner, F.; et al. Global land-use implications of first and second generation biofuel targets. *Energy Policy* **2011**. [CrossRef]
54. Irz, X.; Leroy, P.; Réquillart, V.; Soler, L.-G. Beyond wishful thinking: Integrating consumer preferences in the assessment of dietary recommendations. *PLoS ONE* **2016**, *11*, e0158453. [CrossRef] [PubMed]
55. Rutten, M.; Zimmermann, A.; Havlík, P.; Leip, A.; Heckelei, T.; Achterbosch, T. *D9.1 Modelling Sustainability and Nutrition in Long Run Analyses of the EU Agri-Food system: Work plan for the SUSFANS Toolbox*; SUSFANS, GA No. 633692; SUSFANS: The Hague, The Netherlands, 2016.
56. Kuiper, M.; Oudendag, D.; Bartelings, H.; Shutes, L.; Verma, M.; Tabeau, A. *D9.2 Enhanced Modelling of Sustainable Food and Nutrition Security: Food Consumption and Nutrition Behaviour of European Households;* SUSFANS, GA No. 633692; SUSFANS: The Hague, The Netherlands, 2017.
57. Galafassi, D.; Daw, T.M.; Munyi, L.; Brown, K.; Barnaud, C.; Fazey, I. Learning about social-ecological trade-offs. *Ecol. Soc.* **2017**, *22*, 1–27. [CrossRef]
58. Marshall, F.; Dolley, J.; Priya, R. Transdisciplinary research as transformative space making for sustainability: Enhancing propoor transformative agency in periurban contexts. *Ecol. Soc.* **2018**, *23*, 1–13. [CrossRef]
59. Clapp, J. Food security and food sovereignty: Getting past the binary. *Dialogues Hum. Geogr.* **2014**, *4*, 206–211. [CrossRef]
60. Galafassi, D.; Daw, T.M.; Thyresson, M.; Rosendo, S.; Chaigneau, T.; Bandeira, S.; Munyi, L.; Gabrielsson, I.; Brown, K. Stories in social-ecological knowledge cocreation. *Ecol. Soc.* **2018**, *23*. [CrossRef]

61. Weiler, A.M.; Hergesheimer, C.; Brisbois, B.; Wittman, H.; Yassi, A.; Spiegel, J.M. Food sovereignty, food security and health equity: A meta-narrative mapping exercise. *Health Policy Plan.* **2015**, *30*, 1078–1092. [CrossRef] [PubMed]
62. Landert, J.; Schader, C.; Moschitz, H.; Stolze, M. A holistic sustainability assessment method for urban food system governance. *Sustainability* **2017**, *9*, 490. [CrossRef]
63. Dixon, J.; Omwega, A.M.; Friel, S.; Burns, C.; Donati, K.; Carlisle, R. The health equity dimensions of urban food systems. *J. Urban Health* **2007**, *84*. [CrossRef] [PubMed]
64. Open Data Institute. How Can We Improve Agriculture, Food and Nutrition with Open Data? 2015. Available online: https://www.godan.info/documents/how-can-we-improve-agriculture-food-and-nutrition-open-data (accessed on 15 November 2018).
65. Valin, H.; Sands, R.D.; van der Mensbrugghe, D.; Nelson, G.C.; Ahammad, H.; Blanc, E.; Bodirsky, B.; Fujimori, S.; Hasegawa, T.; Havlik, P.; et al. The future of food demand: Understanding differences in global economic models. *Agric. Econ.* **2014**. [CrossRef]
66. James, S.W.; Friel, S. An integrated approach to identifying and characterising resilient urban food systems to promote population health in a changing climate. *Public Health Nutr.* **2015**, *18*, 2498–2508. [CrossRef] [PubMed]
67. Stirling, A. Keep it complex. *Nature* **2010**, *468*, 1029–1031. [CrossRef] [PubMed]
68. Leach, M.; Scoones, I.; Stirling, A. *Dynamic Sustainabilities: Technology, Environment, Social Justice*; Earthscan: London, UK, 2010; ISBN 1849775060.

sustainability

MDPI

Article

A Reflection of the Use of the Life Cycle Assessment Tool for Agri-Food Sustainability

Oriana Gava *, Fabio Bartolini, Francesca Venturi, Gianluca Brunori, Angela Zinnai and Alberto Pardossi

Department of Agriculture, Food and Environment, University of Pisa, 56124 Pisa, Italy; fabio.bartolini@unipi.it (F.B.); francesca.venturi@unipi.it (F.V.); gianluca.brunori@unipi.it (G.B.); angela.zinnai@unipi.it (A.Z.); alberto.pardossi@unipi.it (A.P.)

* Correspondence: oriana.gava@for.unipi.it

Received: 11 October 2018; Accepted: 20 December 2018; Published: 23 December 2018

check for updates

Abstract: In pursuit of agricultural sustainability and food security, research should contribute to policy-making by providing scientifically robust evidence. Life cycle assessment (LCA) is an excellent candidate for generating that evidence, thereby helping the selection of interventions towards more sustainable agri-food. The purpose of this article is proposing a basis for discussion on the use of the LCA tool for targeting and monitoring of environmental policy interventions in agri-food. The problem of reducing the environmental burden in agri-food can be tackled by acting on the supply and/or demand sides and may benefit from the collaboration of supply chain stakeholders. Agri-food policies that most benefit from LCA-based data concern cross-border pollution, transaction costs following the adoption of environmental standards, adoption of less polluting practices and/or technologies, and business-to-consumer information asymmetry. The choice between the methodological options available for LCA studies (attributional, consequential, or hybrid models) depends on the purpose and scope of the study. The possibility of integrating the LCA with economic and social impact assessments—e.g., under the life cycle sustainability assessment framework—makes LCA an excellent tool for monitoring business or sectoral-level achievements with respect to UN 2030 Sustainable Development Goals.

Keywords: sustainable development goals; sustainability assessment; agricultural sustainability; food security; LCA broadening; LCA deepening

1. Introduction

1.1. Background and Rationale behind the Study

The design and implementation of policy strategies towards agricultural sustainability and food security should base on scientific evidence, to guarantee environmental protection and avoid burden sharing, without reducing the productivity, competitiveness, and profitability of agri-food [1], thereby allowing to meet UN 2030 Sustainable Development Goals (SDGs), especially three interrelated goals, i.e. SDG2, SDG13, and SDG15 [2,3]. The aim of SDG2 (no hunger) is to promote profound changes in the way food is produced and consumed, for feeding the growing world population. Specific targets involve ensuring sustainable food production systems and implementing resilient agricultural practices, while doubling the agricultural productivity and farmers' incomes. SDG13 (climate action) has the objective of limiting the global temperature to rise. To that purpose, national governments are called to implement climate mitigation policies and adaptation policies to promote the resilience and adaptive capacity of socio-ecological systems. SDG15 (life on land) aims at preserving forests and preventing desertification, land degradation, and biodiversity loss. The goal includes the promotion of measures for preserving ecosystems and encourages governments to integrate ecosystem and biodiversity values into national and local planning and development programs.

Among the sources of scientific evidence, information generated via life cycle assessment (LCA) has received growing attention by policy-makers for identifying, selecting, and guiding interventions to reduce the environmental burden of agriculture and food systems [4,5], as well as for setting the objectives and monitoring the impacts of policies [6,7]. LCA is a formalized method (ISO 14040:2006; ISO 14044:2006) (ISO, Geneva, Switzerland) and can be applied for monitoring the achievements of the agricultural sector or single business with respect to SDGs [8]. Policy instruments based on LCAs of traded commodities can mitigate carbon leakage and reduce food losses, by affecting producer and consumer price, thus altering food production and consumption decisions [9,10]. Various governments have set up publicly-available life cycle inventory databases for agricultural activities (e.g., the USA [10], Australia [11], France [12]) and regularly require LCA in their funded research projects [13]. Public investments in LCA research have helped the creation of the scientific knowledge base for evidence-based policy-making [14].

A simple search ("life cycle assessment" AND "agriculture") over major academic citation indexes (viz. Scopus®, Web of Science™) reveals the substantial and recent growth of LCA-based agricultural research, including variously structured literature reviews. Refereed original research has covered the impacts of major agricultural production systems (at least in advanced economies) and has been largely synthesized to highlight, for example, the impacts associated with the production of food [15]—including insects [16] and biofuels [17]—of management practices [18], and of human diets [19], among others. Despite the wide scope for LCA application and development and the growing interest by governments and the research community [20,21], LCA applications targeting the decision-making process in agricultural policy is not so developed, when compared to other economic sectors., with theoretical analyses being almost missing (see [22] for a notable exception). The purpose of the article is not providing a systematic review or meta-analysis of impacts, which are already available from the literature, e.g. [23,24]. Rather this article focuses on needed research development to improve the policy orientation of LCA findings. In a forthcoming study, we will expand on the objective of the present article by delivering a content analysis-based literature review of the academic literature with explicit implications for agricultural policy community, focused on needed research development to improve the policy orientation of LCA findings. Policy interest is among the drivers of LCA popularity in agri-food research [20,25]. Despite that, much more research is needed to adapt LCA studies to policy demand. Key challenges involve adopting innovative perspectives on intervention strategies [26], improving data sources [25], as well as deepening and broadening the LCA technique [27,28]. Drawing on those challenges, this article covers a series of aspects of LCA application, namely research perspectives towards the mitigation of the environmental burden of agricultural and food systems, the scope of the practical use of LCA-based information for regulatory purposes, as well current obstacles to method diffusion and the opportunities for improvement. The aim of the article is to propose a basis for discussion on the use of the LCA tool for targeting and monitoring environmental policy interventions in agri-food, by bringing together and discussing different theoretical and practical elements that should be considered when envisioning LCA studies for agri-food sustainability, some of which were individually addressed by previous research (e.g., [5,20,25,26,28,29]).

The paper aims at inspiring and promoting policy-oriented LCA research, thereby targeting both researchers and policy makers.

1.2. Article Outline and Conceptual Model

A mutual relationship exists between the decision-making process and the evidence generated in LCA studies, with practitioners needing to adjust the methodological approach based on the final use of LCA outputs [30,31]. Figure 1 provides a conceptual representation of different elements of the that relationship.

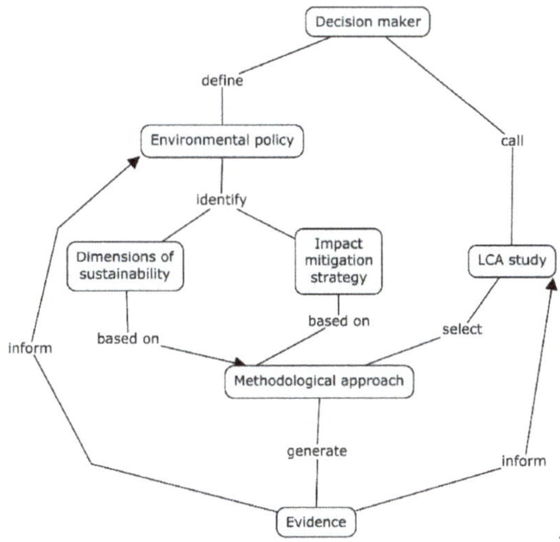

Figure 1. Conceptual model: mutual relationships between the different sections of the article. Source: Authors' own elaboration.

A key determinant of method and data source selection is decision-maker approach to impact mitigation, which identify the stakeholders that are requested to implement actions aimed at impact reduction. For example, technological innovation on firm may allow to reduce the environmental burden of food supply. Contextually, adopting green marketing strategies to inform consumers about the adoption of impact mitigation actions (e.g., food labelling) may increase the consumption of more ecological foods. The type of application of LCA-based information is another determinant of method selection. For example, decision-makers might be interested in learning the impacts of a new technology or in comparing the impacts of two alternatives. When the focus is wider than the business-level, decision makers might be interested in wider impact assessments, including for example economic and social impacts. The role of research is addressing the limitation of the method to improve the reinforcing feedback loops between existing LCA frameworks and decision makers against that background.

Based on the conceptual model above, the present article is structured towards five more sections. The next section provides a synthetic overview of the LCA method, focusing on the key elements that distinguish major methodological approaches. The following section presents the approaches to impact mitigation in agriculture and food systems. Section 4 concentrates on the practical application of LCA-based information for public and private policy. Section 5 addresses method related limitations and presents a series of approaches for dealing with those limitations. The last section presents a discussion of the state-of-the art and delivers some recommendations.

2. Overview of the Method

LCA is the most comprehensive technique for tracing and generating quantitative information about the environmental burdens that originate from the activities and facilities involved in manufacturing, delivering, consuming, and managing the end-of-life of production processes or of the average or marginal output of an industry, i.e., from cradle-to-grave [32]. The assessment has been formalized by the International Standard Organization (ISO) (Geneva, Switzerland). Two ISO (Geneva, Switzerland) rules provide general guidelines for the assessment, thereby allowing great flexibility to practitioners. ISO 14040 (2006) (ISO, Geneva, Switzerland) delivers the principles and the framework for the assessment, including the stepwise procedure, LCA reporting and critical review, limitations of the assessment, and conditions for use of value choices and optional elements. ISO 14044

(2006) (ISO, Geneva, Switzerland) delivers a general guideline to the stepwise procedure, though not providing methodological details about the practical implementation of LCA steps.

LCA is based on an iterative process with four steps, i.e., goal and scope definition, inventory analysis, impact assessment, and interpretation. Based on the goal and scope, the assessment may be narrowed to exclude upstream processes (gate-to-grave), downstream processes (cradle-to-gate) or both (gate-to-gate), or even to exclude the impact assessment step. The latter are more properly referred to as life cycle inventory studies. Life cycle inventories are datasets that account for all material and energy flows and related emissions (via the application of characterization factors), through the various processes within the system boundaries. Input flows refer to consumed natural resources, materials and energy, output flows account for waste, emissions to air, water and soil, and the final goods and services being produced. Input–output flows include variables that should be included in the inventory—e.g., productivity, distance travelled, and type of transport—among others [31]. Based on the part of the system that is under the direct influence of the decision maker, the product system under study can be divided into foreground and background subsystems. The former includes the processes that are directly affected by the decisions made in the study, both in terms of process type and mode of operation; the latter includes the processes aimed at supplying material or energy to the foreground subsystem [30]. When possible, the description of the foreground subsystem should rely on primary data, while secondary or generic data can be used for the background subsystem [31]. Primary data can be directly gathered—e.g., from farmers, food processors or other stakeholders, including consumers—and may be aggregated or concern individual companies. Secondary data can be collected from different sources, such as, for example, scientific literature, business or government reports. In addition, many life cycle inventory databases are available, such as Ecoinvent® and Agrifootprint®. As far as possible, inventories of foreground subsystems should rely on site-specific data [33], though average or generic data may be preferred, depending on research aims [34].

Impact categorization and characterization occurs in the impact assessment step, via the application of impact assessment models. The outputs cover a wide set of problem-oriented (midpoint) or damage-oriented (endpoint), grouped under various impact categories, covering the depletion of natural resources and damages to the environment and human health (an example of recommendations about model and impact category selection is available from [31]). Via contribution analysis, LCA outputs may be used for identifying the activities that most contribute to the environmental impacts of a given product (hot spots), thereby facilitating the targeting of impact mitigation interventions at the sectoral or business-level [29]. Stand-alone LCAs concentrate on a single product, generally for exploratory purpose, while comparative studies aim at supporting product selection [35]. Assessments covering single issues are known as footprint studies. Carbon and water footprints are popular formalized methods (ISO 14067:2018 and ISO 14046:2014, respectively (ISO, Geneva, Switzerland) for calculating and communicating the greenhouse gas emissions and freshwater consumption associated with product life cycles [36]. Footprints are widely used for informing consumers about the environmental performance of foodstuffs via food labelling.

The most widespread software programs for supporting product system modeling and assessment by LCA practitioners are SimaPro® and GaBi®, though other software programs exist, e.g. openLCA® or Umberto®. Software solutions include major life cycle inventory databases and impact assessment models. LCA software tools are generally distributed under a licensing agreement to users, which implies the payment of a fee, though open source software exists, notably openLCA®. Besides publicly available life cycle inventory databases, access to commercial databases (e.g., Ecoinvent®) is generally necessary for gathering data about background subsystems.

Conceptually, LCAs are divided into retrospective and prospective. Retrospective LCAs, or LCAs of the accounting type [30], aim at describing the production of a given product in a system in terms of materials and energy flows, before or after impact mitigation interventions, and the associated environmental impacts; the latter are calculated based on average emissions for producing a unit of the product in the system [37]. Prospective, or change-oriented [30], LCAs consider the environmentally relevant flows in the production of a given product, for describing the consequences of possible impact mitigation interventions (e.g., future policy), i.e., the effects of substitutions in the relevant material and energy flows [37]. In prospective LCAs, system boundaries are expanded, even outside cradle-to-grave system of the product under study, to include the activities introduced via substitution and the associated emissions, with data referring to unitary changes in the relevant flows (marginal data) [30,37]. Prospective LCA studies are especially suited for product development and the ex-ante evaluation of public policies [35].

The division between retrospective and prospective LCA underpins the difference between the attributional (ALCA) and consequential (CLCA) LCA models, respectively. ALCA is the conventional model—formalized in the 1990s (ISO 14040: 1997) (ISO, Geneva, Switzerland)—which provides a static picture of the impacts associated with all the processes included in the boundaries of the system under study. CLCA—formally defined only in 2011 [38]—was developed for creating the connections between environmental and economic models [39]. The purpose of CLCA is to quantify the market-mediated consequences of decisions (measures to mitigate the environmental impact of productive activities) concerning the system under study on other systems [5]. Instead of the processes included in the real (or proposed) production system, CLCA studies, model the systems that are more likely to respond to changes in demand due to the decisions taken to reduce the impacts on the environment (marginal systems) [40]. For example, the diffusion of biogas-to-energy plants based on agricultural biomass can reduce the demand for electricity from the national energy mix [41]. CLCA studies should consider the interactions between agricultural and food policies and changes in consumer behavior [5]. Besides the main product, agricultural supply chains often include intermediate products and by-products (multifunctional processes), for example, calves in the production of cow's milk. ALCA and CLCA differ for the applied method for calculating the emissions associated with multifunctional processes. The attributional model is usually based on allocation, the consequential on system expansion. The allocation method relies on emission distribution among main products, intermediate products and by-products, based on physical properties, such as mass or gross energy, or the market value of the three types of output. The system expansion method considers intermediate products and by-products as resulting from independent (external) production processes, i.e., different from the production process that originate the main product. Those external processes turn to be included within system boundaries, their relative impacts are calculated and subtracted from those of the multifunctional process (saved impacts). CLCA is broader in scope than ALCA, accounting for the horizontal linkages (e.g., competition for a good in alternative applications) at each step in the supply chain, whilst ALCA concentrates on the vertical dependencies throughout the chain [42]. Moving to impact assessment, life cycle impact assessment models for consequential studies model the consequences of one additional unit of a given emission, rather than the average consequences of all emissions [43]. The main drawback of CLCA is the high number of economic assumptions, which grows easily with the number of processes included in the boundaries of the system and may be an important source of variability in study results [44]. Many authors contrasted the key features of attributional and consequential LCAs, in terms of aims of the studies, inputs from and implications for decision-making, and methodological choices, including the way how systems are modeled [30,43,45–48]. The comparative appropriateness and the potential for the complementary use of the two LCA models for political applications are the subject of an open debate within the scientific community [42,44,49,50].

ISO 14040:2006 and 14044:2006 (ISO, Geneva, Switzerland) refer to ALCA and CLCA. Another major approach to LCA exists that has not been covered by ISO (Geneva, Switzerland) rules—i.e., sector-based LCA, known as economic input–output LCA (EIO-LCA). EIO-LCA is an economy-wide assessment of the environmental burdens caused by the business sector, its suppliers and suppliers' suppliers. The assessment adopts a top-down approach by relying on economic input–output analysis for tracing the interdependencies among economic sectors in a given region and giving the transactions an economic value [51]. Sectoral level data from national economic input–output tables are extended by adding a vector of total domestic direct and indirect environmental burdens, e.g., resource use or emissions [52]. Generally, input–output tables do not cover the use and waste management phases of products. Process-based data are often used to cover those life cycle stages. Hybrid LCA links process-based (conventional) LCA with EIO-LCA. EIO-LCA is especially useful when real-world data about given products and processes are not accessible and/or when practitioners are not familiar with the characteristics of a certain industry [50]. However, process-based LCA, i.e. that quantifies the impacts based on physical relations between activities in across product life cycle as standardized in ISO (Geneva, Switzerland) rules, and EIO-LCA have opposite perspectives and different levels of data resolution, which makes the selection of the approach strictly dependent on the objective of the study, as well on its intended audience [51]. The works of [43,51] compare the key features of LCA and EIO-LCA, evaluate their suitability for impact assessment at different level of analysis, and provide recommendation for application.

3. Approaches to Impact Mitigation Strategies in Agriculture and Food Systems

Strategies to reduce the environmental impact of agricultural and food systems involve targeting actions to the supply and/or the demand sides. The former includes the promotion of practices that improve the environmental performance of agriculture, while preventing negative effects on productivity [26]. The purpose of LCA studies is identifying hot spots and comparing management or technological options for improving the environmental performance of single businesses or of the whole sector [13,53]. For example, food packaging has received special attention by the food industry [25]. The life cycle impacts of packaging grow with product water content [54] and, e.g., account for most impacts of wine, especially due to the production and disposal of glass bottles [55]. However, wine consumers may not be willing to accept different packaging materials (e.g., plastics) [56]. So far, supply-side actions have received the greatest attention by researchers [26].

Largely policy-driven, demand-side actions aim at changing consumption patterns to reduce the demand for the most impacting foods [57], e.g., products of animal origin (in particular dairy products and meat of ruminants), those deriving from conventional agriculture [58,59] or that require water-intensive agronomic practices (e.g., rice) [60]. The reduction of food waste is an additional demand-side strategy [57], which, however, has shown a limited impact mitigation potential compared to dietary change [61]. The adoption of impact mitigating practices and/or technologies still occurs in the supply side, but their implementation is mediated by market forces [26]. Regulatory options to promote changes and encourage sustainable consumption generally involve acting on consumer information and education, to raise awareness and policy acceptability [62]. LCA applications involve the comparison of different consumption patterns via the calculation of environmental/nutritional trade-offs associated with the reduction, removal, or substitution of animal-based foods [63]. Compared to supply-side actions, research focusing on the demand side is more recent and has raised the interest and support of more and more research [25]. A key challenge involves addressing the feedback loops between dietary alterations for human health. The LCA literature contributed to answering that challenge mainly by assessing the environmental impacts of diets differing in the content of meat and dairy or by assessing environmental impacts of healthier diets [63]. The former generally compare current consumption patterns (i.e., self-selected diets, generally based on food consumption surveys) against official dietary guidelines and/or other popular type of diets (e.g., vegetarian). The second strand of literature has the nutritional quality of the assessed diets at its core (e.g., [64]). Mainly, the authors propose considering the restriction of the total energy intake as an impact mitigation strategy and to carefully model the choice of meat replacement foods, to avoid burden shifts associated for example with the consumption of greater quantities of food, due to the lower energy density, and of the increased demand of imported or out-of-season foods [65].

Supply and demand side approaches could be combined to promote system-level actions, subject the communication and collaboration among the involved stakeholders [26], such as private businesses (supply side), the public (demand side), governments (resource-use regulation), and researchers and practitioners (impact assessment and monitoring) [66]. To that purpose, the conceptualization and implementation of LCA studies should concern redesigned models of agricultural and food systems (scenarios) that propose sustainable synergetic solutions [61,67]. An example of agri-food system redesign involves the application of the concept of "ecological leftovers" [68], which, though promising, needs more empirical research to be effectively proposed for real world applications, especially for knowing more about its cultural acceptability [69].

4. Policy Applications of LCA Results

Information originating from LCA studies is used in policy-making to face four broad challenges, viz. pollution leakage, ex-post transaction costs of environmental regulation, adoption of environmentally least harmful (technological) options, and business-to-consumer information asymmetry [20]. Pollution leakage is the increase in total pollution outside the policy jurisdiction where the investigated process occurs, in response to a decrease in total pollution within the same jurisdiction, which, e.g., could occur in case of indirect land use change from food to energy cropping [70]. The costs of administration, monitoring, and enforcement of environmental regulation can be high when (potential) polluters are many and heterogeneous, thus making emission monitoring a very costly activity [20]. On one hand, transaction costs may be reduced by monitoring a set of key upstream activities that produce widespread raw materials, e.g., fossil fuels [71]. On the other hand, extended producer responsibility policies have grown in the last two decades, which give producers a significant financial and/or physical responsibility for the treatment or disposal of post-consumer products, in exchange of incentives to prevent waste creation and promote eco-friendly product design [72]. Policy support of the adoption of environmentally least harmful options is widespread across countries (for example, the payments to farmers that adopt practices beneficial for the environment under the Common Agricultural Policy of the EU [73]) and may encompass incentives for developers and/or adopters of innovations [74]. Business-to-consumer information asymmetry occurs when the attributes of products are credence qualities; eliciting information about some of those attributes through environmental labelling reduces information asymmetry, by providing the product with experience qualities, e.g., implemented production practices, emissions occurred during production, type of energy sources, water consumption, among others [75]. However, the effectiveness of environmental labels in conveying the intended information and orienting consumption patterns are subject to consumer education and proper label design [76,77].

To face the four challenges above, [20] classified the application of data generated via LCA (stand alone, retrospective or prospective studies) under three domains, i.e., pure information for decision makers, passive regulation, active regulation. The first type of application is generally intended for providing decision-makers with new knowledge about the impacts of existing or innovative products, practices or technologies. This can be done via stand-alone or comparative LCAs [35]. Passive regulation applications aim at informing the choice among multiple options (e.g., processes, scenarios). Data can be used to target public support towards specific products, technologies, production practices, or end-of-life management via recommendations, labelling or qualification schemes, mandatory targets or incentives to production/consumption of a given product. Lastly, LCA can drive active regulation when highlighting specific parameters that can form the basis, e.g., for an incentive, tax, or subsidy, or for the design of an environmental performance standard. To inform passive or active regulation, LCA studies should be comparative prospective or retrospective [35]. Within the domains of passive and active regulation, policy and business decision makers may also benefit from further elaborations of LCA outputs, e.g., when needing to consider consumer preferences. Food labelling is a typical example, as label design and the way how information about products' emissions (e.g., qualitative or quantitative information) is presented affect consumers' choices [77]. To identify preferences, behavioral economists generally evaluate consumers' willingness to pay for different food/label attributes. This research approach can also evaluate whether adopting a given labelling scheme is an effective differentiation strategy for farmers and what price premiums can be applied to least emitting products [78].

5. Limitations of the Method and Opportunities for Improvement

LCA has evolved from a tool for managing resource use on firm, through a method for monitoring energy use over the production process and for complying with overarching emission reduction mandates, to a metric that drives and is embedded in policy-making [5], including policy evaluation [6,7]. However, a series of improvements may benefit the diffusion of the tool, especially concerning study harmonization, data issues, and the inclusion of economic and social aspects.

5.1. Study Harmonization

Besides data sources, the flexibility of ISO (Geneva, Switzerland) rules has raised concerns about the credibility, transparency, complexity, and communication ability of LCA studies, which can possibly interfere with LCA applications for policy and strategic planning [5]. For example, ISO (Geneva, Switzerland) do not strictly specify how to define the functional units or reference flows and system boundaries, how to select the rules for quantifying the impacts associated with multifunctional processes or how to establish the environmentally relevant impact categories. This wide margin of discretion left to practitioners has been seriously criticized, because the operational choices can generate different and often incomparable results [79]. To overcome those issues, ISO (Geneva, Switzerland) guidelines should be deepened, i.e., revised to include more detailed definitions and stricter rules, to allow greater uniformity among the architecture of different LCA studies, thereby improving the reliability and comparability of findings, as well as widening the scope of results applicability [80]. Trying and answer the need for greater harmonization in the analytical protocols of LCA studies, the European Commission's Joint Research Centre proposed detailed guidelines for the calculation of products' life cycle impacts (Product Environmental Footprint; Recommendation 2013/179/EU). The adoption of a voluntary environmental certification based on LCA (Type III Environmental Declarations ISO 14025:2006, which base on LCA) (ISO, Geneva, Switzerland), also implies the compliance with stricter rules than the ISO (Geneva, Switzerland), providing guidance on data quality requirements and on the selection of environmentally relevant impact categories, among others. The Environmental Product Declaration (EPD®) and the Carbon Footprint of Products are examples of those certifications. Certification adoption by a company allows the producer to provide the product with environmental claims and facilitates comparative assertions.

5.2. Data Issues

LCA is a data-intensive method. Gathering all the necessary data to carry out the assessment is not an easy task. Data originate great concerns within the research community. Primary data collection can be prohibitively expensive [10] and often secondary data cannot cover the lack of information about the processes under study [25]. For agri-food databases, greater transparency is required on data collection, the harmonization of different databases, to facilitate the use of more than one of them, the definition of net boundaries between the technosphere and the ecosphere and the incorporation of spatial variability [25]. The databases should also be updated with innovative and pilot technologies currently available to entrepreneurs [13] and allow their users to address cross-cutting issues, such as food losses along the supply chain, end-of-life treatment technologies [81,82], and the impact of different packaging materials [81,83]. Such improvements are additional aspects of LCA deepening [80].

5.3. Inclusion of Economic and Social Aspects

Another problem of LCA studies is being limited to environmental aspects, omitting economic and social aspects, of great importance to decision makers [79]. The approach for overcoming this problem, known as broadening, involves extending the analysis to the economic and/or social dimensions of sustainability [80]. Research approaches to LCA combinations with various economic assessment and decision-support tools are not recent—e.g., multicriteria decision making analysis [84], stochastic optimization [85], full cost accounting [86]—and have been extensively reviewed [87]. When one can assume a fixed structure of the economy, many authors have turned to environmental input–output-LCA (EIO-LCA) [88–91]. EIO-LCA can eliminate the arbitrariness in system boundary's definition, thereby helping to cope with truncation errors [92]. LCA combination with data envelopment analysis (DEA) is promising, by linking engineering, life sciences, and social sciences, and delivering estimates over the three dimensions of sustainability, thereby improving policy targeting [13,93]. Many examples of LCA combinations with DEA are available from the scientific literature [94]. Compared to other economic sectors, combined DEA-LCA studies in agri-food are few (see [95] for an example). Further improvements may arise from the use of system dynamics modelling to correct model estimates, e.g., with respect to temporal effects, rebound effects, and uncertainty [96]. Compared to economic aspects, LCA combination with social impact assessments has received less attention by researchers, though being key to allow consumer acceptability of innovative food production and consumption systems [69]. The growing popularity of life cycle sustainability assessment (LCSA) has mitigated this knowledge gap, by promoting the combination of LCA with life cycle cost accounting (life cycle costing) and social LCA. LCSA has raised the interest of policy makers, delivering synthetic information on the three dimensions of sustainability that largely base on the same data sources [97,98]. The availability of data on impacts on several dimensions is particularly important in the assessment of debated innovations, for example in the context of the bioeconomy [13]. The adoption of a more holistic methodological approach would also allow to consider the possible effects associated with changes in the use of production factors and yields of production processes, resulting from the large-scale implementation of environmental mitigation measures (rebound effects) [22]. Especially, the adoption of participatory approaches may help with the identification of rebound effects and the possible ways to address them, via the collaboration among supply chain stakeholders, consumers, authorities, and waste-handlers [81]. LCA applications under this perspective need further research, especially for developing indicators against which to measure progress, for incorporating concepts such as human agency and moral responsibility, and for considering the dynamic interactions between economic values and health outcomes of different production systems [26].

6. Discussion and Recommendations

The scientific literature proposes to address the necessary reduction of the environmental impacts agriculture and food systems via the public support to innovation (practices, technologies) to improve resource management efficiency on firm (supply-side), or to facilitate the change in food consumption patterns with indirect effects on food supply (demand-side), or to interventions addressing both the supply and demand-sides that require the collaboration of supply chain stakeholders (system level interventions). LCA is a scientifically robust tool for calculating the impacts of agricultural systems, affected or not by pollution reduction measures. LCA studies investigate and/or compare and immediately communicate the environmental impacts of—e.g., farming methods, input management systems, technologies, or consumer behavior—and can support evidence-based policy making in agriculture and food systems. Information about product impacts can help to provide the basis for establishing legal limits on emissions (e.g., carbon tax) or entry levels to access public tenders (e.g., green public procurement). In agri-food, the policies that benefit most from such information concern cross-border pollution, transaction costs associated with the application of environmental standards, the adoption of less polluting practices or technologies, and the reduction of information asymmetry business-to-consumer. Given the growing consumer sensitivity towards the environmental impacts of agri-food products, environmental certifications and labelling are the most significant example of the opportunities LCA offer to the agribusiness, in terms of product positioning on the domestic and export markets. Concerned consumers are willing to pay a price premium for environmental label on foods [78,99]. However, labels' effectiveness in orienting consumer decisions does not depend just on the type of impact information conveyed [76]. Often, consumers are not aware of the estimates behind the label; moreover, multiple labels exit that can create consumer confusion. Confusion may also arise between environmental declarations (product certifications) and process certifications, e.g., organic farming [100]. Then asymmetric information remain unsolved [101]. Label design is critical to allow the clarity of information delivery and then reduce consumer confusion [102]. Labels should help comparisons within and across food products, use consistent and accessible units of measure, be able to communicate the sustainability criteria behind footprint calculation [77]. Coherent, comprehensive, and cohesive carbon labelling policy, coupled with dedicated social learning campaigns, may help to achieve policy objectives [76,103]. In addition, policy makers should monitor the proliferation of sustainability labels, to play a credible and appropriate role in the development of certification schemes [99].

Improving product positioning is the main drivers for entrepreneurs to adopt an environmental certification [104]. However, certification adoption implies a series of costs associated with the adoption itself and with the requirements the firm need to meet through time to be eligible for the certification scheme, including the adaptation of production facilities [105]. More studies are needed that consider the combined environmental, economic, and social impacts of certification adoption in agribusiness, to evaluate the extent to which the value added associated with certification adoption can be distributed among supply chain stakeholders and to verify if farmers are incentivized to modify their production practices and technologies or are "forced" by the market [106]. Food labels allow the direct comparison of substitutes. However, the extent to which food labels can influence purchasing decisions depends not only on the clarity of the conveyed information, but also on consumers' ability to understand that information [76]. Therefore, entrepreneurs in the agri-food sector wishing to promote their products as "ecological" should pay attention to consumer attitudes, to decide on the type of information and the way how to present it via label design for encouraging purchases [77]. The implementation of consumer education campaigns could improve the effectiveness of demand-side interventions [103]. This is especially relevant when it comes to the promotion of sustainable diets [63,69].

While the high environmental impacts of ruminant meat consumption are recognized, many studies assume that the adoption of more plant-based diets may benefit both health and the environment, without considering the energy density of meat-replacement foods [63]. Besides, decision makers and researchers should consider the potential risk for nutrient deficiencies and the resulting potential increase in the demand for fortified foods or food supplements [107]. To allow the effective promotion of new dietary models, more LCA studies are needed that consider the trade-offs between impact mitigation and health and nutritional outcomes of dietary shifts, also by adopting an epidemiological approach [65]. To inform evidence-based policy, the LCAs of diets should be based on more insightful indicators of nutritional quality, rather than just focusing on macronutrients [65]. More consumer studies are needed to assess the cultural acceptability and affordability of low impact diets [65,69].

Researchers' choices among the several methodological options for carrying out a LCA depend on the purpose and scope of the study. CLCA should be preferred when it is necessary to link environmental and economic models, while EIO-LCA can be useful to correct for method limitations, in specific circumstances [108]. In general, LCA is not without criticism [21] and alternative approaches had many applications in agricultural research, e.g., multi-region input–output analysis [109,110], data envelopment analysis [111], or structural path and structural decomposition analyses [112], among others.

LCA is just one the many sustainability assessment tools and different classifications frameworks exist (see [113] for an overview). For example, indicator-based assessments rely on sustainability indicators, presenting simple information, generally quantitative, about a state of economic, social and/or environmental development in a given region, which may be useful for communicating trends and for non-specialized audience [114]. Monetary assessments, notably cost–benefit, give the costs and benefits of the system under monetary values. The system is evaluated against a benchmark, e.g., some wealth indexes of the measure of the private/public stream of social costs and benefits of alternative systems [115]. Rather than focusing on products, cost–benefit analysis is used for evaluating public or private investment proposals, e.g. by weighing the costs of the project against the expected benefits [114] (see [116] for a detailed comparison among LCA, life cycle costing and cost–benefit analysis). Other methods are at least conceptually close to LCA. For example, being a product-related tool that focuses on the flows associated with the production and consumption of a good, LCA is related to material flow analysis, though concentrating on product-related flows instead of region-related flows [114]. Multicriteria analysis is another widespread tool for sustainability assessment, which can be used when the evaluation needs to consider competing criteria. The method involves the collection of data about the perceived impacts (environmental, economic, social) by supply chain stakeholders [117] and preference synthesis through modeling algorithms [118]. Multicriteria analysis and LCA can usefully complement each other [119].

The sustainable transition towards sustainable agriculture and food systems should involve the evaluation of economic and social aspects. The possibility of integrating LCA within the more comprehensive LCSA framework would improve the suitability of the tool for sustainable development studies, as well as for applications in business, e.g., for supporting and communicating ecological innovation [28]. Greater research efforts to understand farmers' attitudes towards environmentally sound practices and technologies could suggest which interventions are most likely to be undertaken and which policies could be effective in promoting their adoption [120].

Author Contributions: Conceptualization, O.G.; Investigation, O.G., F.B., and F.V.; Writing—original draft preparation, O.G.; Writing—review and editing, O.G., F.B., and F.V.; Supervision, F.B. and A.Z.; Project administration, G.B. and A.P.; Funding acquisition, G.B., A.Z., and A.P.

Funding: This research was funded by the University of Pisa, project SALI (*Strumenti di sostenibilità delle produzioni alimentari*, Sustainability tools for food production) code PRA_2017_34.

Acknowledgments: The authors thank Silvia Coderoni, from Marche Polytechnic University, for useful comments to previous versions of the paper.

Conflicts of Interest: The authors declare no conflict of interest. The funders had no role in the design of the study; in the collection, analyses, or interpretation of data; in the writing of the manuscript, or in the decision to publish the results.

References

1. Van Tongeren, F. *Agricultural Policy Design and Implementation. A Synthesis*; Organisation for Economic Co-operation and Development: Paris, France, 2008; p. 32.
2. About the Sustainable Development Goals. Available online: https://www.un.org/sustainabledevelopment/sustainable-development-goals/ (accessed on 9 October 2018).
3. *Transforming Food and Agriculture to Achieve the SDGs*; Food and Agricultural Organization of the United Nations: Rome, Italy, 2018; ISBN 978-92-5-130626-0.
4. Guinée, J.B.; Heijungs, R.; Huppes, G.; Zamagni, A.; Masoni, P.; Buonamici, R.; Ekvall, T.; Rydberg, T. Life Cycle Assessment: Past, Present, and Future. *Environ. Sci. Technol.* **2011**, *45*, 90–96. [CrossRef] [PubMed]
5. McManus, M.C.; Taylor, C.M. The changing nature of life cycle assessment. *Biomass Bioenergy* **2015**, *82*, 13–26. [CrossRef] [PubMed]
6. Sala, S.; Reale, F.; Cristobal-Garcia, J.; Pant, R. *Life Cycle Assessment for the Impact Assessment of Policies*; Publications Office: Luxembourg, 2016; ISBN 978-92-79-64813-7.
7. Reale, F.; Cinelli, M.; Sala, S. Towards a research agenda for the use of LCA in the impact assessment of policies. *Int. J. Life Cycle Assess.* **2017**, *22*, 1477–1481. [CrossRef]
8. Goedkoop, M.; Martinez, E.V.; de Beer, I. *LCA as the Tool to Measure Progress towards the Sustainable Development Goals*; LIST: Luxembourg, 2017.
9. Huang, H.; von Lampe, M.; van Tongeren, F. Climate change and trade in agriculture. *Food Policy* **2011**, *36*, S9–S13. [CrossRef]
10. Tsiboe, F.; Nalley, L.L.; Durand, A.; Greg, T.; Aaron, S. The Economic and Environmental Benefits of Sheath Blight Resistance in Rice. *J. Agric. Resour. Econ.* **2017**, *42*, 215–235.
11. AusLCI—The Australian Life Cycle Inventory Database Initiative. Available online: http://www.auslci.com.au/index.php/datasets/Agriculture (accessed on 9 October 2018).

12. Agribalyse Program. Available online: https://www.ademe.fr/en/expertise/alternative-approaches-to-production/agribalyse-program (accessed on 9 October 2018).
13. Viaggi, D. Research and innovation in agriculture: Beyond productivity? *Bio-Based Appl. Econ.* **2015**, *4*, 279–300.
14. Sonnemann, G.; Gemechu, E.D.; Sala, S.; Schau, E.M.; Allacker, K.; Pant, R.; Adibi, N.; Valdivia, S. Life Cycle Thinking and the Use of LCA in Policies Around the World. In *Life Cycle Assessment*; Hauschild, M.Z., Rosenbaum, R.K., Olsen, S.I., Eds.; Springer International Publishing: Cham, Switzerland, 2018; pp. 429–463. ISBN 978-3-319-56474-6.
15. Roy, P.; Nei, D.; Orikasa, T.; Xu, Q.; Okadome, H.; Nakamura, N.; Shiina, T. A review of life cycle assessment (Lca) on some food products. *J. Food Eng.* **2009**, *90*, 1–10. [CrossRef]
16. Van Huis, A.; Oonincx, D.G.A.B. The environmental sustainability of insects as food and feed. A review. *Agron. Sustain. Dev.* **2017**, *37*, 43. [CrossRef]
17. Gnansounou, E.; Dauriat, A.; Villegas, J.; Panichelli, L. Life cycle assessment of biofuels: Energy and greenhouse gas balances. *Bioresour. Technol.* **2009**, *100*, 4919–4930. [CrossRef]
18. Gomiero, T.; Pimentel, D.; Paoletti, M.G. Environmental Impact of Different Agricultural Management Practices: Conventional vs. Organic Agriculture. *Crit. Rev. Plant Sci.* **2011**, *30*, 95–124. [CrossRef]
19. Martin, M.; Brandão, M. Evaluating the Environmental Consequences of Swedish Food Consumption and Dietary Choices. *Sustainability* **2017**, *9*, 2227. [CrossRef]
20. Rajagopal, D.; Vanderghem, C.; MacLean, H.L. Life Cycle Assessment for Economists. *Annu. Rev. Resour. Econ.* **2017**, *9*, 361–381. [CrossRef]
21. Freidberg, S. Calculating sustainability in supply chain capitalism. *Econ. Soc.* **2013**, *42*, 571–596. [CrossRef]
22. Benedetto, G.; Rugani, B.; Vázquez-Rowe, I. Rebound effects due to economic choices when assessing the environmental sustainability of wine. *Food Policy* **2014**, *49*, 167–173. [CrossRef]
23. Aleksandrowicz, L.; Green, R.; Joy, E.J.M.; Smith, P.; Haines, A. The Impacts of Dietary Change on Greenhouse Gas Emissions, Land Use, Water Use, and Health: A Systematic Review. *PLoS ONE* **2016**, *11*, e0165797. [CrossRef] [PubMed]
24. Clark, M.; Tilman, D. Comparative analysis of environmental impacts of agricultural production systems, agricultural input efficiency, and food choice. *Environ. Res. Lett.* **2017**, *12*, 064016. [CrossRef]
25. Notarnicola, B.; Sala, S.; Anton, A.; McLaren, S.J.; Saouter, E.; Sonesson, U. The role of life cycle assessment in supporting sustainable agri-food systems: A review of the challenges. *J. Clean. Prod.* **2017**, *140*, 399–409. [CrossRef]
26. Garnett, T. Three perspectives on sustainable food security: Efficiency, demand restraint, food system transformation. What role for life cycle assessment? *J. Clean. Prod.* **2014**, *73*, 10–18. [CrossRef]
27. Onat, N.; Kucukvar, M.; Halog, A.; Cloutier, S. Systems Thinking for Life Cycle Sustainability Assessment: A Review of Recent Developments, Applications, and Future Perspectives. *Sustainability* **2017**, *9*, 706. [CrossRef]
28. Zamagni, A.; Masoni, P.; Buttol, P.; Raggi, A.; Buonamici, R. Finding Life Cycle Assessment Research Direction with the Aid of Meta-Analysis. *J. Ind. Ecol.* **2012**, *16*, S39–S52. [CrossRef]
29. Hellweg, S.; Mila i Canals, L. Emerging approaches, challenges and opportunities in life cycle assessment. *Science* **2014**, *344*, 1109–1113. [PubMed]
30. Tillman, A.-M. Significance of decision-making for LCA methodology. *Environ. Impact Assess. Rev.* **2000**, *20*, 113–123.
31. Hiederer, R. *International Reference Life Cycle Data System (ILCD) Handbook: General Guide for Life Cycle Assessment: Provisions and Action Steps*; Publications Office: Luxembourg, 2011; ISBN 978-92-79-17451-3.
32. Sala, S.; Ciuffo, B.; Nijkamp, P. A systemic framework for sustainability assessment. *Ecol. Econ.* **2015**, *119*, 314–325.
33. Schaltegger, S. Economics of life cycle assessment: Inefficiency of the present approach. *Bus. Strategy Environ.* **1997**, *6*, 1–8. [CrossRef]
34. Zimmermann, P.; Frischknecht, R.; Ménard, M. Background Inventory Data. In *Life Cycle Assessment (LCA) —Quo Vadis?* Schaltegger, S., Braunschweig, A., Büchel, K., Dinkel, F., Frischknecht, R., Maillefer, C., Ménard, M., Peter, D., Pohl, C., Ros, M., et al., Eds.; Themenhefte Schwerpunktprogramm Umwelt; Birkhäuser Basel: Basel, Switzerland, 1996; pp. 39–49, ISBN 978-3-0348-9022-9.

35. Baumann, H.; Tillman, A.-M. *The Hitch Hikers's Guide to LCA: An Orientation in Life Cycle Assessment Methodology and Application*; Studentlitteratur: Lund, Sweden, 2004; ISBN 978-91-44-02364-9.
36. Galli, A.; Wiedmann, T.; Ercin, E.; Knoblauch, D.; Ewing, B.; Giljum, S. Integrating Ecological, Carbon and Water footprint into a "Footprint Family" of indicators: Definition and role in tracking human pressure on the planet. *Ecol. Indic.* **2012**, *16*, 100–112. [CrossRef]
37. Ekvall, T.; Tillman, A.-M.; Molander, S. Normative ethics and methodology for life cycle assessment. *J. Clean. Prod.* **2005**, *13*, 1225–1234. [CrossRef]
38. *Global Guidance Principles for Life Cycle Assessment Databases: A Basis for Greener Processes and Products: "Shonan Guidance Principles"*; United Nations Environment Programme: Nairobi, Kenya, 2011.
39. Earles, J.M.; Halog, A. Consequential life cycle assessment: A review. *Int. J. Life Cycle Assess.* **2011**, *16*, 445–453. [CrossRef]
40. Schmidt, J.H.; Weidema, B.P. Shift in the marginal supply of vegetable oil. *Int. J. Life Cycle Assess.* **2008**, *13*, 235–239.
41. Van Stappen, F.; Mathot, M.; Decruyenaere, V.; Loriers, A.; Delcour, A.; Planchon, V.; Goffart, J.-P.; Stilmant, D. Consequential environmental life cycle assessment of a farm-scale biogas plant. *J. Environ. Manag.* **2016**, *175*, 20–32.
42. Rajagopal, D. Consequential Life Cycle Assessment of Policy Vulnerability to Price Effects: CLCA of Policy Vulnerability to Price Effects. *J. Ind. Ecol.* **2014**, *18*, 164–175.
43. Weidema, B.P.; Ekvall, T.; Heijungs, R. *Guidelines for Application of Deepened and Broadened LCA—Deliverable D18 of Work Packag e 5 of the CALCAS Project*; CALCAS—Co-ordination Action for innovation in Life-Cycle Analysis for Sustainability: Colorado Springs, CO, USA, 2009; p. 49.
44. Rajagopal, D.; Zilberman, D. On market-mediated emissions and regulations on life cycle emissions. *Ecol. Econ.* **2013**, *90*, 77–84. [CrossRef]
45. Weidema, B.P. *Market Information in Life Cycle Assessment*; Danish Environmental Protection Agency: København, Denmark, 2003; p. 147.
46. Weidema, B.P.; Pizzol, M.; Schmidt, J.; Thoma, G. Attributional or consequential Life Cycle Assessment: A matter of social responsibility. *J. Clean. Prod.* **2018**, *174*, 305–314. [CrossRef]
47. Brander, M.; Tipper, R.; Hutchison, C.; Davis, G. *Consequential and Attributional Approaches to Lca: A Guide to Policy Makers with Specific Reference to Greenhouse Gas Lca of Biofuels*; Technical Paper; Ecometrica, 2009. Available online: https://ecometrica.com/white-papers/consequential-and-attributional-approaches-to-lca-a-guide-to-policy-makers-with-specific-reference-to-greenhouse-gas-lca-of-biofuels (accessed on 4 December 2018).
48. Ekvall, T.; Azapagic, A.; Finnveden, G.; Rydberg, T.; Weidema, B.P.; Zamagni, A. Attributional and consequential LCA in the ILCD handbook. *Int. J. Life Cycle Assess.* **2016**, *21*, 293–296. [CrossRef]
49. Creutzig, F.; Popp, A.; Plevin, R.; Luderer, G.; Minx, J.; Edenhofer, O. Reconciling top-down and bottom-up modelling on future bioenergy deployment. *Nat. Clim. Chang.* **2012**, *2*, 320–327. [CrossRef]
50. Baitz, M.; Albrecht, S.; Brauner, E.; Broadbent, C.; Castellan, G.; Conrath, P.; Fava, J.; Finkbeiner, M.; Fischer, M.; Fullana i Palmer, P.; et al. LCA's theory and practice: Like ebony and ivory living in perfect harmony? *Int. J. Life Cycle Assess.* **2013**, *18*, 5–13. [CrossRef]
51. Reimann, K.; Finkbeiner, M.; Horvath, A.; Matsuno, Y. *Evaluation of Environmental Life Cycle Approaches for Policy and Decision Making Support in Micro and Macro Level Applications*; European Commission, Joint Research Centre, Institute for Evironment and Sustainability: Copenhagen, Denmark, 2010; p. 188.
52. Tukker, A.; Huppes, G.; van Oers, L.; Heijungs, R. *Environmentally Extended input-Output Tables and Models for Europe*; Eutopean Commission, Joint Research Centre, Institute for Prospective Technological Studies: Seville, Spain, 2010; p. 120.
53. Garnett, T. Where are the best opportunities for reducing greenhouse gas emissions in the food system (including the food chain)? *Food Policy* **2011**, *36*, S23–S32. [CrossRef]
54. Silvenius, F.; Katajajuuri, J.-M.; Grönman, K.; Soukka, R.; Koivupuro, H.-K.; Virtanen, Y. Role of Packaging in LCA of Food Products. In *Towards Life Cycle Sustainability Management*; Finkbeiner, M., Ed.; Springer Netherlands: Dordrecht, The Netherlands, 2011; pp. 359–370. ISBN 978-94-007-1898-2.
55. Ferrara, C.; De Feo, G. Life Cycle Assessment Application to the Wine Sector: A Critical Review. *Sustainability* **2018**, *10*, 395. [CrossRef]

56. Vázquez-Rowe, I.; Villanueva-Rey, P.; Moreira, M.T.; Feijoo, G. Environmental analysis of Ribeiro wine from a timeline perspective: Harvest year matters when reporting environmental impacts. *J. Environ. Manag.* **2012**, *98*, 73–83. [CrossRef]
57. Godfray, H.C.J.; Beddington, J.R.; Crute, I.R.; Haddad, L.; Lawrence, D.; Muir, J.F.; Pretty, J.; Robinson, S.; Thomas, S.M.; Toulmin, C. Food Security: The Challenge of Feeding 9 Billion People. *Science* **2010**, *327*, 812–818. [CrossRef]
58. Meier, M.S.; Stoessel, F.; Jungbluth, N.; Juraske, R.; Schader, C.; Stolze, M. Environmental impacts of organic and conventional agricultural products—Are the differences captured by life cycle assessment? *J. Environ. Manag.* **2015**, *149*, 193–208. [CrossRef]
59. Herrero, M.; Henderson, B.; Havlík, P.; Thornton, P.K.; Conant, R.T.; Smith, P.; Wirsenius, S.; Hristov, A.N.; Gerber, P.; Gill, M.; et al. Greenhouse gas mitigation potentials in the livestock sector. *Nat. Clim. Chang.* **2016**, *6*, 452–461. [CrossRef]
60. Blengini, G.A.; Busto, M. The life cycle of rice: LCA of alternative agri-food chain management systems in Vercelli (Italy). *J. Environ. Manag.* **2009**, *90*, 1512–1522. [CrossRef] [PubMed]
61. Bryngelsson, D.; Wirsenius, S.; Hedenus, F.; Sonesson, U. How can the EU climate targets be met? A combined analysis of technological and demand-side changes in food and agriculture. *Food Policy* **2016**, *59*, 152–164. [CrossRef]
62. *Policies to Encourage Sustainable Consumption: Full Report*; Publications Office: Luxembourg, 2012; ISBN 978-92-79-25952-4.
63. Garnett, T. Plating up solutions. *Science* **2016**, *353*, 1202–1204. [CrossRef] [PubMed]
64. Vieux, F.; Soler, L.-G.; Touazi, D.; Darmon, N. High nutritional quality is not associated with low greenhouse gas emissions in self-selected diets of French adults. *Am. J. Clin. Nutr.* **2013**, *97*, 569–583. [CrossRef]
65. Perignon, M.; Vieux, F.; Soler, L.-G.; Masset, G.; Darmon, N. Improving diet sustainability through evolution of food choices: Review of epidemiological studies on the environmental impact of diets. *Nutr. Rev.* **2017**, *75*, 2–17. [CrossRef] [PubMed]
66. Friedman, A.L.; Miles, S. *Stakeholders: Theory and Practice*; Oxford University Press: Oxford, UK; New York, NY, USA, 2006; ISBN 978-0-19-926987-7.
67. Dwivedi, S.L.; Lammerts van Bueren, E.T.; Ceccarelli, S.; Grando, S.; Upadhyaya, H.D.; Ortiz, R. Diversifying Food Systems in the Pursuit of Sustainable Food Production and Healthy Diets. *Trends Plant Sci.* **2017**, *22*, 842–856. [CrossRef]
68. Garnett, T. Livestock-related greenhouse gas emissions: Impacts and options for policy makers. *Environ. Sci. Policy* **2009**, *12*, 491–503. [CrossRef]
69. Röös, E.; Patel, M.; Spångberg, J.; Carlsson, G.; Rydhmer, L. Limiting livestock production to pasture and by-products in a search for sustainable diets. *Food Policy* **2016**, *58*, 1–13. [CrossRef]
70. Rajagopal, D. On mitigating emissions leakage under biofuel policies. *GCB Bioenergy* **2016**, *8*, 471–480. [CrossRef]
71. Metcalf, G.E.; Weisbach, D.A. The Design of a Carbon Tax. *Harv. Environ. Law Rev.* **2009**, *33*, 499–556. [CrossRef]
72. Walls, M. *Extended Producer Responsibility and Product Design: Economic Theory and Selected Case Studies*; Resource for the Future: Washington, DC, USA, 2006.
73. Greening. Available online: https://ec.europa.eu/agriculture/direct-support/greening_en (accessed on 9 October 2018).
74. Zilberman, D.; Zhao, J.; Heiman, A. Adoption Versus Adaptation, with Emphasis on Climate Change. *Annu. Rev. Resour. Econ.* **2012**, *4*, 27–53. [CrossRef]
75. Caswell, J.A.; Mojduszka, E.M. Using Informational Labeling to Influence the Market for Quality in Food Products. *Am. J. Agric. Econ.* **1996**, *78*, 1248–1253. [CrossRef]
76. Gadema, Z.; Oglethorpe, D. The use and usefulness of carbon labelling food: A policy perspective from a survey of UK supermarket shoppers. *Food Policy* **2011**, *36*, 815–822. [CrossRef]
77. Leach, A.M.; Emery, K.A.; Gephart, J.; Davis, K.F.; Erisman, J.W.; Leip, A.; Pace, M.L.; D'Odorico, P.; Carr, J.; Noll, L.C.; et al. Environmental impact food labels combining carbon, nitrogen, and water footprints. *Food Policy* **2016**, *61*, 213–223. [CrossRef]
78. Akaichi, F.; Nayga, R.M.; Nalley, L.L. Are there trade-offs in valuation with respect to greenhouse gas emissions, origin and food miles attributes? *Eur. Rev. Agric. Econ.* **2017**, *44*, 3–31. [CrossRef]

79. Boons, F.; Baumann, H.; Hall, J. Conceptualizing sustainable development and global supply chains. *Ecol. Econ.* **2012**, *83*, 134–143. [CrossRef]
80. Jeswani, H.K.; Azapagic, A.; Schepelmann, P.; Ritthoff, M. Options for broadening and deepening the LCA approaches. *J. Clean. Prod.* **2010**, *18*, 120–127. [CrossRef]
81. Halloran, A.; Clement, J.; Kornum, N.; Bucatariu, C.; Magid, J. Addressing food waste reduction in Denmark. *Food Policy* **2014**, *49*, 294–301. [CrossRef]
82. Castellani, V.; Sala, S.; Benini, L. Hotspots analysis and critical interpretation of food life cycle assessment studies for selecting eco-innovation options and for policy support. *J. Clean. Prod.* **2017**, *140*, 556–568. [CrossRef]
83. Venturi, F.; Sanmartin, C.; Taglieri, I.; Xiaoguo, Y.; Andrich, G.; Zinnai, A. The influence of packaging on the sensorial evolution of white wine as a function of the operating conditions adopted during storage. *Agrochimica* **2016**, *60*, 150–160.
84. Shmelev, S.E.; Powell, J.R. Ecological–economic modelling for strategic regional waste management systems. *Ecol. Econ.* **2006**, *59*, 115–130. [CrossRef]
85. Yang, Y.-H.; Lin, S.-J.; Lewis, C. Reduction of acidification from electricity—Generating industries in Taiwan by Life Cycle Assessment and Monte Carlo optimization. *Ecol. Econ.* **2009**, *68*, 1575–1582. [CrossRef]
86. Mouron, P.; Scholz, R.W.; Nemecek, T.; Weber, O. Life cycle management on Swiss fruit farms: Relating environmental and income indicators for apple-growing. *Ecol. Econ.* **2006**, *58*, 561–578. [CrossRef]
87. Settanni, E.; Notarnicola, B.; Tassielli, G. Combining Life Cycle Assessment of food products with economic tools. In *Environmental Assessment and Management in the Food Industry*; Woodhead Publishing Limited: Sawston, UK, 2010; pp. 207–218, ISBN 978-1-84569-552-1.
88. Suh, S. Functions, commodities and environmental impacts in an ecological–economic model. *Ecol. Econ.* **2004**, *48*, 451–467. [CrossRef]
89. Ghertner, D.A.; Fripp, M. Trading away damage: Quantifying environmental leakage through consumption-based, life-cycle analysis. *Ecol. Econ.* **2007**, *63*, 563–577. [CrossRef]
90. Engström, R.; Wadeskog, A.; Finnveden, G. Environmental assessment of Swedish agriculture. *Ecol. Econ.* **2007**, *60*, 550–563. [CrossRef]
91. Cantono, S.; Heijungs, R.; Kleijn, R. Environmental Accounting of Eco-innovations through Environmental Input–Output Analysis: The Case of Hydrogen and Fuel Cells Buses. *Econ. Syst. Res.* **2008**, *20*, 303–318. [CrossRef]
92. Hendrickson, C.T.; Lave, L.B.; Matthews, H.S. *Environmental Life Cycle Assessment of Goods and Services: An Input-Output Approach*; RFF Press Resources for the Future: Washington, DC, USA, 2006; ISBN 978-1-933115-23-8.
93. Sueyoshi, T.; Yuan, Y.; Goto, M. A literature study for DEA applied to energy and environment. *Energy Econ.* **2017**, *62*, 104–124. [CrossRef]
94. Vázquez-Rowe, I.; Iribarren, D. Review of Life-Cycle Approaches Coupled with Data Envelopment Analysis: Launching the CFP + DEA Method for Energy Policy Making. *Sci. World J.* **2015**, *2015*, 1–10. [CrossRef]
95. Dakpo, K.H.; Jeanneaux, P.; Latruffe, L. Greenhouse gas emissions and efficiency in French sheep meat farming: A non-parametric framework of pollution-adjusted technologies. *Eur. Rev. Agric. Econ.* **2017**, *44*, 33–65. [CrossRef]
96. Wang, B.; Brême, S.; Moon, Y.B. Hybrid modeling and simulation for complementing Lifecycle Assessment. *Comput. Ind. Eng.* **2014**, *69*, 77–88. [CrossRef]
97. Kloepffer, W. Life cycle sustainability assessment of products: (with Comments by Helias A. Udo de Haes, p. 95). *Int. J. Life Cycle Assess.* **2008**, *13*, 89–95. [CrossRef]
98. Zamagni, A. Life cycle sustainability assessment. *Int. J. Life Cycle Assess.* **2012**, *17*, 373–376. [CrossRef]
99. Onozaka, Y.; Mcfadden, D.T. Does Local Labeling Complement or Compete with Other Sustainable Labels? A Conjoint Analysis of Direct and Joint Values for Fresh Produce Claim. *Am. J. Agric. Econ.* **2011**, *93*, 693–706. [CrossRef]
100. Sirieix, L.; Delanchy, M.; Remaud, H.; Zepeda, L.; Gurviez, P. Consumers' perceptions of individual and combined sustainable food labels: A UK pilot investigation: Consumers' perceptions of sustainable food labels. *Int. J. Consum. Stud.* **2013**, *37*, 143–151. [CrossRef]
101. Ben Youssef, A.; Abderrazak, C. Multiplicity of Eco-Labels, Competition, and the Environment. *J. Agric. Food Ind. Organ.* **2009**, *7*. [CrossRef]

102. Grunert, K.G.; Hieke, S.; Wills, J. Sustainability labels on food products: Consumer motivation, understanding and use. *Food Policy* **2014**, *44*, 177–189. [CrossRef]
103. Apostolidis, C.; McLeay, F. Should we stop meating like this? Reducing meat consumption through substitution. *Food Policy* **2016**, *65*, 74–89. [CrossRef]
104. Peano, C.; Baudino, C.; Tecco, N.; Girgenti, V. Green marketing tools for fruit growers associated groups: Application of the Life Cycle Assessment (LCA) for strawberries and berry fruits ecobranding in northern Italy. *J. Clean. Prod.* **2015**, *104*, 59–67. [CrossRef]
105. Alberti, M.; Caini, L.; Calabrese, A.; Rossi, D. Evaluation of the costs and benefits of an environmental management system. *Int. J. Prod. Res.* **2000**, *38*, 4455–4466. [CrossRef]
106. Fai Pun, K. Determinants of environmentally responsible operations: A review. *Int. J. Qual. Reliab. Manag.* **2006**, *23*, 279–297. [CrossRef]
107. Van Dooren, C.; Marinussen, M.; Blonk, H.; Aiking, H.; Vellinga, P. Exploring dietary guidelines based on ecological and nutritional values: A comparison of six dietary patterns. *Food Policy* **2014**, *44*, 36–46. [CrossRef]
108. Heijungs, R.; Koning, A.; Suh, S.; Huppes, G. Toward an Information Tool for Integrated Product Policy: Requirements for Data and Computation. *J. Ind. Ecol.* **2006**, *10*, 147–158. [CrossRef]
109. Weber, C.L.; Matthews, H.S. Quantifying the global and distributional aspects of American household carbon footprint. *Ecol. Econ.* **2008**, *66*, 379–391. [CrossRef]
110. Wiedmann, T.; Wilting, H.C.; Lenzen, M.; Lutter, S.; Palm, V. Quo Vadis MRIO? Methodological, data and institutional requirements for multi-region input–output analysis. *Ecol. Econ.* **2011**, *70*, 1937–1945. [CrossRef]
111. Molinos-Senante, M.; Hernández-Sancho, F.; Mocholí-Arce, M.; Sala-Garrido, R. Economic and environmental performance of wastewater treatment plants: Potential reductions in greenhouse gases emissions. *Resour. Energy Econ.* **2014**, *38*, 125–140. [CrossRef]
112. Wood, R.; Lenzen, M. Structural path decomposition. *Energy Econ.* **2009**, *31*, 335–341. [CrossRef]
113. Singh, R.K.; Murty, H.R.; Gupta, S.K.; Dikshit, A.K. An overview of sustainability assessment methodologies. *Ecol. Indic.* **2012**, *15*, 281–299. [CrossRef]
114. Ness, B.; Urbel-Piirsalu, E.; Anderberg, S.; Olsson, L. Categorising tools for sustainability assessment. *Ecol. Econ.* **2007**, *60*, 498–508. [CrossRef]
115. Kopp, R.; Krupnick, A.; Toman, M. Cost-benefit analysis and regulatory reform. *Hum. Ecol. Risk Assess. Int. J.* **1997**, *3*, 787–852. [CrossRef]
116. Hoogmartens, R.; Van Passel, S.; Van Acker, K.; Dubois, M. Bridging the gap between LCA, LCC and CBA as sustainability assessment tools. *Environ. Impact Assess. Rev.* **2014**, *48*, 27–33. [CrossRef]
117. French, S. A manifesto for the new MCDA era. *J. Multi-Criteria Decis. Anal.* **1993**, *2*, 125–127. [CrossRef]
118. Bartolini, F.; Viaggi, D. Recent developments in multi-criteria evaluation of regulations: Recent developments in multi-criteria evaluation. *Qual. Assur. Saf. Crop. Foods* **2010**, *2*, 182–196. [CrossRef]
119. De Luca, A.I.; Iofrida, N.; Leskinen, P.; Stillitano, T.; Falcone, G.; Strano, A.; Gulisano, G. Life cycle tools combined with multi-criteria and participatory methods for agricultural sustainability: Insights from a systematic and critical review. *Sci. Total Environ.* **2017**, *595*, 352–370. [CrossRef] [PubMed]
120. Lefebvre, M.; Espinosa, M.; Gomez y Paloma, S.; Paracchini, M.L.; Piorr, A.; Zasada, I. Agricultural landscapes as multi-scale public good and the role of the Common Agricultural Policy. *J. Environ. Plan. Manag.* **2015**, *58*, 2088–2112. [CrossRef]

sustainability

MDPI

Article

Participative Processes as a Chance for Developing Ideas to Bridge the Intention-Behavior Gap Concerning Sustainable Diets

Leonie Fink [1,2,*], Angelika Ploeger [1] and Carola Strassner [2]

1 Faculty of Organic Agricultural Sciences, Department of Organic Food Quality and Food Culture, University Kassel, 37213 Witzenhausen, Germany; a.ploeger@uni-kassel.de

2 Department of Food Nutrition Facilities, Münster University of Applied Sciences, 48149 Münster, Germany; strassner@fh-muenster.de

* Correspondence: l.fink@uni-kassel.de; Tel.: +49-251-8365614

Received: 30 September 2018; Accepted: 23 November 2018; Published: 27 November 2018

Abstract: Sustainable diets are drivers and results of sustainable food systems. Therefore, they are crucial for improving our global diet-related problems. When trying to adopt sustainable diets, people often struggle with the gap between their good intentions and their actual behavior. Here we see a need for support. To understand people's needs and what could help them, it stands to reason that they can be directly involved in the development processes for appropriate ideas. On that account, we conducted six workshops in different German cities from September to December 2016 with 82 participants in total. We collected data by letting participants generate ideas to bridge the intention-behavior gap. The qualitative data was then coded in internal (168) and external factors (989). Analyzing data shows that the higher numbers of external factors offer a wider range of aspects that contribute to closing the intention-behavior gap from the participant's point of view. We discuss whether the external factors such as availability, advertising, pricing, and education about food and nutrition may be a prerequisite for a broad mass of people to practice a more sustainable diet.

Keywords: sustainable diets; diet adoption; sustainable food system; intention-behavior gap; citizen participation; innovation workshop

---

1. Introduction

Recently the consumption of sustainably produced food, e.g., organic food is increasing [1]. Practicing sustainable diets can promote diet quality and, thus, human health and has—generally speaking—a positive impact on the environment [2]. In order to be clear on the term "sustainable diets" we use the FAO definition, in which "Sustainable Diets are those diets with low environmental impacts which contribute to food and nutrition security and to healthy life for present and future generations. Sustainable diets are protective and respectful of biodiversity and ecosystems, culturally acceptable, accessible, economically fair and affordable; nutritionally adequate, safe and healthy; while optimizing natural and human resources." [3].

Numerous reports on food system related issues, movements towards sustainable lifestyles and, for example, the launch of the United Nations' Sustainable Development Goals are raising public awareness for the importance of sustainable consumption and nutrition [4–9]. These cognitive contact points give people an impulse for their own behavioral improvement [10]. However, a consistent adoption of a sustainable diet seems difficult. To change one's own behavior is no piece of cake. That is why people often struggle with the implementation of their behavioral intentions [11]. This circumstance describes the so-called intention-behavior gap, based on the inconsistency between the behavioral intention and the actual behavior [12]. This gap is a critical aspect of behavior changes, especially when only half of the possible intentions are translated into desired actions [11].

The Theory of Planned Behavior (TPB) describes that human behavior is strongly based on the formed intention (Figure 1). In addition, the model reveals that there are determining factors influencing the formation of an intention. Behavioral beliefs, meaning possible positive or negative consequences of performing the behavior, lead to a specific attitude towards the behavior in question. Normative beliefs represent beliefs about a possible judgment of relevant persons and social pressure. This kind of social influence leads to a specific subjective norm. Control beliefs consider possible internal or external factors that may positively or negatively influence a person's behavior. Furthermore, beliefs on how easy or difficult the performance of the behavior can be, leads to a certain degree of perceived behavioral control [13,14]. The TPB also considers other possible variables called background factors that can influence behavioral, normative and control beliefs. Another variable is feedback loops arising through information from previous behavior. It is assumed that the more the belief that the behavior is under control and the more favorable the attitude and subjective norms are, the more likely it is to form a positive intention towards a particular behavior. Formed intentions don't always lead to the corresponding behavior. This depends on whether a person has actual control over the behavior or whether there are internal (information, skills, abilities, emotions, compulsions) or external factors (opportunity, dependence on others) that are interfering. This actual behavioral control is strongly related to the determinant of perceived behavioral control [14,15].

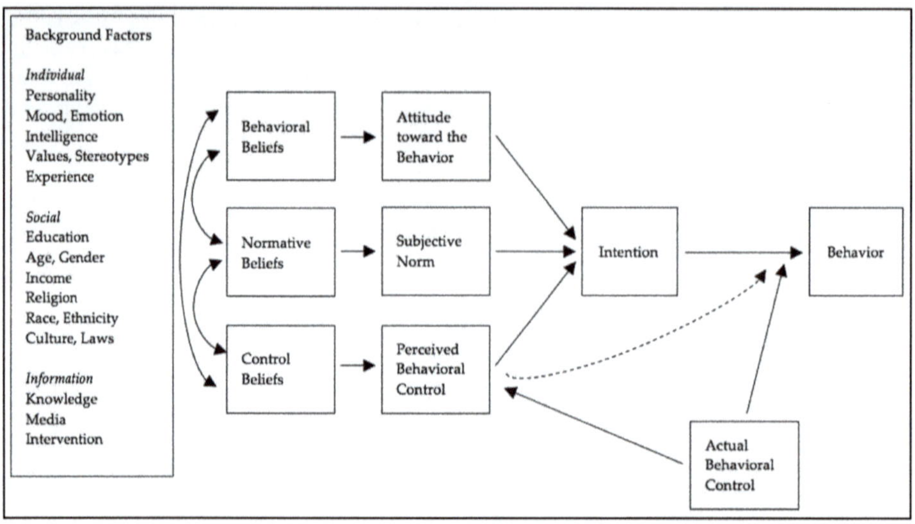

Figure 1. Ajzen's model of the Theory of Planned Behavior (TPB) [16].

The TPB is one of the most widely used social-psychological models for researching human behavior, especially when it comes to examining health behavior or the relation between intentions and behavior [14,17,18]. It offers clearly defined constructs and considers internal and external control factors that can influence the intention-behavior relation [14,18,19]; this is important to our research context.

The imminent need for changing our dietary behavior is argued by many and sustainable dietary consumption is one key driver for the existence of sustainable food systems [20]. At the same time, sustainable diets are also results of sustainable food systems [21]. The question here is how to deal with the difficulties and obstacles associated with the adoption of a sustainable diet. Recommendations for nutritional changes coming from the scientific domain are usually perceived as weak in terms of usability in everyday life. That is why we decided to leave the traditional research paths of nutrition science and go towards innovation to seek solutions for overcoming the intention-behavior gap. With solutions, we mean theoretical and practical innovations that will help us to improve and sustain targeted nutrition behavior [22]. Research, in which responsibilities are shared and collaboration between science and society is promoted, is urgently needed [23].

Therefore, it is clear that we need sustainable development in this area towards a higher adoption rate of sustainable diets. Since sustainable development is an extreme challenge for society [24], it is important that society gets involved in this process of transformation. For a truly sustainable development that requires a change in the current state, innovative strategies are needed and new paths have to be explored [25,26], especially when it comes to dealing with something complex like the intention-behavior gap. In searching for a suitable approach to creating ideas for bridging the intention-behavior gap, we decided to choose the following strategy. To understand people's needs and what could really support the adoption process, it stands to reason that they can be directly involved in the idea generation process.

This paper focuses on the participative idea workshop approach for generating ideas that go along with our research question: How can we close the intention-behavior gap when it comes to adopting sustainable diets? Therefore, the aim is primarily to deliver results in answer to the research question, more precisely: innovative ideas that help deal with the barriers creating the intention-behavior gap. Since the process is open to a certain extent, we cannot commit ourselves in advance to any one type of expected results. In our anticipated range, everything from the necessary framework conditions to practicable approaches and behavioral strategies for everyday nutrition can be contained. Since this research approach in context with our research question is new, it will also be examined for suitability in the context of this research.

This paper is structured as follows. It opens with a methodological view on the development of the research approach and workshop design. Next, we present the data gained from the workshop series. Then we analyze the data within the discussion and finally draw conclusions.

2. Materials and Methods

To develop a suitable workshop design, a comprehensive literature research on participative innovation processes was carried out in advance. Our findings led us to the open innovation process, an approach commonly used in modern product development [27]. Our hypothetical conclusion was that we can derive similar benefits [28] from applying open innovation methods as product developers do. More specifically, we aimed for suitable ideas for bridging the intention-behavior gap, which then led to an increased adoption rate of sustainable diets. Based on our comprehensive research in the field of open innovation we decided that our envisaged involvement of citizens can take place in specially arranged workshops for idea generation, based on the lead-user workshop concept [29–31], which is part of open innovation practices. This concept inspired our research to develop and apply a similar approach. However, we do not claim to call our approach a lead-user workshop, because the recruitment of our participants does not fit with von Hippel's methodology to identify lead-users [32,33].

The aim of conducting the workshops was to develop specific ideas in answer to our research question. We expected to receive ideas in the form of realistic, useful tips that enable people to translate their behavioral intention into actual behavior despite emerging situational barriers. These ideas would then serve as the basis for a kind of toolkit that could promote the adoption of sustainable diets and, along with that, the transformation towards sustainable food systems.

To develop our workshop design, we defined the following criteria: a one-day workshop, allow all interested citizens to participate without special requirements, a fixed participant number between 10 minimum and 20 maximum, an attractive setting as an incentive for participants, a structured workshop led by two moderators and support of the idea generation process by using creativity techniques. In order to gather as many ideas as possible and to achieve a small comparability of the applied methodology, we decided to carry out a series of workshops. To capture different and regional influences regarding existing food systems and communities we decided to choose a major city in the west, north, east, and south of Germany. In particular, these were Dortmund, Hamburg, Berlin, and Freiburg. These workshops were open for every interested citizen. Two additional workshops with students from the University of Kassel with an agricultural background and students from Münster University of Applied Sciences with a nutrition background completed the workshop series.

For the workshops, a time frame of eight hours including lunch and coffee breaks was set. All workshops were scheduled for a Saturday from 9 am to 5 pm. We chose Saturdays because of the higher likelihood of people having more available leisure time. According to our reasoning, we wanted to give ideally everyone, who was interested in participating, the opportunity to do so without checking a particular suitability in advance. We assumed that anyone who signs up for the workshop is interested in the topic and can develop ideas suitable for dealing with the intention-behavior gap in everyday life—at least for themselves. Because nearly everyone deals with this gap in some way, we assumed that no specialized prior knowledge or abilities were necessary. We wanted to attract motivated citizens to participate and not researchers or field experts with alleged solutions. One main challenge was recruiting participants for the workshop. The basic idea was to invite as many people as possible to the workshops in order to enable interested citizens to participate in the workshops. The participants were recruited via general mailing lists, flyers, online event calendars, and writing to companies, associations, and organizations with the request to forward the invitation to all possible contacts (creating a snowball effect). Participation was voluntary and non-remunerated, so the only direct incentive was the workshop topic and format itself. This may have led to selection bias, which we will discuss more detailed in the limitations listed in the discussion. In addition, we had paid extra attention to a pleasant setting and full day catering, where possible, in organic quality. The chosen locations offered a comfortable atmosphere, enough space, modern facilities, and adequate equipment. We decided on a structured workshop led by two facilitators. The role of the facilitators was to introduce new work stages and lead the participants through the workshop, but not interfere with their idea generation processes and be as neutral as possible. The basic conceptual idea for the whole workshop was to generate as many ideas as possible by first stimulating divergent thinking and then enabling thought condensation by convergent thinking.

The workshop procedure provided a combination of different work tasks with different creativity techniques (Table 1). Integrating work tasks offered the possibility to bring a certain structure into the process and to give a certain thematic input of different topic areas. Applying creativity techniques can support the generation of innovative ideas [34,35]. We applied different creativity techniques combined to address the research question from multiple perspectives and to maximize the potential for creative ideas. Since there is no single working formula, we sought to address the different participants by applying various techniques [36].

Depending on the work task, participants worked alone, together with a partner, or in groups of three to five people (Table 2). At the end of the workshop, the idea generation processes led to action-planning in form of developing a concept for implementing the chosen idea.

Table 1. The applied methods; Description and Purpose.

Method	Description	Purpose
Open space [37–40]	For the open space method, we combined two techniques: stimulus picture technique and sentence additions. A total of 8 pairs of pictures and 4 sentence beginnings were hung on brown paper. In this kind of gallery walk, the participants could write down everything they could think of when looking at the pictures and how the picture pairs could relate to each other. The sentences should be supplemented according to spontaneous ideas. Pictures and sentences are closely linked to the topic of sustainable food systems. After the Q&A session participants are supposed to generate ideas in answer to the research question, using the notes from the open space activity.	This methodology allows us to capture impressions before coming up with a thematic input. Based on this, ideas will be generated later in the process. Here, the participants are addressed on a visual and linguistic level.
Idea storage (brainstorming) [37,39–41]	Classic brainstorming, here, called idea storage. After the Q&A session about the workshop's research question, participants are asked to generate ideas they come up with spontaneously in order to answer the question.	Generate intuitive first ideas without using in-depth methods.
Info poster [38–40]	In group work, participants create info posters (illustrations) on a given topic. In the first step, participants depict the topic and all associative aspects. In the next step, the groups walk to a poster from another group and write down everything they can see on the info poster, what they associate with it and what they believe it is supposed to represent. In the final step, the poster illustrations are removed first and only the transcriptions remain available. On this basis, after re-routing the groups, ideas in answer to the research question are generated.	The given topics reflect different eating settings: Nutrition during lunch break Nutrition and rural life Nutrition and out-of-home catering Nutrition and city life Nutrition on the go and during journeys Nutrition and invitations to family or friends The ideas to be generated can address the challenges of implementing a sustainable diet in various settings.
Progressive abstraction [40,41]	The progressive abstraction can be used to generate thematic connections and corresponding measures. Our modified version starts with a given term. In relation to this, question X is answered. Referring to this result, question Y is then answered. Then, in turn, question X related to answer Y is answered. In this system, both questions X and Y will be addressed three times. All three answers to question Y will be used in the next step. Further ideas will be derived from the research question. X: What's good about it? Y: How can this be achieved?	The given terms were economy, ecology, society, culture, and accountability. These are the five dimensions of the Best Practice Guidelines for Agriculture and Value Chains (IFOAM Organics International). Due to this thematic background, common sustainability dimensions (supplemented by culture and accountability) should have a certain impact on the idea generation. Progressive abstraction serves to question how something positive about an aspect can be achieved in other ways.

Table 1. *Cont.*

Method	Description	Purpose
Opposites method (negative brainstorming) [37–41]	The opposites method is a kind of negative brainstorming. This means that the participants are asked questions that are formulated contrary to the actual research question. The next step is to generate ideas in answer to the research question by transferring the previous answers to positive solutions.	The questions bring the participants out of their previous thinking processes by considering the complete opposite of what they did before, e.g., how positive contributions to the implementation of sustainable diets can be prevented: How can we get people not to be interested in their nutrition? How can a worldwide unfair distribution of food be organized? How can we prevent a connection between regional, sustainable agricultural production and human nutrition? How can we discourage people from a healthy diet? How can we reduce the consumption of sustainably produced food? How can we complicate the implementation of sustainable diets? How can people having no access to (healthy and sustainable) food be achieved? How can we prevent a connection between organic farming and consumers of the produced food? How can we convince people not to act according to their knowledge and values regarding their diet? Through the provocative formulation of the questions, a certain kind of creativity should be encouraged that is refreshing and brings variety, in the same way as trying to be negative to get a positive outcome. Based on the results, participants could then generate ideas for answering the research question.
Concept development [38,40]	Based on the selected favorites of the previously generated ideas, a further development of these should be done in group work. Each group received two selected favorite ideas of each idea generation step. Based thereon, concepts for idea implementation should be developed.	In this way, favored ideas should be further developed towards an implementation concept. This also served to give the participants something tangible at the end of the workshop.

Table 2. The workshop design and procedure.

Time	Duration (minutes)	Method	Constellation	Work/Ideas Are Captured on
09:00–09:10	10	Welcome	All together	
09:10–09:30	20	Introduction of participants	All together	
09:30–09:50	20	Open Space method	Individual work	Brown paper
09:50–10:20	30	Input and Q&A session on intention-behavior gap	All together	
10:20–10:30	10	Idea storage (brainstorming) in answer to the research question	Individual work	Yellow cards (1)
10:30–10:40	10	Derive Ideas in answer to the research question (from the Open Space posters)	Individual work	Orange cards (2)
10:40–11:00	20	Info posters, part 1	Group work, 3 participants	Flip chart paper
11:00–11:10	10	Info posters, part 2	Group work, 3 participants	Flip chart paper
11:10–11:25	15	Derive Ideas in answer to the research question (from the info posters)	Group work, 3 participants	Blue cards (3)
11:25–11:40	15	Coffee break		
11:40–11:55	15	Progressive abstraction	Partner work	DIN A3 templates
11:55–12:05	10	Collect and exchange results	Partner work	
12:05–12:20	15	Derive Ideas in answer to the research question (from the progressive abstraction)	Partner work	Green cards (4)
12:20–12:35	15	Selection (1, 2, 3) of favorite ideas	Individual work	Sticky dots
12:35–13:35	60	Lunch break		
13:35–14:20	45	Topic tables	Group work, 3–5 participants	Brown paper
14:20–14:35	15	Opposites method (negative brainstorming)	Partner work	DIN A3 templates
14:35–14:50	15	Rotation & Derive Ideas in answer to the research question (from opposites method)	Partner work	Red cards (5)
14:50–15:05	15	Selection (4, 5) of favorite ideas	Individual work	Sticky dots
15:05–15:20	15	Coffee break		
15:20–16:20	60	Concept development	Group work, 3–5 participants	Brown paper
16:20–16:50	30	Presentation, discussion round	All together	
16:50–17:00	10	Feedback, close the workshop	All together	Feedback questionnaires

The workshops took place from September to December 2016. For the six workshops, we had 82 participants in total, of these 67 were female (81.7%) and 15 male (18.3%; Table 3). People registered themselves by mail. In no case were more than 20 registrations reached for any workshop. Additionally, some registered people did not attend the workshop.

Table 3. The demographic information for the workshop participants.

City	Date	Number of Participants n = 82	Female n = 67 (81.7%)	Male n = 15 (18.3%)
Dortmund	24.09.2016	18	15	3
Kassel	22.10.2016	13	9	4
Hamburg	12.11.2016	11	10	1
Berlin	26.11.2016	10	6	4
Münster	03.12.2016	18	18	0
Freiburg	10.12.2016	12	9	3

3. Results

Before we present the data, we would like to mention that we have tried to avoid the loss of words and meanings by translating the ideas generated from German into English for this article.

For each idea generation step, we had different colored cards, on which participants wrote down their ideas. In this way, we collected the qualitative data during the workshop (these were always ideas in the form of first solutions and not elaborated contributions in the form of directly implementable solutions). Since we collected a large amount of qualitative data, we needed to organize them. For data analysis and coding, all ideas were transferred to a simple spreadsheet software. Analyzing the data, we decided on internal (code 1) and external (code 3) as code-categories for factors influencing the intention-behavior relation as described in the TPB. As we screened data, we expanded these categories by one each (internal+ (code 2) and external+ (code 4)). This additional coding was necessary because of the vague wording of the ideas. From a factual perspective, the coder often had to think one step further or interpret the data in order to assign it to the thematic context of the research question rather than eliminate it.

In total, 1223 ideas were generated during the six workshops that we allocated to five codes (see Supplementary Materials, Table S1). Overall, we can see from Table 4 that we have coded 168 internal (including 142 internal+) and 989 external (including 811 external+) factors. A total of 66 (5.4%) items of the screened data were excluded (code 0). Excluding criteria were (i) pure questions from which no indirect idea can be read and (ii) ideas irrelevant or with no direct link to our research question or (iii) to the subject area of sustainable diets and (iv) sustainable food systems. To illustrate examples of data entries, we listed five ideas for each code (Table 4). It turns out that the external factors outweigh the internal factors, which we will discuss later.

Table 4. The results of data coding, n = 1223.

Factors	Factors	Examples of Participants' Ideas
Internal n = 168, 13.7%	**Internal** n = 26, 2.1%	Taking time for grocery shopping, eating, and cooking Pack food (organic etc.) to go Eating as a conscious time-out Do not buy abundantly, then you can buy healthy things Highlight the individual benefits of sustainable nutrition
	Internal+ n = 142, 11.6%	Gardening yourself (for a little self-supply and a lot of insight) Create Consciousness: What impact has my behavior on other people in the world and in view of that we all share the world? Question trends, do not chase after them Accept more personal responsibility Development of implementation strategies

Table 4. *Cont.*

Factors	Factors	Examples of Participants' Ideas
External n = 989, 80.9%	**External** n = 178, 14.6%	Enable a need-based purchase through unpacked goods (packaging-free shops), and in that way stem food waste (special value packs for fresh products tempt us to by abundantly) Celebrities as an advertising medium for sustainable nutrition (role models) The higher tax rate for animal-based and unhealthy foods, lower for plant-based and healthy foods Apps for information on sustainability when purchasing Explain sustainable diet starting in schools, through courses (cooking class etc.)
	External+ n = 811, 66.3%	Community gardens (as a getting started guide) Increase public advertising Price tag with "real" price → conscience appeal Organic products at every turn More support from the government
Neither internal nor external (not used) n = 66, 5.4%		Individual traffic (cars) → reduce emissions in the cities Stupid people run, smart people wait, wise people go into the garden Repair cafes Start with simple things: "Who likes to shower with dead animals?" Where is the problem in general, when people do not act despite consciousness?

After generating the ideas, we gathered them on a pinboard, where every participant could select two favorites for each color card set. This allowed us to identify the participant's favorite ideas during the workshop and the data analysis. Out of this selection, we generated a top 25 list of favorite ideas (Table 5). Here we point out that not all ideas were available for selection in all workshops.

Table 5. The top 25 favorite ideas, rated by participants during the workshops.

Points	Idea	Code
11	Consume/live consciously and in a resource-saving way	2
9	Direct farm sales, without many processing steps	4
8	More packaging-free grocery stores	3
8	Clarification and education	4
7	Value and advertising-free food (valuation only on content, for example "with x% fruit content" instead of "high content of x")	4
7	Consumers visit farms and discuss animal husbandry, fertilization, etc.	4
7	Nutrition as a subject (cooking, gardening...)	4
7	100% utilization (of things)	4
7	More education in schools	4
6	A product database including true costs (resource consumption etc.) (for example sausages Aldi €0.69 → Earth costs €2.69)	3
6	Food prices must represent the total cost	3
6	Ban factory farming (plus new stricter regulation of fertilizer)	4
6	Create more time resources for people	4
5	Food prices must reflect the full cost	3
5	Higher (value-added-?) tax rate for animal-based and unhealthy foods, lower for plant-based and healthy foods	3
5	Create reward systems for sustainable nutrition	3

Table 5. *Cont.*

Points	Idea	Code
5	Create infrastructure for sustainable strategies (e.g., for neighborhood cooperation) Legal simplifications, e.g., for (food) sharing points, shopping communities; establish jobs that maintain the infrastructure, carry out work	3
5	Sustainability parties in the same way as "Tupperware parties"	4
5	Show an ecological footprint on the product	4
5	Responsible school catering-earning effect for children	4
5	Prohibit or strictly regulate lobbying in agriculture and the food industry	4
5	Make sustainability "noticeable"-price?	4
5	Governmental support	4
5	Organic farming as the only solution for feeding the world	4
5	Integrated production/processing in the city, e.g., urban gardening, rent a field	4

Looking at the results we notice that they are more factors that influence behavior than precise and practicable ideas or useful suggestions to bridge the intention-behavior gap in everyday life.

During the evaluation, it became apparent that a range of topics came up repeatedly. This has led us to an additional inductive data coding, in order to allow a more in-depth discussion of the data. We would like to point out that some data can be assigned to several categories and some data did not fit into the main categories at all. We prefer to use the terminology thematic cluster instead of code categories because we have collected several similar thematic aspects under a generic term (Figure 2).

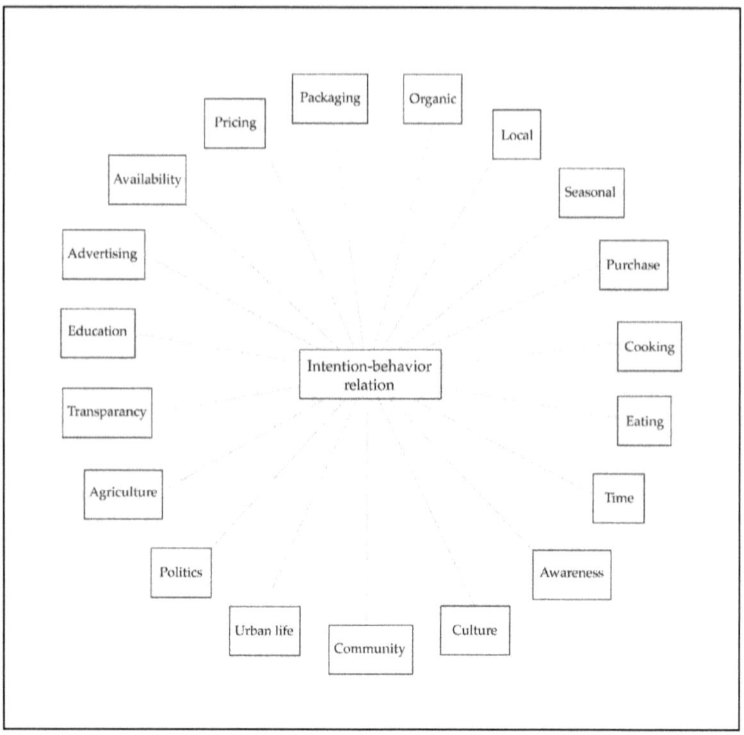

Figure 2. The thematic cluster of factors influencing the intention-behavior relation.

Availability (also meaning the supply, access, and distribution channels) of sustainably produced food with 175 mentions is one large category within the data (example: easier availability → as easy as buying conventional food). Others are nutrition education and education about the food industry including sub-topics like exchange of information, provision of information, and learning (289 mentions; example: creating general knowledge and experiences about food in schools, kitchens, health centers, rural and urban, e.g., nutrition education as a "must"), transparency related to the production and methods as well as to the information on the packed foods by labeling (80 mentions; example: more transparency about the ingredients of food products and understandable declaration) and advertising including all actions that raise public awareness for sustainable nutrition, food, and production. This includes campaigns, positive image building, role model staging and marketing activities (113 mentions; example: promote advertising in favor of organic food and sustainability). It continues with the category community meaning collective action within families and societies (111 mentions; example: introduce collective rituals for the appreciation of food by cooking together, harvest festivals), policies such as promotion, subsidization, prohibition, sanctioning and legislation (178 mentions; example: politically regulate lobbyism in the food industry) and agriculture meaning aspects like food production and the role of farmers (141 mentions; example: promotion of small-scale agriculture with greater consumer involvement). Regarding food products the following aspects were clustered: food pricing (59 mentions; example: offer healthy food more cheaply), organic (113 mentions; example: introduce more organic products into everyday life), local (61 mentions: example: linking regional products and suppliers with the urban world (internet, delivery, "country" shops)) and seasonal (21 mentions; example: supermarkets with seasonal products and a stronger signage). Other clusters concern purchasing (22 mentions; example: avoid emotionally influenced grocery shopping: do I really need it?), cooking (59 mentions; example: cooking, baking, canning, preserving, stockpiling—making it more attractive and imparting craftsmanship), eating (32 mentions; example: develop new eating habits), building awareness including consciousness, mindfulness, and appreciation (79 mentions; example: creating awareness: what effects does my behavior have on other people in the world? Especially considering that we all share the world), food culture (20 mentions; example: (re)establish food culture) and time (33 mentions; example: problem: time pressure → no sustainable nutrition possible; solution: find time and take time to eat). In addition, urban life was often mentioned, especially urban gardening (32 mentions; example: rent a field: urban gardening in open spaces, unused urban areas, parks, green areas) as well as the thematic aspect of food packaging (25 mentions; example: using sleeves instead of plastic packaging for the differentiation of food) concerning the amount and type of packaging material.

4. Discussion

As part of our research, workshops were conducted to collect ideas in answer to the research question. To the best of the authors' knowledge, this was the first attempt of such a method being applied to develop ideas for bridging the intention-behavior gap for adopting sustainable diets.

As described in the introduction, we expected to receive qualitative data in the form of useful ideas that will help people to overcome emerging barriers when dealing with the gap. However, we could not anticipate in advance what kind of ideas we will receive. Looking at the results, they do not seem to provide direct practical or applicable solutions to the addressed problem (no ready-to-use ideas). Within the ideas we coded as internal factors, we have a few that correspond to direct situational solutions.

Most of the data offer important information that we discuss in more detail below, starting with the five most occurring thematic clusters: education, politics, availability, agricultural production, and advertising.

Education is the most mentioned aspect amongst the ideas. Missing knowledge about something can influence the actual behavioral control as an external factor. We can form an intention without having all (necessary) knowledge about aspects concerning our target behavior, which can turn out to be a barrier in the course of the action. This is why lifelong education about nutrition and our food industry is still important. Education may be the one aspect where most research has been done and the most intervention programs have been conducted so far. Many nutrition education approaches are health-related and especially obesity-related [42–46]. Nevertheless, the efficacy of intervention programs for nutrition is still debated [47]. As far as we know, there are no generally applicable guidelines focusing on how to implement an effective intervention program and measures when an intervention is really effective (for example when children help their parents to make more sustainability-led choices). Clearly, educational institutions like schools can directly promote sustainable development by improving the type and quality of the provided food [48,49]. However, it should be noted that social stratification also plays a role in whether children attend school meals or not. Therefore, there is a high probability that those who may need this education cannot be reached [50,51]. In addition, there is some knowledge about different types of emerging resistance when healthy food proposals are implemented in schools [52].

In order to gain knowledge and education, transparency is an important factor and also often mentioned in the data.

The data show that regulatory intervention at the political level is intentional and necessary to have a positive impact on the development of the sustainable food system to enable sustainable nutrition, according to the participants. This mainly concerns subsidies, controls and the labeling of sustainable foods, as well as sanctions for unsustainable economies. This also goes along with the other frequently mentioned aspects such as agriculture, food pricing, organic, local, and seasonal foods and food packaging.

One further observation that seems very important is the frequently emerging aspect of food availability (in grocery stores, restaurants, canteens, catering). Consumption behavior, including that of sustainable nutrition, is based on a decision-making process. Here, the everyday behavior is mainly determined by factors such as habits and convenience that are persistent in terms of possible changes [53,54]. That is why intentions alone can be poor predictors of behavior. In order to change everyday nutritional behavior, external factors play a more important role, especially the availability of what is called sustainable food [53,55]. Without offering such food, people cannot opt for or against food items at the point-of-sale. Moreover, no corresponding consumption patterns can be formed [56]. Food availability can create a direct link to people's behavioral control because everyone can have the intention and behavioral persuasion to practice a sustainable diet but it can be impossible to do so because of a lack of (appropriate) food availability [53]. If availability and accessibility are drivers of food consumption and consumers are drivers of food production [56] then we have a mutually reinforcing, but also interdependent cycle of necessary factors where one cannot exist without the other. Considering that one factor must exist first for this cycle to get going, it is the availability without which there cannot be a corresponding consumption. However, there can be a pure availability of food by agricultural production and corresponding distribution without a demand through consumption (leaving aside economic aspects).

Agricultural food production is, of course, a hot topic considering that the availability of the desired food quality cannot exist without prior sustainable production. The basic premise is that there must be more sustainable agriculture and funding to produce food in the desired quality. In addition, the (sustainable) agricultural production of food is the basis of a (sustainable) food system and human nutrition.

Advertising seems to be a little-used aspect to promote sustainable dietary behavior. The data contain many ideas about marketing strategies, advertising campaigns, and corresponding media communication. The assumption is that a greater media presence of sustainable products is considered necessary to influence the purchasing decisions positively. There are studies about buying behavior successfully influenced by advertising—but mostly for foods that do not promote a healthy and sustainable diet and often targeting children and adolescents [57–60].

The collective action appears to be the way to a new kind of lifestyle, where not everyone undertakes life only individually, but many actions take place at a community level. The mentioned actions are producing, purchasing, cooking and eating food together. Added to this is the demand that people learn to cook and practice this frequently, whether alone or in the community. Another aspect that seems to play a decisive role in here is "time": to spend more time on nutrition decisions, have more time for cooking and eating. Eating also goes hand in hand with the development of awareness of nutrition and consciously deciding for or against food items or behavioral action. This is where the aspect of a food culture also comes in, which should be lived out more distinctively, according to the workshop participants. One of the biggest barriers to sustainable food consumption is the price [61–63]. Within our data, participants suggest that prices for sustainably produced food should be lower than those for conventional products. Another observation we make is, that the workshop contributions also aim at a change in urban space and use. For example, establish a nutrition parliamentary committee in the municipality; edible City: free fruit, nuts, vegetables for everyone (allowed by the municipality); Integrated production/processing in the city, e.g., urban gardening. One reason for the number of urban targeting ideas can be that the workshops took place in bigger cities. Another reason for the focus may be the global challenge of urbanization in particular with food [64–66].

The above discussion shows that all these aspects are highly connected to each other. Altogether, aspects like education, politics, availability, and advertising serve the purpose of bringing sustainable food into the public focus and forming a certain image. Ideally, these transformations would trigger a real trend towards sustainable diets. Previous isolated efforts through nutrition interventions for dietary change and public health could not bring any significant reversal from unhealthy diets [55]. It can, therefore, be assumed that real changes in the sense of improving diets require a large-scale movement of change involving multiple sectors and actors that affect everyday life.

The data indicate what kind of external framework conditions need to be created according to the workshop participants so that there is no gap between intention and behavior regarding the implementation of sustainable diets. Conversely, the data can be used to read perceived aspects that influence dietary behavior (missing availability of health-supporting and sustainably produced food, nutrition education throughout life) or what the potential barriers are (allegedly higher prices of health-supporting and sustainably produced food, lack of time for nutrition and related processes). Individual aspects overlap and can also be found in the literature [53,55,56]. Consequently, we have factors that could be key or at least suitable tools for bridging the intention-behavior gap in decision-making situations.

According to the TPB, intentions are formed and therefore can be influenced by our attitude towards the behavior, our subjective norms, and our perceived behavioral control. In addition, individual, social and informational background factors can have an impact on these variables. Once an intention is formed, the decision-making process to translate the intention into actual behavior can be affected by the person's actual behavioral control consisting of internal and external factors. As we were able to code our results by internal and external factors influencing the behavior, we see how numerous and varied these can be—especially with a complex behavior such as nutrition. Some of the results, e.g., aspects of education, culture, and media also fit the background factors (Figure 1). First, we may need a clearer understanding of or separation between factors influencing the actual behavioral control and background factors. Second, using the TPB, we could also examine within a study which background factor affects which beliefs of attitudes, subjective norms and perceived behavioral control and thereby indirectly influence the intention and behavior [14,16]. This is a very important aspect to consider since the TPB also includes feedback loops coming from behavior to beliefs. Meaning, a performed behavior provides information about consequences, reactions and about how easy or difficult it was to perform the behavior. This information can change people's behavioral, normative, and control beliefs, thereby influencing future intentions and behavior [14]. Combining our results with a suitable follow-up research design may shed light on how our findings may affect the intention-behavior relation. This, in turn, could be used to develop targeted solutions to bridge the intention-behavior gap.

One further interesting aspect we can observe within the results is a kind of rejection of the intention-behavior gap by the participants themselves. The results are, with a few exceptions, never worded as self-referential. Instead, they aim for changes that others have to make, both internal and external. There is mention of individuals who are often called consumers, farmers, politicians, and society, also named community. When screening the data, it seems that the solutions are mainly intended for others and for future generations, rather than for direct optimization for the participants themselves here and now. This is why a large part of the data can be interpreted as targeting a long-term system change that must come from the top down, which is a very interesting aspect to come from applying a bottom-up approach. One indication of this is the high number of external factors (80.9%). In addition, we are able to show a ranking within our dataset. It is also noticeable that among the top 25 selected favorites (Table 5), almost all are external factors (96%). One prominent example of a top-down action that has resulted in a broad change in behavior applies to tobacco. A large-scale strategy that included anti-smoking campaigns, stringent product labeling and offensive media communication of scientific evidence has led to a lifestyle change for many people [55]. If such a multiple offensive can alter a manifested behavior (with an addictive character) such as smoking, this example offers the potential for a similar strategy targeting nutritional behavior.

There are a number of limitations of our study which we want to list here. The first limitation concerns the participants. The absence of a monetary incentive might have selected only people who are who are very committed to sustainable nutrition and related issues. Their assumed high educational status could be one reason why there are so many external factors compared to internal factors within the results. Since we have collected almost no additional data from the participants, we cannot say much about the socioeconomic composition of the sample. However, as stated above we can say that most of our participants are highly educated which may affect the representation of the intention-behavior gap. We raise this in relation to research about the relationship between intentions and behavior affected by the socioeconomic status [67,68]. Social stratification in general also plays an important role, e.g., food consumption is strongly affected by people's economic and cultural resources. Due to the lack of data, we cannot make any statements regarding our results in this respect.

In addition, our sample consists of 81.7% women (Table 3). The gender aspect can also have an impact on health-related behavior. For example, women tend to eat more healthily than men do [69], which in turn could imply that men and women's intention-behavior gaps differ to a great extent. If the proportion of men in our sample were higher, it could have an impact on the results.

Another limitation concerns the definition and interpretation of a sustainable diet. Such a complex construct leaves room for individual interpretations. People's ideas might differ from the scientific definition and its intended meaning. Although we introduced the concept of a sustainable diet at the beginning of each workshop, individual interpretations may have influenced the nature of the results.

A further limitation may be the methodology itself. Because we applied a new approach, we see a need for a subsequent discussion on it, to possibly find out to what extent the type of results was influenced by the applied methods and other parameters. For many people, working with creativity techniques is a new experience and the way of their application can be perceived as abstract and not effective. When using creativity techniques, there are four complex parameters that can be decisive to the success or failure of the creative process. These parameters are processed (problem perception, problem-wording, idea generation, idea evaluation, idea realization), product (what is being worked on), environment (physical and social), and person (attitude, motivation, abilities, personality traits) [70]. Going into detail about these parameters and their individual aspects would go beyond the scope of this discussion. However, we can see here that many factors, some of them uncontrollable, interact, which should be considered when planning to apply such methods. For our research, this means that within this discussion we can take a close look at our workshop design, the research question, and ideas to be generated, possible environmental factors and the participants. The workshop design includes the choice of methods and their combination. Among many methods, we chose those that we considered suitable individually and in combination. The combination of the individual work steps could have been arranged differently. However, there is no standard procedure for a successful workshop of this kind and for this research question. Another aspect is the lack of discussion rounds, which could have also acted as a control for the results. In addition, the interaction between the facilitators or organizer and the participants could have helped to correct or steer the direction in which ideas have been developed. Another consideration is the number of group work tasks. Here the group work and the desired creativity may be contrary. In order to achieve an optimal participation, group dynamic abilities like autonomy, spontaneity, and communication skills have to be learned [70]. Possibly mainly individual work in combination with subsequent discussions could have been more conducive to expose the own ideas of all participants [38,71]. From workshop design to the desired product, the formulation of the research question needs to be discussed. The research question is rather generally worded. One option would be to phrase the question more specifically to situational aspects (e.g., out-of-home catering, grocery shopping) to gain more precise direct situational solutions. However, at the beginning of the workshop, the research question, aim, and expectations were discussed. What remains to be mentioned is that the term 'sustainable food systems', was used in the invitation and the initial explanation. This may have led the participants to generate the ideas in their present form (targeting the whole food system). Further, the participants might not have been able to generate self-related ideas of the kind targeted. They may possibly have a lack of appropriate skills and experience in dealing with creativity techniques or the research question, but this is only an assumption. The structure of the workshop was designed in such a way as to be possible for all people to successfully participate without special prior knowledge; this worked well. First of all, we assumed that every interested participant has experience with the intention-behavior gap and the associated situational barriers. Thinking that anyone could contribute some ideas on how to act in specific situations to bridge the gap was possibly a fallacy. However, we need experts in the continuous practice of sustainable diets. These experts must be found and recruited or at least selected more specifically. Our participants had a variety of backgrounds and we had no further information about their skills and abilities for dealing with the methods and the research question. If we return to the lead-user approach, it is about people who already have solutions. Therefore, it should be considered that a more specific selection of participants and training could be a possible success factor. Within our research, we dealt with a lack of capacities to implement this approach. In addition to the participants, of course, the facilitators also play a major role in a workshop. The facilitators were neutral during the creative process and did not judge the ideas. It seems important to reconsider if a stronger steering would be more effective to release more potential innovative ideas. In a similar way, a lead-user workshop has stakeholders join the workshop and always give clear instructions about what they have in mind. In such cases, the success of the company may very well depend on the developed product. In order to make our applied method more resilient and to find out exactly which aspects have significantly influenced the processes of idea generation, a follow-up study with all participants would be necessary. Though we have not carried this out at this time, what we already have is an evaluated feedback questionnaire distributed to all participants at the end of each workshop. In order to ask for the participant's personal attitude towards the workshop design and other specific aspects, we used a 5-point Likert scale (Table 6).

Table 6. The results of the feedback questionnaire, n = 63.

Selected Questions	Strongly Agree	Agree	Disagree	Strongly Disagree	Neither Agree nor Disagree	No Information
The methods used were suitable and appropriate for the workshop.	50	12	-	1	-	-
The methods used were suitable to support creativity.	45	15	1	1	1	-
The research question is suitable for processing within an idea workshop.	47	14	1	-	1	-
The aim of the workshop was clear after the presentation of the question.	27	33	2	1	-	-
The implementation of several regional workshops makes sense for the research question.	45	15	-	-	1	2
The work materials provided (pens, paper, etc.) were sufficient.	61	2	-	-	-	-
The amount of time was appropriate for the workshop.	45	14	2	-	1	1
The venue was appropriate for the workshop.	58	4	1	-	-	-
The atmosphere during the workshop was pleasant.	58	5	-	-	-	-
Any interested citizen can participate in this workshop without special prior knowledge.	31	24	6	-	1	1
Including citizens in such scientific research makes sense.	50	13	-	-	-	-
By involving citizens in the development of ideas to bridge the intention-behavior gap, ideas can be successful later.	40	18	3	-	-	-
Overall, the facilitators were competent.	56	6	-	-	-	1
The facilitators were clear in the speeches and work tasks.	42	20	-	-	-	1
The facilitators reacted sufficiently to questions.	57	5	-	-	1	-
The facilitators remained neutral within the facilitation.	54	8	1	-	-	-
The facilitators guided the participants well through the workshop.	59	3	-	-	-	1

In total, 63 (76.8%) participants of a total of 82 completed the questionnaire. Since the workshops' procedure was identical, we decided on an overall evaluation without distinguishing between the individual workshops. The discussion of the results below only refers to the participants who completed the questionnaire. It is not possible to make informed statements about the attitudes of the other participants. That is why, despite the generally positive feedback results, we cannot assume that all participants share these opinions. We can see that in total, 62 participants agreed that the applied methods were suitable for the workshop and 60 participants think that the methods were suitable for supporting creativity. In general, we can say that for 61 participants the research question was perceived as suitable for processing within an idea workshop. Additionally, the workshop's aim was clear for the participants (60 participants agreed). Addressing environmental parameters, we can see that in general participants agreed on appropriate time management (59 participants), appropriate venue (62 participants), and pleasant atmosphere (63 participants).

Regarding the question, whether it makes sense to involve citizens in such research, all 63 agreed. However, 6 participants disagreed on the statement that any interested citizen can participate in this workshop without special prior knowledge (whereas 55 agreed on that). Another important point is that 58 participants thought that involving citizens in the development of ideas to bridge the intention-behavior gap can be successful for their later application.

Finally, we cannot say exactly why the results are influencing factors rather than situational solutions and if a different methodological approach would have generated other results that more innovatively address situational barriers, or whether the external factors such as availability, advertising, pricing, and education about food and nutrition are simply a prerequisite for a broad mass of people to practice a sustainable diet. While not a new point, this does seem to reinforce its importance.

5. Conclusions

Sustainable diets cannot exist without sustainable food systems and vice versa. Sustainable food systems wherein environments provide only sustainable food choices evolve slowly, if at all. That is why it is important to support society by researching the adoption and practice of sustainable diets.

The diversity of the gained data and their possible interpretations alone show that our research is part of a highly charged discussion where we are easily crossing set research boundaries. Since the results in themselves are not entirely new ideas, research into innovations that make a decisive contribution needs to continue. All the listed limitations within the discussion could lead us to say that our gained results are rather a list of desiderata of a selected group of people who would like to be able to practice a more sustainable dietary behavior. We have to accept this on the basis of the mentioned limitations, and also because of the sample size and geographical limitation (Germany). However, the interesting aspect of the results is that those participants themselves reject the intention-behavior gap (as we discussed earlier). We see this as an important research topic that deserves further investigation.

Our research did not solve the problem at its first attempt and it seems that we are just at the beginning of the problem-solving process. To achieve our planned goal to research solutions to overcome barriers and bridge the intention-behavior gap we are planning to adjust our methodological approach and our sample, as discussed. It is necessary to collect more socioeconomic data about the participants so that we can later analyze and interpret results in differentiated ways. Ideally, studies will be conducted with particular selected groups of people such as men, woman, mothers, the privileged, the deprived, and experts. It is conceivable to carry out a study that examines the intention-behavior gap of the participants (researching their behavior in vivo). The design should strongly be based on Ajzen's TPB, his published research, and supporting guides on TPB-based questionnaires and interventions [72]. On the other hand, the study design should involve the participants in the solution process. Therefore, we need to develop a methodological approach that leads people to work on ready-to-use ideas for bridging the intention-behavior gap.

We encourage other scientists to apply further methodological approaches that promote participative processes and collaborate with society because studies are required to deliver applicable solutions to people's everyday life. On the basis of our work, we further suggest that future research can use a mixed methodology approach. This allows researchers to collect and distinguish quantitative and qualitative data from participants with different intention-behavior gaps. Consequently, the factors that cause these differences could be more closely analyzed and understood, which in turn could lead to applicable solutions. It remains to be said that, as with many other studies there is no innovation without a risk [73].

Supplementary Materials: The following are available online at http://www.mdpi.com/2071-1050/10/12/4434/s1, Table S1: Complete dataset (Excel chart).

Author Contributions: L.F. developed the workshops design, conducted the workshop series, organized and analyzed the data and wrote the paper. A.P. contributed to the planning of workshops and has significantly revised the paper. C.S. contributed to the theoretical approach of the workshops, significantly revised the paper and did the final language editing. All authors interpreted the results and approved the final version of the paper.

Funding: The workshops were funded by the Department of Organic Food Quality and Food Culture at the University of Kassel, Germany. The costs of this open access publication were covered by the Open-Access-

Publications fund from the University of Kassel which is financed by the German Research Foundation (DFG) and the Library of University of Kassel.

Acknowledgments: We express our thanks to all participants who dedicated their time.

Conflicts of Interest: The authors declare no conflict of interest.

References

1. Willer, H.; Lernoud, J. *The World of Organic Agriculture 2018. Statistics and Emerging Trends.*; Willer, H., Lernoud, J., Eds.; Research Institute of Organic Agriculture (FiBL), Frick, and IFOAM–Organics International: Bonn, Germany, 2018; ISBN 9783940946126.
2. Johnston, J.L.; Fanzo, J.C.; Bogil, B. Understanding Sustainable Diets: A Descriptive Analysis of the Determinants and Processes That Influence Diets and Their Impact on Health, Food Security and Environmental Sustainability. *Adv. Nutr* **2014**, *5*, 418–429. [CrossRef] [PubMed]
3. FAO. *Sustainable Diets and Biodiversity Directions and Solutions for Policy, Research and Action*; Burlingame, B., Dernini, S., Eds.; FAO: Rome, Italy, 2012; ISBN 9789251072882.
4. Runini, L.F.; Ciati, R.; Pratesi, C.A.; Marino, M.; Principato, L.; Vannuzzi, E. Working toward healthy and sustainable diets: the "Double Pyramid Model" developed by the Barilla Center for Food and Nutrition to raise awareness about the environmental and nutritional impact of foods. *Front. Nutr.* **2015**, *2*, 1–6. [CrossRef]
5. Seyfang, G. Shopping for sustainability: Can sustainable consumption promote ecological citizenship? *Env. Polit.* **2005**, *14*, 290–306. [CrossRef]
6. Sachs, J.D. From millennium development goals to sustainable development goals. *Lancet* **2012**. [CrossRef]
7. OECD. *Promoting Sustainable Consumption. Good Practices in OECD Countries*; OECD Publications: Paris, France, 2008.
8. Clark, G. Evolution of the global sustainable consumption and production policy and the United Nations Environment Programme's (UNEP) supporting activities. *J. Clean. Prod.* **2007**, *15*, 492–498. [CrossRef]
9. Toledo, Á.; Burlingame, B. Biodiversity and nutrition: A common path toward global food security and sustainable development. *J. Food Compos. Anal.* **2006**, *19*, 477–483. [CrossRef]
10. Young, W.; Hwang, K.; McDonald, S.; Oates, C.J. Sustainable consumption: Green consumer behaviour when purchasing products. *Sustain. Dev.* **2010**, *18*, 20–31. [CrossRef]
11. Sheeran, P.; Webb, T.L. The Intention–Behavior Gap. *Soc. Personal. Psychol. Compass* **2016**, *10*, 503–518. [CrossRef]
12. Orbell, S.; Sheeran, P. "Inclined abstainers": A problem for predicting health-related behaviour. *Br. J. Soc. Psychol.* **1998**, *37*, 151–165. [CrossRef] [PubMed]
13. Ajzen, I. From Intentions to Actions: A Theory of Planned Behavior. In *Action Control. From Cognition to Behavior*; Kuhl, J., Beckmann, J., Eds.; Springer: Berlin/Heidelberg, Germany, 1985; pp. 11–39. ISBN 978-3-642-69746-3.
14. Ajzen, I. Consumer attitudes and behavior: The theory of planned behavior applied to food consumption decisions. *Rivivista Econ. Agrar.* **2015**, *2*, 121–138. [CrossRef]
15. Ajzen, I. *Attitudes, Personality And Behaviour*, 2nd ed.; Open University Press: Berkshire, UK, 2005; ISBN 978-0335217038.
16. Ajzen, I.; Albarracin, D.; Hornik, R. *Prediction and Change of Health Behavior: Applying the Reasoned Action Approach*; Lawrence Erlbaum Associates: New Jersey, NJ, USA, 2007; ISBN 0-8058-5926-8.
17. Knoll, N.; Scholz, U.; Rieckmann, N. *Einführung Gesundheitspsychologie*, 4th ed.; Ernst Reinhardt Verlag: München, Germany, 2017; ISBN 978-3825247454.
18. Skår, S.; Sniehotta, F.F.; Araújo-Soares, V.; Molloy, G.J. Prediction of behaviour vs. prediction of behaviour change: The role of motivational moderators in the theory of planned behaviour. *Appl. Psychol.* **2008**, *57*, 609–627. [CrossRef]
19. Armitage, C.J.; Conner, M. Efficacy of the Theory of Planned Behaviour: A meta-analytic review. *Br. J. Soc. Psychol.* **2001**, *40*, 471–499. [CrossRef] [PubMed]
20. Mason, P.; Lang, T. *Sustainable diets: How ecological nutrition can transform consumption and the food system*; Routledge: New York, NY, USA, 2017; ISBN 978-0415744720.
21. Meybeck, A.; Gitz, V. Conference on "Sustainable food consumption" Sustainable diets within sustainable food systems. *Proc. Nutr. Soc.* **2017**, *76*, 1–11. [CrossRef] [PubMed]

22. Rothman, A.J. "Is there nothing more practical than a good theory?": Why innovations and advances in health behavior change will arise if interventions are used to test and refine theory. *Int. J. Behav. Nutr. Phys. Act.* **2004**, *1*, 1–7. [CrossRef] [PubMed]
23. Coudel, E.; Devautour, H.; Soulard, C.; Faure, G.; Hubert, B. *Renewing Innovation Systems in Agriculture and Food: How to Go towards More Sustainability?* Wageningen Academic Publishers: Wageningen, The Netherlands, 2013; ISBN 9789086862146.
24. United Nations. *Our Common Future: Report of the World Commission on Environment and Development*; Oxford University Press: New York, NY, USA, 1987; ISBN 978-0192820808.
25. Lowe, P.; Phillipson, J.; Lee, R.P. Socio-technical innovation for sustainable food chains: roles for social science. *Trends Food Sci. Technol.* **2008**, *19*, 226–233. [CrossRef]
26. Ericksen, P.J.; Ingram, J.S.I.; Liverman, D.M. Food security and global environmental change: Emerging challenges. *Environ. Sci. Policy* **2009**, *12*, 373–377. [CrossRef]
27. Chesbrough, H.W. *Open Innovation: The New Imperative for Creating and Profiting from Technology*; Harvard Business School Publishing: Boston, MA, USA, 2003; ISBN 1578518377.
28. Tidd, J.; Bessant, J. *Strategic innovation management*; John Wiley & Sons Inc.: New York, NY, USA, 2014; ISBN 978-1-118-45723-8.
29. Reichwald, R.; Piller, F. *Interaktive Wertschöpfung: Open Innovation, Individualisierung und neue Formen der Arbeitsteilung*; Gabler Verlag: Wiesbaden, Germany, 2009; ISBN 978-3-8349-9440-0.
30. Busse, M.; Siebert, R. The role of consumers in food innovation processes. *Eur. J. Innov. Manag.* **2018**, *21*, 20–43. [CrossRef]
31. von Hippel, E. Democratizing innovation: The evolving phenomenon of user innovation. *J. für Betriebswirtschaft* **2005**, *55*, 63–78. [CrossRef]
32. Von Hippel, E. Lead Users: A Source of Novel Product Concepts. *Manage. Sci.* **1986**, *32*, 773–907. [CrossRef]
33. Lüthje, C.; Herstatt, C. The Lead User method: An outline of empirical findings and issues for future research. *R&D Manag.* **2004**, *34*, 553–568. [CrossRef]
34. Cook, P. The creativity advantage—Is your organization the leader of the pack? *Ind. Commer. Train.* **1998**, *30*, 179–184. [CrossRef]
35. Meinel, M.; Voigt, K.-I. The Application and Impact of Creativity Techniques in Innovation Management. In Proceedings of the XXVII ISPIM Innovation Conference—Blending Tomorrow's Innovation Vintage, Porto, Portugal, 19–22 June 2016.
36. Backerra, H.; Malorny, C.; Schwarz, W. *Kreativitätstechniken. Kreative Prozesse anstoßen, Innovationen fördern*, 3rd ed.; Carl Hanser Verlag: Munich, Germany, 2007; ISBN 978-3446412330.
37. Wack, O.G.; Detlinger, G.; Grothoff, H. *Kreativ sein kann jeder: Kreativitätstechniken für Leiter von Projektgruppen, Arbeitsteams, Workshops und von Seminaren. Ein Handbuch zum Problemlösen*, 2nd ed.; Windmühle Verlag: Hamburg, Germany, 1998; ISBN 978-3922789420.
38. Higgins, J.M.; Wiese, G.G. *Innovationsmanagement. Kreativitätstechniken für den unternehmerischen Erfolg*; Springer: Berlin, Germany, 1996; ISBN 978-3540605720.
39. Gray, D.; Brown, S.; Macanufo, J. *Gamestorming. Ein Praxisbuch für Querdenker, Moderatoren und Innovatoren*; O'Reilly Media Inc.: Köln, Germany, 2011; ISBN 978-3897213265.
40. van Aerssen, B.; Buchholz, C. *Das große Handbuch Innovation. 555 Methoden und Instrumente für mehr Kreativität und Innovation im Unternehmen*; Verlag Franz Vahlen: München, Germany, 2018; ISBN 978-3800656837.
41. Boos, E. *Das große Buch der Kreativitätstechniken*; Compact Verlag GmbH: München, Germany, 2010; ISBN 978-3817451067.
42. Murimi, M.W.; Kanyi, M.; Mupfudze, T.; Amin, M.R.; Mbogori, T.; Aldubayan, K. Factors Influencing Efficacy of Nutrition Education Interventions: A Systematic Review. *J. Nutr. Educ. Behav.* **2017**, *49*, 142–165. [CrossRef] [PubMed]
43. USDA. *A Series of Systematic Reviews on the Effects of Nutrition Education on Children's and Adolescents' Dietary Intake*; U.S. Department of Agriculture: Richmond, VA, USA, 2012.
44. Ajie, W.N.; Chapman-Novakofski, K.M. Impact of computer-mediated, obesity-related nutrition education interventions for adolescents: A systematic review. *J. Adolesc. Heal.* **2014**, *54*, 631–645. [CrossRef] [PubMed]
45. Contento, I.R. Nutrition education: Linking research, theory, and practice. *Asia Pac. J. Clin. Nutr.* **2008**, *17*, 176–179. [CrossRef] [PubMed]

46. Silveira, J.A.C.; Taddei, J.A.A.C.; Guerra, P.H.; Nobre, M.R.C. Effectiveness of school-based nutrition education interventions to prevent and reduce excessive weight gain in children and adolescents: A systematic review. *J. Pediatr. (Rio. J).* **2011**, *87*, 382–392. [CrossRef]
47. Benn, J.; Carlsson, M. Learning through school meals? *Appetite* **2014**, *78*, 23–31. [CrossRef] [PubMed]
48. Morgan, K.; Sonnino, R. *The School Food Revolution: Public Food and the Challenge of Sustainable Development;* Earthscan: London, UK, 2008; ISBN 9781844074822.
49. Sonnino, R. Quality food, public procurement, and sustainable development: The school meal revolution in Rome. *Environ. Plan.* **2009**, *41*, 425–441. [CrossRef]
50. Oncini, F.; Guetto, R. Determinants of dietary compliance among Italian children: Disentangling the effect of social origins using Bourdieu's cultural capital theory. *Sociol. Heal. Illn.* **2017**, *39*, 47–62. [CrossRef] [PubMed]
51. Wills, W.J.; Danesi, G.; Kapetanaki, A.B.; Hamilton, L.K. The Socio-Economic Boundaries Shaping Young People's Lunchtime Food Practices on a School Day. *Child. Soc.* **2018**, *32*, 195–206. [CrossRef]
52. Fletcher, A.; Jamal, F.; Fitzgerald-Yau, N.; Bonell, C. 'We've Got Some Underground Business Selling Junk Food': Qualitative Evidence of the Unintended Effects of English School Food Policies. *Sociology* **2014**, *48*, 500–517. [CrossRef]
53. Vermeir, I.; Verbeke, W. Sustainable Food Consumption: Exploring the Consumer "Attitude-Behavioral Intention" Gap. *J. Agric. Environ. Ethics* **2006**, *19*, 169–194. [CrossRef]
54. Pfeiffer, C.; Speck, M.; Strassner, C. What leads to lunch—How social practices impact (non-)sustainable food consumption/eating habits. *Sustainability* **2017**, *9*, 1437. [CrossRef]
55. Duchin, F. Sustainable consumption of food: A framework for analyzing scenarios about changes in diets. *J. Ind. Ecol.* **2005**, *9*, 99–114. [CrossRef]
56. Kearney, J. Food consumption trends and drivers. *Philos. Trans. R. Soc. B Biol. Sci.* **2010**, *365*, 2793–2807. [CrossRef] [PubMed]
57. Matthews, A.E. Children and obesity: A pan-European project examining the role of food marketing. *Eur. J. Public Health* **2008**, *18*, 7–11. [CrossRef] [PubMed]
58. Story, M.; French, S. Food Advertising and Marketing Directed at Children and Adolescents in the US. *Int. J. Behav. Nutr. Phys. Act.* **2004**, *17*, 1–17. [CrossRef]
59. Chandon, P.; Wansink, B. Does food marketing need to make us fat? A review and solutions. *Nutr. Rev.* **2012**, *70*, 571–593. [CrossRef] [PubMed]
60. Kelly, B.; Hattersley, L.; King, L.; Flood, V. Persuasive food marketing to children: Use of cartoons and competitions in Australian commercial television advertisements. *Health Promot. Int.* **2008**, *23*, 337–344. [CrossRef] [PubMed]
61. Grunert, K.G. Sustainability in the Food Sector: A Consumer Behaviour Perspective. *Int. J. Food Syst. Dyn.* **2011**, *2*, 207–218. [CrossRef]
62. Barosh, L.; Friel, S.; Engelhardt, K.; Chan, L. The cost of a healthy and sustainable diet - Who can afford it? *Aust. N. Z. J. Public Health* **2014**, *38*, 7–12. [CrossRef] [PubMed]
63. Furst, T.; Connors, M.; Bisogni, C.A.; Sobal, J.; Falk, L.W. Food choice: A conceptual model of the process. *Appetite* **1996**, *26*, 247–265. [CrossRef] [PubMed]
64. Satterthwaite, D.; McGranahan, G.; Tacoli, C. Urbanization and its implications for food and farming. *Philos. Trans. R. Soc. B Biol. Sci.* **2010**, *365*, 2809–2820. [CrossRef] [PubMed]
65. Global Panel. *Urban Diets and Nutrition: Trends, Challenges and Opportunities for Policy Action;* Policy Brief No.9; Global Panel on Agriculture and Food Systems for Nutrition: London, UK, 2017.
66. Regmi, A.; Dyck, J. Effects of Urbanization on Global Food Demand. Changing Structure of Global Food Consumption and Trade. 2001. Available online: http://citeseerx.ist.psu.edu/viewdoc/download?doi=10.1.1.738.5776&rep=rep1&type=pdf (accessed on 12 August 2018).
67. Conner, M.; Mceachan, R. Moderating Effect of Socioeconomic Status on the Relationship between Health Cognitions and Behaviors. *Ann. Behav. Med.* **2013**, *46*, 19–30. [CrossRef] [PubMed]
68. Vasiljevic, M.; Ng, Y.L.; Griffin, S.J.; Sutton, S.; Marteau, T.M. Is the intention-behaviour gap greater amongst the more deprived? A meta-analysis of five studies on physical activity, diet, and medication adherence in smoking cessation. *Br. J. Health Psychol.* **2016**, *21*, 11–30. [CrossRef] [PubMed]
69. Oncini, F.; Guetto, R. Cultural capital and gender differences in health behaviours: A study on eating, smoking and drinking patterns. *Health Soc. Rev.* **2017**, *27*, 15–30. [CrossRef]

70. Sikora, J. *Handbuch der Kreativ-Methoden*, 2nd ed.; Katholisch Soziales Institut der Erzdiözese Köln: Cologne, Germany, 2001; ISBN 978-3927566255.
71. Prekel, T.; Sobey, L.-M. Excursions and Metaphors—Getting crazy ideas, and turning them into brilliantsolutions. In *Complex Creativity, The Pathway to Opportunity Finding*; Rajah, K., Ed.; Greenwich University Press: London, UK, 2007; ISBN 978-1861662323.
72. Ajzen, I. Theory of Planned Behavior. Available online: http://people.umass.edu/aizen/tpb.html (accessed on 9 November 2018).
73. Heye, D. Creativity and innovation: Two key characteristics of the successful 21st century information professional. *Bus. Inf. Rev.* **2006**, *23*, 252–257. [CrossRef]

sustainability

MDPI

Article

Participatory Guarantee Systems in Spain: Motivations, Achievements, Challenges and Opportunities for Improvement Based on Three Case Studies

Marta López Cifuentes [1], Christian Reinhard Vogl [1,*] and Mamen Cuéllar Padilla [2]

1 Department of Sustainable Agricultural Systems, University of Natural Resources and Life Sciences Vienna, 1180 Vienna, Austria; m.lopezcifuentes@boku.ac.at
2 Department of Social Sciences and Humanities, University of Córdoba, Plaza del Cardenal Salazar, 14080 Córdoba, Spain; mcuellar@uco.es
* Correspondence: christian.vogl@boku.ac.at; Tel.: +43-1-47654-93312

Received: 18 September 2018; Accepted: 31 October 2018; Published: 7 November 2018

Abstract: The increasing number of producers and consumers of organic products means that there is an increasing need to guarantee the organic characteristics of organic products. Certification is a tool that bridges the information deficit between demand and supply, ensuring that a product complies with the specified standards. Third-party certification (TPC) is the main tool for assessing compliance today. However, there have been criticisms about the suitability of TPC for small-scale producers and alternative certification systems have been developed, such as Participatory Guarantee Systems (PGS). PGS are quality assurance systems in which stakeholders are expected to be involved and assure the quality claims being made by producers. This paper presents three PGS initiatives in Spain. The research methods used in this study were semi-structured and structured interviews. Interviewees felt that their PGS initiatives fulfilled the important motivations of building a community and adding value to their products. The main challenges mentioned were the participation of stakeholders and the efficiency of internal organization. The absence of official recognition for PGS in Spain and insufficient dissemination were also perceived as challenges. Although PGS has the potential for further development in Spain, the interviewees believed that more support and official recognition were still required.

Keywords: certification; organic agriculture; Participatory Guarantee Systems; Spain; alternative certification systems

1. Introduction

In Europe in recent decades, organic agriculture has evolved alongside the regulations that are applicable to it, as well as support programs and subsidies available to the organic sector [1]. Based on these regulations, a control and certification system for organic production is in place in Europe and in an equivalent way for organic imports to Europe as well. Third-party certification (TPC) is currently the main assessment system used to ensure the conformity or compliance of organic agriculture (and its products) with the regulatory framework.

Although required by law and recognized in the literature and in practice for its benefits, there has also been criticism of TPC [2–10]. Participatory Guarantee Systems (PGS) appeared as an alternative to TPC. PGS are expected to provide an organic guarantee, based on the participation of producers and other stakeholders, and are built on participation, trust, social networks and knowledge exchange.

Besides frequent claims made about the possible benefits of PGS, with a few exceptions, the debate on PGS lacks empirical data about their actual performance, benefits and challenges worldwide, including in Spain. The objective of this article is to provide information about three Spanish PGS initiatives in the regions of Valencia, Murcia and Granada in order to identify their members' motivations as well as their achievements, challenges and opportunities for improvement. This paper may serve as an empirical contribution to the global debate on the performance of PGS and the opportunities they present.

1.1. Organic Certification in Europe

In the 1930s and 1940s, farmers and farmers' groups developed the first organic agriculture associations in most industrial countries in response to the industrialization and intensification of agriculture [11,12]. However, it was not until the 1970s that associations of organic farmers started organizing themselves in order to develop their own private organic standards. At the end of the 1980s these associations lost their regulatory importance in certifying their members as governments started regulating organic agriculture themselves [12]. Certification was viewed as a necessity in a growing market to improve the efficiency of the market by creating a "common language" [13].

The first common organic standards in the EU were laid down in 1991 with the aim of protecting organic agriculture and consumers by ensuring fair competition between producers and transparency in the production system. This regulation established the basic guidelines for production, labeling and control [14]. Organic agriculture in the EU is currently regulated by Council Regulation (EC) No. 834/2007, Commission Regulations (EC) No. 889/2008 and No. 1235/2008 [15] and their respective amendments. Inspection and certification in the organic sector are performed by private or public certification bodies in order to ensure that the actors along the organic supply chain follow the European organic agriculture regulations [16]. Certification is a process through which written assurance confirms that a product, process or service conforms to specified requirements [17]. This quality sign depends then on the credibility of the institutions awarding them. And to be credible, quality signs should be given by an external body with no interest in the sales of the products [18]. In the case of organic agriculture, certification is used to guarantee marketing claims for organic quality attributes [4]. TPC is now the main assessment system [13]. According to Deaton (2004) [19], (p. 615) third-party certifiers are "external institutions that assess, evaluate and certify quality claims". TPC bodies are independent and therefore are perceived as objective and transparent [2,13,19]. Furthermore, TPC bodies have to prove their competence, including their capability to carry out inspections, through accreditation based on ISO/IEC 17065:2012 [15].

The importance of TPC is increasing in international quality food markets due to consumers' need for clear and reliable indications of food quality and safety (Anders et al., 2010). Certification of compliance with specific standards has the potential to communicate product claims beyond national boundaries and improve access to markets [20]. TPC reduces the risk of fraud and thus increases confidence in regulatory compliance [2]. It also appears to help producers implement sustainable improvements efficiently in their production practices and legitimizes organic agriculture in the eyes of European consumers [21,22]. Furthermore, TPC's compliance with established EU standards gives producers access to the organic market and economic support or subsidies from member states [23].

Nevertheless, TPC has been criticized for being expensive for producers, which results in consumers paying more for their products, for not being adapted to local circumstances, being hard for small producers to achieve, making agriculture more bureaucratic, simplifying production processes, and for not always being clear about the procedures the TPC bodies are implementing [2–5,7,9]. In addition, TPC does not distinguish between different kinds of organic producers. Thus, small and large producers have to fulfil the same standards in order to certify their products under TPC [10].

In the debate on TPC, it has also been claimed that it is important to develop organic agriculture and its standards in a respectful and coherent way adapted to traditional ecosystems, linked to cultural diversity and including farmers' technological knowledge, aiming at participation, respect and democracy [8,12]. In this context, alternative certification systems that allow producers to create their own locally-based definition of organic agriculture have been developed [23]. Smallholders in the Global South, for example, have developed producer groups with internal control systems or Participatory Guarantee Systems (PGS) [14]. PGS are thought to have first appeared in the 1970s, when groups of organic farmers started certifying themselves using first-party certification, and have reappeared recently in the 21st century as an alternative system following criticisms of TPC [24].

1.2. Participatory Guarantee Systems

The term Participatory Guarantee System was first used on a global level in 2004. PGS initiatives were presented and analyzed during a workshop run by the Latin American Organic Agriculture Movement (MAELA), International Federation of Organic Agriculture Movements (IFOAM) and the Centro Ecológico in Torres, Rio Grande do Sul (Brazil) [25]. PGS were defined as "quality assurance initiatives that are locally relevant, emphasize the participation of stakeholders, including producers and consumers, and operate outside the frame of TPC". The basic elements of PGS, according to IFOAM, are participation, a shared vision, transparency, trust, being a learning process, and horizontality [26].

PGS are expected to be culturally appropriate, require less paperwork than TPC, and are linked to local and alternative commercialization channels (Alternative commercialization channels are considered to be those based on the reconnection and close communication between producer and consumer, building new forms of relationships and governance of the network of actors [27]) [8,26,27]. PGS appear to help producers improve production, achieve political independence, boost livelihoods and establish social networks, and are thought to be appropriate for small producers due to their participatory and horizontal structure allowing a more suitable and less costly system of certification [8]. A PGS initiative should reflect a community's capacity to prove trust through the implementation of diverse social and cultural control instruments in order to provide information to guarantee the integrity of their organic producers [25].

However, PGS are not exempt from mistrust. There is potential for dishonesty, which might raise some doubt concerning PGS [23]. The possible opportunistic behavior and the commitment of members are two of the most common challenges faced by PGS initiatives [28]. The participation of members, especially consumers, is a barrier being faced by PGS initiatives in various countries [23,25,29,30]. This dependence on voluntary work seems to be a significant obstacle to the development of PGS [31].

Despite the obstacles being faced by PGS, there are PGS initiatives operating in Latin America, USA, India, New Zealand, South and East Africa and Europe [26,32]. Even though local conditions and cultural contexts differ, all PGS initiatives appear to share the abovementioned basic elements and principles and also to be similarly organized [33]. In countries such as Brazil and Mexico, PGS work in parallel with organic legislation and TPC [23,34,35], whereas in EU member states producers without TPC are not allowed to sell their products as organic or receive subsidies from organic funds [29], with the exception of Sweden, where the government gives economic support to non-certified organic producers [36]. According to Padel (2010) [14] (p. 70) there are no objective reasons why PGS should be limited to smallholder producers in the Global South. PGS initiatives also have the potential to support organic production based on agroecological ideas and to create fair markets in EU countries [30].

1.3. Participatory Guarantee Systems in the Context of Spain

The first pioneers of organic agriculture in Spain appeared in the 1950s. Nonetheless, it was not until the 1980s that farmers established the first producers' organizations for organic agriculture in the country. In 1988, the name "Organic Agriculture" became official in Spain through Royal Decree 759/1988 and the Regulatory Council of Organic Agriculture (Consejo Regulador de la Agricultura Ecológica, CRAE) was established [6]. Since then, there has been continuous development in this sector, both in terms of the number of producers and area of cultivated land. The area dedicated to organic agriculture in Spain has increased considerably in recent decades, from 4235 ha in 1991 to

1,610,130 ha in 2015 [37]. Nevertheless, the domestic organic market in Spain has not developed in line with its existing productive potential. Around half of the national organic production is destined for export [38].

Cuéllar Padilla (2008) [6] argues that TPC in Spanish organic agriculture could result in the loss of associations and networks that were around at the start of the organic movement in Spain. Moreover, TPC seems to be losing its connection with local conditions in Spain and not taking the socio-political aspects of agriculture into account [23,39]. As a result, there has been a concentration of farms in this sector: in Spain larger farms are certified with TPC, while small and medium-sized farms (farms with less than 10 ha) have been ignored [40,41].

Given the lack of empirical data on PGS, this article attempts to answer the following research questions:

- What are the internal challenges and benefits identified by members of these three PGS initiatives?
- What are the motivations and underlying beliefs about PGS held by stakeholders in these three PGS initiatives?

2. Materials and Methods

2.1. The PGS Initiatives Studied

The PGS initiatives studied are located in the Spanish regions of Valencia (PGS Ecollaures), Murcia (PGS Vecinos Campesinos) and the Valley of Lecrín, Granada (PGS Ecovalle) (Figure 1). They were chosen for their proximity and different characteristics. Based on the authors' own empirical data, they can be characterized as follows:

Figure 1. Map of Spain with its different autonomous communities (grey dotted lines). Drops: well-established Participatory Guarantee Systems (PGS) initiatives in Spain in the year of research (2016); in green: the three PGS initiatives studied: Ecollaures, Vecinos Campesinos and Ecovalle (own creation, Paint, Microsoft Office package, 2010).

Ecollaures was founded by producers in 2010 with the assistance of a student at the University of Valencia (first for his Master's thesis and then for his PhD thesis) to help producers connect with consumers and subsequently improve their participation and communication skills during meetings.

Vecinos Campesinos was founded in 2011 by producers from the region of Murcia who were not interested in TPC and decided to create a PGS initiative. After some meetings and based on a

Master's thesis and the example of Ecovalle's internal regulation, these producers created the PGS Vecinos Campesinos with the involvement of a consumers' association.

Ecovalle started in 2008 by a group of producers in this rural environment who shared an interest in organic and family agriculture. These producers ran workshops, specialist training courses and other activities related to organic agriculture [42,43]. In 2010 they started working on the development of their PGS initiative with the help of a PhD student from the University of Córdoba. Ecovalle is both a production cooperative and a PGS initiative. In 2016 Ecovalle merged with the "vergel de la vega" cooperative. Following the merger, the cooperative changed its name to "Valle y Vega", but the PGS initiative maintained the name of Ecovalle and is independent from the cooperative.

2.2. *Data Collection and Analysis*

In 2016, research began with exploratory fieldwork based on semi-structured interviews with five regional key actors [44], followed by a questionnaire based on the design of Kaufmann [45]. Field research was supplemented by participant observation, i.e., participation in farm visits, meetings, fieldwork and social gatherings, a field diary, pictures and the collection of documents [46,47].

As a next step, within each initiative, snowball and purposive sampling was used [46] to ensure the variety of stakeholders (consumer groups, Non-governmental organizations (NGOs), shops) and areas involved were covered. In total 29 structured interviews were conducted (22 producers, three members of consumer groups, three members of NGOs and one owner of a shop). All semi-structured and structured interviews were recorded using a digital voice recorder (SONY ICD PX333, Sony Corporation, Tokyo, Japan).

The structured interview included pre-coded questions, but also several open questions as well. The answers to these open questions were analyzed as qualitative data. The qualitative data were given numerical values and analyzed quantitatively in just two cases. This was for the use of the word "agroecology" and for statements about the members' political motivations. For the latter, topics related to politics (e.g., food sovereignty, demonstrations, political parties) or the direct use of the words "political" or "politics" were counted as political motivation (Table A1).).

For two questions, interviewees had to rank the importance of a list of proposed statements (about the current status of their PGS initiatives and their motivations for joining PGS) (Table A1)).

All raw quantitative data were recorded in an Excel table (Microsoft Office package, 2010) and analyzed using SPSS (version 21) for Windows (IBM SPSS 2012). The analysis used was descriptive analysis and cross-tabulation. The significance of the association between PGS initiatives and the selected variables was tested using Fisher's exact test, which was carried out at a significance level of 5%. To analyze qualitative information, the recorded interviews were transcribed using Listen N Write (Listen N Write 2016, Elefant Software, Italy) and coded with Atlas.ti (Atlas.ti 2012, Scientific Software Development GmbH, Berlin, Germany). Descriptive and values coding were used to create the codes. A conceptual order display was then created with the qualitative data [46,48,49].

2.3. *"Producer" vs. "Farmer"*

In the literature, the words "farmer" and "producer" often appear in the same text without a clear definition of or differentiation between the terms. Nonetheless, it is considered important to make a distinction between the two terms for this paper. The Oxford dictionary [50] defines producer as "a person, company, or country that makes, grows, or supplies goods or commodities for sale", while a farmer is defined as "a person who owns or manages a farm". According to this definition, while a farmer does not have to be personally involved in the production process (another person might work the land he or she owns), a producer has to "make, grow or supply goods". However, agriculture is implicit in the word "farmer", but "producer" can also be used to talk about non-agricultural products.

Interviewees mainly used the term producer and "producer" rather than "farmer" is always used in the documents collected from the PGS initiatives studied. Moreover, interviewees never talked about farms but about projects. One reason for this might be that not all producers work on a farm, but might produce other food products such as honey (beekeeping) or bread. There are probably several other reasons for the preferred use of the term "producer" over "farmer", but they are not relevant for the purposes of this article.

The term "producer" has been used in this article because the interviewees themselves used it. "Producer" here means people who produce food goods that come either directly from agriculture or from the processing of agricultural products. The term "farmer" is only used in the literature review where it was actually used in the literature. "Farmer" refers to those working in agriculture, whether or not they own the farm.

3. Results

3.1. Main Characteristics of Members of the Three PGS Initiatives

The arithmetic mean age of the interviewees (n = 29) was 40.3, with ages ranging between 27 and 60 years old. A total of 72.7% of the interviewees had a university degree and in the case of Ecovalle none of the members came from Granada (Table A1).

A total of 85.7% of all the members of the three PGS initiatives were producers. Other stakeholders involved in the studied PGS initiatives were NGOs, food cooperatives (FoodCoop), a consumer association and shops. The various stakeholders had their own particular characteristics and functions in the three PGS initiatives, but all stakeholders participated equally in running the PGS initiative (Table 1).

Table 1. Stakeholders involved in the three PGS initiatives studied, their numbers and functions (based on authors' own quantitative and qualitative data).

Stakeholders	PGS Initiatives			Function Based on the Perception of Interviewees; n = 29
	Ecollaures	Vecinos Campesinos	Ecovalle	
Producers *	25	29	6	Production
NGOs	3	0	0	Support (technical, economical ...) Dissemination of PGS Supporting access to public bodies Outside (external) point of view Political lobbying
FoodCoops **	2	0	0	Communication about PGS (networking) Providing the consumers' point of view
Consumer association	0	1	0	Distribution and consumption of produce Voluntary work
Shops	0	4	0	Communication (information about PGS in the shops) Commercialisation of produce Voluntary work
Total	30	34	6	

* The number of producers corresponds to the so-called 'units of production', i.e., more than one person might be involved in a unit of production, but in this study this unit will be counted as one producer. ** A FoodCoop (food cooperative) is an association of consumers that is self-organized and purchases organic and local products that have been produced under fair work conditions. FoodCoops operate a democratic decision-making system based on the participation of members through the distribution of labor and responsibilities. The members seek a direct relationship with producers and thus avoid intermediaries [51,52].

A total of 43% of the interviewed producers had TPC as well as PGS, and their areas under cultivation varied between 3000 m^2 and 13 ha, with an arithmetic mean across the three PGS initiatives of 2.25 ha (Table A1). Of the 22 producers interviewed, five were able to live on agriculture alone. The other 17 producers had another income alongside agriculture, either from their own personal additional income or additional income from their partner.

Agriculture represented 54% of the producers' income (arithmetic mean across PGS initiatives) (Table A1). Producers from Ecollaures had the highest percentage of income from agriculture compared to the other two PGS initiatives (73.5% arithmetic mean, Fisher's exact, n = 22, $p < 0.05$) (Figure 2) (Table A1).

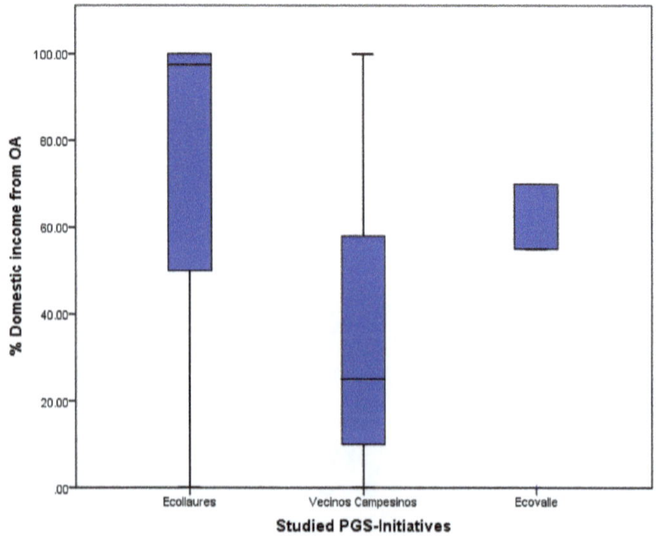

Figure 2. Percentage of domestic income coming from organic agriculture (OA) per PGS initiative studied. The 100% = total family and farm income as perceived by the interviewees. (Ecollaures: $n = 10$; Vecinos Campesinos: $n = 9$; Ecovalle: $n = 5$). Significant difference between PGS initiatives studied (Fisher's exact, $n = 24$, $p < 0.05$) (own data, SPSS (version 21) for Windows).

3.2. Motivations for Joining a PGS Initiative

There was no relationship between the different PGS initiatives studied and the degree of importance attributed by the interviewees to the different statements about why they decided to join a PGS initiative (Fisher's exact, $n = 22$, $p > 0.05$). Interviewees considered PGS to be a tool for social change and a (very) important reason for joining the PGS initiative (Table A1). More than 50% of the interviewed producers did not consider as important the fact that implementing a PGS is cheaper than implementing a TPC (Table A1). Moreover, the possibility of a higher income with the PGS was not a reason for producers to join the PGS initiatives. However, the establishment of a community of producers, consumers and other stakeholders and mutual support were very important to the producers (Table A1) because they shared similar values and ideology. In addition to the proposed statements, interviewees also mentioned other motivations for joining their PGS initiatives such as marketing advantages and less bureaucracy than TPC.

Disagreement with the current certification system was a significant factor in joining/creating their PGS initiatives mainly for producers. Some interviewees perceived TPC to be easier for larger producers with monocultures, and that "it does not meet the needs of small producers". Therefore, the interviewees wanted to differentiate themselves from TPC and "other kinds of organic producers". One member said they did not want to certify products but "certify or guarantee processes, projects and collectives, and people".

Interviewees perceived PGS to be more interactive, real and fair than TPC. The socioeconomic criteria included in the internal regulation of the PGS initiatives studied were an important motivation for the interviewees, especially the fact that the three PGS initiatives studied consider employees' working conditions, support the viability of the projects and include the biodiversity of the fields. Interviewees explained that their PGS initiative was better adapted to their reality and that "PGS go further and give the organic products extra value; PGS show the real work behind organic production".

Other stakeholders of the PGS initiatives (i.e., consumers' groups, shops and NGOs) perceive as main drivers for joining or creating a PGS initiative the support of local producers and the direct relation to them, the promotion of local products and the fact that PGS is a tool for social change. Other important motivations were related to the production methods used by the producers (organic, environmental care) and the quality of the products (variety, healthiness). The implication of NGOs in the PGS initiatives seems especially important to them in order to get to know the reality of producers and consumers in the organic scene, so that they can organize projects or campaigns according to the realities and needs of producers and consumers. Moreover, being part of PGS helps NGOs building bridges between the civil society and the agrarian sector. On the other side, consumers and shops are part of the decision making of the PGS and this way they are able to participate in the decision process linked to their food.

Consumers who join the studied PGS look for organic, local and healthy foods. In the cases of Ecollaures and Vecinos Campesinos, consumers' groups and shops participate in decision making and even were part of the creation of the PGS initiatives. This involvement allows them to adapt the PGS also to their needs, e.g., discuss prices and crops with producers, visit the fields, agree on production method and in general being informed about the food they consume.

All stakeholders in the PGS initiatives discuss together in assemblies and field visits all possible matters related, not only to their internal organization, but also to more general issues such as values, social implications (working conditions), environmental impacts of their production methods or retail, etc.

Some interviewees explicitly expressed a political motivation. The statistical analysis suggested that there was a significant relationship between the PGS initiatives studied and the frequency of terms used that could be attributed to the explicit political motivation of their members (Fisher's exact, $n = 29$, $p < 0.05$). Whereas members of Ecovalle and Vecinos Campesinos did not mention any political motivation during the interviews, 40% of the members of Ecollaures interviewed explicitly perceived PGS to be a political tool (Table A1). While participating in different activities run by Ecollaures and Ecovalle, the members of these PGS initiatives were found to participate or have participated in political activities such as demonstrations, the 11-M movement, talks, events from left-wing parties, etc. Topics such as food sovereignty or farmers' rights came up several times during the research.

The relationship between the use of the word "agroecology" (the number of interviewees who used the term at least once) and the political motivation shown by the interviewees was analyzed. The results indicated a strong and significant relationship between the use of "agroecology" and political motivation (Fisher's exact, $n = 31$, $p < 0.05$). Thus, interviewees who used the word "agroecology" were likely to be politically motivated. There was also a significant relationship between the PGS initiatives studied and the frequency of the use of the word "agroecology" (Fisher's exact, $n = 31$, $p < 0.05$). A total of 80% of the interviewees from Ecollaures (n = 100% = 16) and 60% from Ecovalle (n = 100% = 5) used the word "agroecology" (or derivatives of it) during the interviews, while none of the members of Vecinos Campesinos interviewed (n = 100% = 11) used this word (Table A1). In the words of the interviewees from Ecollaures and Ecovalle, "agroecology is not the same as organic agriculture", "its definition is so broad and deep (. . .) that we have to break it down into its true meaning so the consumer is more aware of it", "agroecology includes social factors" and "it's organic agriculture understood in a broad sense".

3.3. *Achievements of the Three PGS Initiatives Studied*

The interviewees claimed that despite the initial motivations of the members of the PGS initiatives studied, not all of them might be fulfilled. Interviewees perceived the community, network and exchange within the PGS initiative as the most important achievements of their PGS initiatives (Table A1). Interviewees felt supported by other members and that they learned from one other, not only about agricultural techniques but also social skills. Interviewees were satisfied with the group formed in their PGS initiatives, despite the internal issues they might face.

Another achievement perceived by interviewees was the extra (non-monetary) value PGS conferred on their products. These values were established by the members of each PGS initiative and are summarized in their internal regulation. Almost 80% of the interviewees were (very) satisfied with their internal regulation (Table A1), but they considered their own PGS initiative to be an active system that is continuously developing alongside the internal regulation. They mentioned that the internal regulation could be improved and some parts of it should be revised as the first drafts were produced without any experience. Interviewees considered that producers in the PGS differentiate themselves from organic producers with TPC, and that their PGS initiative represents their kind of production, as they expected.

Some other expectations met by the PGS initiatives were product quality, transparency, control, education and personal satisfaction. Transparency and control were achieved through the farm inspections and daily interaction between members. Some members perceived the farm visit to be an important part of their PGS initiative because it is a tool for checking that the others are producing according to the principles established by their PGS initiatives. One member of Ecovalle considered day-to-day life with the other members as fundamental for the trust built within the PGS initiative and assurance of compliance.

The implication of consumers in PGS initiatives increased their interest on eating healthier and organic. For example, one shop (member of a PGS initiative) mentioned that they try to create awareness among consumers about the benefits of organic products for the environment and their own health. They included in their stock different kind of vegetables, not so common among the typical Spanish consumer, and explain to them the benefits of these products and even how to cook them. This was done by other producers and shops too.

Interviewees from Ecollaures and Ecovalle also perceived the commercial opportunities and political battles as achievements. Some markets in Valencia and Granada have started to accept PGS as a guarantee of organic products, with producers allowed to sell them as organic. Interviewees from these PGS initiatives perceived ongoing improvement in the acceptance of PGS in local markets. In the case of Ecollaures, some interviewed members perceived this success in part to be a result of their political work. They also acknowledged the important work of NGOs in fighting for farmers' rights, the widespread use of the word "agroecology" and acceptance of PGS as a valid guarantee.

3.4. Challenges Perceived by PGS Members

Although producers with PGS might be able to sell their products in some Spanish markets, PGS is still not formally accepted as an organic certification and producers are not officially allowed to sell their products as organic. Interviewees were asked about the importance of future legislative recognition of PGS in Spain. Here, interviewees had to evaluate its importance from none to very high (Figure 3). However, before answering the question, several interviewees considered it important to re-define "recognition". Interviewees differentiated between three kinds of recognition:

- legislative recognition: refers to the question asked as to whether PGS are considered a legal certification system as TPC bodies;
- institutional recognition: the acceptance of PGS by public institutions, so that PGS initiatives have a political voice and are allowed to sell their products as organic (at least on a local scale), but without being an official certification body;
- social recognition: consumer acceptance of PGS as a valid guarantee system that consumers can trust.

Interviewees evaluated legislative recognition and some of them pointed out the importance of the other two types of recognition. Interviewees who considered the importance of the legislative recognition of PGS to be (very) low (Table A1) perceived social recognition to be more important than legal recognition. They considered that social recognition could be enough to help producers with commercialization. Other interviewees perceived institutional recognition to be important so that PGS can establish communication on a political level. There were also those who did not consider any of these types of recognition as important and perceived the current independence of PGS as an advantage (Figure 3).

Some interviewed producers mentioned that they were in their PGS initiatives because they preferred group certification. Nevertheless, they had been forced to have TPC because some shops or consumers have demanded TPC in order to reassure consumers that they are buying organic products. These producers stated that as soon as PGS was recognized as a valid organic guarantee, they would no longer certify their products under TPC. These producers also admitted that the high amount of voluntary work and time compared to TPC were major obstacles to adopting PGS.

One interviewee mentioned the difficulty in finding a balance between family duties and the PGS initiative, as the work required for PGS involved a considerable investment of time. In general, time constraints were perceived to be a problem and an obstacle to members' participation in their PGS initiatives. Although 76.6% of interviewees (n = 100% = 29) considered producers' participation to be normal or high (Table A1), they mentioned the participation of all members (not only consumers) as one of the main challenges. The interviewees considered the low participation of their PGS members to be a problem that might eventually lead to just a few members doing all the work (or to "burn out"). Consumer participation was reported as being especially low. This was perceived as a challenge that the interviewees did not know how to address. The interviewees proposed some reasons for low consumer participation such as disinterest, other priorities, lack of need or consumers not finding participation appealing.

"Organization" was mentioned several times as a challenge related to participation. According to the interviewees, "organization" was understood to be the effectiveness of committees and working groups (e.g., in completing tasks), the planning of meetings and farm inspections (e.g., establishing and implementing a calendar of farm inspections), the compilation and formulation of documents, preparation of events and communication of achievements. The perceived low level of participation was thought to influence the organization of PGS initiatives (Table A1). Interviewees perceived that the low participation of members in their PGS initiatives made organizational matters more difficult.

Based on what the interviewees said, efficiency was mainly defined by the performance of meetings (duration, communication between members, capacity for decision-making, participation of members, ability to focus on topics relevant to the PGS initiatives), but also by the quality of the farm inspections and the work undertaken during the year (communication, workshops, publicity, document management, involvement of members etc.). Interviewees perceived the progress of their PGS initiative to be slow. Moreover, interviewees felt that they did not know how to work in a participatory manner, making it difficult to take decisions in a participatory and horizontal way. As communication tools for sharing information and the decisions made, the initiatives and their members mainly used emails and WhatsApp, which were not always perceived as the best methods.

Some interviewees perceived the communication of their PGS initiatives as a challenge that still needed to be addressed (Table A1). Interviewees found publicity of their PGS in the media too low and wanted this to change. Ecollaures was still working on its website, whereas Ecovalle and Vecinos Campesinos had a WordPress site. Interviewees believed publicity could raise awareness among the Spanish population. The lack of professional communication was mentioned as one reason for consumers not participating in PGS initiatives.

Other challenges mentioned were personal issues between the members, the economic viability of the projects and product prices (Table 2). One interviewee mentioned that it is sometimes "difficult to ensure that technical problems do not affect personal relationships". The dispersal of members was perceived to be a problem by the members of Ecollaures and Vecinos Campesinos. Most of the members of Ecollaures lived near Valencia, but some were in villages a long way from the city, even in the southern region of Alicante. The problem was more pronounced in the case of Vecinos Campesinos as all its members were dispersed in the regions of Murcia and Alicante. The maximum distance between members of Vecinos Campesinos was approximately 140 km. However, the members of Ecovalle lived within walking distance of one another in the same village.

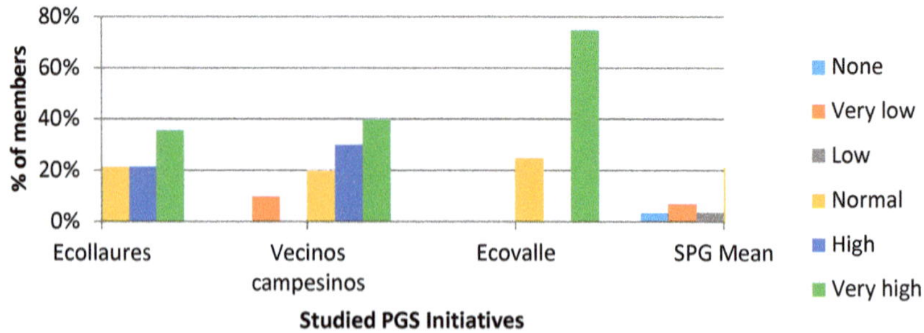

Figure 3. Perception of the interviewed members of the three PGS initiatives about the level of importance of possible future legislative recognition of PGS in Spain (Ecollaures: n = 100% = 14; Vecinos Campesinos: n = 100% = 10; Ecovalle: n = 100% = 4). No significant difference was found between the PGS initiatives studied (Fisher's exact, $n = 29$, $p > 0.05$) (own data, Excel, Microsoft Office package, 2010).

Table 2. Challenges perceived by the interviewees in the three PGS initiatives studied (own creation from qualitative data).

Challenges
Consumer involvement
Economic viability
Efficiency
Legality/official recognition
Dispersal of members
Participation
Personal matters
Prices

3.5. Opportunities for Improvement

Interviewees were aware of the challenges faced by their PGS initiatives and wanted to overcome them in order to improve the initiatives. Several interviewees perceived a need to rely on a hired person who would be in charge of running their PGS initiative (Table 3). They said that producers had too much work and no time to run it. These interviewees thought that an employee would help make progress with their PGS initiative as this person would have more time and the skills needed for the tasks. The PGS initiatives studied were based on voluntary work, but "sometimes it is too much (...) and a level of specialization might be good". However, interviewees were also concerned that hiring somebody might increase the annual fee for the PGS initiatives, result in members participating less in other activities, and that the PGS might require external support (producer associations, NGOs, universities, etc.).

As a way of improving the efficiency of the PGS initiatives studied, members felt that meetings should be more efficient (Table 3). Interviewees from Ecollaures found a workshop on communication skills in participatory-horizontal systems to be particularly helpful and would like to have more events like this. The workshop was about facilitation tools for participative meetings and they have noticed a great improvement in their meetings since the workshop.

Topics discussed during the assemblies were the development and improvement of the internal regulation and farm inspection checklist (Farm inspection checklists are also called "visit guides" locally). A total of 64% of the interviewees were (very) satisfied with their farm inspection checklist (Table A1), but they believed that the farm inspection checklist could still be improved. The general guarantee process (farm inspection checklist plus farm inspection) was also perceived as (very) good (69% of interviewees) (Table A1). In the case of Ecollaures they would like to have different checklists for different products and actors, but they are still working on this (Table 3). A total of 75.6% were (very) satisfied with the internal regulation (Table A1). However, the guarantee process was also perceived to be not good or serious enough by some interviewees (Table A1). Those interviewees perceived that some members did not take the farm inspection seriously and attributed more importance to the social part of it than checking the farm inspection checklist and monitoring that the producer was doing everything right because these members knew the producer and might take it for granted that he/she was working to their standards.

Alongside the potential internal improvements, some producers interviewed were concerned about their production costs (for inputs, machines etc.) and would like to share some costs with other producers in their PGS initiative (Table 3). In the case of Vecinos Campesinos, some interviewed producers perceived production costs to be an important challenge and would like to form an agricultural cooperative and have a distributor selling the products produced by their PGS initiative. They mentioned a wish to have a distributor for more PGS initiatives so that producers from different PGS initiatives could sell their products together. Although this idea was not shared by many interviewees, they all considered it essential to build a PGS network, not for commercialization purposes, but to make the movement stronger in the country (Table 3). PGS in Spain have already started to contact one another. In November 2015 and June 2016 there were two national meetings at which various stakeholders in PGS in Spain met in order to get to know each other and discuss common issues.

Table 3. Aspects perceived by the interviewees that would improve the PGS initiatives studied and suggested tools for improvement (own qualitative data).

Requiring Improvement	Suggestion
Meetings	Respect, good moderator, facilitation workshop
Consumer involvement	Guide for consumers, dissemination, establishing a physical headquarters, fun farm visits
Economic viability/costs *	Collective purchase of inputs, production of seedlings in the PGS initiatives, exchange of products
Publicity	Person in charge of publicity, collective communication work, awareness campaign
Organization	Hired person, farm inspection calendar **
Participation	Monitoring involvement, establishment of minimums
Quality	Different guides for different products
Equality	Different guides for different actors
Relationship with other PGS initiatives	Creation of a PGS network in Spain
Commercialization **	Collective commercialization (cooperative)
Rapid growth ***	Duplication of the PGS initiative, stop admission of new members for a while

* Only producers perceived this to be an area requiring improvement and offered suggestions; ** Only mentioned by members of Vecinos Campesinos; *** Only mentioned by members of Ecollaures.

4. Discussion

4.1. Motivations and Achievements

In other case studies of Latin America, Africa, Asia and Europe most PGS initiatives were initiated by or through the help of other stakeholders, such as NGOs or farmers organizations [32,53]. e.g., in the case of a PGS initiative in Chapingo, Mexico, the initiative was created with support from university and it was gradually taken over by producers [45]. There are also PGS initiatives that were established with the help of a Market (e.g., in South Africa [53]); a network of Markets, such as the Mexican network of tianguis and organic markets; or, as it happens in the Philippines, with the help of an association of sustainable agriculture. These PGS initiatives were mainly established to support small-holder local producers by offering them affordable alternatives to third party certification [53].

In these cases where PGS initiatives were not started by producers, organizers might have had to convince potential member producers of the advantages of PGS [30]. In our case, on the other hand, the studied PGS initiatives were initiated by already-motivated farmers themselves who later on looked for the support of other stakeholders. The high level of education compared to other PGS initiatives might explain why their PGS initiatives were not initiated by another organization. This might also influence their motivations for joining or funding a PGS initiative.

The subject of lower costs of PGS compared to TPC usually appears alongside less paperwork or bureaucracy as the main advantages of PGS in different studies about PGS worldwide [1,6,24–27,29,42,53,54]. Even though producers in the PGS initiatives studied agreed on the importance of low costs and less paperwork, these were not the main reasons for joining the PGS initiatives. This might be due to the members' high level of education and the fact that most producers had additional income alongside agriculture. In addition, almost half of the interviewed producers also had TPC and had to deal with the related paperwork anyway. Moreover, in contrast to other PGS initiatives, e.g., in Mexico, producers of the PGS initiatives studied did not share their costs or their sales (In the case of Ecovalle, interviewees did not consider PGS to be part of the cooperative. They only used the PGS initiative to guarantee their products and the cooperative was separate from it, thus reference is also only made here to the PGS initiative), which explains why the financial and job security motivations found in agricultural cooperatives were not identified in this study [55]. Previous studies about PGS in other countries (e.g., Mexico, Brazil, India, New Zealand, Peru, Philippines, etc. [23,29,34,53]) have not shown that producers who are members of a PGS also have TPC. One exception is Natur et Progrés in France [25]. In the present study producers were found to have both PGS and TPC. This was due to the lack of legislative recognition of PGS in the EU. Producers with only PGS are not allowed to sell their products as organic, except on rare occasions informally established by local authorities [14,56]. Thus, producers of the PGS initiatives studied were not motivated by economic or market access advantages, but by the community building and the extra value added to their products among others. Producers sought to differentiate themselves from other organic producers who did not share their values. PGS members share the values of sustainable agriculture that go beyond an input-substitution production and include a broader definition of organic [23,39]. According to a member of another PGS initiative in Spain: "PGS is not an administrative procedure, PGS is based on trust and takes into account social aspects such as respect for workers, food miles and how farm inputs travel from inputs' suppliers" [56].

This desire for differentiation was also at a political level, which had already been found in Ecovalle by De la Cruz (2015) [42]. In the Global South some politically motivated funding agencies started PGS initiatives because they considered PGS to be more democratic than TPC and decided to make a political statement by supporting PGS. Without previous studies of the local markets or their simultaneous development, the sole political motivation of the funding agencies has been criticized because the reason for starting the PGS initiatives might not have been appropriate for local producers [30]. However, in the present cases, producers themselves instead of external funding agencies were politically motivated and considered PGS to be a tool for social change.

Since 2011, there has been growing discontent with national and European politics and with the global economic system among young people in Spain. Due to the economic crisis that began in 2008, young (mainly high-educated) Spaniards formed the 15-M movement in 2011 to express their disagreement with national politics and the global economic system [57,58]. Members of the PGS initiatives studied here sympathized with the movement and it might have influenced their desire to use PGS as a political tool and differentiate themselves from the certification system dictated by the EU. The PGS initiatives studied are to be seen as a protest movement against modern industrialized agriculture [59].

As a symbol of their protest, Ecollaures and Ecovalle use the word "agroecology" instead of "organic". Although the EU regulation does not directly forbid use of the term "agroecology", it does protect the words "organic", "ecological" and "biological" and their derivatives [14]. As agroecology is actually a derivative of ecological, its use is not allowed in the EU for non-certified organic producers. The members of the PGS initiatives studied would like free use of the term agroecology to sell their products, as happens in many countries in the Global South. For example in Ecuador, producers with PGS are allowed to sell their products at local markets just by using the word agroecological instead of organic [56]. According to the interviewees, agroecology is a better definition of their products than organic. Although, there are many definitions of "agroecology" in the literature, these definitions agree on the common socio-political interpretation of the term agroecology. From the socio-political point of view, agroecology is defined as a tool for the defense, (re)configuration and transformation of rural areas and the key to revitalizing small farming systems and food sovereignty [60–63]. Rover et al. (2016) [35] (p. 11) also found that members of the Brazilian agroecology network, Rede Ecovida, were motivated by an "ideological engagement by transforming the model of rural development, based on agroecology and biodiversity". The PGS initiatives studied and Ecovida are not only interested in production, but in new ways of rural development that consider the reality of each ecosystem and territory [35].

Nonetheless, the fact that the members of the PGS initiatives studied wanted to go beyond TPC might also carry a potential risk that should be considered. In a PGS initiative in New Zealand, there was a trend of making PGS more complex, e.g., by increasing the amount of paperwork for farmers [29]. PGS should be an inclusive system and the desire to include stricter standards than TPC might sometimes be too ambitious, thus making PGS more inaccessible than TPC [23]. The PGS initiatives studied here were aware of this risk and reacted flexibly while including the aspects of agroecology that they consider important.

Apart from the desire to include other aspects of agroecology in PGS, one of the main motivations and achievements was the mutual support and community formed in the PGS initiatives studied. Producers felt supported and satisfied with their colleagues and this might be reflected in trust and guarantee the integrity of their members [25]. PGS are not merely certification systems, but rather communities that are seeking mutual support, local and fair commercialization, improvement in the local ecology and education for the local population [31]. Moreover, the horizontal network has the capability to efficiently transmit information and create incentives to behave in a trustworthy way that strengthens the community [64]. This horizontal and democratic system allows producers to elaborate and control their own standards, which empowers producers, one of the characteristics of PGS suggested by IFOAM [25,65].

The importance of community seems to be a characteristic shared by different alternative food networks that are based on collective and democratic action, such as Community Supported Agriculture (CSA), FoodCoop or networks of farmers' markets [26,66–69]. Nonetheless, in contrast to this study's results, the building of a community is not identified as one of the main motivations for joining a CSA, but as an intrinsic characteristic of CSA [66,70,71]. In their research about CSA in the USA, Brehm and Eisenhauer (2008) [66] (p. 110) found that "strong community attachment clearly has a positive influence on motivations for joining a CSA (...) but community is not always a strong motivation for joining a CSA". In contrast, the main motivations for joining a CSA are the quality of the products, while in the PGS initiatives studied these motivations were not identified [69–71]. One motivation shared by the PGS initiatives studied and CSA is the support of organic local producers and products [70]. While there are several studies about motivation in CSA, few studies address motivations in PGS. More research is needed to identify why producers, consumers and other stakeholders want to join or start a PGS initiative.

4.2. Challenges and Potential Improvements

The present study shows that the three PGS initiatives shared three of the most common challenges reported in the literature: involving consumers in the PGS initiative, gaining recognition and support from authorities, and relying on voluntary work. One or more of these challenges have been found in PGS initiatives in East Africa, Brazil, South Africa, Peru, Mexico, India, Philippines, France and Spain [6,25,31,32,40,42,53].

The results indicate that the low level of consumer involvement might be due to the lack of legislative recognition and limited dissemination of PGS in Spain. Even though many interviewed producers inform their consumers about PGS, there is a high level of ignorance about PGS among

Spanish consumers [6,32]. The lack of legislative recognition make consumers doubt the credibility of the organic products from PGS if they do not know either the producers or the PGS initiative [6]. Credibility is a difficult concept in organic agriculture, so consumers have to trust the producers' quality claims. There are many different quality signs which leads to more confusion rather than an indication for buyers. Besides, there are many situations in which quality claims are weakened due to producers breaking the law [18]. For these reasons, dialogue with consumers to sustain trust can be more important than specific information about traceability or production techniques [72]. In support of this statement, Pole and Gray (2012) [71] (p. 96) found that members of a CSA in New York considered the relationship with the farmer to be an essential aspect of CSA.

Nonetheless, in a broader environment than the PGS initiative itself, public policies in Spain, as in Brazil, could substantially foster PGS, improve its legitimacy and increase the involvement of the state and other public and/or private bodies [9,68]. Cuéllar (2008) [6] argues that a certain degree of legislative recognition of PGS in Spain is necessary to improve the situation of PGS in the country and increase the number of Spanish PGS initiatives. However, legislative recognition presents some challenges when trying to include PGS in a legal framework so that it preserves its core principles [25]. Indeed, the interviewees were aware of these challenges and for this reason proposed institutional and social recognition as alternatives. The interviewees were afraid that legislative recognition of PGS could result in strict hierarchical structures and increase bureaucracy, threatening the essence of PGS and converging with the TPC system [56]. However, in some countries PGS have developed different strategies in recent years to work within the legal framework.

In Mexico, PGS initiatives met the government to design a new regulation of organic products in a participatory way. Although PGS was finally included in organic law, the process demonstrated the difficulty of achieving consensus [23]. Ultimately Mexico, like Chile, recognized PGS as a valid guarantee system for organic products only for small producers and at local markets or through direct marketing (no export) [23,56]. Yet the legal recognition of PGS in Mexico requires that PGS initiatives must comply with the Mexican law for organic products which can be burdensome for the producers and even unachievable. The increase of bureaucracy together with the time constrains of the producers makes it difficult for producer to obtain the legal recognition [45], as the interviewees of our study fear it could happen if PGS are legally recognized in Spain.

In the USA, the US Department of Agriculture allows producers with a gross annual revenue equal to or less than $5000 to sell their products as organic in the local market without TPC. This scenario leaves some space for PGS [14,73]. Thus, a financial threshold that suits the Spanish context and is high enough to include all the desired producers could be defined, making space for PGS to work as a valid guarantee system for organic production in Spain [74].

Nevertheless, in most Latin American countries where PGS are officially recognized, PGS initiatives are overseen by the same regulative body as the TPC system. In Bolivia, Paraguay, Ecuador and Costa Rica small producers are allowed to use PGS as a guarantee of organic production on the national market. However PGS initiatives are then audited by the national Food Safety Authority in most cases [56]. Ecollaures and Ecovalle can be compared to the Peruvian case where PGS are not officially recognised on a national level, but the municipalities or markets support and recognize PGS as a valid guarantee for organic production [56].

Apart from official recognition of PGS, in order to assure consumers of the system's legitimacy, PGS should also be sufficiently formal and members should build trust properly within the PGS initiatives [23]. However, the interviewees admitted the challenge of combining this formality with the characteristic comradeship of PGS. This challenge is typical for cooperatives as the internal tension between efficiency and cooperation are competing views that are always present in collective actions [75]. This and other internal matters seem to be a common challenge in PGS initiatives [6,32,42,53].

Internal matters can lead to conflicts and divisions within the community. In collectives such as PGS, interpersonal issues and differences of opinion among members present a significant challenge [31]. The democratic processes facilitate conflict due to the close proximity and face-to-face decision-making that can lead to an emotional and confrontational environment during meetings [75]. Differences in opinions amongst PGS members and the refusal to compromise and accept others' opinions might result in conflict and separation of PGS initiatives [45]. To avoid conflict and improve the efficiency of PGS initiatives, especially in decision-making, members must establish and uphold standards. These standards have to be clear on issues such as exclusion from the system, monitoring procedures and sanctions for non-compliance [28]. IFOAM (2007) [27] also acknowledges the importance of the clear and prior definition of these issues in order to guide PGS initiatives to put into practice the key elements of PGS as defined by IFOAM.

Another internal issue that affects the PGS initiatives studied was the perceived low level of member participation. In institutions for collective actions such as PGS or agriculture cooperatives, a broad-based commitment and equal degree of participation are the most common challenges faced by members [28,75]. For example, in CSA consumers also acknowledge their wish to volunteer in CSA and support the farmers, "but their busy lives and hectic schedules often prevent participation and active involvement" [71] (p. 97). The present study's results show that the different stakeholders involved in the PGS initiatives also faced time constraints and there was general dissatisfaction with the level of participation by members. One explanation for this dissatisfaction could be the difference in the members' perceptions of their participation and the participation of the other members [76]. The commitment required to participate in meeting can lead to a decline in the quality of democratic decision-making [75]. Time constraints can influence the continuity of the certification process [45] and this might be another reason for the mentioned low efficiency of the PGS initiatives. Nevertheless, a study with a focus on the members' motivation (including consumers) for participation in PGS could help draw more precise conclusions about members' participation and commitment to PGS.

Bouagnimbeck (2014) [53] (p. 58) found that long distances between members of the PGS initiatives and from market to market are a challenge related to time constraints for PGS initiatives in India, Brazil and South Africa. This challenge was not found in the PGS initiatives studied here. Instead the toll on family and free time were considered to be important reasons affecting participation in the PGS initiatives studied. One solution already proposed by producers in the first Andalusian PGS initiatives that is working in some other PGS initiatives (e.g., PGS El Encinar, Granada, Spain) is to have an employee that takes care of the administration, organization and communication of the PGS initiative [6]. Hiring employees for democratic cooperatives has been criticized because it defeats the purpose of the cooperative by creating two types of members [77]. Nonetheless Hernandez (2006) [75] (p. 130) argues for the importance of democracy in the decision-making process of cooperatives, including the choice of whether or not to employ someone. In the PGS initiatives studied here, there is a division of opinion on this matter.

5. Conclusions

Producers of the PGS initiatives studied look for a seal that reflects the work and values behind their products. Their discontent with the current TPC system and the need for a system that assesses their method of production are the main motivations for joining the PGS initiatives studied here. Moreover, the community created through the PGS initiatives is important not only for producers, but for other stakeholders as well.

Currently there are few commercial advantages obtained from being part of the PGS initiatives studied. Instead, the main achievement perceived by members of these PGS initiatives was the community created by their initiatives. The support, knowledge, exchange and social network of the community were some of the benefits of PGS mentioned. The use of PGS as a political tool for improving the situation of organic agriculture and alternative certification systems in Spain was also an important goal of the PGS initiatives studied.

The main challenges perceived by the members of the PGS initiatives were the involvement of consumers, participation of the members, official recognition and time constraints. Moreover, communication between members and internal difficulties seemed to present a challenge for the PGS initiatives studied. In order to overcome these internal challenges, external help is recommended, following the example of Ecollaures. The establishment of a network of PGS in Spain might be helpful to empower PGS and apply pressure for official recognition.

There is a general desire to improve and develop the PGS initiatives studied here. However, until they receive some kind of recognition and the public are given more information, this growth will be limited. The support of public or private organizations, as well as more research, may help PGS develop further in Spain.

PGS might be an example of how people that want to support a different food system to the current one (more ecological, social and fair) come together and create their own initiatives that nowadays have even become a global movement. They reflect a consciousness among consumers, producers and other stakeholder about a healthier food system, not only in terms of human health, but also for the environment.

Author Contributions: Conceptualization and methodology, M.L.C., C.R.V. and M.C.P.; Investigation, M.L.C.; writing—original draft preparation: M.L.C.; writing—review and editing, C.R.V., M.C.P. and M.L.C; Supervision, C.R.V.

Funding: This research received no external funding.

Acknowledgments: We would like to thank the members of Ecollaures, Ecovalle and Vecinos Campesinos for their support, kindness and willingness to collaborate with this project.

Conflicts of Interest: The authors declare no conflict of interest.

Appendix A

Table A1. Summary of results.

Stakeholder	Method	Number of Interviewees	Question			PGS Initiatives Studied: Ecollaures (EL)	PGS Initiatives Studied: Vecinos Campesinos (VC)	PGS Initiatives Studied: Ecovalle (EV)	Arithmetic Mean in the PGS Initiatives STUDIED	Statistical Analysis: Fisher's Exact ($p < 0.05$)
All (producers, FoodCoops, consumers association, NGOs, shops)	Structured and semi-structured interviews	**n = 31** (EL n = 15; VC n = 11; EV n = 5)	Age (years)			36.7	45.5	37.6	40.3	n.s.
		n = 100% = 31 (EL n = 100% = 15; VC n = 100% = 11; EV n = 100% = 5)	**Place of birth**	PGS region		67%	69%	0%	58%	s.
			Studies	University degree		87%	46%	100%	73%	n.s.
			Interviewees who use the word "agroecology"			80%	0%	60%	47%	s.
			Interviewees who use of the word "politics" or related topics			40%	0%	0%	13%	s.
		n = 100% = 24 (EL n = 100% = 12; VC n = 100% = 8; EV n = 100% = 4)	**Level of satisfaction with the internal regulation of their PGS initiative**	Very low		0%	0%	0%	0%	n.s.
				Low		17%	0%	0%	6%	
				Average		58%	0%	0%	19%	
				High		25%	75%	50%	50%	
				Very high		0%	25%	50%	25%	
				No answer (n° interviewees)		3	2	0	-	
		n = 100% = 28 (EL n = 100% = 14; VC n = 100% = 10; EV n = 100% = 4)	**Level of satisfaction with the inspection checklist of their PGS initiative**	Very low		0%	0%	0%	0%	n.s.
				Low		0%	0%	0%	0%	
				Average		43%	30%	25%	33%	
				High		50%	60%	25%	45%	
				Very high		7%	20%	50%	26%	
				No answer (n° interviewees)		1	0	0	-	
		n = 100% = 28 (EL n = 100% = 14; VC n = 100% = 10; EV n = 100% = 4)	**Level of satisfaction with the guarantee process of their PGS initiative**	Very low		0%	0%	0%	0%	n.s.
				Low		7%	0%	0%	2%	
				Average		43%	10%	25%	26%	
				High		50%	90%	50%	63%	
				Very high		0%	0%	25%	8%	
				No answer (n° interviewees)		1	0	0	-	
			Current status of different aspects of the studied PGS initiatives	Prepared statements	Status					
				Feeling of community with the members	Very poor	0%	0%	0%	0%	n.s.
					Poor	0%	0%	0%	0%	
					OK	0%	0%	0%	0%	
					Good	60%	60%	0%	40%	
					Very good	40%	40%	100%	60%	
				Organisation	Very poor	0%	0%	0%	0%	n.s.
					Poor	0%	0%	0%	0%	
					OK	27%	40%	50%	39%	
					Good	73%	60%	50%	61%	
					Very good	0%	0%	0%	0%	
				Communication between members	Very poor	0%	0%	0%	0%	n.s.
					Poor	0%	0%	0%	0%	
					OK	27%	20%	0%	16%	
					Good	73%	80%	75%	76%	
					Very good	0%	0%	25%	8%	

Table A1. *Cont.*

Stakeholder	Method	Number of Interviewees	Question			PGS Initiatives Studied			Arithmetic Mean in the PGS Initiatives STUDIED	Statistical Analysis
						Ecollaures (EL)	Vecinos Campesinos (VC)	Ecovalle (EV)		Fisher's Exact ($p < 0.05$)
				Dissemination of PGS	Very poor	7%	0%	25%	11%	n.s.
					Poor	20%	10%	0%	10%	
					OK	47%	80%	75%	67%	
					Good	27%	10%	0%	12%	
					Very good	0%	0%	0%	0%	
				Possibility to participate in decision-making	Very poor	0%	0%	0%	0%	n.s.
					Poor	0%	0%	0%	0%	
					OK	0%	0%	0%	0%	
					Good	20%	20%	0%	13%	
					Very good	80%	80%	100%	87%	
				Relationship between members	Very poor	0%	0%	0%	0%	n.s.
					Poor	0%	0%	0%	0%	
					OK	0%	0%	0%	0%	
					Good	47%	40%	75%	54%	
					Very good	53%	60%	25%	46%	
				Internal communication methods used	Very poor	0%	0%	0%	0%	n.s.
					Poor	0%	0%	0%	0%	
					OK	33%	30%	0%	21%	
					Good	7%	50%	75%	44%	
					Very good	0%	20%	25%	15%	
				Process of decision-making	Very poor	0%	0%	0%	0%	n.s.
					Poor	0%	0%	0%	0%	
					OK	13%	10%	0%	8%	
					Good	60%	60%	100%	73%	
					Very good	27%	30%	0%	19%	
			Perceived participation of producers in the PGS initiatives studied	Level of participation						n.s.
					Very low	0%	0%	0%	0%	
					Low	8%	11%	0%	6%	
					Moderate	31%	56%	25%	37%	
					High	46%	22%	50%	39%	
					Very high	15%	11%	25%	17%	
			Perceived participation of consumers in the PGS initiatives studied	Level of participation						n.s.
					Very Low	31%	22%	0%	18%	
					Low	54%	56%	0%	36%	
					Moderate	0%	22%	100%	41%	
					High	15%	0%	0%	5%	
					Very high	0%	0%	0%	0%	
			Perceived importance of official recognition of PGS in Spain	Level of importance						n.s.
					None	7%	0%	0%	2%	
					Very low	7%	10%	0%	6%	
					Low	7%	0%	0%	2%	
					Moderate	21%	20%	25%	22%	
					High	21%	30%	0%	17%	
					Very high	36%	40%	75%	50%	

Table A1. *Cont.*

Stakeholder	Method	Number of Interviewees	Question			PGS Initiatives Studied			Arithmetic Mean in the PGS Initiatives STUDIED	Statistical Analysis
						Ecollaures (EL)	Vecinos Campesinos (VC)	Ecovalle (EV)		Fisher's Exact ($p < 0.05$)
Only Producers		**n = 22** (EL n = 10; VC n = 8; EV n = 4)	**Size of land (ha)**			4	1	2	2	n.s.
		n = 100% = 22 (EL n = 100% = 10; VC n = 100% = 8; EV n = 100% = 4)	**% of annual home income from agriculture**			74%	36%	53%	54%	s.
			TPC besides PGS			20%	63%	75%	46%	n.s.
			Motivations for participating in the PGS initiative	Prepared statements	Level of importance					
				Promotion of local products	None	0%	13%	0%	4%	n.s.
					Very low	0%	0%	0%	0%	
					Low	0%	0%	0%	0%	
					Average	0%	0%	0%	0%	
					High	40%	13%	0%	18%	
					Very high	60%	75%	100%	78%	
				Direct relationships with consumers	None	0%	0%	0%	0%	n.s.
					Very low	0%	0%	0%	0%	
					Low	0%	0%	0%	0%	
					Average	10%	0%	0%	3%	
					High	50%	25%	0%	25%	
					Very high	40%	75%	100%	72%	
				Raise awareness among consumers about organic agriculture	None	0%	0%	0%	0%	n.s.
					Very low	0%	0%	0%	0%	
					Low	0%	0%	0%	0%	
					Average	10%	25%	25%	20%	
					High	30%	38%	0%	23%	
					Very high	60%	38%	75%	58%	
				With PGS I can have higher incomes	None	0%	38%	50%	29%	n.s.
					Very low	20%	0%	0%	7%	
					Low	30%	13%	0%	14%	
					Average	20%	38%	50%	36%	
					High	20%	13%	0%	11%	
					Very high	10%	0%	0%	3%	
				This PGS initiative establishes a community of producers	None	0%	0%	0%	0%	n.s.
					Very low	0%	0%	0%	0%	
					Low	10%	0%	0%	3%	
					Average	0%	13%	0%	4%	
					High	30%	13%	50%	31%	
					Very high	60%	75%	50%	62%	
				PGS is cheaper than TPC	None	60%	0%	75%	45%	s.
					Very low	10%	0%	0%	3%	
					Low	10%	0%	0%	3%	
					Average	20%	25%	0%	15%	
					High	0%	38%	25%	21%	
					Very high	0%	38%	0%	13%	

Table A1. *Cont.*

Stakeholder	Method	Number of Interviewees	Question		PGS Initiatives Studied			Arithmetic Mean in the PGS Initiatives STUDIED	Statistical Analysis
					Ecollaures (EL)	Vecinos Campesinos (VC)	Ecovalle (EV)		Fisher's Exact ($p < 0.05$)
			The fact that PGS is a tool for social change	None	0%	0%	0%	0%	n.s.
				Very low	0%	0%	0%	0%	
				Low	0%	0%	0%	0%	
				Average	0%	0%	0%	0%	
				High	40%	13%	0%	18%	
				Very high	60%	88%	100%	83%	
			The opportunity for mutual support with other producers	None	0%	13%	0%	4%	n.s.
				Very low	0%	0%	0%	0%	
				Low	0%	0%	0%	0%	
				Average	0%	0%	0%	0%	
				High	30%	38%	25%	31%	
				Very high	70%	50%	75%	65%	
			The opinion of my family/friends	None	50%	13%	75%	46%	n.s.
				Very low	10%	0%	0%	3%	
				Low	10%	0%	0%	3%	
				Average	20%	38%	25%	28%	
				High	0%	25%	0%	8%	
				Very high	10%	0%	0%	3%	

References

1. Coiduras Sánchez, P.; Díaz Álvarez, J.R.; Porcuna Coto, J.L. Situación y posibilidades de implantación de la certificación participativa en agricultura ecológica. In Proceedings of the VII Congreso SEAE, Zaragoza, Spain, 18–23 September 2006.
2. Tanner, B. Independent assessment by third-party certification bodies. *Food Control* **2000**, *11*, 415–417. [CrossRef]
3. Certification of Organic Products. Available online: http://www.centroecologico.org.br/artigo_detalhe.php?id_artigo=26 (accessed on 10 September 2018).
4. Jahn, G.; Schramm, M.; Spiller, A. The Reliability of Certification: Quality Labels as a Consumer Policy Tool. *J. Consum. Policy* **2005**, *28*, 53–73. [CrossRef]
5. Getz, C.; Screck, A. What organic and Fair Trade labels do not tell us: Towards a place-based understanding of certificatio. *In. J. Consum. Stud.* **2006**, *30*, 490–501. [CrossRef]
6. Cuéllar Padilla, M. Hacia un Sistema Participativo de Garantía para la Producción Ecológica en Andalucía. Ph.D. Thesis, University of Córdoba, Córdoba, Spain, 2008.
7. Albersmeier, F.; Schulze, H.; Jahn, G.; Spiller, A. The reliability of third-party certification in the food chain: From checklists to risk-oriented auditing. *Food Control* **2009**, *20*, 927–935. [CrossRef]
8. Padel, S.; Röcklinsberg, H.; Schmid, O. The implementation of organic principles and values in the European Regulation for organic food. *Food Policy* **2009**, *34*, 245–251. [CrossRef]
9. Velleda Caldas, N.; Sacco dos Anjos, F.; Lozano Cabedo, C. La certificación de productos ecológicos en España y Brasil. *Agrociencia Uruguay* **2014**, *18*, 163–171.
10. Certificación 'de Papel' o de Relaciones Humanas. Los Sistemas de Garantía Participativa como Iniciativas de Soberanía Alimentaria Local. Available online: http://www.academia.edu/15030312/_Certificaci%C3%B3n_de_papel_o_de_relaciones_humanas_Los_sistemas_de_garant%C3%ADa_participativa_como_iniciativas_de_soberan%C3%ADa_alimentaria_local_Certification_on_paper_or_by_the_people_Participatory_guarantee_systems_as_initiatives_for_local_food_sovereignty_in_Spanish_ (accessed on 31 October 2018).
11. Lotter, D.W. Organic Agriculture. *J. Sustain. Agric.* **2003**, *21*, 59–128. [CrossRef]
12. Vogl, C.R.; Kilcher, L.; Schmidt, H. Are Standards and Regulations of Organic Farming Moving Away from Small Farmers' Knowledge? *J. Sustain. Agric.* **2005**, *26*, 5–26. [CrossRef]
13. Hatanaka, M.; Bain, C.; Busch, L. Third-party certification in the global agrifood system. *Food Policy* **2005**, *30*, 354–369. [CrossRef]
14. The European Regulatory Framework and Its Implementation in Influencing Organic Inspection and Certification Systems in the EU. Available online: https://www.researchgate.net/publication/277101680_The_European_Regulatory_Framework_and_its_implementation_in_influencing_organic_inspection_and_certification_systems_in_the_EU (accessed on 31 October 2018).
15. Vogl, C.R.; Axmann, P. Regelungsmechanismen im System des Ökologischen Landbaus. In *Das System Ökologischer Landbau*; Freyer, B., Ed.; UTB Stuttgart: Stuttgart, Germany, 2016.
16. Padilla Bravo, C.; Villanueva Ramírez, I.; Neuendorff, J.; Spiller, A. Assessing the impact of unannounced audits on the effectiveness and reliability of organic certification. *Org. Agric.* **2013**, *3*, 95–109. [CrossRef]
17. Corsin, F.; Funge-Smith, S.; Clausen, J. A Qualitative Assessment of Standards and Certification Schemes Applicable to Aquaculture in the Asia-Pacific region. Asia-Pacific Fishery Commission Food and Agriculture Organization of the United Nations (FAO). Available online: http://www.fao.org/docrep/010/ai388e/ai388e00.htm (accessed on 14 September 2018).
18. Bryla, P. The perception of EU quality signs for origin and organic food products among Polish consumers. *Qual. Assur. Saf. Crop. Foods* **2017**, *9*. [CrossRef]
19. Deaton, B.J. A theoretical framework for examining the role of third-party certifiers. *Food Control* **2004**, *15*, 615–619. [CrossRef]
20. Moser, C.; Hildebrandt, T.; Bailis, R. International Sustainability Standards and Certification. In *Sustainable Development of Biofuels in Latin America and the Caribbean*; Springer: New York, NY, USA, 2014; pp. 27–69. [CrossRef]
21. Hatanaka, M. McSustainability and McJustice: Certification, alternative food and agriculture, and social change. *Sustainability* **2014**, *6*, 8092–8112. [CrossRef]

22. Chkanikova, O.; Lehner, M. Private eco-brands and green market development: Towards new forms of sustainability governance in the food retailing. *J. Clean. Prod.* **2015**, *107*, 74–84. [CrossRef]
23. Nelson, E.; Tovar, L.G.; Gueguen, E.; Humphries, S.; Landman, K.; Rindermann, R.S. Participatory guarantee systems and the re-imagining of Mexico's organic sector. *Agric. Hum. Values* **2015**, *33*, 373–388. [CrossRef]
24. Sistemas Participativos de Garantía. Una Herramienta Clave para la Soberanía Alimentaria. Available online: http://www.redcimas.org/wordpress/wp-content/uploads/2012/10/sistemas_participativos_degarantia.pdf (accessed on 31 October 2018).
25. Källander, I. *Participatory Guarantee Systems—PGS*; Swedish Society for Nature Conservation: Stockholm, Sweden, 2008.
26. How Participatory Guarantee Systems Can Develop and Function. Available online: https://www.ifoam.bio/sites/default/files/page/files/pgs_guidelines_en_web.pdf (accessed on 31 October 2018).
27. IFOAM. *Participatory Guarantee Systems: Shared Vision, Shared Ideals*; IFOAM: Bonn, Germany, 2007.
28. Ostrom, E. *Governing the Commons: The Evolution of Institutions for Collective Action*; Cambridge University Press: New York, NY, USA, 1990.
29. Participatory Guarantee Systems: Case Studies from Brazil, India, New Zealand, USA and France. Available online: https://www.ifoam.bio/sites/default/files/page/files/studies_book_web.pdf (accessed on 10 September 2018).
30. Home, R.; Bouagnimbeck, H.; Ugas, R.; Arbenz, M.; Stolze, M. Participatory guarantee systems: Organic certification to empower farmers and strengthen communities. *Agroecol. Sustain. Food Syst.* **2017**, *41*, 526–545. [CrossRef]
31. Nelson, E.; Gómez Tovar, L.; Schwentesius Rindermann, R.; Gómez Cruz, M.Á. Participatory organic certification in Mexico: An alternative approach to maintaining the integrity of the organic label. *Agric. Hum. Values* **2009**, *27*, 227–237. [CrossRef]
32. Katto-Andrighetto, J. *Participatory Guarantee Systems in East Africa. Case Studies from Kenya, Tanzania and Uganda*; IFOAM: Bonn, Germany, 2013.
33. Torremocha, E. Los Sistemas Participativos de Garantía. Herramienta para la Definición de Estrategias Agroecológicas. Master's Thesis, Internationa University of Andalusia, Seville, Spain, 2012.
34. Sacchi, G.; Caputo, V.; Nayga, R. Alternative Labeling Programs and Purchasing Behavior toward Organic Foods: The Case of the Participatory Guarantee Systems in Brazil. *Sustainability* **2015**, *7*, 7397–7416. [CrossRef]
35. Rover, O.; de Gennaro, B.; Roselli, L. Social Innovation and Sustainable Rural Development: The Case of a Brazilian Agroecology Network. *Sustainability* **2016**, *9*, 3. [CrossRef]
36. Flaten, O.; Lien, G.; Koesling, M.; Loes, A.K. Norwegian farmers ceasing certified organic production: Characteristics and reasons. *J. Environ. Manag.* **2010**, *91*, 2717–2726. [CrossRef] [PubMed]
37. MAPAMA. Agricultura Ecológica: Estadísticas 2015. Available online: http://www.mapama.gob.es/es/alimentacion/temas/la-agricultura-ecologica/estadisticaseco2015connipoymetadatos_tcm7-435957.pdf (accessed on 12 June 2017).
38. López Salcedo, P. Caracterización de la Comercialización y Distribución de Productos Ecológicos a Través de los Canales de Venta Especializados. Available online: http://www.mapama.gob.es/es/alimentacion/temas/la-agricultura-ecologica/estudiocaracterizacioncomercializacioncanalesventaespecializados_tcm7-387574.pdf (accessed on 24 May 2017).
39. Proceso de Construcción y Regulación de un Sistema Participativo de Garantía para la Producción Ecológica en Andalucía. Available online: https://repera.files.wordpress.com/2009/08/sistema_garantia_participativa_andalucia.pdf (accessed on 31 October 2018).
40. Boza Martínez, S. La Agricultura Ecológica como Parte de la Estrategia de Desarrollo Rural Sostenible en Andalucía. Ph.D. Thesis, Universidad Atónoma de Madrid, Madrid, Spain, 2010.
41. Cuéllar Padilla, M.; Calle Collado, Á. Can we find solutions with people? Participatory action research with small organic producers in Andalusia. *J. Rural Stud.* **2011**, *27*, 372–383. [CrossRef]
42. De la Cruz, C. Organizaciones, Sistemas Participativos de Garantía y Procesos Agroecológicos en Andalucía. Ph.D. Thesis, University of Córdoba, Córdoba, Spain, 2015.
43. Ecovalle. Available online: http://ecovalle.org/ (accessed on 11 October 2016).
44. Miles, M.B.; Huberman, A.M.; Saldaña, J. *Qualitative Data Analysis: An Expanded Sourcebook*; SAGE Publications: Houston, TX, USA, 1994.

45. Kaufmann, S.; Vogl, C.R. Participatory Guarantee Systems (PGS) in Mexico: A theoretic ideal or everyday practice? *Agric. Hum. Values* **2017**, *35*, 457–472. [CrossRef]
46. Bernard, R.H. *Research Methods in Anthropology. Qualitative and Quantitave Approaches*; AltaMira Press: Lanham, MD, USA, 2006.
47. Döring, N.; Bortz, J. *Forschungsmethoden und Evaluation in den Sozial- und Humanwissenschaften*; Springer: Berlin, Germany, 2016.
48. Saldaña, J. *The Coding Manual for Qualitative Researchers*; SAGE Publications: Newcastle upon Tyne, UK, 2009.
49. Bühl, A. *SPSS 23—Einführung in die Moderne Datenanalyse*; Pearson: Hallbergmoos, Germany, 2016.
50. Oxford Dictionary. Oxford Living Dictionary. Available online: https://en.oxforddictionaries.com/ (accessed on 12 June 2017).
51. Sense.Lab e.V. *Fair, Bio, Selbstbestimmt: Das Handbuch zur Gründung einer Food-Coop*; Books on Demand GmbH: Hamburg, Germany, 2009.
52. Zitcer, A. Food Co-ops and the Paradox of Exclusivity. *Antipode* **2015**, *47*, 812–828. [CrossRef]
53. Bouagnimbeck, H. *Global Comparative Study on Interaction between Social Processes and Participatory Guarantee Systems*; IFOAM: Bonn, Germany, 2014.
54. IFOAM. *Organic Agriculture and Participatory Guarantee Systems: Marketing and Support for Organic Smallholders*; IFOAM: Bonn, Germany, 2006.
55. Sánchez Hernández, J.L. Redes alimentarias alternativas: Concepto, tipología y adecuación a la realidad española. *Boletín de la Asociación de Geógrafos Españoles* **2009**, *49*, 185–207.
56. Varini, F. How governments regulate PGS and where is it taking us? *Org. Stand.* **2017**, *171*, 3–6.
57. Gentile, A. De vuelta al nido en tiempos de crisis. Los boomerang kids españoles. *Revista Estudios Juventud* **2010**, *90*, 181–203.
58. Santos Ortega, A.; Martín Martín, P. La juventud española en tiempos de crisis: Paro, vidas precarias y acción colectiva. *Sociología Trabajo* **2012**, *75*, 93–110.
59. Alrøe, H.F.; Noe, E. What makes organic agriculture move—Protest, meaning or market? A polyocular approach to the dynamics and governance of organic agriculture. *Int. J. Agric. Resour. Gov. Ecol.* **2006**, *7*, 5–22. [CrossRef]
60. Aportando a la Construcción de la Soberanía Alimentaria Desde la Agroecología. Available online: http://www.ecologiapolitica.info/?p=4848 (accessed on 31 October 2018).
61. Altieri, M.A.; Nicholls, C.I. Agroecología: Única esperanza para la soberanía alimentaria y la resiliencia socioecológica. *Agroecología* **2012**, *7*, 65–83.
62. Rosset, P.M.; Martínez-Torres, M.E. Rural Social Movements and Agroecology: Context, Theory, and Process. *Ecol. Soc.* **2012**, *17*. [CrossRef]
63. Gómez, F.L.; Ríos-Osorop, L.; Eschenhagen, M.L. Epistemological bases of agroecology. *Agrociencia* **2015**, *49*, 679–688.
64. Ostrom, E.; Ahn, T.K. The meaning of social capital and its links to collective action. In *Handbook of Social Capital: The Troika of Sociology, Political Science and Economics*; Gert, T.S., Gunnar, L.S., Northampton, M.E.E., Eds.; Edward Elgar: Bloomington, IN, USA, 2008; pp. 17–35.
65. Alternative Certification and a Network Conformoty Assessment Approach. Available online: https://www.ifoam.bio/sites/default/files/page/files/alternativecertificationandanetworkconformityassessmentapproach.pdf (accessed on 31 October 2018).
66. Brehm, J.M.; Eisenhauer, B.W. Motivations for participating in Community Supported Agriculture and their relationship with community attachment and social capital. *South. Rural Sociol.* **2008**, *23*, 94–115.
67. Brown, C.; Miller, S. The Impacts of Local Markets: A Review of Research on Farmers Markets and Community Supported Agriculture (CSA). *Am. J. Agric. Econ.* **2008**, *90*, 1296–1302. [CrossRef]
68. Zanasi, C.; Venturi, P.; Setti, M.; Rota, C. Participative organic certification, trust and local rural communities development: The Case of Rede Ecovida. *New Med.* **2009**, *2*, 56–64.
69. Papaoikonomou, E.; Ginieis, M. Putting the farmer's face on food: Governance and the producer–consumer relationship in local food systems. *Agric. Hum. Values* **2016**, *34*, 53–67. [CrossRef]
70. Bougherara, D.; Grolleau, G.; Mzoughi, N. Buy local, pollute less: What drives households to join a community supported farm? *Ecol. Econ.* **2009**, *68*, 1488–1495. [CrossRef]
71. Pole, A.; Gray, M. Farming alone? What's up with the "C" in community supported agriculture. *Agric. Hum. Values* **2012**, *30*, 85–100. [CrossRef]

72. Thorsøe, M.H. Maintaining Trust and Credibility in a Continuously Evolving Organic Food System. *J. Agric. Environ. Ethics* **2015**, *28*, 767–787. [CrossRef]
73. United States Department of Agriculture. USDA National Organic Program: Who Needs to Be Cetified? Available online: https://www.ams.usda.gov/sites/default/files/media/who_needs_to_be_certified.pdf (accessed on 18 May 2017).
74. IFOAM. Internal Control Systems (ICS) for Group Certification. Available online: http://www.ifoam.bio/en/internal-control-systems-ics-group-certification (accessed on 17 October 2016).
75. Hernandez, S. Striving for Control: Democracy and Oligarchy at a Mexican Cooperative. *Econ. Ind. Democr.* **2006**, *27*, 105–135. [CrossRef]
76. Österberg, P.; Nilsson, J. Members' perception of their participation in the governance of cooperatives: The key to trust and commitment in agricultural cooperatives. *Agribusiness* **2009**, *25*, 181–197. [CrossRef]
77. Gunn, C.E. *Workers' Self-Management*; Cornell University Press: Ithaca, NY, USA, 1984.

sustainability

MDPI

Article

Scaling-Up Sustainable Development Initiatives: A Comparative Case Study of Agri-Food System Innovations in Brazil, New York, and Senegal

Rositsa T. Ilieva [1,2,*] **and Andreas Hernandez** [3]

1 CUNY Urban Food Policy Institute, CUNY Graduate School of Public Health and Health Policy, City University of New York, New York, NY 10027, USA
2 Parsons School of Design, The New School, New York, NY 10011, USA
3 Department of International Studies, Marymount Manhattan College, New York, NY 10021, USA; ahernandez2@mmm.edu
* Correspondence: Rositsa.Ilieva@sph.cuny.edu

Received: 2 October 2018; Accepted: 29 October 2018; Published: 6 November 2018

Abstract: To effectively address the sustainability crises our planet faces, decision-makers at different levels of government worldwide will have to get a handle on three key challenges: learning from Global North and South initiatives in tandem, taking stock of social innovations alongside technological fixes, and nurturing grassroots sustainable development initiatives next to, or in place of, top-down corporate and government interventions. Current scientific literature and grant-making institutions have often reinforced the compartmentalized fashion in which we learn and draw policy lessons from North/South, social/technical, and bottom-up/top-down sustainability initiatives, including local food system innovations. The strategic levers for global sustainable development lying in-between are thus left out. This paper uses exploratory, multiple case study analysis to address this omission. By concurrently drawing lessons from grassroots innovations in Brazil, New York, and Senegal—three profoundly different socioeconomic and geographic contexts—we identify common pressure points that have enabled local communities to drive system-wide transformations toward climate adaptation, resilience, and sustainability in the agri-food system. The findings of this paper would be of value to scholars, government officials, and community groups engaged in agri-food systems sustainability and interested in the processes of change that have allowed budding innovations to stabilize and scale up.

Keywords: sustainable development; alternative agri-food networks; transition theories; grassroots innovations; socio-technical systems; agroecology; ecovillages; social movements

1. Introduction

Why can some grassroots sustainable development projects scale-up and others cannot? Sustainability transitions are difficult as socio-technical systems like energy, transport, housing, and agri-food are stabilized by lock-in mechanisms that relate to sunk investments, behavioral patterns, vested interests, infrastructure, subsidies, and regulations. Sustainability transitions imply the transformation of these wider technical, social, and economic systems and occur through the emergence, alignment, and scaling up of radical socio-technical innovations.

This research seeks to shed light on the multi-level factors that contribute to the effective scaling up of grassroots sustainable development projects. Our goal is to identify possible transition "pressure points" at multiple levels of community food systems and in multiple sociotechnical domains that may be used to support and guide the effective scaling up of sustainable development initiatives within complex, nonlinear social and technical systems. The aim of this paper is thus to provide theoretically-informed practical recommendations for policymakers seeking to steer community-based sustainability transitions and reform food system governance through, rather than despite of, grassroots innovations. To effectively navigate the sustainability crises our planet faces, decision-makers at different levels of government worldwide will have to get a handle on three key challenges: learning from Global North and South initiatives in tandem, taking stock of social innovations alongside technological fixes, and nurturing grassroots sustainable development initiatives next to, or in place of, top-down corporate and government projects and interventions.

Socio-technical transition studies focusing on community-driven transitions are still limited. There is, however, a growing body of work focusing on "grassroots innovations" and their contribution to different facets of sustainable development [1–6]. To transition the dominant agri-food regime—and dismantle the unjust [7–9], unhealthy [10–12], and inefficient [13,14] systems producing hunger, chronic diet-related diseases, environmental degradation, inhumane treatment of animals, and unfair labor practices—top-down approaches would hardly suffice or even be appropriate. Who should be in charge of sustainable development transitions is, thus, a question in need of urgent investigation. Unsolved dilemmas regarding the role of different societal domains—government, market, civil society, and all intermediary organizations in between—as well as the relationships between efforts to scale up sustainability innovations in Global North and South countries warrant new approaches to the study of grassroots innovations.

To fill this gap in current sustainability research, we undertake a comparative case study, exploring grassroots sustainable development efforts from the Global North and Global South that are transforming wider technical, social, and economic systems. The first case we examine is the Brazilian Landless Movement's (MST) transition to agroecology (ecologically informed sustainable agriculture) focused in the south of Brazil. A group of MST cooperatives have developed one of the most extensive systems of agroecological production globally. The second case we present is the New York City's food movement—a vast and heterogeneous movement of movements which, over the past two decades has advocated for food justice, health equity, environmental sustainability, and fairer labor practices throughout the urban food environment and the food chain more broadly. Lastly, in our third case, we focus on the Ecovillage Movement of Senegal, which is constructing alternative forms of grassroots sustainable development by drawing from West African village life and new green technologies, along with the recuperation of soils.

There has been significant work examining grassroots agri-food movements, global networking and organization, and resistance (see, for example, References [15–18]). There is a necessity for research that seeks to understand the successes and challenges in grassroots agri-food movements in scaling up. This comparative analysis of grassroots sustainable development initiatives calls attention to the coevolution of the different niche, regime, and landscape pressures, and the shared transition levels in each case. We identify crosscutting themes that point to important dimensions of sustainability transitions in each case. We conclude with a systematic summary of the main lessons learnt and outline a set of key recommendations for government officials and policymakers who wish to synergize and scale the innovations emerging from grassroots social movements.

2. Research Methods

2.1. Theoretical Framework

Radical transformations of systems of provision—such as food, water, and transportation—are best understood as the outcome of both social and technological innovations. Neither engineers alone nor policymakers or non-expert citizens can singlehandedly shift complex matrixes of infrastructures to a more equitable or sustainable regime of operation. Rather, it is the concurrent interplay, and alignment, of changes in multiple societal domains that can, under certain circumstances, bring about radical system transformation. Technological innovation examples, such as cellular phones and personal computers, are well known, as are social innovations such as microcredit loans, sharing economies, community-owned renewables, and food cooperatives. We argue that both social and technological innovations are fundamentally socio-technical, and that this view is essential if we are to guide, rather than simply react to, socio-technical transitions in the near future.

However, what helps innovations endure, scale up, and transform mainstream institutions, infrastructures, and social norms? Transition theories such as the multi-level perspective (MLP) [19–21] provide some cues into that. The strength of this perspective lies, in fact, in the integration of social and technical understandings of sustainability transitions and the emphasis on multiple levels of stability, or path-dependency, of a socio-technical system. In brief, according to the MLP, socio-technical *regimes*, which are locked in and stabilized through mainstream infrastructures and institutions, can change when there is an alignment between a disruption (e.g., climate change, peak oil, obesity, economic recession) in the *landscape*, the highest level of system stability, and the level of *niches* of grassroots innovations which offer solutions to such disruptions and a promise for a reinstated stability.

The MLP has its roots in science, technology, and society studies [22] and evolutionary economics [23], and has been widely adopted to theorize historical transitions and the upsurge of nineteenth and twentieth-century innovations, such as steam ships [19], sewer systems [24], digital computers [25], and rock 'n' roll music [26], to mention a few, but also more recent cases of innovative municipal waste management [27], renewable energy [28], alternative food networks [29], and low-carbon transportation [30]. Drawing from the rich literature of historical case studies, contemporary theorists of sociotechnical transitions have also used the MLP to devise intervention-oriented theoretical frameworks, such as transition management [31–33], focusing on the governance of transition processes, and strategic niche management [34,35], offering insights into the distinct internal characteristics that successful niche innovations exhibit.

Yet the MLP, and its companion management frameworks, has also been faced with criticisms [36]. Close observers of grassroots innovations and social movements argue that current transition theories are limited in their ability to comprehend the full range of action of grassroots sustainability initiatives. Scholars caution that the focus of strategic niche management on novelty limits the ability to capture the scope of civil society action [37], that niche theory is not able to adequately explain the transformational power of social movements, and that this framework must take on a more critical capacity [1]. In fact, application of the growth-oriented approach of strategic niche management to grassroots innovations may not be able to encompass the complexity of diverse and conflicted realities on the ground [38]. Additionally, analyses of community currencies [39] reveal that grassroot innovations diffuse differently from conventional innovations, and that the MLP and niche theories require adaptation for this context.

To counter some of these limitations, we extend the MLP framework by embedding it in a spatially informed understanding of political economy, while simultaneously focusing on the agency of the collectivities constructing sustainable development initiatives.

2.2. Data Collection and Analysis

We utilize qualitative data collection and analysis research methods to gain an in-depth understanding of each case and conduct a comparative analysis. We conceived each of the cases as exploratory case study [40]. The MST and Senegalese Ecovillage Movement cases rely on ethnographic research including visual ethnography, photo-elicitation techniques, filming, and participant observation. The New York City food movement case relies primarily on secondary data sources and the analysis of peer-reviewed articles, government reports, and policy evaluations by nongovernmental organizations. Unstructured interviews with academic experts, government officials, and practitioners provided further insights into the key issues and turning points in each transition process. The extended MLP framework guided the transition analyses and thematic coding and analysis methods were used to identify shared themes across the cases. The final set of themes, or transition levers, were identified through discussion between the two authors and comparison of intermediate findings on each of the three cases.

3. Cases Overview

3.1. MST Agroecological Cooperatives

From its modest roots in Southern Brazil, the *Movimento dos Trabalhadores Rurais Sem Terra* (Landless Rural Workers Movement, or MST) gradually grew into the largest nationally-based social movement in Brazilian history and is widely recognized as the most organized, dynamic, and influential mass movement in Latin America today [15,41–43]. Through organizing landless families to occupy unproductive agricultural land, the MST has pressured Brazilian governments into enacting the Constitution and redistributing more than seven million hectares of unproductive agricultural land on which one and a half million members are now growing food [15].

The MST has developed some of the largest scale agroecological systems on the planet. MST cooperatives have used agroecological techniques to delink from agribusiness and banks, produce more food at a higher quality and lower cost, and recuperate their soils. Perhaps most notably is the *Grupo Gestor de Arroz* (Rice Management Group) in the south of Brazil, which has created several intertwined cooperatives bringing together 501 families, across 16 municipalities, who are cultivating rice using diverse agroecological methods. The democratically owned and managed *Grupo Gestor* stores, processes, packages, and markets an estimated 500,000 sacks of rice per year, over 5513 hectares in several regions of the state of Rio Grande do Sul. In addition to internal organization (including building large scale food processing plants), at the core of the scaling up of the *Grupo Gestor* has been the construction on institutional markets, which the Movement has worked to create at every scale through agreements and through policy. The *Grupo Gestor* provides 1000's of livelihoods and provides food for families, the region, government institutions around the country, and for global export. They have built a large scale, horizontal, and democratic food system, demonstrating that agroecological methods are an effective option for peasant farmers to stay on the land and feed their regions. While there are numerous agroecological settlements and cooperatives in the MST, here we focus on the *Grupo Gestor* in the context of the wider Movement.

3.2. New York City Food Movement

The New York City food movement is a vast and heterogeneous movement of movements which, over the past two decades has advocated for food justice, health equity, environmental sustainability, and fairer labor practices throughout the urban food environment and the food chain more broadly. While the city's food system is far from having radically transitioned to sustainability, it has effectively been reconfigured, both in institutional and physical infrastructure terms. Today, New York City has a dedicated Office of the Director of Food Policy, has released over twenty different reports on food policy-related matters [44], issues yearly food metrics reports, and has made the right to free lunch accessible to all public-school students in the city. The city has also witnessed the scaling up of many innovations in its local food system: it has more than 900 food producing gardens, over a dozen rooftop farms, more than 140 farmers markets of which more than half are located in high-poverty neighborhoods, a pilot curbside food scraps collection program already reaching over one million people, a network of more than one thousand upstate farmers engaged in sustainable watershed management practices, and a regional food hub supporting local food producers.

3.3. Senegalese Ecovillage Movement

The Senegalese Ecovillage Movement brings together hundreds of villages in a heterogenous network that seeks community-led development by taking the best of West African village life and combining this with green technologies and recuperation of soils and forests. Movement leaders assert they have flipped the Northern ecovillage model on its head, saying that West African villages already have strong community, cooperation, and spiritual systems that link them to nature, but that they need green technologies. The Movement began in the traditional fishing village of Yoff in coalition with the Ithaca Ecovillage and the third international EcoCity Conference which was held there in 1995.

Through the internal successes of what became EcoYoff, the ecovillage framework began to spread organically to villages in ecologically diverse regions of Senegal. Government officials, including a President, took note in the early 2000's and launched the Ministry of Ecovillages, which later became the National Agency for Ecovillages (ANEV) with the project of transitioning half of the country's 28,000 villages into ecovillages. With this institutionalization, funds from the UNDP and other international donors became available, creating a split in the Movement. The complex network now has a wing linked to the government and a wing focused on grassroots community-led development (although also linked with international donors). Although the goal of 14,000 villages remains distant, hundreds of villages are adopting aspects of the African ecovillage model, often in coalition with ANEV or NGOs, creating one of the most successful grassroots development efforts on the continent. The model is spreading to neighboring countries such as Mali and Democratic Republic of Congo.

Villages have developed projects as diverse as: solar power grids, extensive permaculture gardens, biogas and solar cookers not reliant on scarce wood fuel, reforestation, reintroduction of dry crops such as millet, and water pumps and tanks that extend growing seasons. According to the UNDP [45], in at least one ecovillage, years of outmigration have reversed as young people return to new opportunities in villages.

4. Results

The cross-case thematic analysis of the three case studies led to the identification of eleven main crosscutting themes, which we argue offer insights into possible levers for socio-technical transitions to sustainability.

4.1. Environmental Pressure and Drive

Each of these cases is responding to very real environmental pressures that disrupt established forms of production, distribution and consumption in the agri-food system. It is in relationship to these environmental pressures, that the movements began to, in part, redefine their understandings and practices in the food system.

MST (Landless Rural Workers Movement): Most farmers and settlements transitioned to agroecology in great part as they were not able to produce on highly-degraded land, which was redistributed through the state and federal government. This was coupled with health complications from pesticide use (especially among children and those in the fields), the high costs of purchasing chemical inputs, as well as the high cost of interest through public and private banks. These dynamics conditioned settlements to look for alternatives to improve soil health and intensify soil capacity. Through trial and error, they began practicing a constellation of soil intensification techniques using materials available on their settlement, such as animal and crop rotation, organic inputs, and ground covers, to build their soil's capacities. These agroecological techniques recuperated degraded soils while delinking with expensive chemical inputs, and farmers report improved health conditions.

NYCFM (New York City Food Movement): Social justice and environmental concerns, rather than economic development motifs, have been at the heart of most threads of the food movement in New York City as well. Derelict and dilapidated urban spaces in the late 1970s triggered community groups organizing to convert them in quality green spaces. Most recently, the deepening health inequalities between New Yorkers of different socioeconomic statuses have also been a central driver for food justice activism and community food system innovations such as affordable community supported agriculture (CSAs) (e.g., Corbin Hill Food Project) and youth-run farmers markets. Additionally, the urban agriculture movement has gained further support in the aftermath of extreme weather events, such as Hurricane Sandy in 2011, and risks of flooding and environmental degradation.

SEM (Senegalese Ecovillage Movement): Villages face dire environmental conditions, which are intertwined with difficult social conditions. In the north of the country, the Sahara is arriving where forests existed 60 years ago. Deforestation by colonial powers, villages, and companies have left impoverished landscapes. Organizations such as USAID and the Chinese Government have advocated and subsidized chemical and water intensive rice production to sell nationally and for export, poisoning rivers and mining soils. This constellation of factors has impoverished villages and contributed to hunger, outmigration, and social breakdown. A key dynamic of successful ecovillages has been recuperating the local environments on which they rely, particularly through improving and intensifying soils and through reforestation.

4.2. Reframing Innovations as Political Tools

A key tenet in the MLP on transitions is that regime actors perceive the radical niches of innovation as advantageous and consequently take action to transform current institutions and practices. This need for compatibility between mainstream and innovative practices poses a paradox, but is also a key pressure point for transforming entrenched sociotechnical systems such as energy, food, and housing. Successful transition initiatives in each of the three cases examined have been able to take advantage of this by effectively reframing the social benefits (or the challenge) their innovation is a means to addressing.

MST: The MST has challenged the assumption that large-scale, chemically-intensive industrial agriculture is the only, or the most efficient, way to feed the world, and that the peasant farmer is outmoded. Through their practice, they posit that the peasant farmer is the best steward of the land for intergenerational use, and that agroecological cooperatives, such as the *Grupo Gestor*, can provide high quality, low cost food for their regions, provide livelihoods, and recuperate the earth. Agroecological methods are referred to as technologies that intensify soil, social, and political capacities. The MST has reframed agroecology as a *political tool* for peasants to stay in the countryside. Hannah Wittman [46] has theorized the MST project as agrarian citizenship, in which "political participation, local food production, and environmental stewardship redefine the ongoing constitution of the relationship between land, state, and rural society".

NYCFM: Many of the successes of the New York City food movement are attributable to the timely and effective reframing of the key issue at stake and how it links to the highest priorities on the mayoral agenda. Examples include the reframing of urban agriculture as a tool for social justice [47], environmental and nutrition education, and green infrastructure for climate resiliency; the re-envisioning of farmers markets as a tool for community development and public health; and sustainable regional farming as a tool for safeguarding the city's drinking water. The reconceptualization of food as an urban system and of food justice goals as part of the responsibilities of local government are arguably two of the most consequential shifts in local political discourse over the past decade and a half.

SEM: This Movement has reframed the notion of ecovillages coming from the Global North, and in the process also reframed ideas of West African rural development. The Movement seeks community-led development by taking the best of West African village life and combining this with green technologies and the recuperation of soils and forests. Movement leaders assert that Northern ecovillages are often focused on creating community and ecologically viable worldviews and spiritual systems. African villages, they argue, already possess these social and cultural resources, and seek to bring in "clean modern technologies to uplift living conditions" (interview with Ousmane Pame, July 2016) while recuperating the environments upon which villages depend. Leaders report that the holistic framework of ecovillages is highly resonant with West African traditional worldviews and provides an effective tool for development that respects traditional village culture while opening to the world and introducing technology.

4.3. Openness to Experimentation

Transitions are complex, coevolutionary processes defying any attempt to plan and implement them in a linear fashion. Successful grassroots innovations and movements, as those discussed in this paper, have been able to circumvent this challenge by remaining open to new ideas and experimentation and timely making adjustments in response to changing socioeconomic and political conditions, or internal struggles.

MST: Early in the Movement's history, MST leadership implemented cooperatives based on the Cuban model of agricultural modernization, cultivating monocultures with investment into machinery and chemical inputs. Many early settlements failed, due to high costs and increasing debt on equipment and chemical packages, difficulties in accessing markets and credit, and soil deterioration. A grassroots rebellion in the Movement forced the leadership to adopt more open-ended approaches, with settlements taking initiative and following multiple pathways towards effective production and livelihoods [15]. Through successful experiments at the settlement level, and later regional level with the *Grupo Gestor*, agroecology emerged as one of the most effective new pathways and was adopted as a pillar of the Movement in 2000. The open-endedness of agroecology itself, which proposes the holistic engagement of constellations of social and ecological relationships, has provided a fluid and agile tool for innovation and scaling up.

NYCFM: In New York, examples of the openness of food system entrepreneurs and policymakers to experimentation include pilot initiatives to test different models of curbside composing, demonstration urban farms at a public housing sites, forging new links between local farmers and preschool centers, call centers for food and nutrition assistance benefits, and online school food programs enrollment. Examples in the nongovernmental domain include developing alternative, healthy school food meal deliveries (e.g., Red Rabbit), pop-up drop-off sites for food scraps (e.g., Lower East Side Ecology Center), green jobs for youth through green roof construction, culinary education, and urban agriculture (e.g., Green Bronx Machine [48]), youth-managed farmers markets (e.g., GrowNYC Youthmarkets), and the conversion of industrial buildings' rooftops into food-producing farms (e.g., Eagle Street Rooftop Farm, Brooklyn Grange, Gotham Greens).

SEM: The ecovillage framework was initially adopted by a village being surrounded and subsumed by Dakar, which had been sprawling since at least the 1970s, to defend livelihood and culture. This framework set the foundations for innovative responses, outside of both traditional village modalities as well as mainline development pathways. Village leaders express how the ecovillage model provides a framework to engage the interrelations of culture, economy, technology, and environment, to promote materially and culturally better ways of living over the long term. The ecovillage framework they say is not prescriptive but orients innovative approaches to protracted problems. For example, the ongoing issue of food insecurity is being addressed the village of Mbackombel through installing solar powered microgrids. Among numerous other benefits, this grid powers pumps to store water, and thus expands the growing season, and creates new permaculture gardens, reforestation, and fish ponds. It also frees up young girls charged with getting water to go to school.

4.4. Partnerships and Coalition Building

Coalition building is essential for the alignment and scaling up of niches of innovation. Links between participants with different powers and roles across government and market institutions are also key for the translation of niches' value in terms that can be seen as advantageous by mainstream actors in the socio-technical system.

MST: MST settlements, and the Movement as a whole, realized early on that they needed partnerships to survive politically and physically; as a movement they needed to challenge existing patterns of private property with direct links to the colonial era. The development of the agroecological systems of the *Grupo Gestor* has been accomplished through partnerships with universities, agronomists, religious organizations, and other organic farmers, among others. Beyond this, creating new markets required partnerships with city and state governments, other social movements, and technical support. The MST was a founding member of the global network Via Campesina, the world's largest social movement. Via Campesina has transformed global debate on food and agriculture, introducing democratic principles. Their idea of food sovereignty asserts the rights of peoples to define and control ecologically sound food systems, rather than the demands of international commodity markets and corporations.

NYCFM: While the food movement in New York City is effectively a movement of movements and largely diverse and fragmented, partnerships have played an important role in both stabilizing grassroots innovations and influencing mainstream businesses and policies. Examples of coalitions include the NYC Community Gardens Coalition, which was key in preserving community gardens threatened from development, the now defunct Brooklyn Food Coalition, the New York City Coalition Against Hunger, City Harvest and its Community Action Networks, the NYC Agriculture Technology Collective, and the New York City Food Assistance Collaborative, among others. Cross-sectoral coalitions, such as the New York City Food and Fitness Partnership—a collaborative effort between City Harvest, Brooklyn Rescue Mission, and Transportation Alternatives—have also been essential in scaling up school food and food access initiatives throughout the city. New York City is also part of the cross-city Urban School Food Alliance (established in 2012) together with Orlando, Dallas, Miami, Los Angeles, Chicago, and Fort Lauderdale, and since 2017, Las Vegas, Philadelphia, Baltimore and Boston as well. The Alliance has been successful in mandating antibiotic-free chicken and compostable trays across school districts, collectively influencing more than 3 million meals and thousands of school cafeterias.

SEM: The ecovillage movement was born out of partnership with Northern ecovillages in the U.S. and Europe, as well as the Global Ecovillage Network. These exchanges continue, bringing the strengths of the African and Northern experiences to bear upon each other. The Ecovillage framework was at the center of the development of several coalition organizations, including the Senegal Ecovillage Network, GEN Africa, and village led NGOs, such as REDES, which coordinates the development of five regional villages. Ecovillages work with international organizations, such as the UNDP, Gaia Education, and IFAD, as well as the National Ecovillage Agency (ANEV), which is discussed below, making a new more engaged and effective relationship with the Senagalese state.

4.5. Building and Maintaining Autonomy

Historically, social movements have often been weakened through a mixture of cooptation and coercion, sapping the movements of independent and creative action. Each of these three cases has consciously fought to remain relatively autonomous, while in interaction with state, civil society, and NGO stakeholders. Autonomy has made it possible for the movements to continue to innovate, have policy impact, and scale up their projects.

MST: MST agroecological farmers and cooperatives, challenge the idea of growing food for money (and export), and then using money to buy food. High costs and poor soil quality catalyzed the development of farming methods that intensify soil with what is available on settlements, and to delink from the high costs of chemical inputs from agribusiness (and the high cost of bank credit in Brazil). The MST cooperatives have sought to first build their own self-sufficiency and autonomy (soil inputs, seeds, food, etc.), before extending to build wider exchanges. They argue that this provides independence and stability from varying macroeconomic conditions, as well as a core space of strength in which to act within wider social and political systems. The Movement has been successful at building capacity on the settlements, often in partnership with sympathetic organizations, to train settlement members in areas such as accounting, machine operation, and repair, and perhaps critically, political analysis.

NYC: As innovations scale up, one key dilemma is how to maintain their independence from the government agencies and private companies they are trying to resist and provide an alternative to. The recent rise of a commercial strand of the urban agriculture movement can potentially be coopted by mainstream food businesses and community-based composting initiatives are now gradually being "phased out" via the new city-led pilot programs. Yet, changes in mainstream practices are occurring because the pioneer initiatives were able to be sustainable on their own first. Spaces like Farm School

NYC, kitchen incubators, and the new urban agriculture business incubator continue to provide movement entrepreneurs with the skills and tools to build and maintain their autonomy.

SEM: The Senegal Ecovillage Network (GENSEN) was born out of the first ecovillages in the late 1990's. This network fell apart and the movement split into two heterogenous wings as the federal government became involved with first the Ministry of Ecovillages, and then ANEV. One part of the Movement asserts that the community-led dimensions of ecovillage development are essential, and direct government intervention weakens community agency, creating a situation that looks like other government-led development efforts. The other part of the Movement insists that Government and international aid provides access to crucial and expensive technologies (such as solar power) and infrastructures (such as irrigation), and that villages remain agents in this relationship, participating in decisions of what interventions or resources will be provided. The Global Ecovillage Network (GEN) created GEN Africa, which has become the overarching and unifying organization to which most ecovillages may relate.

4.6. Creating New Markets

The three movements found substantial barriers in the market systems in which they were situated. Each movement worked to create new market relationships, and often value-added enterprises, which became key infrastructures to their economic viability. Importantly, many of the new markets were developed by cooperatives, and in some cases between cooperatives.

MST: Foundational for the scaling up for MST agroecological systems has been the creation of institutional markets at every scale through direct agreement and through policy. Perhaps the most notable policy has been the Food Acquisition Program (PAA), which requires municipal governments to procure up to 30% of their food from family farms for city operations. Other institutional markets include the military, universities, and prisons, and importantly the National Program for School Meals (PNAE). Some cooperatives and organizations, such as the *Grupo Gestor*, also process and package their own brands (rice, milk, sauces, etc.) which are available in MST stores, grocery stores, and are exported. Farmers markets, organized in partnership with city governments and other institutions, and with other organic farmers, have emerged as critical spaces for MST farmers to gain dignified livelihoods by selling their production directly to consumers.

NYCFM: While far from replacing mainstream food production, procurement, retail, and disposal, the multiple streams of food justice activism in New York City have effectively reconfigured the marketplace. Currently, there are over 140 farmers markets, multiple links between the city's over 900 urban gardens and farmers markets, dozens of CSAs (including Community Supported Fisheries [CSFs]), food co-ops, farm to preschool programs, and new regional food hubs (e.g., Greenmarket Co., Lucky Dog Food Hub) now in operation. The pilot city compost collection and recycling program has also effectively been scaled up to now reach over one million New Yorkers. Other new businesses related to food waste, such as the recycling of used cooking oil into biodiesel, have also changed the local food and energy market and established themselves as viable local businesses (e.g., TriState Biodiesel, Grease Lightning).

SEM: In many villages, recuperating soils, creating permaculture gardens, and increasing growing seasons remains the focus, within the context of food insecurity. However, women's groups particularly, as they find success with permaculture methods, are able to gain increasing income. Villager farmers are building on existing institutional markets, such as selling produce to local and regional schools. As value added enterprises are launched (discussed below), these are also creating new market opportunities.

4.7. Mobilization of Women's Groups

Women's groups played decisive roles in the three grassroots social movements. Often, it was women or women's groups that created key new practices and infrastructures, which supported the movements and helped scale up movement projects.

MST: Although the Movement continues be led at the highest levels disproportionately by men, women have organized effectively within the Movement to create greater gender balance. For example, all elected coordinator positions from the settlement to the national level must be composed of one man and one woman. This gender balance within the movement organization has been foundational on the settlement level for experimenting with agroecological practices, which were often proposed and first implemented by women who sought to protect their families from sickness and economic hardship. One example was with the transition of a dairy operation to agroecology in the settlement COOPAVA. The first change proposed was to treat the cows with kindness, instead of with the historic rough treatment using dogs, horses, and whips. Women on the settlement embraced this proposal and led the initiative. University technicians report that changing the treatment of the cows increased milk production by 25% within one month.

NYCFM: Women and women's groups have been a powerful driving force behind the NYC food movement. The city's first community garden was initiated by Liz Christy in 1973, Christine Datz spearheaded the first community composting program and founded the Lower East Side Ecology Center in the 1980s, Annie Novak (together with Ben Flanner) co-founded the city's (and U.S.'s) first commercial food producing rooftop farm in 2009; Onika Abraham directs the first urban farming training program in the city Farm School NYC (co-founded by Ursula Chanse, Lorrie Clevenger and others); Karen Washington founded Black Urban Growers (BUGS) and directed the NYC Community Garden Coalition; and Linda Goode Bryant founded Project EATS—Active Citizen Project, to mention a few. Other noteworthy women-led initiatives, which have ignited the NYCFM, include Hot Bread Kitchen, an ethnic breads company allowing immigrant women an opportunity to start their own businesses; La Finca del Sur, a women-led community farm in the South Bronx; the Harvest Home farmers markets network, led by Maritza Owens; and Community Food Advocates, co-founded by Kathy Goldman (previously founder of the Community Food Resource Center 1980–2003) and Agnes Molnar who, together with Liz Accles, Jan Poppendieck (co-founder of the New York City Food Policy Center and the CUNY Urban Food Policy Center and author of *Free For All: Fixing School Food in America* [49]), and others, led a successful campaign for universal free school lunch in New York City.

SEM: Women and women's groups have often led the way in transforming food production in the ecovillages. For example, in the village of Djara, women's groups have created extensive permaculture gardens, providing the majority of village food, while building soil health. Each garden is a mixture of collective and family plots. The men of the village continue with chemical and water-intensive rice cultivation, which has had at best mixed success financially while adding to significant health problems in the village due to significant pesticides in their only water sources. Many villagers report symptoms of pesticide toxicity, such as joint pain and stomach problems. Women have also used the ecovillage framework to assert women's agency in formal village life, and in making direct relationship with Northern ecovillages and NGO's.

4.8. Mobilizing Public Institutions (While Maintaining Autonomy)

Each of these movements has not only affected policy but has been able to mobilize public institutions at critical intersections. The movements, while all founded as agri-food movements, expanded their understandings of what is required for alternative agri-food systems. In these very diverse cases, these understandings are articulated differently, but each calls for the support of public and private institutions to build what might be loosely termed as citizenship rights.

MST: Through occupations, advocacy, and politics, the movement has been able mobilize the redistribution of land for almost a million members. This is less than the land reform initially envisioned when the movement began. They have been more successful for the *struggle on the land*, mobilizing city, state, and federal institutions to provide citizenship rights for settlements including schools, healthcare, roads, electricity, and other infrastructure. Many have argued the focus on the *struggle on the land* is part of what has made the MST more successful than many other landless movements around the world. The Movement has sought (and sometimes struggled) to remain autonomous while actively participating in formal politics. The MST was a founding force of the Workers Party (PT), which has held power at all levels and the majority of members continue to support. There have been numerous MST members elected to political office at local, state, and national levels. In some regions, where settlements are concentrated, the MST has taken electoral control of rural towns. The Movement has struggled internally with how much to push sympathetic governments (for instance, through land occupations) and how much to work with them (building infrastructure, new markets, etc.).

NYCFM: In December 2017, New York City passed a bill (Intro 1661-A) to create the city's first centralized digital hub for urban agriculture. This is just one example of how the alignment of bottom-up innovations, in this case commercial urban agriculture companies, can mobilize institutions to change the rules. Other examples include the expansion of the universal free lunch program, achieved through joint efforts by nonprofit advocacy organizations, like Community Food Advocates, and government officials like the City's Public Advocate. Under pressure of environmental groups in the food movement, the city also recently carried out a comprehensive food system resiliency study (2016), which assessed its degree of disaster preparedness.

SEM: As discussed above, GENSEN was able to help create the world's first Ministry for Ecovillages. This achievement was recognized with the GEN meeting held in Dakar in 2014. Although government involvement remains divisive, many villages report that they are able to mobilize financial and technical resources from ANEV and other international organizations, while continuing to be community determined.

4.9. *Affecting and Participating in Policy*

These successful movements were all able to, in different ways, begin to affect policy in ways that then fed back into their practices to support scaling their work. These policies also served to support transformation in consciousness of wider communities interlinked with the movements. Critically, beyond social and political pressure, each movement was able to utilize or create ongoing processes to propose policy and actively participate in policymaking.

MST: A key explanatory factor for the success of MST cooperatives in transitioning to, and scaling up agroecological systems, has been their ability to affect, participate in, and create policy at municipal, state, and federal levels. The Movement's success is in part due to its ability to participate in building policy that sets the stage for expanding its political and agricultural projects. The MST helped pressure social dimensions of the 1989 Constitution, which legitimized their struggle. Perhaps most crucial in the scaling of agroecology to regional levels has been the creation of policy for institutional markets, guaranteeing large purchases of food from family farms, such as the Food Acquisition Program (PAA) and the National School Food Program (PNAE). These programs that provide high quality food to schools, hospitals, and other public institutions through government purchases were designed as guaranteed and stable markets for agroecological cooperatives.

The Movement has worked with state governments to transform the industrial bias of agricultural support. For example, in the State of Rio Grande do Sul, subsidies were put in place for the support of organic and agroecological farming, including organic fertilizer, technical support, and infrastructure, such as irrigation, and support for building local markets. The Movement also has participated in the design of state and federal educational policy, building government-funded technical schools with specialties in areas such as agroecology and cooperative management. The MST voice is present in global forums through the Via Campesina and the food sovereignty perspective (see above).

NYCFM: Over the past two decades, the different strands of the New York City food movement have been able to reconfigure part of the local food system regime through concerted and sustained activism and coalition-building. Community groups have effectively prevented community gardens to be sold out for development in the 1990s and, more recently, the NYC Community Gardens Coalition saved nearly 70% of the community gardens that were threatened by affordable housing development. Other policy changes include the Zone Green amendment incentivizing rooftop greenhouses, the introduction of universal free lunch for all public-school students, the increase of the minimum wage for fast food workers, the introduction of food procurement standards for city agencies, the ban on trans fats, and the requirement for calorie and sodium labeling for chain restaurants.

Many of the successes and the expansion and scaling up of local food initiatives are attributable to a blend of tactics that have enabled community food advocates to participate decision-making processes. Among these are community board meetings, participatory budgeting, demonstrations, legislative hearings chaired by City Council and the state, and electoral forums as the precedent-setting 2013 Mayoral Forum on Food Policy. Recently, food justice and food access advocates testified before City Council on how to revise the city's Food Retail Expansion to Support Health (FRESH) program offering tax and zoning incentives to developers to integrate fresh and healthy food retail in designated high-need neighborhoods.

SEM: The expansion of the Ecovillage Model has been driven both through the grassroots, as well as through government initiatives which villages both inspired and participated in building. Perhaps most important was the development of the Ministry of Ecovillages, following a visit of by the Country's President to a series of ecovillages in the late 2000s. This Ministry was set up with the task to transition half the country's 28,000 villages into ecovillages. A few years later, this project was moved to a new National Ecovillage Agency (ANEV) under the environment Ministry. ANEV seeks to involve and support the villages with development assistance that villages request. This includes interventions such as implementing solar power, providing seeds, infrastructure for irrigation, and technical support. The formal power structures of villages vary between elected mayors and hereditary chiefs. In both cases, villages have taken on the ecovillage framework usually with the leadership, or at least with the strong support of, these formal village positions. Thus, government resources are leveraged directly towards ecovillage development at the village level, as villages make this a political focus. Village leadership is also then able to formally interact with federal organs, particularly with ANEV.

4.10. *Access to Land and Land Tenure*

The three movements had different relations to land and land access. The thread weaving through the three movement histories is that questions of land tenure were political from the beginning. Whether through occupations, or in defense of traditional lands, the movements encountered powerful resistance, often by some of the most powerful segments of their societies, to establishing their agricultural practices.

MST: The MST was born through the desire of farmers displaced by the Green Revolution and Military Government to gain land tenure. The Movement continues to pressure governments through advocacy, occupations, and politics to fulfill the constitutional mandate to redistribute unproductive land, as well as to provide citizenship rights on the settlements, such as education, roads, electricity, and healthcare. The main tool has been to occupy unproductive land as a kind of rural strike to force the federal government to fulfill the constitutional mandate.

NYCFM: Community gardens are often under pressure from more lucrative commercial and residential land uses. While the market overwhelmingly favors built-up spaces, the NYC food movement has been successful in institutionalizing a formal Garden Review Process (since 2010) that requires developers and the city to seek alternative sites for the relocation of existing community gardens (The Rules of the City of New York, Title 56, Chapter 6–05). Most importantly, several community-led land trust groups like the Brooklyn Queens Land Trust (which helped prevent over 120 gardens from being auctioned in the 1990s), the Bronx Land Trust, and the Manhattan Land Trust have been essential in helping urban gardeners stay on the land. The City's Green Thumb program and the nonprofit 596 Acres (which ceased operations in Summer 2018, after seven years of sustained advocacy) have also been playing a central role in facilitating access to public land and scaling up community food system initiatives.

SEM: The Movement began in part as resistance to land grabbing by public and private entities as Dakar expanded to encircle and subsume traditional villages. Farther away from urban centers, most villages have access to lands, but are often historically degraded by deforestation, poor farming practices, and overgrazing.

5. Discussion

The results of this paper confirm and extend existing theories of socio-technical transitions in three main ways. First, the manifold set of levers, or pressure points, uncovered through the comparative analysis corroborates the hypothesis that radical system-wide transitions occur when a rich and diverse set of strategies are deployed, and transition champions engage with the concurrent transformation of different segments of the mainstream regime—markets, government policies, physical infrastructures, and social norms and practices [19]—through multiple, sustained reiterations in time.

Second, it also echoes prior findings on the importance of the interplay between different levels of power—relational, dispositional, and structural [50]—in sociotechnical transitions. According to MLP theorists [51], niche innovators spur system transformation by leveraging their *relational powers* stemming from their connectedness and unity with other grassroots entrepreneurs while regime players, in turn, bring about (or resist) change through their *dispositional powers* by using existing legislative and regulatory mechanisms. Finally, landscapes, or the aggregate of niche and regime groups, and the economic, ideological, and environmental settings they operate in are used to collectively entrench (or disrupt) existing systems through their *structural powers*. As our cases reveal, strong networks of grassroots innovators can, thus, mobilize mainstream institutions and businesses to use their dispositional powers and change formal rules and policies, which, in turn, can upset mainstay beliefs and conventions of what is, or should be, normal.

Third, the analysis of the three cases also reconfirmed the already known paradox [29] in transition processes, or the need for radical niche innovations to exhibit some degree of compatibility with the dominant systems and sociotechnical regimes that seek to overturn or reconfigure. In fact, building and maintaining autonomy is a key trait of successful sustainability innovations, but even more so the ability to mobilize existing institutions while maintaining autonomy.

One of the limitations of the research presented in this paper is that, while drawing on a markedly diverse and rich set of cases, we cannot make the claim that our findings are generalizable beyond the three cases analyzed. This was a purposefully qualitative, exploratory case study that afforded an in-depth examination of emerging themes and shared traits across three cases, yet these findings remain grounded in the specific study settings we chose to study. An additional caveat is that agri-food systems are socio-technical and inherently complex and dynamic, and so even local policymakers and activists seeking to apply the findings from this paper need to proceed with caution. The systems that yielded those findings may well no longer exist—coalitions disband and reform, values may shift, and technologies are rapidly morphing into new infrastructures and services. Yet, while developing a universal theory of scaling up niche innovations was beyond the scope of this paper, this does not preclude the possibility that the findings and recommendations we put forward are relevant for, or applicable to, current circumstances or geopolitical contexts other than those we examined.

Future research would benefit from delving deeper into questions about the role of political entrepreneurs in steering niche innovations, the relationship between different streams of funding and the trajectories and longevity of sustainable development initiatives, as well as the tactics that transformative niches of innovations use to cope with failure and seemingly intractable challenges as they seek to transform entrenched systems of production and consumption. An overt examination of the possible downsides of scaling up grassroots innovations or connecting them would also afford a clearer understanding of the possible limitations of approaches seeking to replicate and normalize place-based solutions.

6. Conclusions: Pressure Points and Policy Recommendations

To effectively address the sustainability crises our planet faces, including those stemming from a behemoth yet fragile global agri-food system, we have suggested that it is necessary for decision-makers at different levels of government worldwide to engage three challenges: learning from Global North and South initiatives in tandem, taking stock of social innovations alongside technological fixes, and nurturing grassroots sustainable development initiatives next to, or in place of, top-down corporate and government projects and interventions. In this research we sought to address the question of what key pressure points for guiding socio-technical transitions to sustainability exist and what is the scope for learning from success cases from Global North and Global South countries in tandem. We addressed these questions by exploring the accomplishments of three distinct social movements: the Senegalese ecovillage movement, MST agroecological cooperatives, and the New York City food movement. Our findings reveal that the successes of those movements in reconfiguring dominant systems of production and consumption lie in a rich amalgam of factors, which all point to the importance of movement's "soft skills" and the ability to build robust social infrastructures alongside transformations of the physical environment. Specifically, among these skills are the movement's ability to:

- reframe the key issues at stake;
- remain open to experimentation;
- forge diverse cross-sectoral partnerships and coalitions;
- amass political support and affect policy;
- create self-sustaining new markets;
- nurture and encourage women leadership;
- secure access to land and land tenure;
- build and maintain autonomy from mainstream systems and institutions;
- mobilize public institutions to change rules and practices;
- be actively engaged and participate in policymaking processes.

Policy Recommendations

Based on the findings from our cross-case analyses, we offer a set of recommendations for government decision-makers at all scales interested using pressure points to steer socio-technical transitions to sustainability. The emphasis on steering and pressure points over command and control and execution of blueprints is key in that contemporary societies, and the systems of provision they rely on, such as the agri-food system, are increasingly complex and tend to evolve rapidly in a nonlinear fashion. Thus, while comprehensive plans, targets, and indicators are essential tools in planning for local and global sustainability, implementation relying on 20th-century theories of change is unlikely to succeed.

An agile planning and implementation, acknowledging the impossibility to have complete information about the systems we seek to transform and their complexity, is therefore a more promising and, in light of the findings of this paper, we argue, necessary approach. Further, to make global sustainability targets and indicators meaningful to local administrations as diverse as the ones we explored in this paper, a granular understanding of the levers (and barriers) that have accompanied agri-food transitions already underway is warranted. Our analysis of sustainability innovations in Brazil, New York, and Senegal revealed a series of pressure points that local actors have acted on in seeking transformative change and durable transitions to healthier, more equitable, and environmentally sound systems and communities. We suggest that future sustainability planning and implementation will thus benefit from grounding action in emergent evidence of what bundles of actions work here and now in addition to, or despite of, institutionalized rules and practices, which often become obsolete by the time a plan gets to the stage of implementation. In particular, we suggest that governments consider the five strategies listed below, with the caveat that they are conceivable only if existing norms, expectations, and institutions change as well:

1. Where resources are scarce, governments, instead of continuing a path-dependent momentum, should support movement innovation and alternatives, which are embedded and responding to local physical and cultural geographies;
2. Include innovative movements in debate and policy making and support movement-led policy that builds new movement capacity and innovation.
3. Support movement autonomy, through supporting conditions for self-sufficiency. By investing first in movement self-sufficiency, this provides a foundation to nurture or strengthen innovations;
4. Support movement value-added ventures, even if value-added alternatives challenge regulations or path dependencies of the present system;
5. Support the design of flexible, territorially-sensitive policies and plans. Rigidity of policies and indicator frameworks, both local and global, is one of the most frequent reasons for their failure and even rejection by local communities.

Yet, while these five strategies were key in scaling up the grassroots innovations we investigated, it would be naive to suggest that local governments can pursue them in the absence of conducive institutions. New political spaces [52] and creative bureaucrats [53,54] open to working at the margins of routinized practices and comfortable collaborations are core prerequisites for engaging with any of these tactics. Organizational innovations, such as the Ministry for Ecovillages in Senegal or the Office of the Director of Food Policy in New York, are two among many other examples of the importance of nonconventional spaces and leaders within existing regimes for scaling up grassroots innovations.

Ultimately, our research reveals that grassroots sustainable development initiatives are advancing some of the most creative system-wide transformations and transitions toward climate adaptation, resilience, and sustainability in the agri-food system. While not universal, and contingent upon foresighted, open-minded governments, we suggest that these five broad strategies, implemented with the participation of grassroots social movements, and embedded in local social and ecological conditions, may help catalyze the creative innovations needed to create socially and ecologically resilient and sustainable forms of collective life.

Author Contributions: Both authors contributed to the conceptualization and writing of this manuscript and the ideas it contains.

Funding: This research received no external funding.

Acknowledgments: We are deeply thankful for the time and energies given by members and allies of the Brazilian Landless Movement, New York City Food Movement, and the Senegalese Ecovillage Movement. Without their openness and interest in the critical investigation of their projects, this research would not have been possible. We are also grateful for the valuable insights and constructive feedback provided by the three anonymous reviewers on earlier versions of the manuscript.

Conflicts of Interest: The authors declare no conflict of interest.

References

1. Smith, A.; Seyfang, G. Constructing grassroots innovations for sustainability. *Glob. Environ. Chang.* **2013**, *23*, 827–829. [CrossRef]
2. Hielscher, S.; Seyfang, G.; Smith, A. Grassroots innovations for sustainable energy: Exploring niche-development processes among community-energy initiatives. In *Innovations in Sustainable Consumption: New Economics, Socio-Technical Transitions and Social Practices*; Cohen, M.J., Szejnwald Brown, H., Vergragt, P.J., Eds.; Edward Elgar: Cheltenham/Northampton, UK, 2013; pp. 133–158.
3. Monaghan, A. Conceptual niche management of grassroots innovation for sustainability: The case of body disposal practices in the UK. *Technol. Forecast. Soc. Chang.* **2009**, *76*, 1026–1043. [CrossRef]
4. Feola, G.; Nunes, R. Success and failure of grassroots innovations for addressing climate change: The case of the Transition Movement. *Glob. Environ. Chang.* **2014**, *24*, 232–250. [CrossRef]
5. Seyfang, G.; Haxeltine, A. Growing grassroots innovations: Exploring the role of community-based initiatives in governing sustainable energy transitions. *Environ. Plan. C Gov. Policy* **2012**, *30*, 381–400. [CrossRef]

6. Kirwan, J.; Ilbery, B.; Maye, D.; Carey, J. Grassroots social innovations and food localisation: An investigation of the Local Food programme in England. *Glob. Environ. Chang.* **2013**, *23*, 830–837. [CrossRef]
7. *Cultivating Food Justice: Race, Class, and Sustainability*; Alkon, A.H., Agyeman, J., Eds.; MIT Press: Boston, MA, USA, 2011; ISBN 0262016265.
8. Broad, G. *More Than Just Food*; University of California Press: Oakland, CA, USA, 2016.
9. Allen, P. Mining for justice in the food system: Perceptions, practices, and possibilities. *Agric. Hum. Values* **2008**, *25*, 157–161. [CrossRef]
10. Feng, J.; Glass, T.A.; Curriero, F.C.; Stewart, W.F.; Schwartz, B.S. The built environment and obesity: A systematic review of the epidemiologic evidence. *Health Place* **2010**, *16*, 175–190. [CrossRef] [PubMed]
11. Lee, H. The role of local food availability in explaining obesity risk among young school-aged children. *Soc. Sci. Med.* **2012**, *74*, 1193–1203. [CrossRef] [PubMed]
12. Pan, L.; Sherry, B.; Njai, R.; Blanck, H.M. Food insecurity is associated with obesity among US adults in 12 states. *J. Acad. Nutr. Diet.* **2012**, *112*, 1403–1409. [CrossRef] [PubMed]
13. Robertson, G.P. Greenhouse Gases in Intensive Agriculture: Contributions of Individual Gases to the Radiative Forcing of the Atmosphere. *Science* **2000**, *289*, 1922–1925. [CrossRef] [PubMed]
14. Garnett, T. Where are the best opportunities for reducing greenhouse gas emissions in the food system (including the food chain)? *Food Policy* **2011**, *36*, S23–S32. [CrossRef]
15. Karriem, A. The rise and transformation of the Brazilian landless movement into a counter-hegemonic political actor: A Gramscian analysis. *Geoforum* **2009**, *40*, 316–325. [CrossRef]
16. Wittman, H. Reworking the metabolic rift: La Vía Campesina, agrarian citizenship, and food sovereignty. *J. Peasant Stud.* **2009**, *36*, 805–826. [CrossRef]
17. Holt Giménez, E.; Shattuck, A. Food crises, food regimes and food movements: Rumblings of reform or tides of transformation? *J. Peasant Stud.* **2011**, *38*, 109–144. [CrossRef] [PubMed]
18. Patel, R. *Stuffed and Starved: The Hidden Battle for the World Food System*; Melville House Pub: Brooklyn, NY, USA, 2012; ISBN 1612191274.
19. Geels, F.W. Technological transitions as evolutionary reconfiguration processes: A multi-level perspective and a case-study. *Res. Policy* **2002**, *31*, 1257–1274. [CrossRef]
20. Geels, F.W. Processes and patterns in transitions and system innovations: Refining the co-evolutionary multi-level perspective. *Technol. Forecast. Soc. Chang.* **2005**, *72*, 681–696. [CrossRef]
21. Geels, F.W. Ontologies, socio-technical transitions (to sustainability), and the multi-level perspective. *Res. Policy* **2010**, *39*, 495–510. [CrossRef]
22. Hughes, T.P. The evolution of large technological systems. *Soc. Constr. Technol. Syst. New Dir. Sociol. Hist. Technol.* **1987**, *82*, 86–89.
23. Silverberg, G.; Dosi, G.; Orsenigo, L. Innovation, Diversity and Diffusion: A Self-Organisation Model. *Econ. J.* **1988**, *98*, 1032. [CrossRef]
24. Geels, F.W. The hygienic transition from cesspools to sewer systems (1840–1930): The dynamics of regime transformation. *Res. Policy* **2006**, *35*, 1069–1082. [CrossRef]
25. Van den Ende, J.; Kemp, R. Technological transformations in history: How the computer regime grew out of existing computing regimes. *Res. Policy* **1999**, *28*, 833–851. [CrossRef]
26. Geels, F.W. Analysing the breakthrough of rock 'n' roll (1930–1970) Multi-regime interaction and reconfiguration in the multi-level perspective. *Technol. Forecast. Soc. Chang.* **2007**, *74*, 1411–1431. [CrossRef]
27. Uyarra, E.; Gee, S. Transforming urban waste into sustainable material and energy usage: The case of Greater Manchester (UK). *J. Clean. Prod.* **2013**, *50*, 101–110. [CrossRef]
28. Raven, R.P.J.M. Towards alternative trajectories? Reconfigurations in the Dutch electricity regime. *Res. Policy* **2006**, *35*, 581–595. [CrossRef]
29. Smith, A. Green niches in sustainable development: The case of organic food in the United Kingdom. *Environ. Plan. C Gov. Policy* **2006**, *24*, 439–458. [CrossRef]
30. Geels, F.W. A socio-technical analysis of low-carbon transitions: Introducing the multi-level perspective into transport studies. *J. Transp. Geogr.* **2012**, *24*, 471–482. [CrossRef]
31. Rotmans, J.; Kemp, R.; van Asselt, M. More evolution than revolution: Transition management in public policy. *Foresight* **2001**, *3*, 15–31. [CrossRef]

32. Loorbach, D.; Rotmans, J. Managing Transitions for Sustainable Development. In *Understanding Industrial Transformation: Views from Different Disciplines*; Olsthoorn, X., Wieczorek, A.J., Eds.; Springer: Dordrecht, The Netherlands, 2006; pp. 187–206.
33. Kemp, R.; Loorbach, D. Governance for Sustainability through Transition Management. Work. Paper. 2003, pp. 1–27. Available online: http://sedac.ciesin.columbia.edu/openmtg/docs/kemp.pdf (accessed on 6 November 2018).
34. Schot, J.; Geels, F.W. Strategic niche management and sustainable innovation journeys: Theory, findings, research agenda, and policy. *Technol. Anal. Strateg. Manag.* **2008**, *20*, 537–554. [CrossRef]
35. Kemp, R.; Schot, J.; Hoogma, R. Regime shifts to sustainability through processes of niche formation: The Approach of Strategic Niche Management. *Technol. Anal. Strateg. Manag.* **1998**, *10*, 175–195. [CrossRef]
36. Geels, F.W. The multi-level perspective on sustainability transitions: Responses to seven criticisms. *Environ. Innov. Soc. Transit.* **2011**, *1*, 24–40. [CrossRef]
37. Hargreaves, T.; Haxeltine, A.; Longhurst, N.; Seyfang, G. *Sustainability Transitions from the Bottom-Up: Civil Society, the Multi-Level Perspective and Practice Theory*; CSERGE Working Paper 2011-01; EconStor: Kiel, Germany, 2011.
38. Smith, A.; Hargreaves, T.; Hielscher, S.; Martiskainen, M.; Seyfang, G. Making the most of community energies: Three perspectives on grassroots innovation. *Environ. Plan. A* **2016**, *48*, 407–432. [CrossRef]
39. Seyfang, G.; Longhurst, N. Desperately seeking niches: Grassroots innovations and niche development in the community currency field. *Glob. Environ. Chang.* **2013**, *23*, 881–891. [CrossRef]
40. Yin, R. *Case Study Research: Design and Methods*, 2nd ed.; Sage: Thousand Oaks, CA, USA, 1994.
41. Kay, C. Reflections on rural violence in Latin America. *Third World Q.* **2001**, *22*, 741–775. [CrossRef]
42. Robles, W. The landless rural workers movement (MST) in Brazil. *J. Peasant Stud.* **2001**, *28*, 146–161. [CrossRef]
43. Branford, S.; Rocha, J. *Cutting the Wire: The Story of the Landless Movement in Brazil*; Latin America Bureau: London, UK, 2002; ISBN 1899365516.
44. Freudenberg, N.; Cohen, N.; Poppendieck, J.; Willingham, C. *Food Policy in New York City Since 2008: Lessons for the Next Decade*; CUNY Graduate School of Public Health and Health Policy: New York, NY, USA, 2018.
45. United Nations Development Programme (UNDP) "L'Ecovillage de Mbackombel sort de l'ombre". Available online: http://www.sn.undp.org/content/senegal/fr/home/ourwork/environmentandenergy/successstories/-l_ecovillage-de-mbackombel-sort-de-lombre.html (accessed on 29 October 2018).
46. Wittman, H. Reframing agrarian citizenship: Land, life and power in Brazil. *J. Rural Stud.* **2009**, *25*, 120–130. [CrossRef]
47. Reynolds, K.; Cohen, N. *Beyond the Kale: Urban Agriculture and Social Justice Activism in New York City*; University of Georgia Press: Athens, GA, USA, 2016.
48. Ritz, S. *The Power of a Plant: A Teacher's Odyssey to Grow Healthy Minds and Schools*; Rodale Inc.: Emmaus, PA, USA, 2017.
49. Poppendieck, J. *Free for All: Fixing School Food in America*; University of California Press: Berkeley, CA, USA, 2011.
50. Grin, J.; Rotmans, J.; Schot, J. *Transitions to Sustainable Development. New Directions in the Study of Long Term Transformative Change*; Routledge: New York, NY, USA; Abingdon, UK, 2010.
51. Grin, J. The Governance of Transitions: An Agency Perspective. In *Transitions to Sustainable Development: New Directions in the Study of Long Term Transformative Change*; Grin, J., Rotmans, J., Schot, J., Eds.; Routledge: New York, NY, USA, 2010; pp. 265–284.
52. Boudreau, J.-A. Making new political spaces: Mobilizing spatial imaginaries, instrumentalizing spatial practices, and strategically using spatial tools. *Environ. Plan. A* **2007**, *39*, 2593–2611. [CrossRef]
53. Landry, C. *The Creative City: A Toolkit for Urban Innovators*; Routledge: New York, NY, USA; Abingdon, UK, 2012; ISBN 1136554483.
54. Block, T.; Paredis, E. Urban development projects catalyst for sustainable transformations: The need for entrepreneurial political leadership. *J. Clean. Prod.* **2013**, *50*, 181–188. [CrossRef]

sustainability

MDPI

Article

Strategy for the Sustainability of a Food Production System for the Prosperity of Low-Income Populations in an Emerging Country: Twenty Years of Experience of the Peruvian Poultry Association

Alejandro Fontana [1,*], Ignacio de los Ríos Carmenado [2], Johan Villanueva-Penedo [1], José Ulloa-Salazar [1] and Denisse Santander-Peralta [1]

1 PAD Escuela de Dirección, Universidad de Piura, Av. Aldebarán 160, Lima 15023, Peru; j.villanueva.penedo@gmail.com (J.V.-P.); jose.ulloa.salazar@gmail.com (J.U.-S.); dsantander@mba.pad.edu (D.S.-P.)

2 GESPLAN Research Group. School of Agricultural, Food and Biosystems Engineering, Universidad Politécnica de Madrid, Calle Ramiro de Maeztu, 7, 28040 Madrid, Spain; ignacio.delosrios@upm.es

* Correspondence: afontana@pad.edu; Tel.: +51-999-166-073

Received: 29 September 2018; Accepted: 30 October 2018; Published: 3 November 2018

Abstract: This research shows a business initiative that has been able to integrate into an environmentally sustainable food production system, such as poultry farming, a positive impact on food security and public health patterns of low-income populations in an emerging country. For a process that took 20 years, the adopted strategy has become a positive experience of sustainability and prosperity in low-income populations in Peru. The objective of the research is to conceptualize and identify the key elements of this experience so that its replication in other food production systems to impact favorably the prosperity of such vulnerable population. The Working With People (WWP) model, a validated methodology for analyzing the sustainability and prosperity of rural areas in Europe, is used for the analysis of this experience. The analysis shows that the presence of the three dimensions of this model (political-contextual, technical-business, and ethical-social) ensure the sustainability of a food production system that has an impact on the prosperity of low-income populations in emerging countries. This balance is important to enrich the connections between sustainability and prosperity, with other concepts such as core values in companies, public-private cooperation, food safety, inclusion and consumption patterns.

Keywords: sustainability; food systems; nutrition; public-private cooperation; corporate values; sanitary control; food safety; prosperity

1. Introduction

The case study shows a private initiative that has integrated the perspective of an environmentally friendly food production, such as poultry farming [1], into a positive impact on consumption and social inclusion, and a contribution to public health standards.

The Peruvian Poultry Association, founded in 1938, is a group of companies dedicated to raising poultry products, and which brings together the whole domain of companies in the sector. Currently, the eleven most important companies represent 85% of market share. On the initiative of the board of directors of this association, a system of poultry meat production has been developed in Peru, which has allowed to families who had had minimal access to any kind of animal protein to increase it steadily in the last 20 years.

During this period of time, the actions promoted by the Peruvian Poultry Association (APA) have generated production to grow from 400 thousand tons annuals to almost 2 million tons annually. In parallel, the annual consumption of low-income families has been increased from 10 kg per year to 35 kg per year. On the other hand, this initiative has also improved the health safety of poultry meat.

The objective of the research has been to analyze this case study with the purpose of extracting from it some lessons that can be recommended for the creation of food production systems that have an impact on the prosperity of populations with low resources in emerging countries. In this sense, the research question has been formulated as: What strategy can ensure the sustainability of food production systems that impact the prosperity of populations with fewer resources from emerging countries?

To analyze this experience, we have used the Working With People (WWP) framework [2], a methodology previously validated for the analysis of the sustainability of production systems and its contribution to the prosperity of rural populations [3,4]. This methodology allows identifying the critical variables of the success of an experience through the analysis of its three dimensions: territorial-contextual; technical-business; and ethical-social.

The analysis of the case study under the WWP methodology shows that an initiative of an association of companies can generate a system of food production that benefits the low-income people of an emerging country. This analysis underscores the objective of the business initiative must focus on the prosperity of society.

At the same time, this case validates the assertion that the strategy of a food production system that contemplates the three dimensions of the WWP model ensures the sustainability of a system that impacts the prosperity of low-income populations in an emerging country.

The article has been structured collecting, in the first place, some experiences on the sustainability of food production systems; then the methodology used has been described, including a detailed description of the case, its conceptualization and the elements that demonstrate its sustainability; thirdly, the case analysis under the WWP model is presented under its three dimensions.

1.1. Sustainability: Relevance and Scope

Two important controversial issues impact the present times: on one hand, the rapid increase of environmental challenges caused by the climate change and a greater awareness of the environmental impact; and on the other hand, the ethical requirement to contribute to prosperity of low-income population: poverty eradication and hunger [5]. Also, this dilemma needs to be solved in the medium term. At current growth rates, by 2050, an increase of between 60% and 110% in global food production will be needed [6]. And this increase will take place in a context of climate change and without compromising resources needed to serve future generations. Parallelly, according to the World Bank, the number of people under the poverty line of US$ 1.90/day in the world has increased in 2.5% between 2016 and 2017: increasing to 768 million of people [7].

For Rockström et al. [6], the solution to this dilemma must focus on the transformation of intensive agricultural systems, so that agriculture ceases to be the biggest global change factor to become a key contributor in the transition to a sustainable world. However, as Brklacich et al. presented, the concerns about the sustainability of food production systems is not only be confined to agriculture systems and their impact in the environment degradation [8].

These authors presented in 1991 a detailed and exhaustive review of the academic literature about different dimensions for the concept of sustainable food production systems [8]. Thus, they found out that six major themes underpinning sustainability: environmental accounting, sustained yield, carrying capacity, production unit viability, product supply and security, and equity [8].

Environmental accounting refers to biophysical limits for agricultural systems and has been presented as the most important theme for sustainability [8]. Sustained yield describes the group of conditions that would stabilize crop yields from year to year [9]. This theme was extended to agricultural activities from the forest sector.

Carrying capacity refers to the maximum population size who could use a site without causing permanent damage to the natural environment. However, as Brklacich et al. pointed out, the estimation of population that may be supported by a particular site is affected by the interactions among a wide range of socioeconomic, behavioral, and biophysical factors. This way, "estimating population size which can be supported by a particular region is central to measuring the sustainability of food production systems" [8] (p. 7).

Production unit viability focuses on the performance and viability of individual farms. "The viability of farms is an important concern within the broader context of sustainable food production systems. Without adequate economic rewards for agricultural producers, it will be increasingly difficult for farms to remain in food production system" [8] (p. 8).

Product supply and security means the adequacy of food supply levels to meet the population minimal requirements for nutrition and satisfy cultural demands or dietary preferences. In fact, food security cannot be reduced to just one economic activity: food is a complex matrix that demands a holistic approach to capture the complexity of nutrition and its impact on health [10,11].

Equity refers "to fair access to production opportunities and a balanced distribution of production costs, goods, and services associated with resource use" [8] (p. 9). This way, it is a common requirement that enhancing prospects for sustainable food production must include conditions for improving living conditions for the least advantaged groups within society: "increasingly, a secure food supply that is accessible to all members of society is being recognized as a vital component of a sustainable food production system" [8] (p. 9).

There are authors who recognize there cannot be sustainability without inclusion. A system of food production cannot be sustainable if it is not equitable: if it does not meet the basic needs of all, and at the same time, it opens opportunities to improve their quality of life. Not doing so would mean being exposed to the problems of major migrations, geopolitical crises, and social conflicts [12].

Equity has also an intergenerational level of application. The concern is "meeting the needs of the present without compromising the ability of future generations to meet their own needs" [13] (p. 43). This is also recognized by Food and Agriculture Organization (FAO): the sustainability of an agricultural system implies ensuring food security and nutrition for all now and in the future [14].

Brklacich et al. propose a definition of sustainable food production system: "an agri-food sector that over the long term can simultaneously: maintain or enhance environmental quality; provide adequate economic and social rewards to all individuals and firms in the production system; and produce a sufficient and accessible food supply" [8] (p. 10).

Keoleian and Heller [15] argue that improving the sustainability of food production systems requires a clear understanding of the relationships between the behavior of population consumption, food processing, distribution activities, and production practices. Thus, these authors conclude that an adequate understanding of the life cycle of the product is a very useful tool to understand the existing relationships between social needs, and social and natural processes. These relationships allow understanding those needs and the environmental impact of the satisfaction of said needs.

Another factor mentioned in the academic literature for the sustainability of the productive system is the impact of food on public health. The current concern is given by the appearance or reappearance of human diseases that are of zoonotic origin. Even the term One Health is used to refer to the essential relationship that exists between human health and animal health. In this sense, a practical recommendation is the reduction of the use of antibiotics in the care of animals [10].

This multiplicity of social and productive factors implies that responsibility for the sustainability of food production systems does not rest solely with the state. Van Meijl, Ruben and Reinhard [12] argue that sustainability is everyone's job and that the business sector is called to participate in it, making inclusion and sustainability part of the company's core values. On the other hand, Verburg et al. [10] argue that food security is linked to government policies, that is, a strategic vision that looks at the long term, and therefore demands a governance approach: an adequate interaction between the state, the market and the civil society [14,16], and effective institutions to obtain good results in sustainable agricultural development strategies [14].

1.2. Prosperity

Prosperity has been conventionally defined as the economic growth, but this definition has not been able to account for the negatives environmental consequences, serious material impacts that compromise future possibilities and lack of opportunities for all. A prosperity focuses on few people, and founded on ecological degradation and steadily social injustice is not appropriated for a fair society [17].

From this point of view, Jackson [18] has popularized a new approach of prosperity: the human beings flourishing, the achievement of greater social cohesion and higher levels of well-being while reducing their material impact on the environment.

Indeed, he accepts that prosperity has material dimensions to ensure adequate supply of food, shelter, clothing, water, etc. without an environmental degradation. But, he also explains that a human being is also affected by psychological and social dimensions. Thus, everyone needs a meaning and purpose in life, and to participate in building on the common good of one's community: working usefully, enjoying respect to other people, or voluntarism. In other words, a sense of community and meaningful work is essential: "ensures nutritional health, takes part in the life of the community, uses their education, finds worthwhile jobs, appears in public without shame, visit friends and relations" [19] (p. 15).

This vision of prosperity may serve us to precise the most well-known and well-used dimensions of sustainability for food production systems: economic, social, and environmental. "The economic, social and environmental have been the most well-known and well-used pillars of sustainability" [20] (p. 13). For example, when Castellini [21] compared three poultry production systems, he defined the social dimension as labor safety, this is, he only took into account the impact of the different production processes on worker health, and not how those systems are socially affecting the whole society.

2. Methodology

The methodology used was the analysis of a case study that occurred in an emerging country. The access to official information and also the access to the most relevant actors of this case has allowed assuring the internal and external validity. In fact, each step of the analysis process has been documented to ensure its reliability [22]. The APA case study is located in Peru, a country in South America with an area of more than 1,200,000 km^2 and also with a complex geography (Figure 1).

The objective of the research has been to find out the variables that can ensure the sustainability of a food production system that contributes to the prosperity of low-income families of emerging countries, and the methodology used for this analysis has been the WWW framework.

WWP proposes a 'new approach' for sustainability and prosperity in postmodernity in rural and food production systems. Key to the WWP model is 'planning as social learning' and a 'new postmodern sensibility' [2,23,24]. The name Working With People was chosen to convey the need to overcome the traditional technical-economic vision of prosperity, and the need to focus on individuals' behavior and the context in which they work.

In 2016, the WWP framework has been validated as an instrument to analyze rural development, rural prosperity and resilience in a world of growing demands and limited resources. This validation was done under the international RETHINK research program supported by European Commission and funding bodies in 14 countries under the FP7 and the RURAGRI ERA-NET called RETHINK [3,4]. "The research members of Rethink formed a multi-sectorial and inter-disciplinary Expert Panel to create the reporting guidelines on each theme. The Expert Panel provided research advice on effective strategies and ways of rural prosperity and established the common guidelines for the creation of questionnaires based on the methodological framework Working With People (WWP)" [3] (p. 87).

Figure 1. The political map of Peru. Source: National Geographic Institute.

The WWP model seeks to integrate expert knowledge with experienced knowledge, and also connect knowledge with action, but incorporating the values of the different participants of the flow of goods or services of any project [3].

In this sense, this model considers behavioral competences and a solid ethical basis; contextual competencies; and technical competencies as key elements for the sustainability of a system that has an impact on the prosperity of a society [3]. This methodological framework goes beyond the technical-economic view, and integrates social learning processes in the analysis and the building of prosperity from three dimensions: ethical-social, technical-entrepreneurial and political-contextual. These three components interact through social-learning processes (Figure 2), and include the four fields of a social-relationship system, as defined by Friedmann [25]: the political field, the public administration field, the private and entrepreneurial field, and finally the civil society field. The apparent simplicity of WWP involves a large social complexity [2] given the richness of the relationships and lessons that occur between the three components of the model.

From the Political-contextual dimension, organizations involved ought to adapt their priorities and their projects in the context in which they work to achieve sustainable management [3]. This dimension covers the ability to make relations with political organizations and with the different public-administrations that enhance sustainable development and thereby foster the prosperity of rural areas and its agricultural communities. The design of modern strategies for the sustainability of a food production system requires social integration and anticipating the long-term trends affecting rural areas. The configuration of WWP strategies for the sustainability must ensure that organizational change processes and structural processes are generated to allow adaptation to the priorities of involved people, also working with actors from the political and public administration fields. WWP organization has, therefore, an instrumental character, to serve the population, and it is flexible and changing according to the learning and the new information generated. This way, the WWP organization becomes a living entity, which transmit values to society—from its ethical component—and is capable of influencing and changing political priorities and to work together to achieve sustainable management. This political-contextual dimension builds on the understandings of prosperity as capabilities for flourishing [18,26].

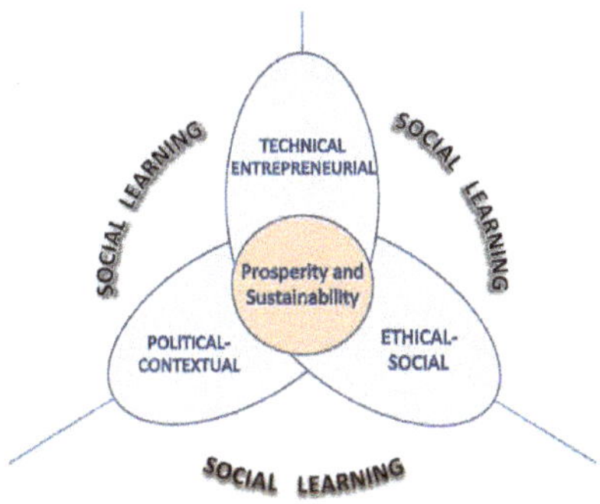

Figure 2. Dimensions of sustainability and prosperity from the WWP model. Source: [2]. With authors' permission.

The Technical-business dimension is oriented towards the generation of sustainable products and services for the society based on quality standards and providing environmental, social and economic benefits. From the point of view of social relations, this dimension corresponds to the private-entrepreneurial field, which comprises all activities of private initiative. The WWP project adopts a "business function"—as mobilizing human, economical, public, private resources—leading to the arrangement and negotiation between various actors and involves a commitment to assume and manage risk. This technical-business dimension exceeds economic imperative and include social and environmental criteria [27,28] and multi-stakeholder business governance to creates rich social networks that allow new sustainable strategies [29], recognizing that—given a diminishing marginal utility of goods—"more is not always better". Given this view, the merely commercial and financial aspects of WWP project are exceeded; it serves not only to achieve "tangible" benefits to care about the "intangible" benefits in the form of expansion of knowledge, and social and cultural aspects to develop sustainable solutions [2].

And from the Ethical-social dimension considers the context of behaviors, attitudes, and values of people who interact throughout production processes. This component is identified with the social subsystem, consisting of all interpersonal relationships that are taking place within society. The ground of the social system that surrounds WWP project is to cover conduct and moral behavior of stakeholders and it sets out the "foundations" to make people, both from private and public fields, come to work together, with commitment, confidence and personal freedom [2]. The incorporation of ethics, means considering the project as not 'neutral', but based on an ideal of service and guided by values. This dimension integrates behavioral competences with ethics and values as appropriate elements to overcome potential moral conflicts related to the interested parties involved in the project [30]. The WWP project seen from this dimension tries to achieve the best for a greater number of people, especially the poorest people with few resources and the participation of the beneficiaries for the effective sustainable development and modernizing the farming practices, like sustainable food systems.

Finally, the integrating dimension of the three dimensions is social learning [25]. These dimension to ensure social learning processes among the different subsystems, which lead to learn from the real agents of change. The social learning process runs with the main assumption that all effective learning comes from the experience of reality change. The population affected by the project actively participates in planning, with their own behaviors, attitudes and values—ethical-social component—to promote and manage the WWP project. Therefore, it requires to generate actions directed to integrate the experienced knowledge [31] of the affected population, along with the planner's expertise, providing mutual learning. To do so, reliability is essential among the agents of the different social subsystems, generated from responsibility, a proper behavior, rigor, trustworthiness, and an open and consistent attitude [30]. Similarly, to ensure these social learning areas and processes, it requires to have a proper appreciation of values, defined as the ability to understand the inherent qualities of others and understand their points of view. This leads us again to say that ethics and behavior of people involved are the basis of WWP project.

According to the principles set up within the WWP framework, the analysis has been made from two types of data: qualitative and quantitative data. First, during the last six years there have been several interviews with the key actors of this intervention. In 2003, an interview was conducted with the Director of National Agrarian Health Service (SENASA), who previously was the head of Animal Health in 1996, and the person who represented SENASA at the beginning of the relationship with APA. Also, an interview was conducted with the person of APA who represented the association in that relationship. Additionally, another interview was undertaken with the CEO of APA. Then, to ensure full and complete participation of the different company sized members of APA, a fourth interview was carried out with one of the CEOs of the smallest company members of APA. In 2015, the CEO of APA was interviewed again. All of these interviews were recorded.

Then, in 2018, and in order to update the data for this research, a focus group was carried out with the President of APA; the CEO of APA; the person in charge of the market access project of APA (who was the first head of the national program of Poultry Health by SENASA, in 2004); and the head of the poultry and egg committee of APA. Then, a phone interview was conducted with the CEO of the leading company of the sector, and finally a last interview was carried out with the President and the CEO of APA.

For all of these in-depth interviews, semi-structured questionnaires were prepared. The questions were divided into the three dimensions of the WWP framework: territorial-contextual, technical-business, and ethical-social in order to recognize the key variables of the APA initiative on each one of these dimensions. The interviews were conducted face to face with the stakeholders involved.

Information was also obtained from previous investigations, statistics and official documents of the government of Peru, which were collected to examine the evolution of APA actions and their impact on the sustainability of a food production system that benefits the population of fewer resources in the country.

To ensure the internal validity of the case, a quantitative analysis was carried out based on the National Household Survey (ENAHO), which the National Institute of Statistics and Informatics (INEI) performs each year at the national level with the intention of measuring, individually and by family, different attributes of the population. This survey has modules on employment, health, income, expenses, consumption, among others, which have representation at the national, regional, rural, urban or geographical levels.

Specifically, Household Expenses module and the SUMMARY module for expenditures has been used. The data analysis software used has been STATA 15.0, and the methodology has had two parts: processing and calculation. The processing included the review and identification of the variables to be used and the household identifiers; the unification of the databases; and the creation of the necessary variables for doing the calculation. It also involved the estimation of the values of the different variables. In this way, as the sample is representative at a national level, the evolution of the average annual consumption of poultry by families between 2004 and 2017 were found for three segments: extremely poor, poor, and non-poor.

Regarding the limitations of the quantitative analysis, information is only available since 2004. On the other hand, the poverty categorization used by the INEI is based on the calculation of a poverty line linked to the monetary and non-monetary value of a basket of consumption appropriate to the circumstances of the respondent's region.

The Case of the Peruvian Poultry Association (APA)

In 1998, the APA consisted of 13 contributing companies that accounted for more than 70% of the poultry production in Peru, however it included all the producer companies of poultry products in the country as members of the association. As part of its activities, the association held the representation of the companies of the sector before the government and organized, also collectively, several commercial campaigns. The factor that had most influenced the collective action was the success of the commercial campaign to spread the quality of the chicken raised in industrial farms.

In the years prior to 1998, companies in the sector had to face several crises: from a market of prices controlled by the State in the 1970s and 1980s, the market had switched to a system of free competition; the sector had stopped being exonerated the 18% of value added tax, an amount that had to be absorbed by the companies themselves, and that caused the bankruptcy of several of them. And finally, in 1996, the business association and twelve companies in the sector were accused of price agreements.

In 1998, the per capita consumption of poultry meat per year in Peru was 19.5 kg, and represented the main source of animal protein consumed by Peruvians: 46% of animal protein. In addition, the sector represented 1.16% of the national GDP, and 14.8% of the agricultural sector [32].

It was in these circumstances that the APA board realized that poultry production and marketing was more than a simple exchange of consumer goods for an amount of money. If the need they were satisfying was a component of the prosperity of the Peruvian population, then they should act seeking the food and nutritional security of the entire population. With these concerns, they realized the importance of working to improve their productivity. Their main cost overrun was the expense caused by chicken diseases: drop in feed conversion, losses of chickens, and a high expenditure on medicines for chickens focused in carrying out a health control program. The APA board also considered that competition between companies in the sector would ensure the transfer of said savings to the consumer (Interview with the CEO of the leading company of the sector and interview with a member of the APA Board of directors in 1996).

In this way, the APA board decided to approach the National Agrarian Health Service (SENASA), the decentralized entity of the Ministry of Agriculture in charge of sanitary control. However, as SENASA did not have the poultry sector in its operational plan, it did not have the financial resources to run a vaccination program for domestic poultry, fowl and migratory birds. Therefore, the program had to be financed by private companies until public resources generated by a national sanitary regulation could be available. In this sense, it was also necessary to work on this regulation, which was estimated to go into effect in less than two years (Interview with the head of animal health of SENASA in 1996).

In these circumstances, the eight largest companies in the sector decided to finance this program for three and a half years, until the sanitary standard set a rate for all industrial poultry breeders (Interview with a member of the APA Board of directors in 1996).

3. Results: The Analysis of the Case Study under the WWP Framework

In order to demonstrate the research hypothesis of this work, the case study is analyzed under each one of the WWP dimensions: political-contextual; technical-business; and ethical-social. This analysis allows to identify the key variables and also to organize them systematically, so the model could be replicated in emerging countries. For this reason, first of all, the intervention's strategy has been drawn. Secondly, the key variables emerged from the different interviews conducted and the public information collected have been written down under the appropriate WWP dimension.

3.1. Political-Contextual Dimension: Conceptualization of the Intervention Strategy

A first key element of the intervention strategy of this case (Figure 3) has been the adequate definition of the problem: the main cost overrun is due to the lack of sanitary control in poultries that are outside industrial farms.

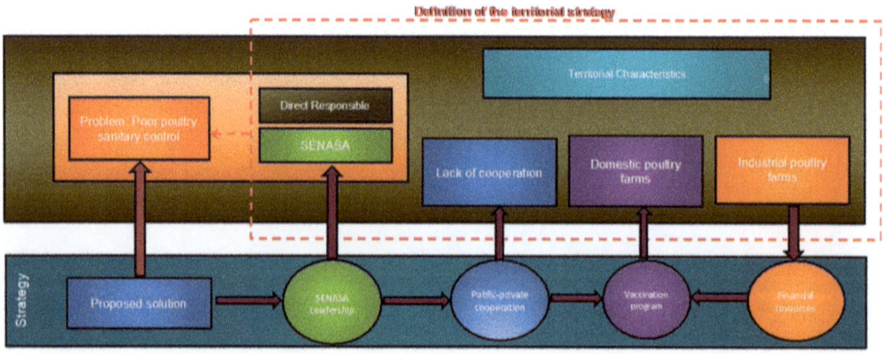

Figure 3. The definition of the intervention's strategy. Source: Own elaboration. SENASA: National Agrarian Health Service.

A second key element was the definition of the territory. An important aspect of the sanitary control program is the definition of the scope. Initially, the poultry companies considered necessary only the attention of the farms adjoining their farms. However, the SENASA staff helped APA to understand that the risk came from any poultry that was in the country, so that the measures had to be extended to the entire national territory, including domestic farms which raise poultries for self-consumption or for small commercialization scale, fowl, and migratory birds (interview with the head of animal health of SENASA in 1996).

A third key variable of this dimension has been the horizontal cooperation between the private companies in the sector. In 1998, APA had been in operation for 60 years, and the owners of the 13 associated poultry companies took an active part in the association. This horizontal cooperation was also manifested in the joint decision of the eight main companies in the sector to finance the vaccination program for three and a half years. "We started by asking the leader for his contribution, and when we had it, we told each of the rest if we could count on theirs because we already had the leader's contribution. Then each one said yes" (interview with an APA member of the Board of Directors in 1996).

Another key variable has been the public-private cooperation between APA and SENASA. Until 1996, there was no articulation between the companies in the sector and SENASA. Each one did his tasks according to his best criteria. The industrial farms worried about the running of their business, and they did not allow any public official to enter their facilities or share their information. In addition, as the international cooperation agencies did not grant loans for assistance to the poultry sector, in the annual plans of SENASA there was no such sector. And since the evaluation of public officials is based on compliance with the objectives and goals included in their annual plans, there was no reason to think about a health control program for the poultry sector. This cooperation required several attempts and efforts of the directors of both institutions. The Head of Animal Health of SENASA in 1996 commented: "It took two years of meetings and constant conversations between APA and SENASA to build trust and find a common vision for the future" (interview with a member of APA Board of directors in 1996 and the Head of Animal Health of SENASA in 1996).

Other key element of the intervention strategy was to give prominence to the direct people in charge: SENASA, who was in charge of the direction and management of the vaccination program for domestic and combat poultries, and migratory birds. Table 1 shows the number of animals vaccinated between 2003 and 2012.

Table 1. Number of vaccinated domestic poultry per year.

Province	2003	2004	2005	2006	2007	2008	2009	2010	2011	2012
Ancash							1155			
Piura									23,048	11,711
Ica	2814	775,684	713,066	565,871	119,143	259,155	330,551	269,893	304,294	291,667
Cajamarca							6	16,023	3877	3135
Arequipa	142,428	300,671	321,554	273,950	39,923	245,205	202,111	125,490	83,466	217,537
Lambayeque							310			17,477
La Libertad	321,137	156,763	353,217	706,255	138,115	392,331	444,430	299,736	464,396	519,172
Madre de Dios				1071		1304	4443		9530	348
Moquegua							2349			
Lima y Callao	30,639	519,266	536,805	828,100	69,905	308,007	514,716	137,949	592,642	793,016
Puno					1297	17,268				
Tacna	13,584	4501	2677	5853	8541		62,135	44,606	76,831	57,842
Apurimac							1960		4132	11,602
Total	510,602	1,756,885	1,927,319	2,381,100	376,924	1,223,270	1,564,166	893,697	1,562,216	1,923,507

Source: SENASA data. Own elaboration.

The last key perspective of this dimension has been the articulation of the four social actors: the business sector, public administration, the political sphere, and civil society. This articulation was essential for the design and promulgation of the national sanitary control system with an international standard, that was given in 2007. The promulgation of this norm signified an articulation plan, which had as its origin, the work between APA and SENASA (Interview with the general director of Animal Health of SENASA in 1996). The Figure 4 shows the most important events during all the process.

Event	Beginning of conversations between APA-SENASA	Signature of the agreement between APA-SENASA	Beginning of the vaccination program	Beginning of the National Health Poultry Program	Declaration as free avian influenza country	Regulation of the Poultry Health System
Year	1996	2000	2002	2004	2005	2007

Figure 4. The relation of the most important events during the process. Source: SENASA, 2010. The management of SENASA for the poultry health system: roles and obligations. Own elaboration. APA: Peruvian Poultry Association.

In this way, from the perspective of the planning of the public sphere, the intervention has been characterized by having two complementary movements. A bottom-up movement: the private initiative at the beginning of the intervention, typical of Social Learning planning model [25] and a top-down movement: the securing of public recourses and the action of SENASA, typical of Policy Analysis planning model [25]. The WWP model [2] proposes that in order to ensure the sustainability of an intervention in a given territory, both planning models are required.

3.2. Technical-Business Dimension: Results and Sustainability of the Production System

Regarding the technical-entrepreneurial dimension, the relevant variables have been the generation of a sanitary control system and a series of actions to improve the productivity of the sector. For the Head of Animal Health of SENASA in 1996, the most significant result of this intervention has been the generation of a sanitary control system that was defined with the new poultry health regulation, which was enacted in November 2007.

For the generation of this health control system, the initial financing of the largest companies in the sector played a relevant role. However, the sustainability of this system depends in a determining way on public resources that are generated from a mandatory rate for all industrial farms.

The improvement of productivity is also a consequence of geo-referencing all of the 1500 poultry farms in the country; the nationwide enumeration of the 45,000 domestic farms and fowl breeders; the annual vaccination of approximately 2 million domestic poultries; the surveillance of poultries in small farms, domestic breeding and migratory birds of 11 wetlands.

Another relevant technical variable for the sustainability of this production system has been a feature of the system that companies in the sector have known how to hold: the short value chain. The Peruvian market has always been characterized by consuming only the chicken that has been benefited on the same day. This causes that in Metropolitan Lima, a city of 9 million inhabitants, daily benefit and distribute between 750,000 and 1,000,000 chickens in the 7195 poultry sale stalls, in a window of time that only goes from 3:30 a.m. at 8:00 a.m. (interview with the President of APA in 2018; interview with a poultry distributor and owner of sale stalls) [33,34].

In relation to 1998, the production of poultry meat has multiplied by 5.8, as shown in Figure 5. At the same time, annual per capita consumption has gone from 19.5 kg in 1998 to 50.52 kg in 2017 (Figure 6); and in that period, it has been the animal protein that has shown the most internal consumption (Figure 7).

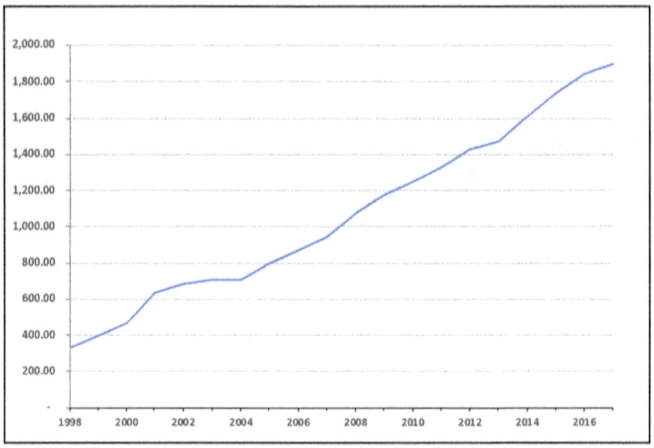

Figure 5. The evolution of poultry meat production. Source: Agrarian Regional Directions – Direction of Agrarian Information. Ministry of Agriculture and Irrigation (MINAG). Own elaboration.

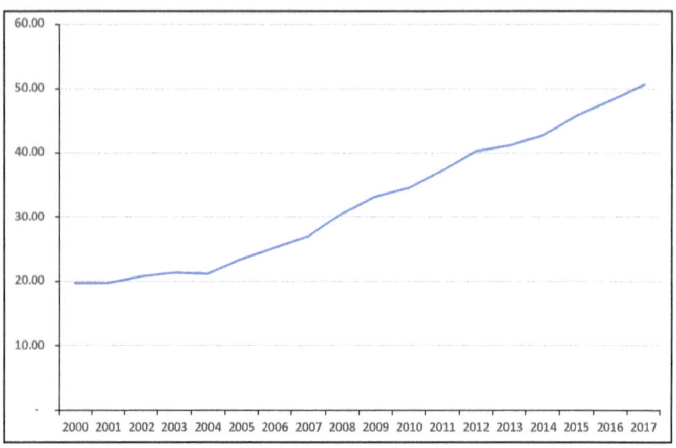

Figure 6. The evolution of internal demand of poultry meat. Source: Agrarian Regional Directions—Direction of Agrarian Information. MINAG. Own elaboration.

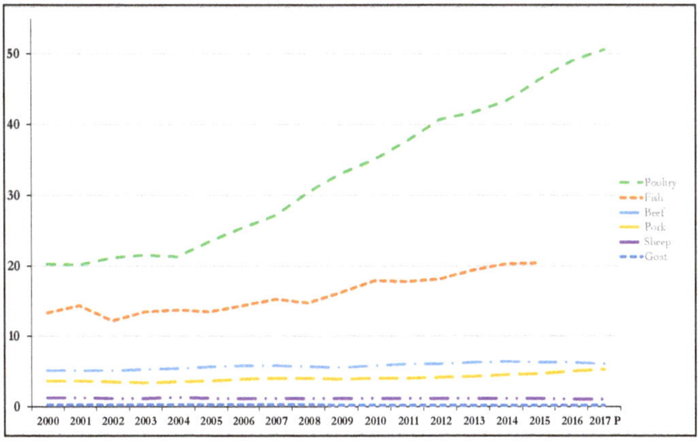

Figure 7. The evolution of internal demand for different types of source of animal protein. Source: Regional Directories of Agriculture; Ministry of Production (Produce); FAO; MINAG. For 2017, some results were still projected. Own elaboration.

On the other hand, in Figure 7 it can also be seen that the greater demand for poultry meat has not been at the expense of the consumption of the other sources of animal protein: it is a higher consumption for people who already consumed poultry or the consumption by people who previously had not consumed any type of animal protein.

Regarding the impact that this intervention has had on populations with fewer resources in the country, Table 2 shows the comparison of the evolution of the average consumption of poultry meat for extremely poor, poor, and non-poor households between 2004 and 2017. The three segments substantially increase their consumption: the extremely poor segment has an increase of 250%; the one of poor, 27%; and the one of non-poor, 20%. The greatest impact of this production system is precisely in improving the diet of the population with the least resources in the country.

Table 2. Comparison of the evolution of poultry meat consumption in extremely poor, poor and non-poor families in Peru (kg/per family per year).

Year	Extremely Poor			Poor			Non Poor		
	Average	Inf. Limit	Sup. Limit	Average	Inf. Limit	Sup. Limit	Average	Inf. Limit	Sup. Limit
2004	10.40	9.24	11.55	55.52	53.31	57.72	85.37	82.77	87.97
2005	12.14	10.42	13.86	60.20	57.43	62.97	89.46	86.79	92.12
2006	8.83	7.80	9.87	52.04	49.64	54.43	93.62	90.92	96.33
2007	8.83	7.67	9.99	49.98	47.58	52.38	95.10	92.79	97.40
2008	29.08	25.23	32.93	71.68	68.30	75.05	113.81	111.46	116.16
2009	33.81	28.66	38.95	68.88	65.73	72.02	112.31	109.91	114.70
2010	29.46	26.65	32.28	66.40	63.08	69.73	111.89	109.49	114.28
2011	27.33	24.67	29.98	67.89	64.67	71.12	111.00	108.87	113.13
2012	30.08	26.48	33.69	67.44	64.42	70.46	109.38	107.35	111.42
2013	40.22	36.65	43.78	77.98	74.14	81.82	106.23	104.55	107.90
2014	28.73	25.50	31.96	69.85	66.76	72.94	106.41	104.79	108.02
2015	30.56	27.58	33.55	67.05	63.65	70.45	103.51	101.94	105.08
2016	28.81	25.96	31.66	67.27	64.10	70.45	101.47	100.06	102.88
2017	35.87	29.17	42.58	70.34	67.03	73.66	102.28	100.86	103.70

Source: Own elaboration since National Household Survey (ENAHO) data.

In the extremely poor segment, the positive impact has also occurred in the subsistence family economies of the rural areas of the country. These families usually raise poultry in order to have a product that they can commercialize in case of facing an illness or a special celebration. As the sanitary control program has had a national scope, these families have been also benefited, because the program has reduced the mortality of their chickens (Interview with a member of the APA Board of directors in 1996).

On the other hand, Figure 8 shows the evolution of the problem of anemia and child malnutrition, detailing its effect in the rural area. The improvement and positive evolution of the ratios of anemia and chronic child malnutrition in Peru are due to a series of activities carried out by the State to improve the living conditions of children: supplementing the infant diet with micronutrients; better control of their development; vaccination and education campaigns. But to these activities, greater access to food must also be added. The Demographic and Family Health Survey shows that in 2004 the percentage of infants who received some type of animal protein was 61%; the year 2013, 65.8%; and the year 2017, 69.8% in nursing infants and 94% in non-nursing infants [35].

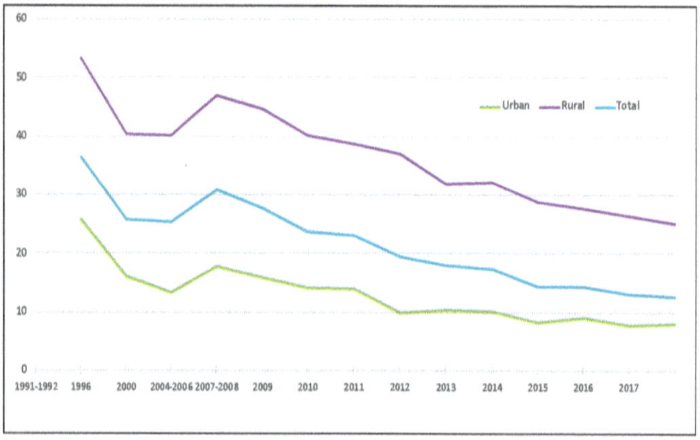

Figure 8. The evolution of infant malnutrition in children under five years of age according to area of residence in Peru (%). Source: National Statistics and Information Institute (INEI). Demographic and family health survey.

It should also be mentioned that the national health system implemented by the APA initiative has resulted in a decrease in the consumption of antibiotics to combat poultry diseases. This decrease in the use of antibiotics improves the safety of this protein source (Interview with the President of APA 2018).

The Head of Animal Health of SENASA in 1996 also commented that thanks to this joint work between APA and SENASA, avian influenza did not enter Peru, despite having appeared in Chile and Colombia. As a result of this, the declaration of an avian influenza-free country was given in June 2005, and the possibility of entering the Japanese market in December of that same year.

Carranza [36], former Minister of Economy of Peru, pointed out that there have been few sectors in the Peruvian economy that have gained this level of productivity. In a national conference, he explained the poultry sector has been able to manage external shocks, to improve the food safety conditions and to work together with the public sector.

The sustainability of this food production system is manifested in the results obtained in the production of poultry meat in the last 20 years; in the increase of the animal protein consumption of the populations of fewer resources; in improving the safety of this protein source; and in the impact it has had on the prosperity of society.

3.3. Ethical-Social Dimension

This dimension has been present in a transversal way throughout the intervention, however, the social awareness that has had the largest companies in the sector has been essential. Several authors suggest that business associations are instruments to face similar problems in several companies [37–42].

On the other hand, as the cooperation between them in a common objective to the sector, it does not mean that each company does not maintain its own competitive objectives, these common interests can respond to strategies that benefit only the dominant companies in the industry [43].

In this sense, the performance of the APA has had a different objective. Moved by a social sensibility, the association has acted focused in the search of the prosperity of the Peruvian population with fewer resources. The CEO of the leading company commented: "If we could do something to contribute to prosperity of people with fewer resources, we had to act on it. It is an important food for people" (Personal phone interview).

So that they asked themselves how they should manage this business. What objectives should companies pursue in the sector? and what specific actions could be promoted from the APA to contribute to the improvement of families with few nutrition resources? (Interview with the President of the APA in 2018).

This interest for contributing to prosperity of the national population also was concerned to some of the SENASA officials. The head of Animal Health of SENASA in 1996 commented that he was one of those who considered that public administration should be involved with the private sector in the sanitary control of the poultry sector: "the agrarian health is a public good, and it claims pretty much that all of us work for it. It is about serving society as a whole" (Interview with the head of Animal Health in 1996).

On the other hand, this cooperation did not alter competition among companies in the sector: Carranza [36], argues that it has been a highly competitive sector. In addition, these measures did not harm small companies in the sector either. Referring to the mandatory rate for industrial farms, the general manager of one of the small companies of APA commented: At the beginning, there were many interpretations. Some of the directors of small businesses thought it would be restrictive for small companies. However, it was finally seen as a self-assessment, in order to regulate ourselves, and be more competitive. And he added: "the disease eradication program was not a measure of the dominant companies in the sector to prevent the growth of small ones. There was no bad intention. SENASA regulated and ordered the sector" (Interview with the general manager of a small associated company).

4. Discussion

The objective of this research has been to analyze the case of the APA initiative in order to draw the key variables which have to be taken in account to create a sustainable food production system, that impacts positively on the prosperity of low-incomes families in emerging countries. For this reason, the research question was set as what kind of strategy can assure the sustainability of a food production system which impacts the prosperity of low-resources people in an emerging economy.

The literature review has suggested to start out our analysis for the sustainability of a food production system using the definition proposed by Brklacich et al.: "an agri-food sector that over the long term can simultaneously: (1) maintain or enhance environmental quality, (2) provide adequate economic and social rewards to all individuals and firms in the production system, and (3) produce a sufficient and accessible food supply" [8] (p. 9).

Also, this analysis considers the definition of prosperity proposed by Jackson [18] and the theoretical framework of WWP. Thereby, the analysis of this study case allows plotting the main features of a strategy for the sustainability of a food production system targeting the prosperity of low-income populations in an emerging country.

Now, we are going to discuss these features of this strategy. First, the strategy starts out from a food production system which is itself environmentally friendly. Broiler poultry is the most environmentally efficient livestock among lamb, beef cattle, eggs, swine, dairy (milk), and turkey. "chicken broiler is the most efficient with an input of 4 kcal of fossil energy per 1 kcal of broiler protein produced" [1]. Beef cattle varies between 20:1 and 40:1 and lamb raises up to 57:1 [1].

Also, producing 1 kg of beef requires about 43,000 L of water. In contrast, 1 kg of broiler chicken can be produced with 2.6 kg of grain, which requires approximately 3500 L of water [1].

Comparing the animal protein consumption of fossil energy to the plant protein consumption (1 kcal of animal protein requires approximately 10 times the energy expended to produce 1 kcal of plant protein) [1], Pelletier criticizes the statement of environmentally friendly given to the poultry livestock, and suggests to promote the consumption of plant [44]. However, as Pimentel also mentions, "it should be noted that animal protein is 1.4 times more nutritious for humans than plant protein" [1]. Thereby, for emerging countries where there are high levels of malnutrition, especially in rural areas, the broiler chicken should be accepted as an adequate supply of animal protein.

According to Van Mejil, Ruben and Reinhard [12], sustainability should be everyone's job and the business sector should be called to participate in it. The study case shows precisely this participation from the beginning of the initiative. Also, the companies' core values have been the engine of this initiative which has pursued the prosperity of the low-income people of the country.

Thereby, the study case shows that an initiative of business companies can lead to the creation of a sustainable food production system. This private initiative should satisfy some conditions:

First, sustainability and prosperity require a change of mentality in business men that allows them to establish public–private partnerships and links with civil society. Herrero et al. state that "business as usual investments in agriculture, although necessary are unlikely to deliver sustainable solutions as the world rapidly change" [45] (p. 822). The aim of that initiative should be the prosperity of the whole society, especially the low-income resource people.

Thereby, the social dimension of sustainability should concern the prosperity of society. For example, when Castellini [21] presents his comparative analysis of three poultry production systems, he considers labor safety as the social dimension of sustainability. Instead, the link between sustainability and prosperity allows understanding that a sustainable food production system cannot be built over an unfair social system [17], but over one which allows the human flourishing of the whole society [18].

Second, the initiative should accomplish adequately three dimensions: ethical-social, technical-business and political-contextual.

From the ethical-social dimension, it can be affirmed that in the strategy of the APA, this component plays an essential role both in the concept of the initiative and in its execution. Interpersonal relationships and behaviors lay the foundations for the managers and partners of the APA—together with other actors from the public and private spheres—to work and advance together towards the design of a strategy that ensures the sustainability of the production system of foods that impact on the prosperity of populations with fewer resources. These processes allow improving people's abilities and competencies, with ethics and values as fundamental elements to overcome possible moral conflicts and to work as a team.

This ethical requirement is according to the Brklacich, Bryant and Smit statement of their three-dimensional definition of sustainable food production system (SFPS): "by casting this three-dimensional definition for a SFPS in the context of long-term goals for agricultural land use and food production, the equity perspective transcends all the three dimensions [8] (p. 10).

APA has developed a strategy that influences the technical-business dimension of sustainability and prosperity, allowing the creation of a solid business structure that generates goods and services to society and the most vulnerable populations. The initiative has allowed to the companies to reduce production costs by cutting inefficient expenditures in medicines for poultry, and also by decreasing the number of death birds by pests. And this savings has been translated to consumer by decisions of the main companies and the high competition among all the companies in the sector. Thereby, the strategy has achieved a considerable increase of the broiler chicken market: the annual consumption has raised from 19.5 kg/hab in 1998 to 50 kg/hab.

Thus, the strategy of the APA has firstly ensured the accomplishment of the second dimension of the Brklacich et al. definition of SFPS: provide adequate economic and social awards to all individuals and firms in the production system; and secondly, to fulfill simultaneously the third dimension of their definition: produce a sufficient and accessible food supply.

The study case, thereby, has accomplished the three components proposed in the Brklacich, Bryant and Smit definition of SFPS. However, the analysis of the APA experience with the theoretical WWP framework permits finding out unnoticed key variables for the Brklacich et al. definition of SFPS. This is the political-contextual dimension of the WWP framework.

The strategy of APA to ensure the sustainability of food production systems and generate prosperity also has an important political-contextual dimension, contributing to a strategic vision of its activity of territorial scope. From this dimension, the APA improves the sustainability of its food production system by complementing the actions of the business sector and public administration. These synergies and complementarities are especially effective when carried out from alliances based on trust, commitment, and reliability among partners, creating favorable environments for good governance of policies. The capacities of the managers of these associations to establish interrelations and negotiations between agents (public and private) of different levels (regional, national or international) and to form strategic alliances, is a key factor for sustainability and prosperity. The experience of the APA has generated horizontal cooperation between companies and cooperation between the business sector and the public sector, which have been relevant to the sustainability of their food production system.

In this way, the results confirm the need for a balance between the different dimensions of the strategy: ethical-social, technical-business, and political-contextual to ensure the sustainability of food production systems that affect the needs of populations with fewer resources.

Finally, the results also confirm the need for a balanced action among the different social actors to lead to effective solutions to problems. In general, the case shows that a private initiative, business or social, can solve a social problem. Therefore, from an academia perspective, it would be advisable to propose the generation of more private initiatives that focus on solving social problems.

5. Conclusions

The sustained success of this business association is achieved when, in addition to generating profitability in the associated companies, other values are generated that affect the improvement of the quality of life of people. The companies incorporate social and environmental values, and guarantee a means of subsistence for producers and businessmen, and cover the needs of the population.

The case of APA provides lessons of experience for the transformation of the conventional systems of government of the business associations, towards a greater sustainability. In other words, when businesses focus on the prosperity of low-income populations in an emerging country, a sustainable system of food production can be generated, and then positively impact the prosperity of these populations. The formation of a business association (APA) from the WWP model has managed to generate a food production system that affects the three dimensions necessary to ensure sustainability and prosperity.

The case study shows that an initiative of an association of companies can generate a sustainable system of food production when considering the three dimensions of the WWP model: political-contextual, technical-business, and ethical-social. In this way, it tries to answer the challenge posed by Brklacich et al.: "the design and implementation of analytical systems that can readily consider environmental, cultural, and economic conditions of particular region and simultaneously measure the implications of alternative futures for environmental quality, food supplies, and the well-being of individual producers" [8] (p. 11).

Author Contributions: Conceptualization and methodology, A.F. and I.d.l.R.C.; software, J.V.-P.; validation and formal analysis, A.F., I.d.l.R.C. and J.V.-P.; investigation, A.F. and J.V.-P.; resources, A.F., I.d.l.R.C., J.V.-P., J.U.-S., D.S.-P.; writing—original draft preparation, A.F.; writing—review and editing, A.F. and I.d.l.R.C.

Funding: This research was funded by Universidad de Piura. PAD Escuela de Dirección. We propose to APA to fund the APC.

Acknowledgments: We would like to thank Apolonio Suárez and Mario Berrocal for their valuable collaboration. We also are grateful with Stephen Pereira for his style review.

Conflicts of Interest: The authors declare no conflict of interest. The funders had no role in the design of the study; in the collection, analyses, or interpretation of data; in the writing of the manuscript, or in the decision to publish the results.

References

1. Pimentel, D.; Pimentel, M. *Food, Energy and Society*; CRC Press: Boca Raton, FL, USA, 2008.
2. Cazorla, A.; De Los Ríos, I.; Salvo, M. Working with People (WWP) in Rural Development Projects: A Proposal from Social Learning. *Cuad. Desarro. Rural* **2013**, *10*, 131–157.
3. De los Ríos, I.; Rivera, M.; García, C. Redefining Rural Prosperity through Social Learning in the Cooperative Sector: 25 Years of Experience from Organic Agriculture in Spain. *Land Use Policy* **2016**, *54*, 85–94. [CrossRef]
4. Rivera, M.; Knickel, K.; de los Rios, I.; Ashkenazy, A.; Pears, D.Q.; Chebach, T.; Šūmane, S. Rethinking the Connections between Agricultural Change and Rural Prosperity: A Discussion of Insights Derived from Case Studies in Seven Countries. *J. Rural Stud.* **2018**, *59*, 242–251. [CrossRef]
5. Deaton, A. Health, Inequality, and Economic Development. *J. Econ. Lit.* **2003**, *41*, 113–158. [CrossRef]
6. Rockström, J.; Williams, J.; Daily, G.; Noble, A.; Matthews, N.; Gordon, L.; Wetterstrand, H.; DeClerck, F.; Shah, M.; Steduto, P.; et al. Sustainable Intensification of Agriculture for Human Prosperity and Global Sustainability. *Ambio* **2017**, *46*, 4–17. [CrossRef] [PubMed]
7. Ferreira, F. The 2017 Global Poverty Update from the World Bank. Available online: http://blogs.worldbank.org/developmenttalk/2017-global-poverty-update-world-bank (accessed on 29 October 2018).
8. Brklacich, M.; Bryant, C.R.; Smit, B. Review and Appraisal of Concept of Sustainable Food Production Systems. *Environ. Manag.* **1991**, *15*, 1–14. [CrossRef]
9. Altieri, M.A. *Agroecology: The Science of Sustainable Agriculture*; CRC Press: Boca Raton, FL, USA, 2018.
10. Verburg, P.H.; Mertz, O.; Erb, K.H.; Haberl, H.; Wu, W. Land System Change and Food Security: Towards Multi-Scale Land System Solutions. *Curr. Opin. Environ. Sustain.* **2013**, *5*, 494–502. [CrossRef] [PubMed]
11. Dwivedi, S.L.; Lammerts van Bueren, E.T.; Ceccarelli, S.; Grando, S.; Upadhyaya, H.D.; Ortiz, R. Diversifying Food Systems in the Pursuit of Sustainable Food Production and Healthy Diets. *Trends Plant Sci.* **2017**, *22*, 842–856. [CrossRef] [PubMed]
12. Van Meijl, H.; Ruben, R.; Reinhard, S. *Towards an Inclusive and Sustainable Economy*; Wageningen University & Research: Wageningen, The Netherlands, 2017.
13. WCED. *Report of the World Commission on Environment and Development: Our Common Future*; Oxford University Press: Oxford, UK, 1987.
14. FAO. *Desarrollo Agrícola Sostenible Para La Seguridad Alimentaria y La Nutrición: ¿qué Función Desempeña La Ganadería?* FAO: Rome, Italy, 2016.
15. Heller, M.C.; Keoleian, G.A. *Life Cycle-Based Sustainability Indicators for Assessment of the U.S. Food System*; Center for Sustainable Systems, University of Michigan: Ann Arbor, MI, USA, 2000.
16. Kooiman, J. *Governing as Governance*; Sage: Thousand Oaks, CA, USA, 2003.
17. Jackson, T.; Anderson, V. *Redefining Prosperity-Essays in Pursuit of a Sustainable Economy*; Earthcan: London, UK, 2009.
18. Jackson, T. *Prosperity without Growth: Economics for a Finite Planet*; Routledge: London, UK, 2009.
19. Darnhofer, I.; De los Ríos, I.; Knickel, K.; Koopmans, M.; Lamine, C.; Olsson, G.A.; de Roest, K.; Rogge, E.; Šūmane, S.; Tisenkopfs, T. *Rethinking the Links between Farm Modernisation, Rural Development and Resilience in a World of Increasing Demands and Finite Resources*; European Commission: Brussels, Belgium, 2014.
20. Xu, L. Integrated Assessment of Sustainable Agri-Food Production System in Shaanxi Province, China. Ph.D. Dissertation, University of Lethbridge, Lethbridge, AB, Canada, 2017.
21. Castellini, C.; Boggia, A.; Cortina, C.; Dal Bosco, A.; Paolotti, L.; Novelli, E.; Mugnai, C. A Multicriteria Approach for Measuring the Sustainability of Different Poultry Production Systems. *J. Clean. Prod.* **2012**, *37*, 192–201. [CrossRef]
22. Yin, R.K. Case Study Methodology. In *Case Study Research: Design and Methods*, 3rd ed.; Sage: Thousand Oaks, CA, USA, 2003.

23. Cazorla-Montero, A.; de los Ríos-Carmendo, I.; Díaz-Puente, J.M. La Iniciativa Comunitaria Leader Como Model de Desarrollo Rural: Aplicación a La Región Capital de España. *Agrociencia* **2005**, *39*, 697–708.
24. De los Ríos-Carmenado, I.; Díaz-Puente, J.M.; Cadena-Iñiguez, J. La Iniciativa Leader Como Modelo de Desarrollo Rural: Aplicación a Algunos Territorios de México. *Agrociencia* **2011**, *45*, 609–624.
25. Friedmann, J. Planning in the Public Domain: Discourse and Praxis. *J. Plan. Educ. Res.* **1989**, *8*, 128–130. [CrossRef]
26. Sen, A. The Living Standard. *Oxf. Econ. Pap.* **1984**, *36*, 74–90. [CrossRef]
27. Davidson, J.; Lockwood, M. Partnerships as Instruments of Good Regional Governance: Innovation for Sustainability in Tasmania? *Reg. Stud.* **2008**, *42*, 641–656. [CrossRef]
28. Grin, J. The Politics of Transition Governance in Dutch Agriculture. Conceptual Understanding and Implications for Transition Management. *Int. J. Sustain. Dev.* **2012**, *15*, 72–89. [CrossRef]
29. Heley, J. Soft Spaces, Fuzzy Boundaries and Spatial Governance in Post-Devolution Wales. *Int. J. Urban Reg. Res.* **2013**, *37*, 1325–1348. [CrossRef]
30. IPMA. *Bases Para La Competencia En Dirección de Proyectos: Versión 3.1.*; IPMA: Valencia, Spain, 2010.
31. Hulme, D. Learning and Not Learning from Experience in Rural Project Planning. *Public Adm. Dev.* **1989**, *9*, 1–16. [CrossRef]
32. Saldarriaga, M.A.; Structural policies specialist Central Reserve Bank of Peru. email message to authors. 17 May 2013.
33. MINAG. *Comercialización de Pollos y Gallinas En Los Centros de Acopio de Lima Metropolitana y Callao*; MINAG: Lima, Perú, 2018.
34. INEI. *Censo Nacional de Mercados de Abasto*; INEI: Lima, Peru, 2016.
35. INEI. *Encuesta Demográfica y de Salud Familiar*; INEI: Lima, Peru, 2017.
36. Carranza, L. Contexto Nacional e Internacional y el Sector Avícola Peruano. Presented at Semana del Avicultor Conference, Lima, Peru, Junio 2011.
37. Aldrich, H.E.; Staber, U.H. Organizing Business Interests: Patterns of Trade Association Foundings, Transformation, and Deaths. In *Ecological Models of Organization*; Carrol, G., Ed.; Ballinger: Cambridge, MA, USA, 1988; pp. 111–126.
38. Oliver, C. Determinants of Interorganizational Relationships: Integration and Future Directions. *Acad. Manag. Rev.* **1990**, *15*, 241–265. [CrossRef]
39. Hemphill, T.A. Self-Regulating Industry Behavior: Antitrust Limitations and Trade Association Codes of Conduct. *J. Bus. Ethics* **1992**, *11*, 915–920. [CrossRef]
40. David, P.; Baron, D.P. The Nonmarket Strategy System. *Sloan Manag. Rev.* **1995**, *37*, 73.
41. Hillman, A.J.; Hitt, M.A. Corporate Political Strategy Formulation: A Model of Approach, Participation, and Strategy Decisions. *Acad. Manag. Rev.* **1999**, *24*, 825–842. [CrossRef]
42. Barnett, M.L. One Voice, but Whose Voice? Exploring What Drives Trade Association Activity. *Bus. Soc.* **2013**, *52*, 213–244. [CrossRef]
43. Williamson, O. Wage Rates as a Barrier to Entry: The Pennington Case in Perspective. *Q. J. Econ.* **1968**, *82*, 85–116. [CrossRef]
44. Pelletier, N. Environmental Performance in the US Broiler Poultry Sector: Life Cycle Energy Use and Greenhouse Gas, Ozone Depleting, Acidifying and Eutrophying Emissions. *Agric. Syst.* **2008**, *98*, 67–73. [CrossRef]
45. Herrero, M.; Thornton, P.K.; Notenbaert, A.M.; Wood, S.; Msangi, S.; Freeman, H.A.; Bossio, D.; Dixon, J.; Peters, M.; Van De Steeg, J.; et al. Smart Investments in Sustainable Food Production: Revisiting Mixed Crop-Livestock Systems. *Science* **2010**, *327*, 822–825. [CrossRef] [PubMed]

sustainability

MDPI

Article

Offshore Wind Farms as Potential Locations for Flat Oyster (*Ostrea edulis*) Restoration in the Dutch North Sea

Pauline Kamermans [1,*], Brenda Walles [1], Marloes Kraan [1,2], Luca A. van Duren [3], Frank Kleissen [3], Tom M. van der Have [4], Aad C. Smaal [5] and Marnix Poelman [1]

1 Wageningen Marine Research, Wageningen University and Research, P.O. Box 77, 4400 AB Yerseke, The Netherlands; brenda.walles@wur.nl (B.W.); marloes.kraan@wur.nl (M.K.); marnix.poelman@wur.nl (M.P.)
2 Environmental Policy Group, Wageningen University and Research, P.O. Box 8130, 6700 EW Wageningen, The Netherlands
3 Deltares, P.O. Box 177, 2600 MH Delft, The Netherlands; luca.vanduren@deltares.nl (L.A.v.D.); frank.kleissen@deltares.nl (F.K.)
4 Bureau Waardenburg, P.O. Box 365, 4100 AJ Culemborg, The Netherlands; t.m.van.der.have@buwa.nl
5 Aquaculture and Fisheries Department, Wageningen University and Research, P.O. Box 17 338, 6700 AH Wageningen, The Netherlands; aad.smaal@wur.nl
* Correspondence: pauline.kamermans@wur.nl

Received: 29 August 2018; Accepted: 24 October 2018; Published: 30 October 2018

check for updates

Abstract: The "Dutch Energy Agreement" motivates governments and industries to invest in renewable energy sources, of which offshore wind energy is one of the solutions to meet the agreed target of 16% of the total energy budget from renewable resources by 2023. An option for the multi-use of wind farms is nature-inclusive building, in which the design and construction of wind farms make use of the potential for co-design with oyster bed restoration. This can support the government's ambitions, for the Dutch North Sea, to achieve biodiversity goals, restore ecosystem functions, and enhance ecosystem services, including future seafood production. For the recovery of flat oyster (*Ostrea edulis*) beds, knowledge is required about the conditions under which active restoration of this species in the North Sea can be successfully implemented. This paper gives a framework and presents results to determine suitability of wind farms for flat oyster restoration, and provides recommendations for pilot studies. Our analysis showed that a number of wind farms in the Dutch section of the North Sea are suitable locations for development of flat oyster beds. Combining oyster restoration and oyster culture, as a protein source, is a viable option worth investigating.

Keywords: *Ostrea edulis*; native oyster restoration; North Sea; site selection; pilot study; offshore wind farms

1. Introduction

The "Dutch Energy Agreement", a document by the Dutch government and dozens of organizations and interest groups, presents the energy ambitions and targets up until 2023 and beyond. This agreement motivates governments and industries to invest in renewable energy sources, of which offshore wind energy is one of the solutions to meet the agreed 16% increase in the share of renewable energy by 2023. Significant North Sea space is required to meet the future goals of the energy transition, resulting in competition between current users, such as exploitation of oil and gas, fisheries, nature, shipping, and the new wind farms. Current insights in multi-use options are still in an exploratory phase, and policies are currently developed to support the multi-use of space.

Besides the goals set for energy arrangements, there is an increasing global demand for protein and carbohydrate sources. The world population is likely to grow up to 9 billion people by 2050, resulting in the need to provide proteins for all, of which fish and marine products can be part of [1]. Furthermore, alongside its contribution to protein intake, the value of fish is even stronger for poor consumers because of its micro-nutrient and lipid content [1]. This trend requires a more flexible food provision system and new sources. On the one hand, the current production, production methods, and value chains should be assessed, improved, and optimized. This needs to coincide with the arrangement of adapted locations and suitable space, as well as with an adequate governance strategy. On the other hand, production of food and proteins needs to expand and/or convert to alternative sources. Aquaculture is "projected to remain the fastest growing food commodity sector" [1]. As part of this development, aquaculture in open marine systems is seen as a potential large-scale expansion possibility [2,3]. The main reasons for global demand for seafood is the increase in world-population, the increase of consumption per capita, as well as the increase in income [1,4]. Because limits in the wild capture fisheries have been reached, with demand unable to be supported, aquaculture is considered the major future seafood contributor [5,6].

Mariculture has a wide history in nearshore and inshore areas and estuarine environments. The physical conditions and the relatively easy accessibility of these areas are the driving forces behind their use. However, there are growing concerns about the impact on the environment [7,8], and nearshore and inshore competition for space is setting limitations for the expansion or introduction of aquaculture [9,10]. Therefore, the "blue revolution" (the remarkably fast expansion of marine aquaculture and marine use in recent years) is set up to expand aquaculture practices in more exposed open sea and offshore areas. New technologies provide further impulse for this expansion. For example [11,12], investigated the potential of aquaculture for offshore seaweed and mussel farming in the Dutch North Sea.

Gerty [13] reported that the potential of marine aquaculture is demonstrated in many parts of the world. Worldwide the potential of development is far greater than the seafood demand, and expansion for space is not limited in many offshore locations. However, locally the options are not always directly suitable. The North Sea is a crowded and intensively used sea. Combinations of functions are therefore welcome. Currently, the potential for multi-use functions of offshore platforms, including wind farms, are under investigation worldwide [14,15]. One of the options is nature-inclusive construction, in which the design and construction of wind farms include the potential to enhance biodiversity and natural resources.

Until about a century ago, flat oyster beds constituted an important habitat in the North Sea. The flat oyster was a key species that once existed in the North Sea ecosystem, a fact that has almost disappeared from public memory. According to field surveys conducted in the 19th and early 20th century, the North Sea harbored substantial areas of oyster beds in that time (over 25,000 km^2, [16] in [17,18]. However, by the end of the 19th century the flat oyster fishery became too intensive, which caused the oyster population to decline rapidly [17,18]. By the beginning of the 20th century, the oyster beds were already decimated [17]. Later, other types of bottom trawling fisheries also impacted on the remaining beds. Other reasons for the decimation are not documented. As a result, the oyster community, including related species, disappeared from the North Sea.

For the recovery of flat oyster (*Ostrea edulis*) beds, knowledge is required about the conditions under which active restoration of this species in the North Sea can be successfully implemented. This paper gives a framework and presents results to determine suitability of wind farms for flat oyster restoration, and provides recommendations for pilot studies. The information is used to support the marine spatial planning of wind farm and nature restoration.

1.1. Flat Oyster Bed Ecosystem Services in the North Sea

Several factors motivate flat oyster restoration in the North Sea. Flat oyster beds are a threatened species and habitat (OSPAR, European Union (EU) Habitat Directive, biogenic reefs, EU Red List of Species and Habitats). They once constituted a key element of a rich North Sea. Oyster beds are some of the most striking structures in soft-sediment environments, providing many ecosystem services, including the provision of habitat, which is important for biodiversity [19,20], pelagic–benthic ecosystem coupling, shoreline stabilization, water quality regulation, and the enhancement of fishery production [21–23]. There is a lack of knowledge on the ecosystem services provided by deep-water flat oyster beds, as most studies have focused on the intertidal reefs of the Pacific oyster. Despite this, some mechanisms are assumed to be generic for both species and environments. [24], for example, demonstrated that the species facilitations function of the intertidal native European oyster bed is similar to the invasive Pacific oyster (*C. gigas*) reefs.

Beds of the European oyster *Ostrea edulis* function as a habitat for many species (Figure 1). Beds provide settlement substrate for epibenthic flora and fauna, food and shelter to mobile invertebrates (e.g., crab, lobster, and shrimps), and fish can use the beds as shelter (particularly in the juvenile stage) and spawning grounds (e.g., herring). The hard substrate, formed by the *O. edulis* shells, increases in time through recruitment and growth, harboring higher species diversity than non-living hard substrate [25].

Figure 1. Shells of the European oyster *Ostrea edulis* provide substrate for epibenthic flora and fauna. Pictures by Youri van Es and Aad Smaal.

Oyster reefs, as ecosystem engineers, impact the benthic environment by altering the biogeochemical and physical properties of the sediment, affecting community composition, abundance, and species richness. As filter feeders, oysters enhance the pelagic–benthic ecosystem coupling through the production of fecal and pseudofecal deposits, followed by mineralization [26,27], which may stimulate phytoplankton turnover, enhancing primary production [28]. In addition, reefs of filter feeders can concentrate and trap organic matter, which fuel local food webs (Figure 2). Concentration of organic matter has been demonstrated for intertidal mixed blue mussel (*Mytilus edulis*) and Pacific oyster reefs [29], deep reef structures of the cold-water coral *Lophelia pertusa* [30], and tubeworm *Lanice conchilega* reefs [31]. De Smet et al. (2016) [31] demonstrated that *L. conchilega* can significantly concentrate and trap organic matter, produced in an area no less than 15 times the reef area, within their reef food web, resulting in a higher macrofaunal biomass and a more diverse food web than in the absence of the reef.

Large beds of bivalves can also have a major effect on the local fine sediment dynamics [32,33]. The large expanses of oyster beds in the 19th century must have had a large-scale effect on turbidity in the North Sea. The area of the old "oyster grounds" is characterized by relatively fine sediment [34], which, under high wave conditions, can resuspend into the water column affecting light penetration. It is to be expected that the presence of 25,000 km^2 of oyster bed has affected these processes significantly. To what extent the changes in light penetration may have also affected local primary production is unknown.

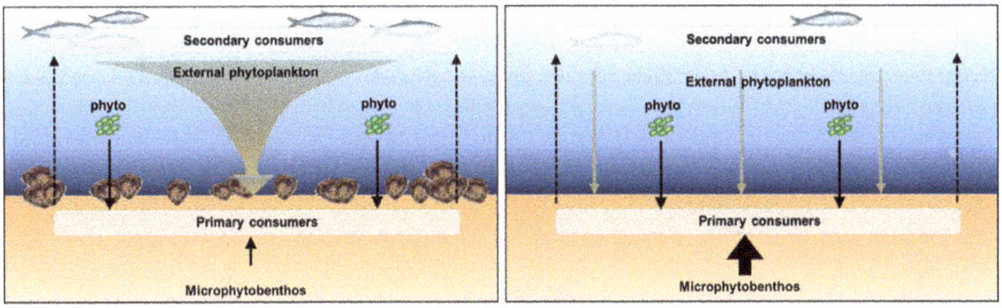

Figure 2. In the presence of the filter feeder *Ostrea edulis*, in situ-produced and laterally advected external phytoplankton dominate the carbon input into the bed, supporting a diverse-bed food web. In bare sand areas, the carbon input is much lower and mainly based on in situ primary production by microphytobenthos. Figure adapted from [31] De Smet et al. (2016).

Habitat formed by the oysters, combined with biodeposition, increases prey abundance, which is the driving force for enhanced fish production. [22] quantitatively estimated enhanced fish and mobile invertebrate production by oyster reefs. In the Gulf of Mexico, gross production was increased with 397 g m^{-2} $year^{-1}$ in the presence of oyster reefs [22]. Korringa (1954)] [35] reported about 250 species living in association with or on oyster beds. Hence, restoration would provide opportunities for ecosystem services, such as the commercial exploitation of fish and mobile invertebrates, as well as flat oysters themselves; provided that exploitation methods are developed that leave the flat oyster beds intact. As a consequence, the loss of oyster beds leads to a less productive and less diverse ecosystem and the loss of ecosystem services. In the Wadden Sea, the disappearance of the flat oyster beds resulted in a less diverse and productive mudflat ecosystem [36].

1.2. Policy

Restoring oyster beds will contribute to fulfil Dutch obligations, following from the Marine Strategy Framework Directive (MSFD). The recovery of natural hard substrates, including flat oyster beds, is an objective stated in the policy documents: Nature Ambition Large Water Bodies [37]; and the Implementation Agenda Natural Capital [38], under the title of shellfish beds in general; and in recent OSPAR recommendations (OSPAR, 2013). Sustainable energy supply, robust nature, and climate proof food supply are part of the Dutch Development Strategy for the North Sea (Noordzee 2050 Gebiedsagenda), and can potentially be realized with flat oyster restoration in wind farms.

Based on the European Common Fisheries Policy (CFP), a National Strategic Aquaculture Plan 2014–2020 has been described for the stimulation of aquaculture in the Netherlands. Among the prospects, the multi-use of space and offshore aquaculture have been identified as potentials for development.

1.3. Food and Nutrition Security

The utilization of the ocean for food is not a new development. Fish and aquatic products are long since the most traded food commodity in the world, whereby many seafood markets are global markets (https://www.wto.org/english/res_e/reser_e/ersd201003_e.htm). In the last couple of years, attention to fish and aquaculture products, and the contribution they make to food and nutrition security, has grown [1,39,40]. In particular, special attention has been given to the contribution of small-scale fisheries [41–43] and the importance of (small) fish for feeding the poor [44,45]. Furthermore, in Europe, attention to the contribution of fish and aquatic products for food security has risen, mainly as part of what has been called "blue growth". The EU focusses on aquaculture when discussing blue growth, as catches in fisheries are regarded as stabilized. The EU is currently the fourth largest fish producer, after China, Indonesia, and India, and the largest trader (in value) of fishery and aquaculture products in the world. Accordingly, the EU has quite an impact on fish trade and production (European Market Observatory for Fisheries and Aquaculture Products EUMOFA). It also is the largest importer of fish products, buying 60% of the fish and aquaculture products consumed from other countries in the world (https://www.euractiv.com/section/global-europe/opinion/closing-the-net-the-eu-must-step-up-enforcement-of-seafood-import-controls/). Although, by doing so, the EU contributes, indirectly, to the food security of people elsewhere in the world whose livelihoods rely on fisheries, aquaculture, and trade. Due to the rising prices of seafood and the EU's higher financial power, the EU will eventually (and has already in some places (https://www.pri.org/stories/2018-03-30/senegalese-women-turn-exporting-fish-spite-local-shortages)), by safeguarding its own food security, impact on the direct food security of people in other countries [46]. Thus, the more the EU can produce itself, the better.

There are a couple of advantages of fish and aquatic products for food and protein, compared to other sources, in terms of ecological impact: Fish and aquatic products have a relatively high energy intensity compared to other food items [47]; the environmental costs, in terms of water use, fertilizer use, pesticide use, antibiotic use, and soil loss, as well as greenhouse gasses per kg, are much lower for marine resourced food compared to dairy and meat [48]; and bivalves show a reduction of eutrophication (kg PO_4 equivalent) compared to terrestrial derived meat and crops [49] (Hall et al., 2011). In terms of impacting biodiversity, [48] has argued that capture fisheries are sustained by maintaining semi-natural ecosystems, whereas agriculture relies on replacing natural ecosystems with highly productive exotic species, and aquaculture falls somewhere in between (https://rayhblog.files.wordpress.com/2012/10/hilborn-world-fisheries-congress.pdf). Thus, when seeking opportunities to increase production of food, many turn to the ocean. It has also been established that, until now, mostly the top of the aquatic trophic pyramid has been harvested, which in fact is inefficient and less sustainable [50]. We could make much more use of the food potential of the ocean if we also, but not solely, consume fish and marine products at lower levels in the food chain. Flat oysters currently have a very high market value, due to the relatively low supply. If restoration programs become a success, in terms of growth, survival, and reproduction of oyster beds, the return of direct or indirect commercial exploitation may become possible. Currently an estimated 3000 km^2 of wind farm area is accommodated for wind farm production (up to 2023). If 0.1% of this area could be used for oyster restoration, in combination with oyster production, a total of 300 ha is then available for restoration in combination with food production (current production area is ~2.000 ha in The Netherlands).

1.4. Wind Farm Suitability

In September 2014, the Dutch government designated three areas for the development of offshore wind farms over the years to come: Borssele, Noord-Holland, and Zuid-Holland (Figure 3). In the North Sea 2050 Spatial Agenda, the Ministries of Economic Affairs and Infrastructure and the Environment expressed their ambition for the combined use of offshore space. This is included in the environmental regulations and design regulations of the Wind Farm Site Decisions for next generation wind farms. More specifically, the Ministry of Economic Affairs stated their wish to establish whether the areas recently designated as offshore wind farm sites offer opportunities for the development of flat oyster beds. The fact that wind farms are in the current regulatory framework, free from seabed-disturbing activities, is regarded as a major precondition for the restoration of flat oyster beds [17,51]. Oyster bed restoration appears to be a viable option to support these ambitions in the Dutch North Sea, to simultaneously achieve biodiversity goals and restore and enhance ecosystem services, including future seafood production.

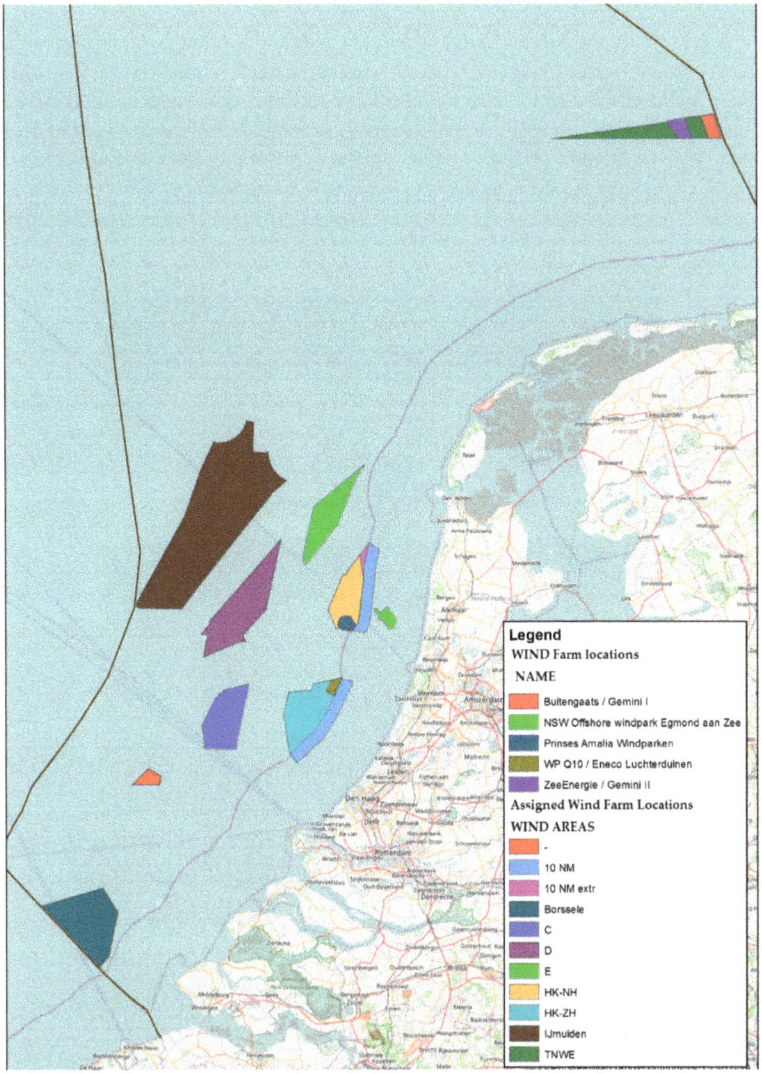

Figure 3. Locations of existing and planned wind farms in the Dutch section of the North Sea (DSC: Dutch Continental Shelf). The present study focusses on Borssele, Buitengaat/Gemini, offshore wind farm Egmond aan Zee (OWEZ), Dutch coast South and North (HK-NH, HK-ZH), Luchterduinen, Princes Amalia, and Zee-Energie. These wind farms are either in operation or are soon to be realized from [52].

Current conditions may favor the return of the flat oyster in the North Sea. The flat oyster has survived in estuaries around the North Sea (e.g., Limfjorden in Denmark, Lake Grevelingen and Oosterschelde in the Netherlands, plus various inlets on the coast of the British Isles, Norway, and Sweden). Recent records of individuals on shipwrecks, buoys, and marine wind farms in the North Sea show that it can still survive, reproduce, and disperse in the open sea [53]. Newly installed marine protected areas, wind farms, and offshore installations could provide sheltered areas that are free from bottom trawling fisheries. Yet, without human assistance the oyster beds may not return on a large scale. In the southern North Sea, the sea floor consists mainly of sand and silt, whereas rocks are uncommon. Oyster larvae have a limited dispersal range and need hard substrate to settle on [54,55], but without oysters, very little natural hard substrate has remained on the North Sea floor. So, once the oyster beds are gone, they will probably not return on their own because the source populations are too distant, even if the conditions are favorable. For the recovery of the flat oyster beds, knowledge of the conditions under which the active restoration of this species in the North Sea can be successfully implemented is required. To determine suitability of wind farms for flat oyster restoration we developed a framework, which considered the life history and habitat preferences of the species (Figure 4).

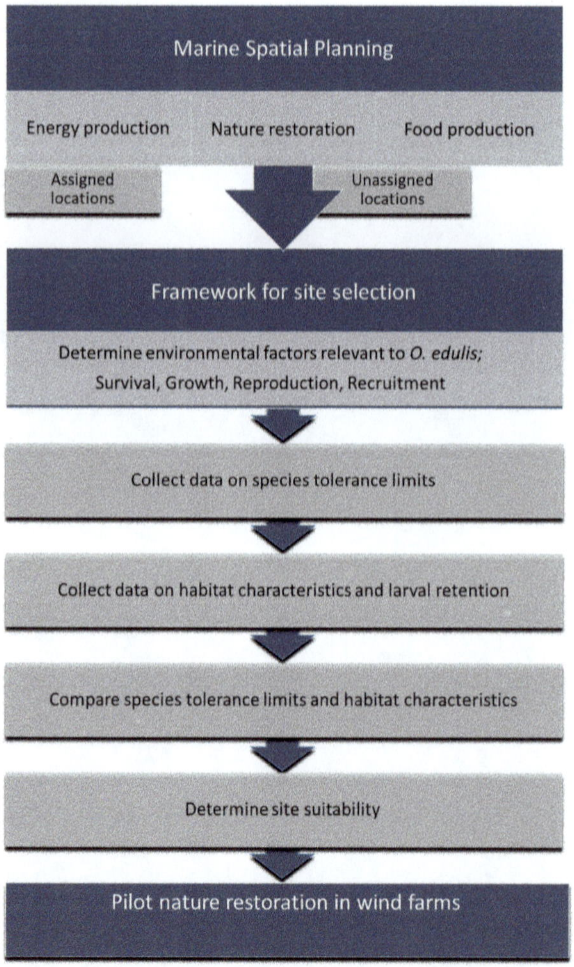

Figure 4. Schematic overview of the framework for site selection applied in the marine spatial planning of nature restoration, with special application of *Ostrea edulis* in wind farms.

1.5. Life History

Mature flat oysters can switch between sexes to be male or female [56]. Generally, flat oysters start as males and become females as they grow older. Older oysters can spawn twice during one season, once as a male and once as a female. Sperm cells, clustered in spermatozeugmata [57], are expelled through the exhalant siphon. Eggs remain in the mantle cavity of the female, where they are fertilized and develop into larvae with two shells in a period of one to two weeks. When they are released, the shell length of the larvae is 170–190 μm. During their free-swimming stage (another 8–30 days: [58] the length increases to 290–360 μm. Metamorphosis, from swimming larvae into sessile spat, depends on food availability for the larvae. Settlement occurs when a suitable location is detected. A drop of cement is produced and the left valve is glued to the surface. As a result of the relative short free-swimming stage, compared to other bivalve species, the dispersal distance of *O. edulis* is limited (on average 1 km [59]), although longer distances are occasionally possible during favorable conditions (up to more than 10 km [60]). Oyster spat settles on hard substrates, such as stones, shell fragments, or oyster shells. They adhere or fix themselves to the substrate and do not disperse further. For spat collection, calcified tiles have been employed in many areas and are still in use in Arcachon Bay (France). Oyster shells in existing oyster beds are a preferred settling substrate for oyster spat. Oyster bed development is therefore a self-reinforcing process, due to the positive feedback of existing oysters on successful recruitment and settlement (e.g., Eastern oyster [61]). It implies that oyster beds have a critical mass below which recruitment may fail, due to limited substrate availability in relation to the number of larvae produced [60,62].

2. Materials and Methods

The conditions for the long-term development of a flat oyster bed were largely determined by four life-history variables: Survival, growth, reproduction, and recruitment. Reproduction referred to the capacity to produce larvae; recruitment denoted the successful settlement of larvae in a specific site. These four variables were influenced by a range of abiotic and biotic factors, which are listed in Table 1. Survival depended on environmental factors such as large- and small-scale sea bed dynamics, oxygen content, salinity, and predation. Growth was mainly determined by phytoplankton and the concentration of suspended particles. Reproduction required a parent population in suitable age classes, and the right water temperature for gonadal development and spawning. Recruitment depended on water temperature, the quantity of larvae in a specific area, and the presence of suitable substrate for settlement. Recruitment was determined by the size of the parent population that produced the larvae, and served as larval settlement substrate, and by the water motion that determined larval retention in a specific area (see also [63,64]).

Table 1. Environmental factors, relevant to the various life-history processes of *O. edulis*, used as site selection criteria. The symbol (+) indicates a positive effect and (–) a negative effect.

	Survival	Growth	Reproduction	Recruitment
Sea bed shear stress	–			
Sea bed motion	–			
Concentration of suspended particles		–		
Larval retention			+	
Coarse sediment				+

Tolerance ranges for abiotic and biotic environmental factors were derived from the literature to evaluate the suitability of already present and planned wind farm areas in the Dutch North Sea for flat oysters (Table 2). The wind farms under study were within the limits for the factors: Salinity, water depth, water temperature, current velocity, food supply, and oxygen concentration [52]. The selection criteria used were: Sea bed motion, bed shear stress, sediment composition, type of substrate, concentration of suspended particles, and rate of larval dispersal and retention. Information on the historical range of flat oysters in the North Sea (e.g., [16,17,51]) was used as an extra "habitat suitability indicator" (Figure 5).

Table 2. Tolerance range and optimum of *O. edulis* for abiotic and biotic environmental factors. Adapted from [65].

Environmental Factor	Response Variable	Range	Optimum	Reference
Temperature (°C)	Survival adults	<30		[66]
	Growth adults	6–19	7–14	[66–68]
			17	[69]
	Gonad development	7–14		[68]
	Survival and growth larvae	10–31	25–27	[70]
	Spat fall	>18.5		[71]
	Survival spat	>3		[72]
Salinity (PSU)	Survival and growth larvae	20–39.5	25–35	[73]
	Growth adults	>19		[74]
Water depth (m)	Survival adults	<–80		[75]
Current velocity (m/s)	Survival adults	<0.25	0.03	[76]
Sea bed motion (cm/day)	Survival spat and adults		<0.8	[77]
Sediment composition	Survival adults		Firm silty sand or silty gravel with shells and stones	[78]
Suspended matter concentration (mg/L)	Food intake	<90		[79,80]
Food supply (chla ug/L)	Gonad development	1.68		[81]
Oxygen concentration (mg/L)	Survival adults	>0.5		[82]

Figure 5. Map from Olsen's Piscatorial Atlas [16], with the oyster range represented in brown.

2.1. Sea Bed Motion and Bed Shear Stress

Seabed dynamics can be described on different levels of scale. Therefore, a distinction was made between bed mobility morphodynamics, which concerns bed dynamics on a larger scale, and bed shear stress, as a measure of the force exerted by currents and waves on the bed. Bed shear stress is the driving force for the formation of sand waves [83–85]. It also has a direct influence on the settlement probability of shellfish larvae [86–88], as well as on the stability of the substrate on which they settle. The threshold is determined by sediment grain size, sediment density, and the influence of biota on sediment cohesion [89,90]. In a laboratory experiment, Grant et al. [77] demonstrated that low concentrations of suspended sediment, when swirled up 0.1 cm/day, can contribute to nutrient intake and, as such, have a positive effect on the growth speed of *O. edulis*, but that sediment churned up to 0.8 cm per day reduces growth. Tolerance of sea bed motion of up to 0.8 cm/day was therefore assumed in this study.

The southern North Sea, and particularly the Dutch continental shelf, is characterized by the occurrence of sand waves [91,92]. These features can range in height from less than 1.5 m up to 10 m, and can have a wave length of 100 to 1000 m. Under the influence of residual currents and waves, these waves can move over the bed surface of the North Sea with speeds up to several meters per year [93]. Although the average annual movement of sand waves is generally less than 0.8 cm/day, movement is not constant throughout the year and is often mediated by strong wave conditions. Many areas with migrating sand waves will, therefore, regularly experience sedimentation or erosion speeds that exceed the tolerance levels of flat oyster beds. Figure 6 shows the sea bed structure and sea bed motion features of the North Sea, as derived from the North Sea Atlas [94]. The effect of waves tends to be the most powerful in shallow waters. In deep waters, waves do not reach the seabed. Wave energy is often expressed as orbital velocity: the velocity of water just above the seabed caused by waves. The extent to which wave energy reaches the seabed depends on the height of the waves and the depth of the water. Waves are lower in coastal waters, but because these waters are shallow the orbital velocity is high. Offshore waves are higher, but under normal conditions their effect only reaches the seabed in shallower areas (such as the Dogger Bank). Orbital velocity and current velocity were both taken into account in the Delft 3D model of the Southern North Sea, often used to calculate fine sediment dynamics [95,96]. Comparison with the historical distribution of *O. edulis* (Figure 5) allowed us to assume that areas where sea bed shear stress was less than 0.6 N/m^2 were to be deemed suitable for the development of flat oyster beds. Sea bed shear stress levels are presented in Figure 7.

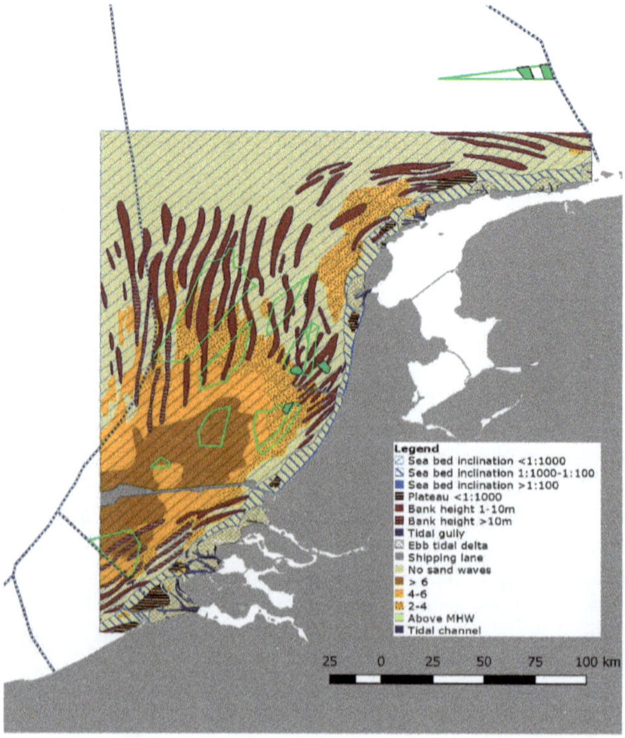

Figure 6. Sea bed structure and motion (source: North Sea Atlas, [95]), and wind farm sites in the Dutch section of the North Sea.

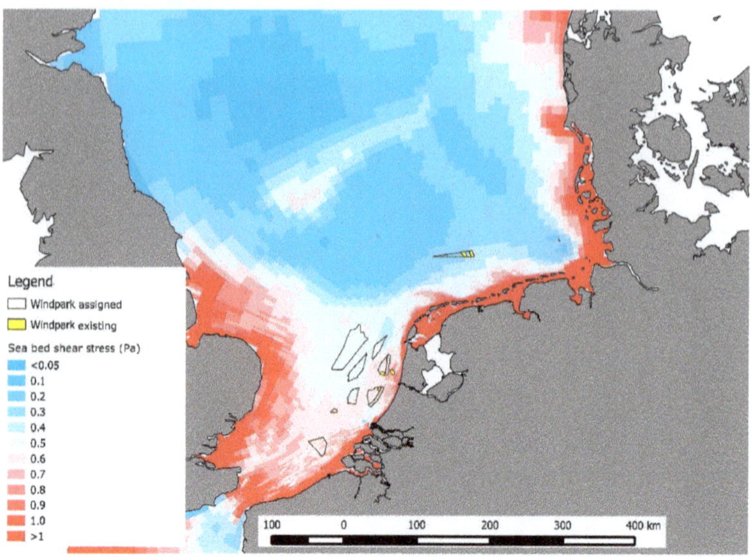

Figure 7. Average sea bed shear stress (in Pascal) in the North Sea, with combined effects of currents and waves. Wind farm sites are indicated.

2.2. Concentration of Suspended Particles and Sediment Composition

The concentration of suspended particles is an important factor in oyster growth. Oysters filter suspended matter from the water; this includes phytoplankton, detritus, and inorganic matter. The oysters use the phytoplankton to fuel their growth, but they have no use for the inorganic material in their metabolism. Higher concentrations of inorganic material will reduce the oysters' capacity for growth. This applies to most filter feeders, including flat oysters (e.g., [97]). Experiments have confirmed that suspended particle concentration levels in excess of 90 mg/L strongly inhibit growth in Pacific oysters (*Crassostrea gigas*) [80]. To our knowledge, no such experiments have been performed for European oysters. Measurements in a field set-up by [98] have shown that the total concentration of suspended matter just above the seabed (60 mg/L) is significantly higher than 80 cm above the seabed (40 mg/L). This resulted in significantly lower filtration rates for *O. edulis*, although no significant effect was found on the condition of the oysters after 15 months. Concentration of suspended particles near the seabed was obtained from model calculations [96] and are presented in Figure 8.

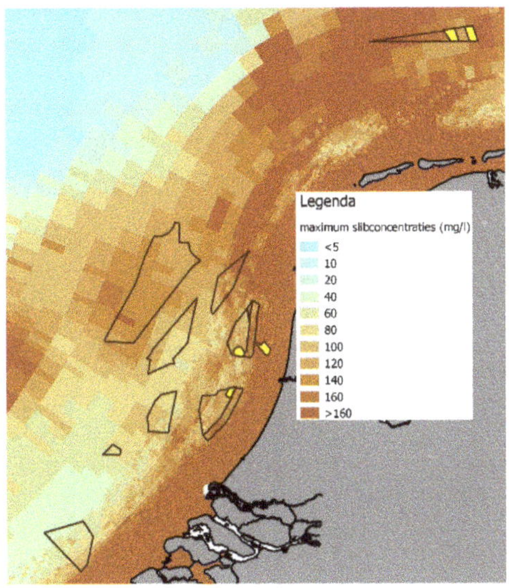

Figure 8. Maximum concentrations of inorganic suspended particles (mg/L) near the seabed in Dutch coastal waters, based on model simulation data. Wind farm sites are indicated.

Sediment composition is important as it determines substrate suitability for recruitment. First of all, shells and existing oyster beds will promote recruitment. In the absence of shells and oyster beds, sediment grain size is the most commonly used parameter. Sediment grain size is easy to measure. Spatial patterns in the range of shellfish often correlate with sediment grain size, but there is little evidence that this is the primary factor; other factors that co-vary with grain size are probably far more important [99]. Water motion can have an impact on sediment composition because heavy particles settle sooner than light ones. This is why silt tends to accumulate in sheltered spots, whereas coarser sediment particles are found in exposed areas. As a rule, grain size varies according to the average local current. [100] made a reconstruction of sediment composition along the Belgian coast, as it was a hundred years ago, based on the results of the Gilson surveys of 1899 to 1914. In their reconstruction, the seabed that accommodated extensive oyster beds before 1860 was described as follows: A heterogeneous gravel field partially covered by a thin layer (<15 cm) of sand through which bits of stone protruded. [101] studied factors that could explain the dispersal of Olympia oysters (*Ostrea lurida*) in an estuary in California. Burial under a layer of fine sediment impeded the oysters' survival on small hard substrates. In areas with large amounts of silt, the oysters could only be found on large rocks. In the northwestern part of the Mediterranean, oyster growth on shipwrecks is dominated by *Ostrea edulis* [102]. Based on this dispersal data, coarse sand (grain size >210 µm) was classified as unsuitable, fine sand (>63 µm) as moderately suitable, and firm silty sand, or silty gravel with shells and stones (not defined in terms of grain size), was considered suitable for oyster growth [78]. Seabed structure is presented in Figure 8.

2.3. *Food Availability*

Growth and survival of adult and larval oysters clearly depends on food supply. For benthic filter feeders, food availability is driven by primary production as well as by transport (i.e., hydrodynamics). Even in areas with very high productivity, stratification can limit the access of benthic filter feeders to food supply [103,104]. The current wind farms, and the current potential wind farm areas, are in areas relatively close to the coast with large nutrient input and, hence, high levels of primary production. Additionally, in these areas, water is not stratified for prolonged periods of time, apart from close to the western coast of the Netherlands, where the fresh water plume from the rivers Rhine and Meuse have an influence [105]. However, these areas are shallow and highly productive. In the locations under investigation in this study, food supply as well as oxygen depletion were not limiting for the formation of oyster beds, although there may be limits on the carrying capacity of wind farm sites for large beds. Other parts of the North Sea are characterized by prolonged periods of stratification [105], which often coincide with oxygen depletion [106,107]. For these locations food availability as well as oxygen dynamics need to be considered specifically.

2.4. *Larval Dispersal and Retention*

For an estimate of potential oyster larvae dispersal from various locations and the retention rate within a certain area, a particle dispersal model was used (with the particles representing the larvae) that was driven by water motion data from the Delft3D-FLOW model [95,108]. The three-dimensional water motion was derived from the ZUNO-DD model [96]. Hydrodynamic data were available for the years 2003 through 2011 on the following features: Water levels, current velocity, water temperature, and salinity. The forcing functions for simulations, using the hydrodynamic model, were based on actual data, such as river outflows, water temperature, and wind data. Because this data covered a period of nine years, it covered the annual and inter-variability in hydrodynamic conditions. In this study, the larvae were modelled as passive particles. The transport of these particles was driven by water motion and by dispersal and diffusion processes. In the model, the age at which the larvae settled was assumed to be ten days from the moment the larvae leave the mother oyster [109]. There was no mortality until then. Larvae that are less than ten days old were unable to settle. Larvae were removed from the model as soon as they were more than ten days old, and they played no further role in the simulation. The particles entered into the model in a way that reflected data from actual larval counts during the season [110], with the release of larvae from the middle of June to the middle of August, and highest concentrations in the middle of July. The model did not present absolute concentration figures, but offered proportional figures of larval concentration relative to the total number of larvae generated in a specific period (Figure 9). The dispersal maps presented a picture of larval dispersal in relative concentrations over the months June–August. Using colors, ranging from dark green to yellow, to represent decreasing larval concentration levels, the maps showed that larval retention in Borssele and the two northerly wind farms, Buitengaats and Zee-Energie, was much higher than on the other wind farms along the Zuid- and Noord-Holland coast, which were subject to the outflow of the Rhine and the northward current. Though the Borssele and Gemini sites did experience tidal currents, the net current was relatively weak.

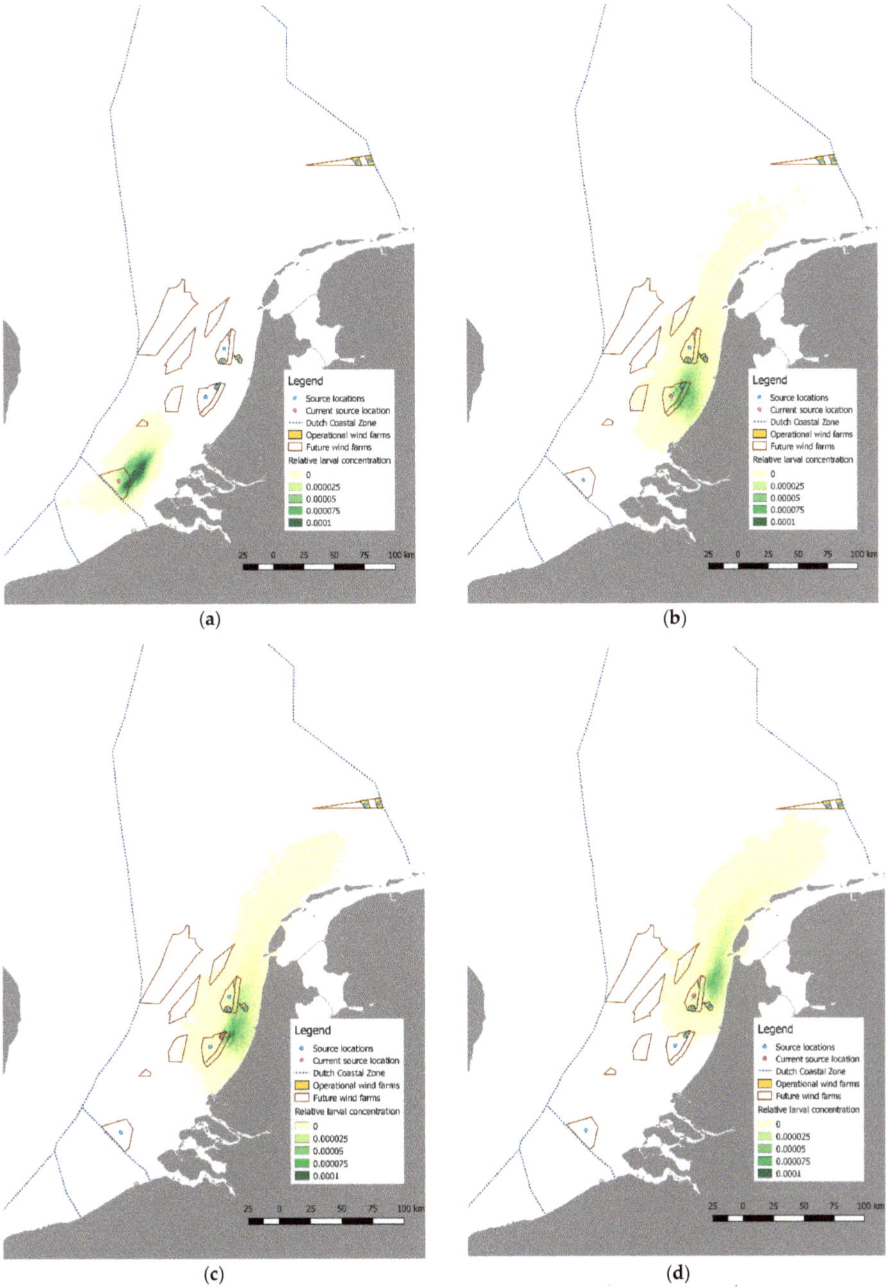

Figure 9. *Cont.*

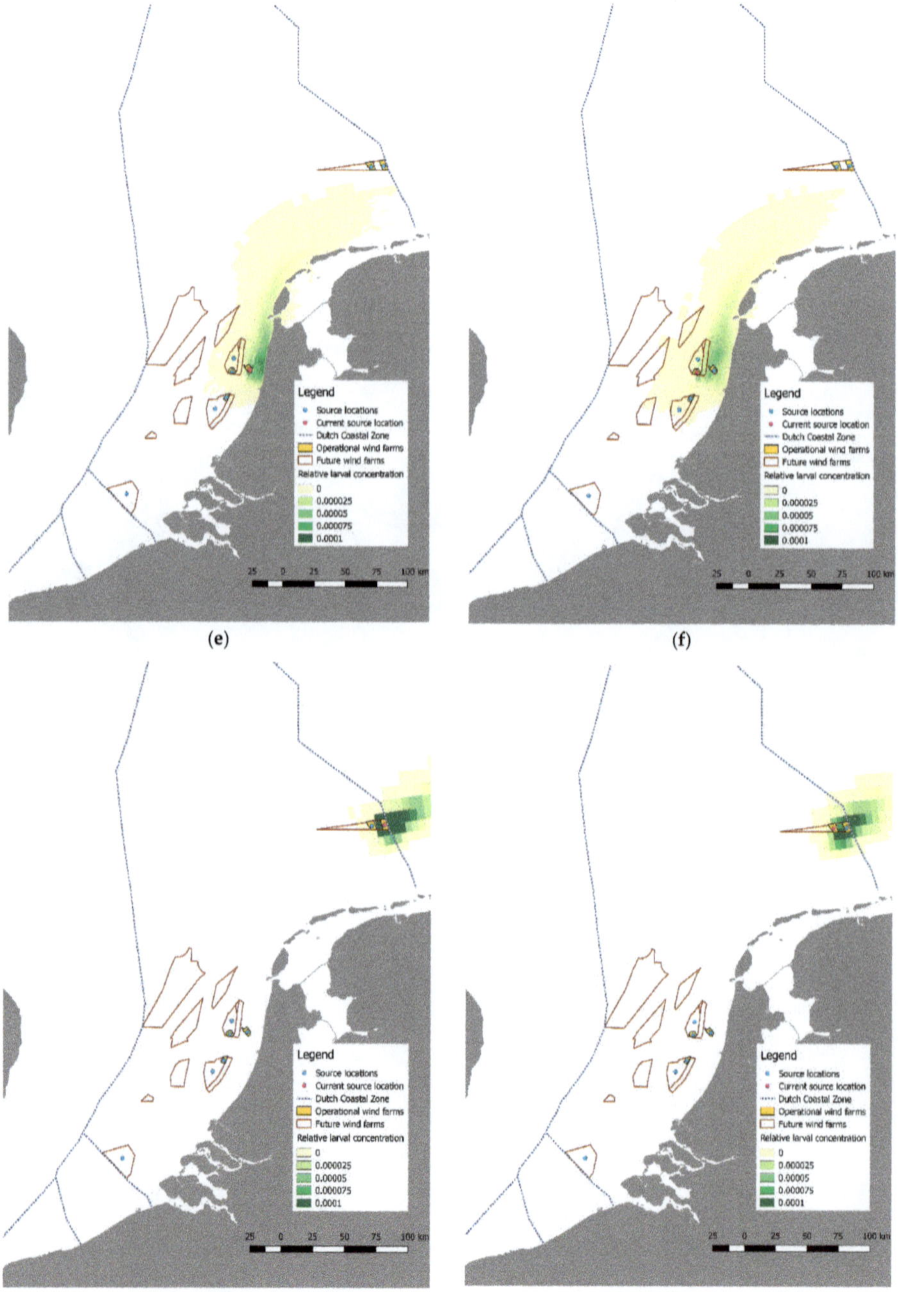

Figure 9. Simulated larval dispersal over a 10-day period at wind farms: (**a**) Borssele; (**b**) HK-Zuid; (**c**) Luchterduinen; (**d**) HK-Noord; (**e**) OWEZ; (**f**) Prinses Amalia; (**g**) Buitengaats; and (**h**) Zee-Energie. The legend presents relative concentration values. This means that the figure presents dispersal rather than absolute concentration figures.

2.5. Size of Parent Stock

"Critical mass" is defined as the quantity of oysters required to ensure a larval abundance that is sufficient for potentially successful spat settlement and survival in the bed of origin. Estimates in the literature as to the required critical mass are quite diverse. According to [60], a very large oyster population in the North Sea is required—many tens of millions of individual oysters—if successful recruitment is to be achieved in the North Sea and the Wadden Sea. This is linked to the openness of these systems and the origin of the larvae, which, according to these authors, originate principally from the English Channel. However, Smyth et al. (2016) [55] have demonstrated that following a period of over-exploitation, a flat oyster population in Strangford Lough recovered on the basis of a limited critical mass of flat oysters.

3. Results and Discussion

3.1. Wind Farm Selection

The various sites experienced different levels of sea bed shear stress, with the lowest values found in the Gemini area and on the Zee-Energie and Buitengaats wind farms (Table 3). The values for the habitat characteristics were retrieved from the maps shown in Figures 5–8, and were compared to the tolerance range and optimum range for *O. edulis* presented in Table 2. This resulted in a score of two (green, more suitable) when the values were within the optimal range, or one (orange, less suitable) when the values were within the tolerance range but not optimal. For the historical occurrence, 2 indicated that it was within the historical range, and 1 indicated that it was not within the historical range in that area. For larval retention, the locations scored a two when the map showed a relatively small area of larval distribution, whereas the other locations scored a one.

At most wind farms, bed shear stress corresponded with sea bed sediment composition; the seabed at Zee-Energie contained slightly larger amounts of silt. With respect to survival, the highest scores were for Zee-Energie, Buitengaats, and Borssele, which were all relatively close to sites where flat oysters occurred in the past (Table 3). The coastal sites attracted lower scores in terms of sea bed shear stress, sediment composition, and historical range. Growth and reproduction mainly depended on the availability of food, which was determined by the presence of phytoplankton and suspended particles. Though phytoplankton was not seen as a limiting factor in the potential wind farm sites analyzed in this study, suspended particles did qualify as an inhibitor. For that reason, to what extent the levels of near-seabed suspended particles differ and have limiting effects was determined among the various wind farm sites. Such an effect was found only at offshore wind farm Egmond aan Zee (OWEZ), the site which was closest to the coast (Table 3). Regarding the chance of settlement, model simulations showed that Buitengaats qualified for potentially successful recruitment, especially from a relatively small parent population at Zee-Energie, aided by proximity and the net eastward current (Table 3). The simulations also showed that the Prinses Amalia site should be able to serve as a source for spat settlement in OWEZ. With a moderate population, the Borssele wind farm could see successful recruitment within its boundaries; the same applies to Luchterduinen.

The analysis showed that Zee-Energie combined with Buitengaats were relatively suitable for the development of oyster beds; Borssele was also deemed suitable, followed by Luchterduinen. Luchterduinen also qualified as a pilot site because it was already in use and was favorably situated from a logistics point of view. The shallower coastal sites also offered opportunities for developing flat oyster beds, but the seabed in these locations was slightly less stable and they experienced more larval outflow without immediately benefiting neighboring sites.

Table 3. Overview of relative suitability of wind farm sites for the development of flat oyster beds. Numbers indicate relative suitability, 2 is more suitable than 1. The last column is the sum of all scores.

Wind Farm	Sea Bed Shear Stress	Suspended	Matter	Sediment	Historical Occurrence	Larval Retention	Sum
	avg	max	avg				
Borssele	1	2	2	1	2	2	10
HK-Zuid	1	2	2	1	1	1	8
OWEZ	1	1	2	1	1	2	8
HK-Noord	1	2	2	1	1	1	8
Luchterduinen	1	2	2	1	1	2	9
Princes Amalia	1	2	2	1	1	1	8
Buitengaats	2	2	2	1	2	1	10
Zee-energie	2	2	2	2	2	2	12

3.2. Pilot Studies

Given the habitat characteristics of several wind farms in the Dutch section of the North Sea, some of those sites, in principle, provide opportunities for the development of flat oyster beds. However, information on the abundance of predators is lacking. In addition, it is unclear which critical population size is required for a self-sustaining oyster bed. To verify actual suitability of offshore wind farms for flat oyster restoration, empirical tests are needed before large-scale restoration efforts are started.

Uncertainties can be reduced by conducting pilot studies in the field to identify the most relevant factors. Experiences gained in the pilot studies provide a basis for development of pilots in wind farms in the Dutch North Sea, these include: Studies conducted by nature conservation organizations ARK and WWF at the Dutch Voordelta [110–112], at Borkum Reef, and at the wind farm Gemini in the Dutch North Sea (www.platteoester.nl); experiments carried out with flat oysters around the North Sea (RESTORE project in Germany, https://www.awi.de/en/science/biosciences/shelf-sea-system-ecology/main-research-focus/european-oyster.html; Solent Oyster Restoration project https://www.bluemarinefoundation.com/project/solent/); the Essex Native Oyster Restoration Initiative (ENORI) (https://www.zsl.org/regions/uk-europe/thames-conservation/native-oyster-restoration), both of which were carried out in England; the DEEP project in Scotland (http://www.theglenmorangiecompany.com/about-us/deep/); and the Bivalve Project in Sweden.

The objective of such a pilot study could be formulated as follows: Conducting practical tests of life history variables to establish (1) the extent to which flat oysters are able to survive, grow, and reproduce on the chosen site; (2) whether the oyster bed is able to sustain itself through recruitment (larval production and sufficient substrate for settlement); (3) the extent to which the oyster bed can serve as a habitat for other species; and (4) which features within wind farms are most suitable to use for restoration efforts.

In outline, the following approach could be pursued: (1) Create a source of larvae; and (2) provide suitable substrate. To generate a source of larvae it is necessary to establish a minimum flat oyster population with different age (= size) classes, of which a part is protected against predators. Oyster development will have to be monitored during the growth and spawning seasons for survival, growth, and gonad development. An effective way to study survival, growth, and reproduction is to install cages with oysters for monitoring. During the spawning season, water samples can be taken to establish the larval abundance. Spat collectors can be used to monitor larval settlement. Clean shell materials (e.g., mussels and oysters) can be deposited at the pilot site during the period of peak larval abundance to serve as substrate to promote the settlement of flat oyster larvae.

Additionally, morphodynamic developments need to be monitored. The scour protection around turbines provides an area with hard substrate that may be suitable (or can be made suitable) for oyster settlement. However, these constructions can often induce edge scour along the edges [113]. These scour holes can be very dynamic and may prevent the expansion of an oyster bed. Whether this is a limiting issue or not will depend on the local dynamics within wind farm locations. Another option in a wind farm may be to utilize the protective layers over cable crossings. However, there remains a major discussion on the requirements to decommission any anthropogenic structures in the North Sea (OSPAR), which conflicts with the concept of long-term nature development strategies.

A combination of oyster culture and oyster restoration is an option. Off-bottom oyster growth in the open sea has been demonstrated to be possible [114]. The oysters will spawn at least once before they reach market size and are harvested. The larvae that are produced may be able to settle on substrate placed on the bottom. This will enlarge the population of the natural bed. When the bed has become larger than is needed to sustain itself, controlled harvest from the bottom can be investigated. Oyster farmers may be interested in explorative experiments. Apart from off-bottom farming systems, an area set aside for such activities and more robust ships are needed. At present, farmers are reluctant to invest in equipment and new ships without guarantees for the long-term (e.g., 10 years) use of an area. Possibilities for the government to allow such long-term experiments, and combined use of wind farm service vessels, need to be explored.

4. Conclusions

Our analysis showed that a number of wind farms in the Dutch section of the North Sea are suitable locations for flat oysters and for the development of flat oyster beds. This offers opportunities for multi-use in the form of nature-inclusive construction and exploitation. It can support the achievement of biodiversity goals, the restoration of ecosystem functions, and the enhancement of ecosystem services, including future seafood production. To verify actual suitability of off shore wind farms for flat oyster restoration, empirical tests are needed before large-scale restoration efforts can be started. Pilots should focus on estimation of life history variables of the oysters, including survival, growth, reproduction, and recruitment. In addition, expansion of the bed and self-sustainability, by the positive feedback created from the provision of new settlement substrate by the newly established bed, should be studied, as well as the enhancement of biodiversity in general and the increased production of fish and large, mobile crustaceans in particular. Combining oyster restoration with oyster culture for commercial purposes is a viable option worth investigating. It could be very productive, as cultivation offers broodstock for larvae, which can expand and maintain the natural bed, and harvest adds to food security.

Author Contributions: P.K. lead the writing of the manuscript and provided, together with T.M.v.d.H and A.C.S., biological information on *O.edulis* and its habitat. L.A.v.D. provided information and wrote on physical habitat conditions and larval dispersal. F.K. provided larval dispersal model results. B.W. provided information and wrote on goods and services of oyster beds. M.P. provided information and wrote on oyster aquaculture and M.K. provided information and wrote on food security. The authors declare no conflict of interest.

Funding: Part of this study was funded by the Dutch Ministry of Economic Affairs, project code: BO-11018. Writing of the manuscript was supported by the Wageningen Research project KB22.

Conflicts of Interest: The authors declare no conflict of interest.

References

1. Béné, C.; Barange, M.; Subasinghe, R.; Pinstrup-Andersen, P.; Merino, G.; Hemre, G.; Williams, M. Feeding 9 billion by 2050 Putting fish back on the menu. *Food Secur.* **2015**, *7*, 261–274. [CrossRef]
2. Lovatelli, A.; Aguilar-Manjarrez, J.; Soto, D. (Eds.) *Expanding Mariculture Farther Offshore—Technical, Environmental, Spatial and Governance Challenges*; FAO Technical Workshop, FAO Fisheries and Aquaculture Proceedings No. 24; Food and Agriculture Organization of the United Nations (FAO): Rome, Italy, 2013; p. 73.
3. Rubino, M.C. *Offshore Aquaculture in the United States: Economic Considerations, Implications and Opportunities*; Technical Memorandum NMFS F/SPO-103; National Oceanic and Atmospheric Administration: Silver Spring, MD, USA, 2008.

4. Godfray, H.C.J.; Beddington, J.R.; Crute, I.R.; Haddad, L.; Lawrence, D.; Muir, J.F.; Pretty, J.; Robinson, S.; Thomas, S.M.; Toulmin, C. Food Security: The Challenge of Feeding 9 Billion People. *Science* **2010**, *327*, 812–818. [CrossRef] [PubMed]
5. Costello, C.; Ovando, D.; Clavelle, T.; Strauss, C.K.; Hilborn, R.; Melnychuk, M.C.; Branch, T.A.; Gaines, S.D.; Szuwalski, C.S.; Cabral, R.B.; et al. Global fishery prospects under contrasting management regimes. *PNAS* **2016**, *113*, 5125–5129. [CrossRef] [PubMed]
6. Goldberg, R.; Naylor, R. Future seascapes, fishing and fish farming. *Front. Ecol. Environ.* **2005**, *3*, 21–28. [CrossRef]
7. Borja, Á.; Rodríguez, J.G.; Black, K.; Bodoy, A.; Emblow, C.; Fernandes, T.F.; Forte, J.; Karakassis, I.; Muxika, I.; Nickell, T.D.; et al. Assessing the suitability of a range of benthic indices in the evaluation of environmental impact of fin and shellfish aquaculture located in sites across Europe. *Aquaculture* **2009**, *293*, 231–240. [CrossRef]
8. Read, P.; Fernandes, T. Management of environmental impacts of marine aquaculture in Europe. *Aquaculture* **2003**, *226*, 139–163. [CrossRef]
9. Lester, S.E.; Stevens, J.M.; Gentry, R.R.; Kappel, C.V.; Bell, T.W.; Costello, C.J.; Gaines, S.D.; Kiefer, D.A.; Maue, C.C.; Rensel, J.E.; et al. Marine spatial planning makes room for offshore aquaculture in crowded coastal waters. *Nat. Commun.* **2018**, *9*, 945. [CrossRef] [PubMed]
10. Sarà, G.; Mangano, M.C.; Johnson, M.; Mazzola, A. Integrating multiple stressors in aquaculture to build the blue growth in a changing sea. *Hydrobiologia* **2018**, *809*, 5–17. [CrossRef]
11. Van den Burg, S.W.K.; Kamermans, P.; Blanch, M.; Pletsas, D.; Poelman, M.; Soma, K.; Dalton, G. Business case for mussel aquaculture in offshore wind farms in the North Sea. *Mar. Policy* **2017**. [CrossRef]
12. Jansen, H.M.; van den Burg, S.; Bolman, B.; Jak, R.G.; Kamermans, P.; Poelman, M.; Stuiver, M. The feasibility of offshore aquaculture and its potential for multi-use in the North Sea. *Aquacult. Int.* **2016**. [CrossRef]
13. Gentry, R.R.; Froehlich, H.E.; Grimm, D.; Kareiva, P.; Parke, M.; Rust, M.; Gaines, S.D.; Halpern, B.S. Mapping the global potential for marine aquaculture. *Nat. Ecol. Evol.* **2017**, *1*, 1317–1324. [CrossRef] [PubMed]
14. Pirlet, H.; Claus, S.; Copejans, E.D.C.E.; Guanche García, R.; Møhlenberg, F.; Rappé, K.; Schouten, J.-J.; Zanuttigh, B. *The Mermaid Project—Innovative Multi-Purpose Offshore Platforms*; Flanders Marine Institute (VLIZ): Ostend, Belgium, 2014; p. 20.
15. Quevedo, E.; Delory, E.; Castro, A.; Llinás, O.; Hernández, J. Modular multi-purpose offshore platforms, the TROPOS project approach. In Proceedings of the 4th International Conference on Ocean Energy, Dublin, Ireland, 17–19 October 2012.
16. Olsen, O.T. *The Piscatorial Atlas of the North Sea, English and St. George's Channels*; Grimsby: London, UK, 1883.
17. Gercken, J.; Schmidt, A. *Current Status of the European Oyster (Ostrea edulis) and Possibilities for Restoration in the German North Sea 2014*; BFN Report; Bundesamt für Naturschutz: Bonn, Germany, 2014.
18. Houziaux, J.-S.; Kerckhof, F.; Degrendele, K.; Roche, M.; Norro, A. *The Hinder Banks Yet an Important Region for the Belgian Marine Biodiversity?* Final Report HINDERS Project; Belgian Science Policy Office: Brussels, Belgium, 2008; p. 249. Available online: http://www.belspo.be/belspo/home/publ/pub_ostc/EV/rappEV45_en.pdfS (accessed on 28 August 2018).
19. Bruno, J.F.; Stachowicz, J.J.; Bertness, M.D. Inclusion of facilitation into ecological theory. *Trends Ecol. Evol.* **2003**, *18*, 119–125. [CrossRef]
20. Bouma, T.J.; Olenin, S.; Reise, K.; Ysebaert, T. Ecosystem engineering and biodiversity in coastal sediments: Posing hypotheses. *Helgol. Mar. Res.* **2009**, *63*, 95–106. [CrossRef]
21. Lenihan, H.S.; Peterson, C.H. How habitat degradation through fishery disturbance enhances impacts of hypoxia on oyster reefs. *Ecol. Appl.* **1998**, *8*, 128–140. [CrossRef]
22. zu Ermgassen, P.S.E.; Grabowski, J.H.; Gair, J.R.; Powers, S.P. Quantifying fish and mobile invertebrate production from a threatened nursery habitat. *J. Appl. Ecol.* **2016**, *53*, 596–606. [CrossRef]
23. Baggett, L.P.; Powers, S.P.; Brumbaugh, R.; Coen, L.D.; DeAngelis, B.; Greene, J.; Hancock, B.; Morlock, S. *Oyster Habitat Restoration Monitoring and Assessment Handbook*; The Nature Conservancy: Arlington, VA, USA, 2014; 96p, Available online: http://www.oyster-restoration.org/wp-content/uploads/2014/01/Oyster-Habitat-Restoration-Monitoring-and-Assessment-Handbook.pdf (accessed on 28 August 2018).
24. Zwerschke, N.; Emmerson, M.C.; Roberts, D.; O'Connor, N.E. Benthic assemblages associated with native and non-native oysters are similar. *Mar. Pollut. Bull.* **2016**, *111*, 305–310. [CrossRef] [PubMed]

25. Smyth, D.; Roberts, D. The European oyster (*Ostrea edulis*) and its epibiotic succession. *Hydrobiologia* **2010**, *655*, 25–36. [CrossRef]
26. Newell, R.I.E. Ecosystem influences of natural and cultivated populations of suspension-feeding bivalve molluscs: A review. *J. Shellfish Res.* **2004**, *23*, 51–61.
27. Kellogg, M.L.; Cornwell, J.C.; Owens, M.S.; Paynter, K.T. Denitrification and nutrient assimilation on a restored oyster reef. *Mar. Ecol. Prog. Ser.* **2013**, *480*, 1–19. [CrossRef]
28. Dame, R.F.; Prins, T. Bivalve carrying capacity in coastal ecosystems. *Aquat. Ecol.* **1997**, *31*, 409–421. [CrossRef]
29. Van der Zee, E.; Heide, T.; Donadi, S.; Eklöf, J.; Eriksson, B.; Olff, H.; Veer, H.; Piersma, T. Spatially extended habitat modification by intertidal reef-building bivalves has implications for consumer-resource interactions. *Ecosystems* **2012**, *15*, 664–673. [CrossRef]
30. Van Oevelen, D.; Duineveld, G.; Lavaleye, M.; Mienis, F.; Soetaert, K.; Heip, C.H.R. The cold-water coral community as a hot spot for carbon cycling on continental margins: A food-web analysis from Rockall Bank (northeast Atlantic). *Limnol. Oceanogr.* **2009**, *54*, 1829–1844. [CrossRef]
31. De Smet, B.; van Oevelen, D.; Vincx, M.; Vanaverbeke, J.; Soetaert, K. *Lanice conchilega* structures carbon flows in soft-bottom intertidal areas. *Mar. Ecol. Prog. Ser.* **2016**, *552*, 47–60. [CrossRef]
32. Van der Heide, T.; Tielens, E.; van der Zee, E.M.; Weerman, E.J.; Holthuijsen, S.; Eriksson, B.K.; Piersma, T.; van de Koppel, J.; Olff, H. Predation and habitat modification synergistically interact to control bivalve recruitment on intertidal mudflats. *Boil. Conserv.* **2014**, *172*, 163–169. [CrossRef]
33. Ysebaert, T.; Hart, M.; Herman, P.M.J. Impacts of bottom and suspended cultures of mussels Mytilus spp. on the surrounding sedimentary environment and macrobenthic biodiversity. *Helgol. Mar. Res.* **2009**, *63*, 59–74. [CrossRef]
34. Puls, W.; Sündermann, J. Simulation of Suspended Sediment Dispersion in the North Sea. In *Residual Currents and Long-Term Transport*; Coastal and Estuarine Studies; Springer: Berlin, Germany, 1990; Volume 38.
35. Korringa, P. The shells of *Ostrea edulis* as a habitat. *Arch. Neerl. Zoöl.* **1954**, *10*, 32–146. [CrossRef]
36. Reise, K.; Beusekom, J.E.E. Interactive effects of global and regional change on a coastal ecosystem. *Helgol. Mar. Res.* **2008**, *62*, 85–91. [CrossRef]
37. Ministry of Economic Affairs. 2014 Nature Ambition Large Water Bodies. Available online: www.minez.nl (accessed on 28 August 2018).
38. Ministry of Economic Affairs. 2013 Implementation Agenda Natural Capital. Available online: www.minez.nl (accessed on 28 August 2018).
39. High Level Panel of Experts. *Sustainable Fisheries and Aquaculture for Food Security and Nutrition: A Report by the High Level Panel of Experts on Food Security and Nutrition of the Committee on World Food Security*; FAO: Rome, Italy, 2014.
40. Golden, C.D.; Allison, E.H.; Cheung, W.W.L.; Dey, M.M.; Halpern, B.S.; McCauley, D.J. Nutrition: Fall in fish catch threatens human health. *Nature* **2016**, *534*, 317–320. [CrossRef] [PubMed]
41. Allison, E.H.; Béné, C.; Andrew, N.L. Poverty reduction as a means to enhance resilience in small-scale fisheries. In *Small-Scale Fisheries Management—Frameworks and Approaches for the Developing World*; Pomeroy, R.S., Andrew, N.L., Eds.; CABI: Wallingford, UK, 2011; pp. 216–238.
42. Béné, C.; Arthur, R.; Norbury, H.; Allison, E.H.; Beveridge, M.; Bush, S.; Campling, L.; Leschen, W.; Little, D.; Squires, D.; et al. Contribution of fisheries and aquaculture to food security and poverty reduction: Assessing the current evidence. *World Dev.* **2016**, *79*, 177–196. [CrossRef]
43. Beveridge, M.C.M.; Thilsted, S.H.; Phillips, M.J.; Metian, M.; Troell, M.; Hall, S.J. Meeting the food and nutrition needs of the poor: The role of fish and the opportunities and challenges emerging from the rise of aquaculture. *J. Fish Boil.* **2013**, *83*, 1067–1084. [CrossRef] [PubMed]
44. Kolding, J.; van Zwieten, P.; Marttin, F.; Poulain, F. *Fisheries in the Drylands of Sub-Sahara. Fish Come with the Rains*; Building Resilience for Fisheries Dependant Livelihoods to Enhance Food Security and Nutrition in the Drylands; FAO Report FIAF/FIAP/C1118(En).7; Food and Agriculture Organization of the United Nations (FAO): Rome, Italy, 2016.
45. Bogard, J.R.; Thilsted, S.H.; Marks, G.C.; Abdul Wahab, M.; Hossain, M.A.R.; Jakobsen, J.; Stangoulis, J. Nutrient composition of important fish species in Bangladesh and potential contribution to recommended nutrient intakes. *J. Food Compos. Anal.* **2015**, *42*, 120–133. [CrossRef]

46. Kurien, J. *Responsible Fish Trade and Food Security*; FAO Fisheries Technical Paper 456; Food and Agriculture Organization of the United Nations (FAO): Rome, Italy, 2005.
47. Pelletier, N.; Audsley, E.; Brodt, S.; Garnett, T.; Henriksson, P.; Kendall, A.; Kramer, K.J.; Murphy, D.; Nemecek, T.; Troell, M. Energy Intensity of Agriculture and Food Systems. *Annu. Rev. Environ. Resour.* **2011**, *36*, 223–246. [CrossRef]
48. Hilborn, R.; Tellier, P. *The Environmental Cost of N. Z. Food Production*; New Zealand Seafood Industry Council Ltd.: Wellington, New Zealand, 2012; ISBN 978-0473-17824-6.
49. Hall, S.J.; Delaporte, A.; Phillips, M.J.; Beveridge, M.; O'Keefe, M. *Blue Frontiers: Managing the Environmental Costs of Aquaculture*; The WorldFish Center: Penang, Malaysia, 2011.
50. Garcia, S.M.; Kolding, J.; Rice, J.; Rochet, M.J.; Zhou, S.; Arimoto, T.; Beyer, J.E.; Borges, L.; Bundy, A.; Dunn, D. Reconsidering the consequences of selective fisheries. *Science* **2012**, *335*, 1045–1047. [CrossRef] [PubMed]
51. Smaal, A.C.; Kamermans, P.; van der Have, T.M.; Engelsma, M.; Sas, H.W.J. *Feasibility of Flat Oyster (Ostrea edulis) Restoration in the Dutch Part of the North Sea*; IMARES Report C028/15; IMARES Wageningen UR: Ijmuiden, The Netherlands, 2015.
52. Smaal, A.; Kamermans, P.; Kleissen, F.; van Duren, L.; van der Have, T. *Flat Oysters on Offshore Wind Farms. Opportunities for the Development of flat Oyster Populations on Existing and Planned Wind Farms in the Dutch Section of the North Sea*; Wageningen Marine Research Rapport C052/17; Wageningen Marine Research: Yerseke, The Netherlands, 2017.
53. Kerckhof, F.; Coolen, J.W.P.; Rumes, B.; Degraer, S. Recent findings of wild European flat oysters *Ostrea edulis* (Linnaeus, 1758) in Belgian and Dutch offshore waters: New perspectives for offshore oyster reef restoration in the southern North Sea Belgian. *J. Zoöl.* **2018**. [CrossRef]
54. Burke, K.; Bataller, É.; Miron, G. Spat collection of a non-native bivalve species (European oyster, *Ostrea edulis*) off the Eastern Canadian coast. *J. Shellfish Res.* **2008**, *27*, 345–353. [CrossRef]
55. Smyth, D.; Mahon, A.M.; Roberts, D.; Kregting, L. Settlement of *Ostrea edulis* is determined by the availability of hard substrata rather than by its nature: Implications for stock recovery and restoration of the European oyster. *Aquat. Conserv. Mar. Freshw. Ecosyst.* **2018**. [CrossRef]
56. Walne, P.R. *Culture of Bivalve moLluscs: 50 Years of Experience at Conwy*; Fishing News Books Ltd.: Farnham, UK, 1974; 189p.
57. O'Foighil, D. Role of spermatozeugmata in the spawning ecology of the brooding oyster *Ostrea edulis*. *Gamete Res.* **1989**, *24*, 219–228. [CrossRef] [PubMed]
58. Muus, B.J.; Dahlstrom, P. *Meeresfische*; BLV: Munchen, Germany, 1973.
59. Jackson, A. *Ostrea edulis. Native oyster. Marine Life Information Network: Biology and Sensitivity Key Information Sub-Programmem [on-line]*; Marine Biological Association: Plymouth, UK, 2007.
60. Berghahn, R.; Ruth, M. The disappearance of oysters from the Wadden Sea: A cautionary tale for no-take zones. *Aquat. Conserv. Mar. Freshw. Ecosyst.* **2005**, *15*, 91–104. [CrossRef]
61. Jordan-Cooley, W.C.; Lipcius, R.N.; Shaw, L.B.; Shen, J.; Shi, J. Bistability in a differential equation model of oyster population and sediment volume. *J. Theor. Boil.* **2011**, *289*, 1–11. [CrossRef] [PubMed]
62. Kennedy, R.J.; Roberts, D. Commercial oyster stocks as a potential source of larvae in the regeneration of *Ostrea edulis* in Strangford Lough, Northern Ireland. *JMBA* **2006**, *86*, 153–159. [CrossRef]
63. Vera, M.; Carlsson, J.; Carlsson, J.E.L.; Cross, T.; Lynch, S.; Kamermans, P.; Villalba, A.; Culloty, S.; Martinez, P. Current genetic status, temporal stability and structure of the remnant wild European flat oyster populations: Conservation and restoring implications. *Mar. Biol.* **2016**, *163*, 239. [CrossRef]
64. Smyth, D.; Kregting, L.; Elsäßer, B.; Kennedy, R.; Roberts, D. Using particle dispersal models to assist in the conservation and recovery of the overexploited native oyster (*Ostrea edulis*) in an enclosed sea lough. *JSR* **2016**, *108*, 50–59. [CrossRef]
65. Capelle, J.J. *Restoration of the Oyster Ostrea edulis L. in the Wadden Sea*; Wageningen IMARES, Rapport; IMARES: Yerseke, The Netherlands, 2008; p. 61.
66. Haure, J.; Penisson, C.; Bougrier, S.; Baud, J.P. Influence of temperature on clearance and oxygen consumption rates of the flat oyster *Ostrea edulis*: Determination of allometric coefficients. *Aquaculture* **1998**, *169*, 211–224. [CrossRef]

67. Newell, R.C.; Johnson, L.G.; Kofoed, L.H. Adjustment of the components of energy balance in response to temperature change in *Ostrea edulis*. *Oecologia* **1977**, *30*, 97–110. [CrossRef] [PubMed]
68. Lubet, P. Ecophysiologie de la reproduction chez les mollusques lamellibranches. *Haliotis* **1976**, *7*, 49–55.
69. Buxton, C.D.; Newell, R.C.; Field, J.G. Response-Surface Analysis of the Combined Effects of Exposure and Acclimation Temperatures on Filtration, Oxygen Consumption and Scope for Growth in the Oyster. *Mar. Ecol. Prog. Ser.* **1981**, *6*, 73–82. [CrossRef]
70. Davis, H.C.; Calabrese, A. Survival and growth of larvae of the European oyster (*Ostrea edulis* L.) at different temperatures. *Boil. Bull.* **1969**, *136*, 193–199. [CrossRef] [PubMed]
71. Hoek, P.P.C. *Rapport Over de Oorzaken van den Achteruitgang in Hoedanigheid van de Zeeuwsche oester*; Ministerie van Waterstaat, Handel en Nijverheid: 's-Gravenhage, The Netherlans, 1902.
72. Child, A.R.; Laing, I. Comparative low temperature tolerance of small juvenile European, *Ostrea edulis* L.; Pacific oysters, *Crassostrea gigas* Thunberg. *Aquac. Res.* **1998**, *29*, 103–113. [CrossRef]
73. Davis, H.C.; Ansell, A.D. Survival and growth of larvae of the European oyster (*Ostrea edulis* L.) at lowered salinities. *Boil. Bull.* **1962**, *122*, 33–39. [CrossRef]
74. Hutchinson, S.; Hawkins, L.E. Quantification of the physiological responses of the European flat oyster, *Ostrea edulis* L. to temperature and salinity. *J. Molluscan Stud.* **1992**, *58*, 215–226. [CrossRef]
75. Hayward, P.J.; Ryland, J.S. *Handbook of the Marine Fauna of North-West Europe*; Oxford University Press: New York, NY, USA, 1998.
76. Drinkwaard, A.C. Current velocity as an ecological factor in shell growth of *Ostrea edulis*. *ICES Shellfish Committee* **1961**, *47*, 1–3.
77. Grant, J.; Enright, C.T.; Griswold, A. Resuspension and growth of *Ostrea edulis*: A field experiment. *Mar. Boil.* **1990**, *104*, 51–59. [CrossRef]
78. Tillin, H.M.; Hull, S.C. *Tools for Appropriate Assessment of Fishing and Aquaculture Activities in Marine and Coastal Natura 2000 Sites*; Report VII: Intertidal and Subtidal Reefs, Report No. R. 2074; ABPmer for the Marine Institute: Galway, Ireland, 2013.
79. Hawkins, A.J.S.; Bayne, B.L.; Bougrier, S.; Héral, M.; Iglesias, J.I.P.; Navarro, E.; Smith, R.F.M.; Urrutia, M.B. Some general relationships in comparing the feeding physiology of suspensionfeeding bivalve molluscs. *J. Exp. Mar. Boil. Ecol.* **1998**, *219*, 87–103. [CrossRef]
80. Barillé, L.; Prou, J.; Heral, M.; Razet, D. Effects of high natural seston concentrations on the feeding, selection, and absorption of the oyster *Crassostrea gigas* Thunberg. *JEMBE* **1997**, *2012*, 149–172. [CrossRef]
81. Millican, P.F.; Helm, M.M. Effects of nutrition on larvae production in the European flat Oyster, *Ostrea edulis*. *Aquaculture* **1994**, *123*, 83–94. [CrossRef]
82. Davis, J.C. Minimal Dissolved Oxygen Requirements of Aquatic Life with Emphasis on Canadian Species: A Review. *J. Fish. Board Can.* **1975**, *32*, 2295–2332. [CrossRef]
83. Borsje, B.W.; Bouma, T.J.; Rabaut, M.; Herman, P.M.J.; Hulscher, S.J.M.H. Formation and erosion of biogeomorphological structures: A model study on the tube-building polychaete *Lanice conchilega*. *Limnol. Oceanogr.* **2014**, *59*, 1297–1309. [CrossRef]
84. Buijsman, M.C.; Ridderinkhof, H. Long-term evolution of sand waves in the Marsdiep inlet. II: Relation to hydrodynamics. *Cont. Shelf Res.* **2008**, *28*, 1202–1215. [CrossRef]
85. Dorst, L.L.; Roos, P.C.; Hulscher, S.J.M.H.; Lindenbergh, R.C. The estimation of sea floor dynamics from bathymetric surveys of a sand wave area. *J. Appl. Geodesy* **2009**, *3*, 97–120. [CrossRef]
86. Abelson, A.; Weihs, D.; Loya, Y. Hydrodynamic impediments to settlement of marine propagules, and adhesive-filament solutions. *Limnol. Oceanogr.* **1994**, *39*, 164–169. [CrossRef]
87. Crimaldi, J.P.; Thompson, J.K.; Rosman, J.H.; Lowe, R.J.; Koseff, J.R. Hydrodynamics of larval settlement: The influence of turbulent stress events at potential recruitment sites. *Limnol. Oceanogr.* **2002**, *47*, 1137–1151. [CrossRef]
88. Hendriks, I.E.; van Duren, L.A.; Herman, P.M.J. Turbulence levels in a flume compared to the field. *J. Sea Res.* **2006**, *55*, 15–29. [CrossRef]
89. Grabowski, R.C.; Droppo, I.G.; Wharton, G. Erodibility of cohesive sediment: The importance of sediment properties. *Earth-Sci. Rev.* **2011**, *105*, 101–120. [CrossRef]
90. Wright, L.D.; Schaffner, L.C.; Maa, J.P.-Y. Biological mediation of bottom boundary layer processes and sediment suspension in the lower Chesapeake Bay. *Mar. Geol.* **1997**, *141*, 27–50. [CrossRef]

91. McCave, I.N. Sand waves in the North Sea off the coast of Holland. *Mar. Geol.* **1971**, *10*, 199–225. [CrossRef]
92. Van Dijk, T.A.G.P.; Kleinhans, M.G. Processes controlling the dynamics of compound sand waves in the North Sea, Netherlands. *J. Geophys. Res. Earth Surf.* **2005**, *110*. [CrossRef]
93. Damen, J.M.; van Dijk, T.A.G.P.; Hulscher, S.J.M.H. Spatially Varying Environmental Properties Controlling Observed Sand Wave Morphology. *J. Geophys. Res.* **2018**, *123*, 19. [CrossRef]
94. ICONA. *Noordzeeatlas Voor Het Nederlandse Beleid en Beheer*; ICONA Den Haag; Stadsuitgeverij: Amsterdam, The Netherlands, 1992.
95. Gerritsen, H.; Vos, R.J.; Van Der Kaaij, T.; Lane, A.; Boon, J.G. Suspended sediment modelling in a shelf sea (North Sea). *Coast. Eng.* **2000**, *41*, 317–352. [CrossRef]
96. Van der Kaaij, T.; van Kessel, T.; Troost, T.; van Duren, L.A.; Villars, M.T. *Modelondersteuning MER Winning Suppletie- en Ophoogzand Noordzee 2018–2027*; Modelvalidatie 1230888-002; Deltares: Delft, The Netherlands, 2017.
97. Laing, I.; Walker, P.; Areal, F. *A Feasibility Study of Native Oyster (Ostrea edulis) Stock Regeneration in the United Kingdom*; CARD Project Report FC1016; Centre for Environment, Fisheries & Aquaculture Science: Lowestoft, UK, 2005.
98. Sawusdee, A.; Jensen, A.C.; Collins, K.J.; Hauton, C. Improvements in the physiological performance of European flat oysters *Ostrea edulis* (Linnaeus, 1758) cultured on elevated reef structures: Implications for oyster restoration. *Aquaculture* **2015**, *444*, 41–48. [CrossRef]
99. Snelgrove, P.V.R.; Butman, C.A. Animal-sediment relationships revisited: Cause versus effect. *Oceanogr. Mar. Boil. Annu. Rev.* **1994**, *32*, 111–177.
100. Houziaux, J.-S.; Fettweis, M.; Francken, F.; van Lancker, V. Historic (1900) seafloor composition in the Belgian–Dutch part of the North Sea: A reconstruction based on calibrated visual sediment descriptions. *Cont. Shelf Res.* **2011**, *31*, 1043–1056. [CrossRef]
101. Wasson, K. Informing Olympia Oyster Restoration: Evaluation of Factors That Limit Populations in a California Estuary. *Wetlands* **2010**, *30*, 449–459. [CrossRef]
102. Peirano, A. Wrecks on the Bottom: Useful, Ecological Sentinels? *Mar. Technol. Soc. J.* **2013**, *47*, 118–127. [CrossRef]
103. Petersen, J.K.; Maar, M.; Ysebaert, T.; Herman, P.M.J. Near-bed gradients in particles and nutrients above a mussel bed in the Limfjorden: Influence of physical mixing and mussel filtration. *Mar. Ecol. Prog. Ser.* **2013**, *490*, 137–146. [CrossRef]
104. Wiles, P.J.; van Duren, L.A.; Haese, C.; Larsen, J.; Simpson, J.H. Stratification and mixing in Limfjorden in relation to mussel culture. *J. Mar. Syst.* **2006**, *60*, 129–134. [CrossRef]
105. Van Leeuwen, S.; Tett, P.; Mills, D.; van der Molen, J. Stratified and nonstratified areas in the North Sea: Long-term variability and biological and policy implications. *J. Geophys. Res.* **2015**, *120*, 4670–4686. [CrossRef]
106. Greenwood, N.; Parker, E.R.; Fernand, L.; Sivyer, D.B.; Weston, K.; Painting, S.J.; Kröger, S.; Forster, R.M.; Lees, H.E.; Mills, D.K.; et al. Detection of low bottom water oxygen concentrations in the North Sea; implications for monitoring and assessment of ecosystem health. *Biogeosciences* **2010**, *7*, 1357–1373. [CrossRef]
107. Queste, B.Y.; Fernand, L.; Jickells, T.D.; Heywood, K.J.; Hind, A.J. Drivers of summer oxygen depletion in the central North Sea. *Biogeosciences* **2016**, *13*, 1209–1222. [CrossRef]
108. Cronin, K.; Plieger, R.; Gaytan Aguilar, S.; de Lima Rego, J.R.; Blaas, M. *MoS2-II: Deterministic Model Calibration: Updates of the ZUNO-DD Hydrodynamic and SPM Model Deltares Report*; Rep. No. 1204561; Deltares: Delft, The Netherlands, 2013.
109. Korringa, P. Relations between the Moon and Periodicity in the Breeding of Marine Animals. *Ecol. Monogr.* **1947**, *17*, 347–381. [CrossRef]
110. Sas, H.; Kamermans, P.; van der Have, T.; Lengkeek, W.; Smaal, A. *Shellfish Reef Restoration Pilots, Voordelta, The Netherlands*; Annual Report; ARK Natuurontwikkeling: Nijmegen, The Netherlands, 2016.
111. Sas, H.; Kamermans, P.; van der Have, T.; Christianen, M. *Shellfish reef Restoration Pilots, Voordelta, The Netherlands*; Annual Report; ARK Natuurontwikkeling: Nijmegen, The Netherlands, 2017.
112. Christianen, M.J.A.; Bergsma, J.H.; Coolen, J.W.P.; Didderen, K.; Dorenbosch, M.; Driessen, F.M.F.; Kamermans, P.; Reuchlin-Hugenholtz, E.; Sas, H.; Smaal, A.; et al. Return of the native facilitated by the invasive? Population composition, substrate preferences and epibenthic species richness of a recently discovered shellfish reef with native European flat oysters (*Ostrea edulis*) in the North Sea. *Mar. Boil. Res.* **2018**, *14*, 590–597. [CrossRef]

113. Petersen, T.U.; Mutlu Sumer, B.; Fredsøe, J.; Raaijmakers, T.C.; Schouten, J.J. Edge scour at scour protections around piles in the marine environment—Laboratory and field investigation. *Coast. Eng.* **2015**, *106*, 42–72. [CrossRef]
114. Pogoda, B.; Buck, B.H.; Hagen, W. Growth performance and condition of oysters (*Crassostrea gigas* and *Ostrea edulis*) farmed in an offshore environment (North Sea, Germany). *Aquaculture* **2011**, *319*, 484–492. [CrossRef]

sustainability

MDPI

Article

Macroeconomic Impacts of Climate Change Driven by Changes in Crop Yields

Shinichiro Fujimori [1,2,*], Toshichika Iizumi [3], Tomoko Hasegawa [2], Jun'ya Takakura [2], Kiyoshi Takahashi [2] and Yasuaki Hijioka [2]

1 Department of Environmental Engineering, Kyoto University, Kyoto 615-8540, Japan

2 Center for Social and Environmental Systems Research, National Institute for Environmental Studies, Tsukuba 305-8506, Japan; hasegawa.tomoko@nies.go.jp (T.H.); takakura.junya@nies.go.jp (J.T.); ktakaha@nies.go.jp (K.T.); hijioka@nies.go.jp (Y.H.)

3 Institute for Agro-Environmental Sciences, National Agriculture and Research Organization, Tsukuba 305-8604, Japan; iizumit@affrc.go.jp

* Correspondence: sfujimori@athehost.env.kyoto-u.ac.jp; Tel.: +81-75-383-3367

Received: 10 September 2018; Accepted: 10 October 2018; Published: 14 October 2018

Abstract: Changes in agricultural yields due to climate change will affect land use, agricultural production volume, and food prices as well as macroeconomic indicators, such as GDP, which is important as it enables one to compare climate change impacts across multiple sectors. This study considered five key uncertainty factors and estimated macroeconomic impacts due to crop yield changes using a novel integrated assessment framework. The five factors are (1) land-use change (or yield aggregation method based on spatially explicit information), (2) the amplitude of the CO_2 fertilization effect, (3) the use of different climate models, (4) socioeconomic assumptions and (5) the level of mitigation stringency. We found that their global impacts on the macroeconomic indicator value were 0.02–0.06% of GDP in 2100. However, the impacts on the agricultural sector varied greatly by socioeconomic assumption. The relative contributions of these factors to the total uncertainty in the projected macroeconomic indicator value were greater in a pessimistic world scenario characterized by a large population size, low income, and low yield development than in an optimistic scenario characterized by a small population size, high income, and high yield development (0.00%).

Keywords: agricultural impacts; climate change impacts; integrated assessment model; CGE model

1. Introduction

The economic impact of climate change on key economic sectors has been studied for a long time. The latest findings in this research area were summarized by working group II in Chapter 10 of the Fifth Assessment Report of the Intergovernmental Panel on Climate Change (IPCC AR 5) [1] and a recent literature review [2]. With respect to the economic impact on the agricultural sector, the information reported by IPCC AR5 is limited [3], and a review of studies using global economic models found in Chapter 7 only provides the impacts on food prices and food security, but not those on the macroeconomy. The same can be said of the Agricultural Model Inter-comparison Project (AgMIP) exercises [4,5].

There are some studies that deal with the macroeconomic implications of agricultural climate change effects. For example, Reilly et al. [6] analyzed the economic impacts of reduction in agriculture production using a partial equilibrium model, and estimated which regions of the world would be winners or losers under climate change. Recently, on a global scale, Ren et al. [7] conducted a similar analysis using a computable general equilibrium (CGE) model, and concluded that the macroeconomic impact would be small in absolute terms (less than 1% of gross domestic product (GDP)). Roson and Damania [8] made an assessment with multiple socioeconomic scenarios from the point of view of water scarcity. Another study used a partial equilibrium model and estimate changes in agricultural welfare due to climate change [9]. The magnitude of welfare changes in trade liberalization scenarios was reported to be 0 to −0.5%. Ciscar, et al. [10] investigated the economic effects of climate change in Europe, and found that a macroeconomic loss of about 0.3% would occur in most global warming scenarios when using a CGE model. As seen, the order of magnitude of the projected global agricultural economic losses due to climate change is small (about 0–1% of GDP). Therefore, it is reasonable to use food prices and population at risk of hunger as indicators rather than GDP change when assessing climate change impacts on the agricultural sector.

However, a quantification of the GDP impacts in the agricultural sector is important as it enables one to compare the climate change impacts across multiple sectors using a single macroeconomic indicator. Ultimately, this can enable us to make comparisons in total cost between mitigation and adaptation feasible, like in the literature [11]. However, the quantification of the economic impacts on the agricultural sector involves many uncertain factors, including the use of different yield aggregation methods and GCMs, and the use of different assumptions of the CO_2 fertilization effect, global warming level, and socioeconomic conditions.

We evaluated the relative contributions to the uncertainty in the projected GDP impacts in agricultural sector and identified largest factors to change the GDP impact associated with agricultural climate change impact. Here our scope is to understand the macroeconomic responses to the yield changes; we thus apply a CGE model rather than a partial equilibrium model. The factors that we have taken into account are the use of (1) different aggregation methods of gridded yield information, (2) different assumptions of the CO_2 fertilization effect, (3) GCMs, (4) level of mitigation policy and (5) socioeconomic assumptions associated with Shared Socioeconomic Pathways (SSPs).

2. Methods

We use an integrated assessment model AIM (Asia-Pacific Integrated Model), which includes a crop model (Crop Yield Growth Model with Assumptions on climate and socioeconomy: CYGMA), a land use allocation model (AIM/PLUM: Asia-Pacific Integrated Model/integration Platform for Land-Use and environmental Modelling) model in AIM/CGE model) [12], and a global computable general equilibrium model (AIM/CGE) [13] were used as the main tools (see Supplementary Materials, Figure S1). In addition, the Dynamic Integrated model of Climate and the Economy (DICE) optimization model was used to derive the global greenhouse gas (GHG) emissions constraints for mitigation scenarios. AIM/CGE first performed a simulation (reference case) that had 17 aggregated regions (see Table S1). As with losses in other sectors (e.g., health effects, and flood damage), the impacts in the agricultural sector affect macroeconomic economies through changes in production factor inputs such as labor, capital, and land. CGE models are suitable for the analysis of such influences, and they have been used in many of the studies mentioned above compared to partial equilibrium models.

Forms of land use, such as cropland (see Table S2), were then allocated using AIM/PLUM, which handles land use on a grid basis. The yield potential map generated by CYGMA was then aggregated into 17 AIM/CGE regions using either the current gridded harvest area or AIM/PLUM (depending on scenario assumptions). Finally, the simulation covered the world from 2005 to 2100. The potential yield of crops was based on estimates from CYGMA. However, CYGMA simulates only rice, wheat, soybean, and maize yield. Therefore, for other crops (e.g., sugar crops), we used the climate change impact on yield, generated by the Lund-Potsdam-Jena managed Land Dynamic Global Vegetation and Water Balance Model (LPJmL) [14,15]. For the no climate change (NoCC) yield change of other crops, we adopted the assumptions used in an earlier study [16] (Figure S2). For each model's performance, see more in Fujimori et al. (2017) [17], Hasegawa et al. (2017) [12] and Iizumi et al. (2017) [18] for the CGE, land-use, and crop models, respectively.

Regarding the yield aggregation methods, we considered two methods. One was the time-fixed cropland pattern and the other was time-varying cropland pattern simulated by a land-use allocation model (there is an attempt to do two time-fixed crop model comparison [19]). This is an important source of uncertainty as aggregated yield changes differ depending on gridded cropland patterns [20] in which similar approach has been implemented by earlier studies [21,22]. Biophysical crop models in general have a high geographical resolution (e.g., a 0.5-degree grid size) and therefore the spatial aggregation of crop model output is required when it is used as the input to CGE models with the spatial resolution of 10 to 20 regions at the global level. Although some agricultural economic models explicitly considered changes in the geographic land-use pattern (e.g., MAgPIE [23] and GLOBIOM [24]), a CGE or regionally aggregated equilibrium approach does not treat land-use changes in this way (agro-ecological zone (AEZ), which is now used in many CGEs in the agricultural economic assessments partly allows such geographic land-use pattern). For the CO_2 fertilization effect, we compared the cases with and without the CO_2 fertilization effect on agricultural yields. For GCMs, we used a set of GCMs that was the same with ISI-MIP (The Inter-Sectoral Impact Model Intercomparison Project). Recently developed SSPs and Representative Concentration Pathways (RCPs) framework allows us to comprehensively investigate the effects of climate change and socioeconomic development patterns [25].

2.1. *Model Brief Description*

2.1.1. AIM/CGE

The AIM/CGE model includes 42 industrial classifications (see Table S3). The production sectors are assumed to maximize profits under multi-nested constant elasticity substitution (CES) functions for each input price. Household expenditures on each commodity are described by a linear expenditure system (LES) function of which income elasticities for agricultural products are show in Table S7. The parameters adopted in the linear expenditure system function are recursively updated by income elasticity assumptions. The saving ratio is endogenously determined to balance saving and investment, and capital formation for each good is determined by a fixed coefficient. The Armington assumption is used for trade (CES and constant elasticity of transformation are assumed), and the current account is assumed to be balanced.

The AIM/CGE model has a land-nesting strategy, similar to the approach taken in earlier studies [26]. Land is categorized into one of three ecological zones, and there is a land market for each zone. The allocation of land by sector is formulated as a multi-nominal logit function to reflect differences in substitutability across land categories with land rent. As such, the function assumes that land owners in each region and AEZ decide on land sharing among options, with the land rent depending on the production of each land unit (i.e., crops, livestock, and wood products). This is validated in Fujimori et al. [17]. More details of this model can be found in the model documentation [13].

The agricultural yield impacts associated with climate change increase the cost of production. The corresponding agricultural sectors attempt to expand harvesting area and demand in response to the high price, which decreases the consumption. Compared with the conditions where labor, capital and land resources are optimally allocated in terms of welfare maximization, the climate change moves demand and supply of all goods, serves and primary factors to different points which causes macroeconomic costs.

CGE model parameters are usually calibrated based on a single year's social accounting matrix. Our social accounting matrix is based on the reconciled various statistical data (e.g., GTAP [27], National accounts [28], and so on.

2.1.2. Global Gridded Crop Model (CYGMA)

CYGAM is a biophysical global gridded crop yield model that can explicitly consider changes in agronomic inputs to the production system, technology and management associated with economic growth, and changes in the biophysical response of a crop to environmental conditions. The model operates on a grid cell basis, with a grid interval of 0.5°. CYGAM has an advantage to other crop models in the way that it can explicitly deals with socioeconomic changes in yield evolutions. Crop growth is simulated on a daily basis and the influences on crop productivity (yield) regarding the crop's thermal requirement (this determines crop duration), sowing date and availability of heat, water, and nitrogen, are considered. Growth stresses associated with nitrogen shortage, heat, cold, water shortage, and water excess are also considered. The amount of annual nitrogen input is parameterized as a function of per capita GDP and per capita agricultural area, country by country. All stress types considered here are a function of the knowledge stock of agricultural technologies and climatic conditions. The knowledge stock improves according to economic growth and increases the use of improved varieties and associated agronomic management in farm fields, which allows simulated crops to have an increased tolerance to stresses. For the base year calibration, the crop-specific coefficients that represent tolerance to abiotic stresses and yield response to nitrogen deficit were calibrated based on the high-resolution (5-min arc or 10-km) average actual and potential yield data in 2000 (Monfreda et al. 2008 [29], Mueller et al. 2012 [30]). The similar method that uses a time-constant, spatial yield dataset for the model calibration can be seen in Deryng et al. (2011) [31]. The coefficients are crop-specific, but universal across locations. The verification of the model and its application to the future climate and socioeconomic conditions are described in Iizumi et al. [18].

2.1.3. AIM/PLUM

The AIM/PLUM is a global land-use allocation model used to downscale the AIM/CGE's aggregated regional land-use projections into a spatial gridded land-use pattern for the interactive assessment of human activities and biophysical elements. Regional-scale land demand estimated by AIM/CGE (17 regions) was fed into the AIM/PLUM land-use allocation model and was spatially distributed into grid cells ($0.5° \times 0.5°$). The cropland and afforestation area was allocated based on optimization (profit maximization), where a landowner was assumed to decide the mix of land-uses to obtain the highest profit for a given biophysical land productivity condition (e.g., crop yield production per unit area). Because the optimization was solved for each region that had the same regional classification as that used in AIM/CGE, land transactions across the regions were not allowed. There were seven crop types, with or without irrigation. Land for harvested wood was excluded from the model framework. The bioenergy crop yield and forest carbon sequestration were based on estimates from the Vegetation Integrative Simulator for Trace Gases (VISIT) [32]. Please see Hasegawa, Fujimori, Ito, Takahashi and Masui [12] for more details of the model description and validation.

2.2. Scenarios

We computed scenarios considering the five factors "crop yield aggregation method," "CO_2 fertilization," "socioeconomic assumptions," "mitigation policy," and "multi-GCMs." All scenarios that were quantified are shown in Table 1. There were two ways to aggregate the crop model CYGMA gridded information into the aggregated area of AIM/CGE. The first method was to calculate the average yield of 17 regions by fixing the gridded land use or harvest area to the current situation ("base" in Table 1). The other method was a case in which there was a land-use change option to change gridded cultivated land according to yield using AIM/PLUM ("Change" in Table 1). The second factor was that of consideration, or not, of the CO_2 fertilization effect, which is considered in CYGMA with the Representative Concentration Pathways (RCPs) CO_2 concentration information. The third factor was a socioeconomic assumption using the SSPs (SSP1, SSP2, and SSP3). Social and economic conditions such as GDP, population, and food preference also followed the SSPs [16,25].

For each case, we conducted a run with and without climate change cases. The cases without climate change used current climate conditions. For the climate change cases, RCP 8.5 and RCP2.6 [33] were used. Although SSP1, SSP2, and SSP3 do not reach such a high level of forcing [25], considering the comparability of scenarios and relevancy of the climate impacts, the RCP8.5 climate condition was the most appropriate for this study. Moreover, the uncertainties of using multiple GCMs were incorporated into the five GCMs, which were also used in ISIMIP (Table S4). The mitigation policy approximately corresponded to a 450 ppm CO_2 concentration stabilization. We capped the global total GHG emissions constraint for AIM/CGE derived from a modified DICE model [34] of which emissions pathway might be slightly different from original RCP pathways but they would be close enough for this paper's analysis. Here we did not consider near-term (2025 to 2030) policies such as Paris Agreement, but for the agricultural macroeconomic impact this position would not change the major findings because the climate change impact becomes severe after 2030. For the agricultural markets, GDP and population are well known representative socioeconomic indicators but also dietary preferences and the degree of openness of trade are considered. They are described in Popp et al. (2017) [35]. More details of how we implemented the mitigation scenario are provided in Fujimori, et al. [16].

We ran selected combinations of each of the factors rather than all possible combinations to appropriately address the main research questions, as shown in Table 1. The main aims of this study were to clarify the macroeconomic impact due to climate change. Therefore, the basic strategy was to compare NoCC and climate change cases (e.g., RCP8.5). To determine the impact of differences in the yield aggregation method, we considered the differences between selected scenarios; for example (scenario 3–scenario 2) and (scenario 5–scenario 4) (scenario numbers are shown in Table 1). Another example is to assess the impact of CO_2 fertilization we considered the differences between (scenario 3–scenario 4) and (scenario 6–scenario 5). Here, two things had to be addressed. First, using the current harvest area was adopted as a pivot for the yield aggregation method. Second, scenario 10 was a hypothetical scenario in which the climate was stabilized at a low CO_2 concentration, without mitigation efforts. However, we computed this scenario to derive the pure climate change effect in the RCP2.6 climate condition (scenario 10–scenario 11).

Table 1. List of scenarios.

Scenario Number	Socioeconomic Condition	Climate Condition	Yield Aggregation Method *	CO_2 Fertilization Effect	Mitigation
1	SSP2	NoCC	Base		
2	SSP2	NoCC	Change		
3	SSP2	RCP8.5	Change		
4	SSP2	RCP8.5	Change	X	
5	SSP2	RCP8.5	Base	X	
6	SSP1	NoCC	Base		
7	SSP1	RCP8.5	Base		
8	SSP3	NoCC	Base		
9	SSP3	RCP8.5	Base		
10	SSP2	RCP2.6	Base		
11	SSP2	RCP2.6	Base		X

* If the yield aggregation is based on the current harvest gridded map, it shows Base. If the gridded land use change is taken, it is "Change."

3. Results

3.1. Macroeconomic Losses

The global total macroeconomic impact (rate of GDP change) in 2100 due to changes in agricultural yield, relative to the corresponding NoCC cases (e.g., scenario 3–scenario 1), is shown in Figure 1. In total, the range of the GDP change was from 0.00% to −0.57%. For the scenario 1 case, which is without gridded land-use change, has no CO_2 fertilization, and used SSP2 and RCP8.5, in 2100 the median change is 0.04%. The yield aggregation method produces about half (0.02%) of the change of the no gridded land use case (panel a). However, considering the uncertainty of GCMs, this difference is ambiguous. When focusing on the existence of the CO_2 fertilization effect (panel b), the median change is about 0.02%, which is similar to the yield aggregation method effect. The climate change condition is a stronger factor than either the yield aggregation method effect or the CO_2 fertilization effect in terms of the macroeconomy. The use of RCP2.6 results in a worldwide rate of GDP loss of −0.01%, which represents an overall positive effect, with all GCMs producing a positive change (panel c). This positive effect would be caused by modest warming. Note that these scenarios do not consider mitigation costs. If mitigation is considered, the mitigation costs would dominate the GDP changes, resulting in clear losses and the climate change impact on yield being much smaller than the mitigation cost (see Figure S3). On the other hand, it can be seen from (panel d) in Figure 1 that the difference depending on the socioeconomic conditions is remarkable in SSP3, with the digits differing by one order of magnitude. The use of SSP3 produces a difference of 0.57%, while the use of SSP1 has little influence.

Figure 2 shows time series information about each climate change level, with regional information. From a worldwide perspective, the rate of GDP loss (negative GDP change) increases rapidly after 2050. The range of uncertainty also increases over time. The uncertainty range in RCP8.5 is larger than that in RCP2.6, which is basically driven by crop model outcome (shown later in Figure 4).The increase is not a unique characteristic across the five large regions. For example, the OECD median value is almost stable, while the range of uncertainty increases in the latter half of the century. The median in Asia is also stable and the range of uncertainty is almost constant over time across the scenarios. In contrast, Africa (MAF) and Latin America (LAM) are relatively low-latitude zones that could be sensitive to the effects of global warming. In these countries, RCP8.5 resulted in a higher GDP loss than RCP2.6. Reforming regions (REF; mostly the former Soviet Union) are also likely to have a negative impact in RCP8.5, but the range of uncertainty is large. More detailed regional results indicate that there are some regions or countries that the order of the magnitude is higher than those shown in Figure 2 (e.g., Rest of Asia; XSA in Figure S4). Note that regional heterogeneity is also apparent across the yield aggregation method options and CO_2 fertilization assumption differences (Figures S5 and S6).

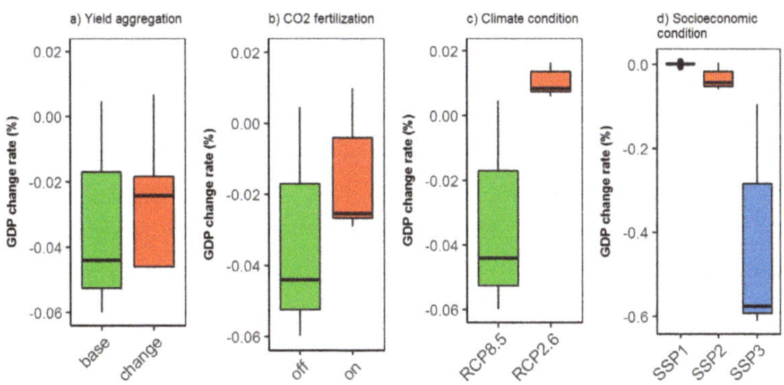

Figure 1. Global total macroeconomic impact due to changes in agricultural yield in the year 2100 considering: (**a**) yield aggregation method differences ("Base" considers a fixed grid-level harvest map and "Change" considers land-use change at grid level, with both under SSP2), (**b**) CO_2 fertilization (on or off), (**c**) climate condition, and (**d**) socioeconomic condition. Boxplots represent the uncertainty of five general circulation models (GCMs). For panels (**a**,**b**,**c**) left, and the center of (**d**) boxplots are identical because they are the same scenario (RCP8.5), with no CO_2 fertilization and no land-use change cases.

Interestingly, the added agricultural value has much more visible changes than GDP, and the sign is positive, which means an increase in agricultural added value (Figure 3). The food consumption response is much lower than yield change (Figures S7 and S8) due to the price elasticity far less than 1 (Table S6), the basic reactions of the CGE model to the yield changes are expanding the cultivated area (Figure S7). Therefore, the yield negative effect will require additional labor and capital, which will increase the production price and added value of the agriculture sector. On the other hand, these additional labor and capital in agriculture sectors would decrease the resource availability in other industries that would have relatively higher productivity than agriculture sectors, which eventually generates GDP loss as shown above in spite of an increase in added agricultural value. The order of the magnitude is around 10-30% of changes compared to no climate change cases, which are remarkably large.

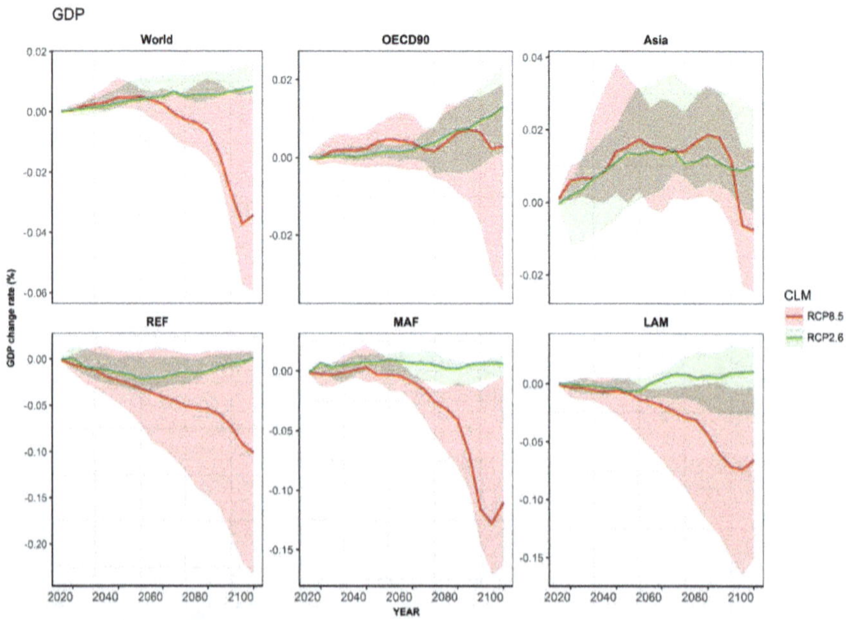

Figure 2. Macroeconomic impact due to changes in agricultural yield in SSP2 in five regions and as a global total (positive means increase compared with baseline). The colors represent climate change level (RCPs). The shaded area is the general circulation model (GCM) range of uncertainty (five GCMs for RCP8.5, and RCP2.6). The lines are the median of five GCMs for each RCP2.6 and RCP8.5. The scenarios in which CO_2 fertilization and land-use change are both included are shown. Regional codes are OECD9, OECD regions; Asia, Asia; REF, Reforming region; MAF, Middle East and Africa; and LAM, Latin America. (The mapping procedure from 17 regions is shown in Table S1.)

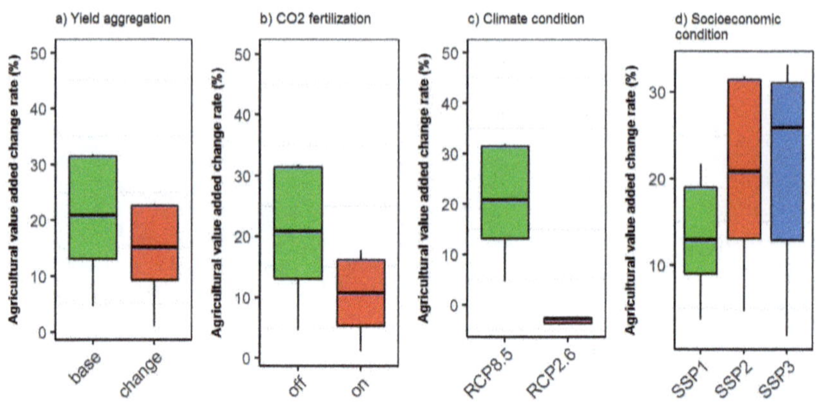

Figure 3. Global agricultural value added impact due to changes in agricultural yield in the year 2100 considering (positive means increase compared with baseline): (**a**) yield aggregation method differences ("Base" considers a fixed grid-level harvest map and "Change" considers land-use change at grid level, with both under SSP2), (**b**) CO_2 fertilization (on or off), (**c**) climate condition, and (**d**) socioeconomic condition. Boxplots represent the uncertainty of five general circulation models (GCMs). For panels (**a**,**b**,**c**) left, and the center of (**d**) boxplots are identical because they are the same scenario (RCP8.5), with no CO_2 fertilization and no land-use change cases.

3.2. Changes in Agricultural Yield Associated with Climate Change and Its Consequences

Figure 4 (panel a) shows the average yield of five crops (rice, wheat, other cereals, oil crops, and sugar crops) for AIM/CGE in 17 regions and the global total (see individual crops in Figures S9–S12). The yields are clearly different among the cases with different climate change conditions, i.e., RCP2.6, RCP8.5, and NoCC. The mean global mean yield is currently around 3 t/ha. Because of technological progress, which is mainly caused by the growth of income, the yield increases to about 5 t/ha in the scenario where climate conditions are kept at the current level (NoCC). The RCP2.6 case is similar to the NoCC case. This trend is apparent in the global total, but varies regionally in some cases. For example, the yield in India is about 25% lower for the RCP2.6 case than NoCC, while in China it is slightly larger than NoCC (less than 10%). Although the impact on global average yield is negative in RCP2.6 there is a positive impact on GDP. This is mainly due to the weighting used to calculate the average value. Regional GDP, the impact on yield, and agricultural production has different weighting, and the way that the average is calculated causes discrepancies.

In the RCP8.5 case, there is a substantial decrease in yield. The difference in the global mean decrease from the NoCC decrease is not dissimilar in the first couple of decades of the 21st century, before the influence of climate change becomes apparent but, over time, the deviation becomes large. In 2100, it is almost 3 t/ha, which is almost the same as in 2005. Although there are always uncertainties associated with the use of multiple GCMs, the negative effect on yield is clearly seen. In particular, the remarkable tendency for a deviation from the yield decrease in the NoCC case mentioned above is significant in developing countries such as India (IND), other Asia (XSA), and other Africa (XAF). Even when CO_2 fertilization is considered, a similar pattern is apparent, although the negative effect is modest (see Figure S5).

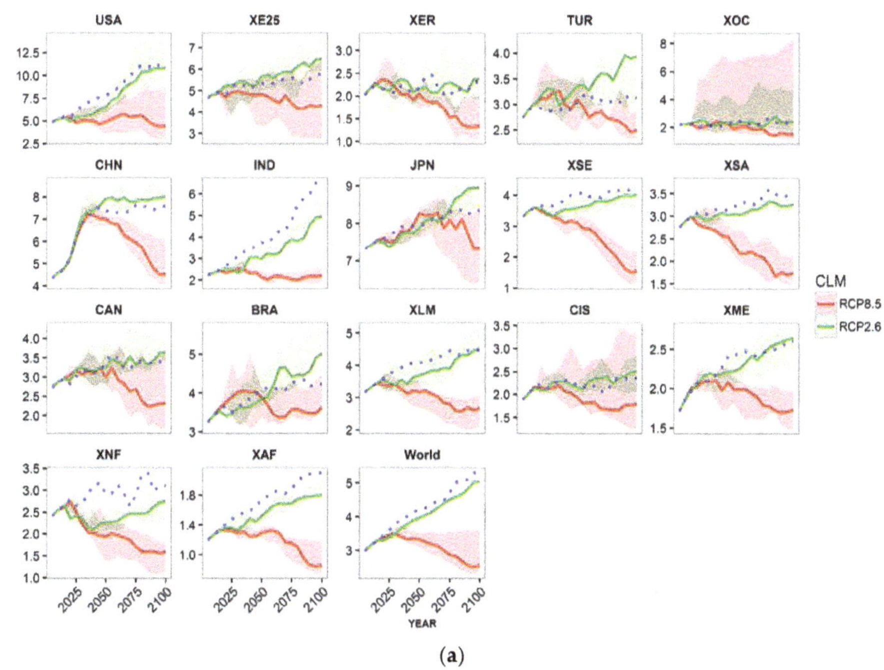

(a)

Figure 4. *Cont.*

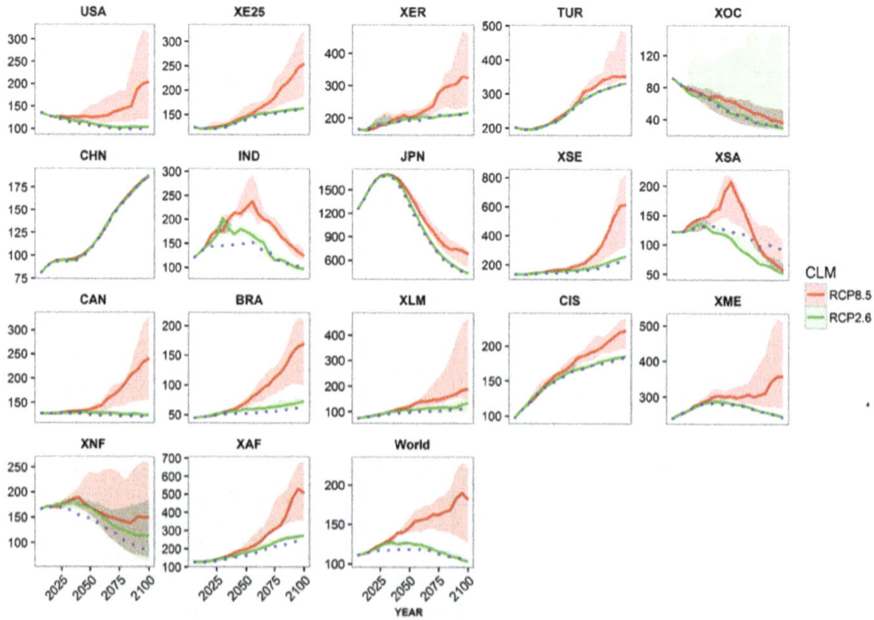

(b)

Figure 4. (**a**) Mean changes in a) yield and (**b**) price of five major crops for 17 regions (Table S1) and the global average in two climate change cases (green; RCP2.6 and red; RCP8.5) and the no climate change (NoCC) case (blue dot). The ribbon indicates the range of uncertainty for five general circulation models (GCMs). The units are t/Ha and 1000$/t, respectively.

As can be seen from Figure 4 (panel b), the yield decreases lead to a food price increase, which is also discussed in IPCC AR5 and AgMIP literature [3–5]. In 2100, the global average price of five major crops in the RCP8.5 case is much higher than in the NoCC and RCP2.6 cases. The RCP2.6 and NoCC cases are similar, and RCP8.5 is around 70% (range of 30 to 120%) higher than NoCC. This price response is higher than that reported in the AgMIP study (0 to 60%). Since the yield information, and the models are different from AgMIP, it is not fully comparable, but one of the reasons could be that AgMIP's focus is the year 2050, in which the impact of climate change is not projected to be as severe as in 2100.

3.3. Gridded Yield Aggregation Method Effect

We addressed the changes in GDP associated with aggregation method of gridded yield information in Section 3.1. The next question with respect to this land-use change treatment is how much significant changes are generated in the yield between these two methods. The Figure 5 illustrates yield differences and some regions are quite overlapping, which means land-use change treatment do not change in the macroscale results. However, interestingly, many regions show quite different yield trajectories across the land use treatment. In particular, North Africa (XNF) and the Middle East (XME) are remarkably different regions. Crops specific results are more diverse. These results imply that an analysis that requires regional and crop-specific focuses should deal with the spatial explicit land-use change appropriately.

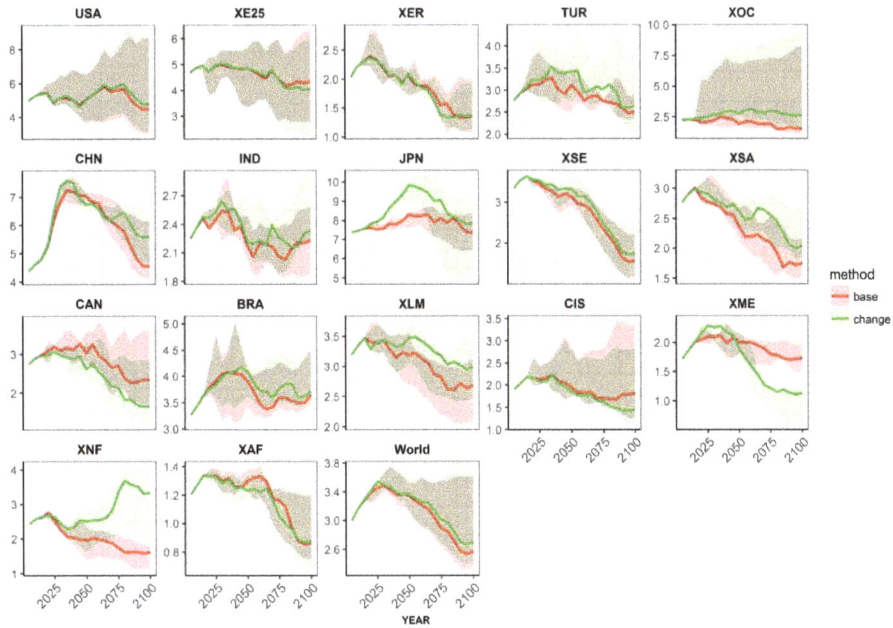

Figure 5. Mean yield of five major crops for 17 regions (Table S1) and the global average in two yield aggregation method for land-use change treatment cases (red; fixed as base year harvested area and red; changed dynamically) under RCP8.5 and non-CO_2 fertilization cases. The ribbon indicates the range of uncertainty for five general circulation models (GCMs). The units are t/Ha.

4. Discussion

The socioeconomic differences cause a relatively large impact on the global macroeconomy in terms of agricultural yield differences. There are two implications from this. First, in the situation where climate mitigation failed, it would be desirable for societal development to be directed toward SSP1 like world in the SSPs context so that the impact of climate change can be kept to a minimum locally and globally. Second, the fact that socioeconomic conditions are a major factor means that further detailed research is required in this area.

One of the focuses of this study is how much the spatially explicit land-use change responded to changes in crop yield due to the influence of climate change on the macroeconomy. Conventional CGE studies have not treated these issues appropriately. From the results of this study, this is found to be a significant factor. It should be noted that the GCM uncertainty could hide the significance of the results of GDP. However, considering the regional variation, the methodology aggregating yield information by using future spatial land-use change is important at the regional or crop specific level. This would imply that methodological improvement may need to be considered for the CGE modeling which might also be better to reconsider the model intercomparison exercise [36].

The macroeconomic impact of changes in agricultural yield associated with future climate change is found to be small compared with the economic analysis of climate effects in other sectors where the large sectors would be 0 point several percentages or even higher [37,38]. For example, in IPCC AR5, [1] reported that a 3.0 °C increase in global mean temperature would lead to around 1 to 3% welfare loss. The most severe climate change case in our study is RCP8.5, which has an increase in global mean temperature of more than 4 °C. The macroeconomic loss in agriculture due to climate change is one order of magnitude less than that of the total impact across all sectors. The magnitude of this loss is similar to that reported in earlier studies [7,9,10]. There are several possible reasons for this. First, the value added by the agricultural sector is small relative to the GDP of the economy as a whole.

The current share of agricultural value added is 3.7% of total GDP (Table S5). The GDP in 2100 is projected to increase by about seven times compared to the current value in SSP2, but because most of the growth would originate from secondary and tertiary sectors, the ratio of value added in agriculture in 2100 to GDP is 1.3%. Second, depending on the region, there are some areas where the influence on the macroeconomy is projected to be positive (indicated by negative GDP losses in the figure), which are offset in the global values. However, this offset effect is not significant. Third, the adjustment effect due to international trade could be a factor. However, the value added by the agricultural sector is considered to be the primary reason. Although here we focused on macroeconomic implications in order to obtain climate change impact information that is comparable across sectors, we should note that it does not mean agricultural climate change is less important despite small macroeconomic changes. Rather, they are related to human basic needs and may entail a human health effect [39,40].

With regard to the regional variation, no specific trend could be identified from the regional information, and it is not clear in which region climate change would have a negative influence. It is clear that low latitude regions experienced relatively high impacts, which could be due to a pure climate change effect, but could also be due to the low GDP per capita. In these regions, the value added of agriculture to GDP is higher than developed countries and the economic impact of climate change seems to be higher as well, which suggests that the impacts on GDP could be larger in developing countries.

5. Limitations

There were several limitations of this research. First, we only used one crop model and one economic model, and the results might have been different if multiple models had been used. We believe that additional model inter-comparison experiments are, therefore, necessary. Second, there are some constraints in the use of CYGMA information. (1) The wheat sector yield is biased because CYGMA wheat currently only deals with spring wheat. This aspect of the model should be improved in the future. (2) The yields of crops other than those treated by CYGMA were obtained from LPJml and the two models are not fully consistent. Because the trends in the four major crops were dealt with appropriately, the major outcomes would be unaffected by this treatment. This crop-cover issue should also be resolved in future studies. (3) We experimented limited scenario combinations and there could be much more possibilities such as SSP4 and SSP5, or different climate levels (e.g., RCP45). (4) The study attempts to account for all relevant uncertainties as much as possible considering current model capability, but there are still some missing elements (e.g., the geopolitical international situation and the possibility of transgenic species). As in studies of the impact of climate change in other sectors, particularly for economic analyses, it is difficult to cover all aspects of the subject, although the modeling framework attempts to capture the main factors.

6. Conclusions

This study reveals the macroeconomic impact of climate change in the agriculture sector using an integrated assessment modelling framework. By considering various factors, such as mitigation policy, the yield aggregation method associated with gridded land-use change, GCM uncertainty, and CO_2 fertilization, the macroeconomic impact is found to be small (0.02–0.06%), even in the case with the highest level of warming. However, when the socioeconomic condition is changed to SSP3 from SSP2, the scale of the impact increases by an order of magnitude. This study highlights the importance of consideration of future socioeconomic conditions in the agricultural economic implications associated with climate change impacts.

Supplementary Materials: The following are available online at http://www.mdpi.com/2071-1050/10/10/3673/s1, Figure S1: Asia-Pacific Integrated Model (AIM) modeling framework, Figure S2: Ten-year mean annual yield growth rate for the SSP1, SSP2, and SSP3 baseline scenarios, Figure S3: The macroeconomic impact due to mitigation policy (RCP2.6 equivalent) and agricultural yield change for RCP2.6 for five regions and the global total, Figure S4: Macroeconomic impact due to changes in agricultural yield in SSP2 in five regions and as a global total, Figure S5: The macroeconomic impact due to changes in agricultural yield for RCP8.5 with and without

CO_2 fertilization for five regions and the global total, Figure S6: The macroeconomic impact due to changes in the agricultural yield for RCP8.5 with and without consideration of land-use change for five regions and the global total, Figure S7: Global main indicators' changes for RCP2.6 (right) and RCP8.5 (left) in 2050 (top) and 2010 (bottom), Figure S8: Food consumption as colorific in take per capita, per day change for the 17 regions of the Asia-Pacific Integrated Model/Computable General Equilibrium (AIM/CGE) and the global average in two climate change cases (green; RCP2.6 and red; RCP8.5) and the no climate change (NoCC) case (blue dot), Figure S9: Yield change of coarse grains for the 17 regions of the Asia-Pacific Integrated Model/Computable General Equilibrium (AIM/CGE) and the global average in two climate change cases (green; RCP2.6 and red; RCP8.5) and the no climate change (NoCC) case (blue dot), Figure S10: Yield change of oil seeds for the 17 regions of the Asia-Pacific Integrated Model/Computable General Equilibrium (AIM/CGE) and the global average in two climate change cases (green; RCP2.6 and red; RCP8.5) and the no climate change (NoCC) case (blue dot), Figure S11: Yield change of rice for the 17 regions of the Asia-Pacific Integrated Model/Computable General Equilibrium (AIM/CGE) and the global average in two climate change cases (green; RCP2.6 and red; RCP8.5) and the no climate change (NoCC) case (blue dot), Figure S12: Yield change of wheat for the 17 regions of the Asia-Pacific Integrated Model/Computable General Equilibrium (AIM/CGE) and the global average in two climate change cases (green; RCP2.6 and red; RCP8.5) and the no climate change (NoCC) case (blue dot), Figure S13: Yield change of five major crops for the 17 regions of the Asia-Pacific Integrated Model/Computable General Equilibrium (AIM/CGE) and the global average in a climate change (RCP8.5) case and the no climate change (NoCC) cases (blue dot), Figure S14: Rice yield for 17 regions in the Asia-Pacific Integrated Model/Computable General Equilibrium (AIM/CGE) and the global average in two gridded yield aggregation method associated with land-use change (red; fixed as base year harvested area and red; changed dynamically) under RCP8.5 and non-CO_2 fertilization cases, Figure S15: Wheat yield for 17 regions in the Asia-Pacific Integrated Model/Computable General Equilibrium (AIM/CGE) and the global average in two gridded yield aggregation method associated with land-use change (red; fixed as base year harvested area and red; changed dynamically) under RCP8.5 and non-CO_2 fertilization cases, Figure S16: Coarse grain yield for 17 regions in the Asia-Pacific Integrated Model/Computable General Equilibrium (AIM/CGE) and the global average in two gridded yield aggregation method associated with land-use change (red; fixed as base year harvested area and red; changed dynamically) under RCP8.5 and non-CO_2 fertilization cases, Table S1. Regional classifications., Table S2: Land use classification, Table S3: Industrial classification, Table S4: List of general circulation models (GCMs), Table S5: Share of primary sector's value added in total GDP for SSP1, SSP2, and SSP3, Table S6: Price elasticity in 2050 baseline case. The numbers are derived from LES (Linear Expenditure System) consumption function, and Table S7: Income elasticity in 2050 baseline case. The numbers are derived from LES (Linear Expenditure System) consumption function.

Author Contributions: S.F., T.I. and T.H. designed the research; S.F., T.I. and T.K. carried out the simulations; S.F. carried out the analysis of the modeling results; S.F. led the writing of the paper; S.F. created figures; all authors contributed to the discussion and interpretation of the results.

Funding: This research is supported by JSPS KAKENHI Grant Number JP16K18177, JSPS Overseas Research Fellowships, and Environment Research and Technology Development Fund S-14-5 of the Environmental Restoration and Conservation Agency.

Conflicts of Interest: The authors declare no conflicts of interest.

References

1. Arent, D.; Tol, R.; Faust, E.; Hella, J.; Kumar, S.; Strzepek, K.M. Key economic sectors and services. In *Climate Change 2014: Impacts, Adaptation, and Vulnerability. Part A: Global and Sectoral Aspects*; Contribution of Working Group II to the Fifth Assessment Report of the Intergovernmental Panel on Climate Change; Field, C.B., Barros, V.R., Dokken, D.J., Mach, K.J., Mastrandrea, M.D., Bilir, T.E., Chatterjee, M., Ebi, K.L., Estrada, Y.O., Genova, R.C., et al., Eds.; Cambridge University Press: Cambridge, UK, 2014.
2. Tol, R.S.J. The Economic Impacts of Climate Change. *Rev. Environ. Econ. Policy* **2018**, *12*, 4–25. [CrossRef]
3. Porter, J.R.; Xie, L.; Challinor, A.J.; Cochrane, K.; Howden, S.M.; Iqbal, M.M.; Lobell, D.B.; Travasso, M.I. *Chapter 7: Food Security and Food Production Systems*; Cambridge University Press: Cambridge, UK, 2014.
4. Nelson, G.C.; van der Mensbrugghe, D.; Ahammad, H.; Blanc, E.; Calvin, K.; Hasegawa, T.; Havlik, P.; Heyhoe, E.; Kyle, P.; Lotze-Campen, H.; et al. Agriculture and climate change in global scenarios: Why don't the models agree. *Agric. Econ.* **2014**, *45*, 85–101. [CrossRef]
5. von Lampe, M.; Willenbockel, D.; Ahammad, H.; Blanc, E.; Cai, Y.; Calvin, K.; Fujimori, S.; Hasegawa, T.; Havlik, P.; Heyhoe, E.; et al. Why do global long-term scenarios for agriculture differ? An overview of the AgMIP Global Economic Model Intercomparison. *Agric. Econ.* **2014**, *45*, 3–20. [CrossRef]
6. Reilly, J.; Hohmann, N.; Kane, S. Climate change and agricultural trade. *Glob. Environ. Chang.* **1994**, *4*, 24–36. [CrossRef]

7. Ren, X.; Weitzel, M.; O'Neill, B.C.; Lawrence, P.; Meiyappan, P.; Levis, S.; Balistreri, E.J.; Dalton, M. Avoided economic impacts of climate change on agriculture: Integrating a land surface model (CLM) with a global economic model (iPETS). *Clim. Chang.* **2016**, *146*, 517–531. [CrossRef]
8. Roson, R.; Damania, R. The macroeconomic impact of future water scarcity: An assessment of alternative scenarios. *J. Policy Model.* **2017**, *39*, 1141–1162. [CrossRef]
9. Stevanović, M.; Popp, A.; Lotze-Campen, H.; Dietrich, J.P.; Müller, C.; Bonsch, M.; Schmitz, C.; Bodirsky, B.L.; Humpenöder, F.; Weindl, I. The impact of high-end climate change on agricultural welfare. *Science Adv.* **2016**, *2*, e1501452. [CrossRef] [PubMed]
10. Ciscar, J.C.; Iglesias, A.; Feyen, L.; Szabo, L.; Van Regemorter, D.; Amelung, B.; Nicholls, R.; Watkiss, P.; Christensen, O.B.; Dankers, R.; et al. Physical and economic consequences of climate change in Europe. *Proc. Natl. Acad. Sci. USA* **2011**, *108*, 2678–2683. [CrossRef] [PubMed]
11. van der Mensbrugghe, D. Chapter 14—Modeling the Global Economy–Forward-Looking Scenarios for Agriculture. In *Handbook of Computable General Equilibrium Modeling*; Dixon, P.B., Jorgenson, D.W., Eds.; Elsevier: Amsterdam, The Netherlands, 2013; Volume 1, pp. 933–994.
12. Hasegawa, T.; Fujimori, S.; Ito, A.; Takahashi, K.; Masui, T. Global land-use allocation model linked to an integrated assessment model. *Sci. Total Environ.* **2017**, *580*, 787–796. [CrossRef] [PubMed]
13. Fujimori, S.; Masui, T.; Matsuoka, Y. *AIM/CGE [Basic] Manual*; Version 2012-01; Center for Social and Environmental Systems Research, National Institute Environmental Studies: Tsukuba, Japan, 2012; pp. 1–87.
14. Bondeau, A.; Smith, P.C.; Zaehle, S.; Schaphoff, S.; Lucht, W.; Cramer, W.; Gerten, D.; Lotze-Campen, H.; MÜLler, C.; Reichstein, M.; et al. Modelling the role of agriculture for the 20th century global terrestrial carbon balance. *Glob. Chang. Biol.* **2007**, *13*, 679–706. [CrossRef]
15. Rosenzweig, C.; Elliott, J.; Deryng, D.; Ruane, A.C.; Müller, C.; Arneth, A.; Boote, K.J.; Folberth, C.; Glotter, M.; Khabarov, N.; et al. Assessing agricultural risks of climate change in the 21st century in a global gridded crop model intercomparison. *Proc. Natl. Acad. Sci. USA* **2014**, *111*, 3268–3273. [CrossRef] [PubMed]
16. Fujimori, S.; Hasegawa, T.; Masui, T.; Takahashi, K.; Herran, D.S.; Dai, H.; Hijioka, Y.; Kainuma, M. SSP3: AIM implementation of Shared Socioeconomic Pathways. *Glob. Environ. Chang.* **2017**, *42*, 268–283. [CrossRef]
17. Fujimori, S.; Dai, H.; Masui, T.; Matsuoka, Y. Global energy model hindcasting. *Energy* **2016**, *114*, 293–301. [CrossRef]
18. Iizumi, T.; Furuya, J.; Shen, Z.; Kim, W.; Okada, M.; Fujimori, S.; Hasegawa, T.; Nishimori, M. Responses of crop yield growth to global temperature and socioeconomic changes. *Sci. Rep.* **2017**, *7*, 7800. [CrossRef] [PubMed]
19. Babel, M.S.; Deb, P.; Soni, P. Performance Evaluation of AquaCrop and DSSAT-CERES for Maize Under Different Irrigation and Manure Application Rates in the Himalayan Region of India. *Agric. Res.* **2018**, 1–11. [CrossRef]
20. Porwollik, V.; Müller, C.; Elliott, J.; Chryssanthacopoulos, J.; Iizumi, T.; Ray, D.K.; Ruane, A.C.; Arneth, A.; Balkovič, J.; Ciais, P.; et al. Spatial and temporal uncertainty of crop yield aggregations. *Eur. J. Agron.* **2017**, *88*, 10–21. [CrossRef]
21. Ruth, D.; Gernot, K.; Florian, Z.; Wolfram, M. Global economic–biophysical assessment of midterm scenarios for agricultural markets—biofuel policies, dietary patterns, cropland expansion, and productivity growth. *Environ. Res. Lett.* **2018**, *13*, 025003. [CrossRef]
22. Mauser, W.; Klepper, G.; Zabel, F.; Delzeit, R.; Hank, T.; Putzenlechner, B.; Calzadilla, A. Global biomass production potentials exceed expected future demand without the need for cropland expansion. *Nat. Commun.* **2015**, *6*, 8946. [CrossRef] [PubMed]
23. Humpenöder, F.; Popp, A.; Stevanovic, M.; Müller, C.; Bodirsky, B.L.; Bonsch, M.; Dietrich, J.P.; Lotze-Campen, H.; Weindl, I.; Biewald, A.; et al. Land-Use and Carbon Cycle Responses to Moderate Climate Change: Implications for Land-Based Mitigation? *Environ. Sci. Technol.* **2015**, *49*, 6731–6739. [CrossRef] [PubMed]
24. Havlík, P.; Schneider, U.A.; Schmid, E.; Böttcher, H.; Fritz, S.; Skalský, R.; Aoki, K.; Cara, S.D.; Kindermann, G.; Kraxner, F.; et al. Global land-use implications of first and second generation biofuel targets. *Energy Policy* **2011**, *39*, 5690–5702. [CrossRef]
25. Riahi, K.; van Vuuren, D.P.; Kriegler, E.; Edmonds, J.; O'Neill, B.C.; Fujimori, S.; Bauer, N.; Calvin, K.; Dellink, R.; Fricko, O.; et al. The Shared Socioeconomic Pathways and their energy, land use, and greenhouse gas emissions implications: An overview. *Glob. Environ. Chang.* **2017**, *42*, 153–168. [CrossRef]

26. Wise, M.; Calvin, K. *GCAM3.0 Agriculture and Land Use: Technical Description of Modeling Approach*; Pacific Northwest National Laboratory: Richland, WA, USA, 2011.
27. Dimaranan, B.V. *Global Trade, Assistance, and Production: The GTAP 6 Data Base*; Dimaranan, B.V., Ed.; Center for Global Trade Analysis, Purdue University: West Lafayette, IN, USA, 2006.
28. United Nations, (UN). *National Accounts Main Aggregates Database*; United Nations: New York, NY, USA, 2013.
29. Monfreda, C.; Ramankutty, N.; Foley, J.A. Farming the planet: 2. Geographic distribution of crop areas, yields, physiological types, and net primary production in the year 2000. *Glob. Biogeochem. Cycles* **2008**, *22*. [CrossRef]
30. Mueller, N.D.; Gerber, J.S.; Johnston, M.; Ray, D.K.; Ramankutty, N.; Foley, J.A. Closing yield gaps through nutrient and water management. *Nature* **2012**, *490*, 254–257. [CrossRef] [PubMed]
31. Deryng, D.; Sacks, W.J.; Barford, C.C.; Ramankutty, N. Simulating the effects of climate and agricultural management practices on global crop yield. *Glob. Biogeochem. Cycles* **2011**, *25*. [CrossRef]
32. Ito, A.; Inatomi, M. Water-Use Efficiency of the Terrestrial Biosphere: A Model Analysis Focusing on Interactions between the Global Carbon and Water Cycles. *J. Hydrometeorol.* **2012**, *13*, 681–694. [CrossRef]
33. van Vuuren, D.P.; Edmonds, J.; Kainuma, M.; Riahi, K.; Thomson, A.; Hibbard, K.; Hurtt, G.C.; Kram, T.; Krey, V.; Lamarque, J.-F.; et al. The representative concentration pathways: An overview. *Clim. Chang.* **2011**, *109*, 5–31. [CrossRef]
34. Fujimori, S.; Su, X.; Liu, J.-Y.; Hasegawa, T.; Takahashi, K.; Masui, T.; Takimi, M. Implication of Paris Agreement in the context of long-term climate mitigation goals. *SpringerPlus* **2016**, *5*, 1–11.
35. Popp, A.; Calvin, K.; Fujimori, S.; Havlik, P.; Humpenoeder, F.; Stehfest, E.; Bodirsky, B.L.; Dietrich, J.P.; Doelmann, J.C.; Gusti, M.; et al. Land-use futures in the shared socio-economic pathways. *Glob. Environ. Chang. Hum. Policy Dimens.* **2017**, *42*, 331–345. [CrossRef]
36. Hasegawa, T.; Fujimori, S.; Havlík, P.; Valin, H.; Bodirsky, B.L.; Doelman, J.C.; Fellmann, T.; Kyle, P.; Koopman, J.F.L.; Lotze-Campen, H.; et al. Risk of increased food insecurity under stringent global climate change mitigation policy. *Nat. Clim. Chang.* **2018**, *8*, 699–703. [CrossRef]
37. Park, C.; Fujimori, S.; Hasegawa, T.; Takakura, J.Y.; Takahashi, K.; Hijioka, Y. Avoided economic impacts of energy demand changes by 1.5 and 2 °C climate stabilization. *Environ. Res. Lett.* **2018**, *13*, 045010. [CrossRef]
38. Takakura, J.Y.; Fujimori, S.; Takahashi, K.; Hijioka, Y.; Hasegawa, T.; Honda, Y.; Masui, T. Cost of preventing workplace heat-related illness through worker breaks and the benefit of climate-change mitigation. *Environ. Res. Lett.* **2017**, *12*, 064010. [CrossRef]
39. Hasegawa, T.; Fujimori, S.; Takahashi, K.; Yokohata, T.; Masui, T. Economic implications of climate change impacts on human health through undernourishment. *Clim. Chang.* **2016**, *136*, 1–14. [CrossRef]
40. Springmann, M.; Godfray, H.C.J.; Rayner, M.; Scarborough, P. Analysis and valuation of the health and climate change cobenefits of dietary change. *Proc. Natl. Acad. Sci. USA* **2016**, *113*, 4146–4151. [CrossRef] [PubMed]

MDPI
St. Alban-Anlage 66
4052 Basel
Switzerland
Tel. +41 61 683 77 34
Fax +41 61 302 89 18
www.mdpi.com

Sustainability Editorial Office
E-mail: sustainability@mdpi.com
www.mdpi.com/journal/sustainability

www.ingramcontent.com/pod-product-compliance
Lightning Source LLC
Chambersburg PA
CBHW041751220926
51847CB00014B/170
* 9 7 8 3 0 3 8 9 7 8 1 4 5 *